Nanoscale Devices

Nanoscale Devices

Physics, Modeling, and Their Application

Edited by
Brajesh Kumar Kaushik

CRC Press is an imprint of the
Taylor & Francis Group, an **Informa** business

CRC Press
Taylor & Francis Group
6000 Broken Sound Parkway NW, Suite 300
Boca Raton, FL 33487-2742

First issued in paperback 2020

ISBN 13: 978-0-367-57072-9 (pbk)
ISBN 13: 978-1-138-06034-0 (hbk)

Library of Congress Cataloging-in-Publication Data

Names: Kaushik, Brajesh Kumar, editor.
Title: Nanoscale devices : physics, modeling, and their application / [edited by] Brajesh Kumar Kaushik.
Description: Boca Raton : Taylor & Francis, a CRC title, part of the Taylor & Francis imprint, a member of the Taylor & Francis Group, the academic division of T&F Informa, plc, 2019. | Includes bibliographical references and index.
Identifiers: LCCN 2018026865| ISBN 9781138060340 (hardback : acid-free paper) | ISBN 9781315163116 (ebook)
Subjects: LCSH: Nanoelectronics. | Nanoelectromechanical systems. | Nanotechnology.
Classification: LCC TK7874.84 .N3835 2019 | DDC 621.381--dc23
LC record available at https://lccn.loc.gov/2018026865

Visit the Taylor & Francis Web site at
http://www.taylorandfrancis.com

and the CRC Press Web site at
http://www.crcpress.com

To my Father, Late Mr. Jai Prakash Kaushik, and my Mother, Mrs. Karuna Kaushik, for their affection and untiring efforts in my upbringing. Also dedicated to my wife, Supriya, and our loving son, Partha Kaushik. Life has been really enjoyable because of the efforts of all my friends and students. Dedicated to all of them.

Brajesh Kumar Kaushik

Contents

Section IV Graphene and Carbon Nanotube Transistors and Applications

Section V Modeling of Emerging Non-Silicon Transistors

Section VI Emerging Nonvolatile Memory Devices and Applications

Preface

The pace of advances in nanoscale materials and devices has led to a gamut of developments in the emerging areas of science and technology that has permeated to almost all applications. The *rapid growth* of the electronics industry can be attributed by and large to the continuous progress in nanotechnology driven by a relentless pursuit of innovation. With the new prospects from emerging nanomaterial and devices, a frontier challenge is to gain insight into their behavior and properties. The aim of this book is to integrate developments in the area of nanoelectronics devices and systems, evolving from a unifying set of principles and practices. This book is a collective effort of well-known pioneers and many leading researchers in the field of nanoelectronics, nanomaterials, plasmonics, and related areas, providing extraordinary breadth and depth of coverage. To comprehend this book, a basic knowledge of semiconductor physics is assumed. In 17 chapters, grouped under 6 subareas, the book eloquently covers the major advancements in the field of nanoelectronics. The book presents an interdisciplinary perspective of nanoelectronics and can serve as a comprehensive reference for researchers and academicians working in the area of nanotechnology.

The continuous miniaturization of transistors has been considered a primary reason behind the progress in the electronics industry. New approaches based on quantum mechanics and a multiphysics perspective are now required to accurately model these nanoscale transistors. The first section of this book provides a comprehensive overview of quantitative understanding of the operation of nanoscale transistors. It begins with an introduction to nanoscale transistors, their current state of development, and various challenges associated with commercialization of the technology. Chapter 1 of the book deals with the recent development in modeling and simulation of nanoscale transistors. A multiphysics approach has been used to capture the relevant physics at the atomic level while taking quantum effects into account. The major challenge to the present and next-generation nanoscale device is the variability that requires serious attention. Considering this fact, Chapter 2 discusses the major sources of variability in nanoscale devices and circuits. To increase the robustness of a design, the impact of process, voltage, and temperature variation has been modeled and analyzed using technology computer-aided design (TCAD) simulations. An epitaxial delta-doped channel (EδDC) MOS transistor is a promising approach for extending the scalability of bulk metal oxide semiconductor (MOS) technology for low-power system-on-chip applications. The subsequent part of the chapter discusses the structure and characteristics of EδDC MOS transistors and compares them with a conventional MOS transistor.

For nanoscale semiconductor field-effect transistor (FET) devices, several efforts have been made to reduce the short-channel effects (SCEs). For controlling short-channel effects, a ground plane can be formed in the silicon substrate underneath the buried oxide. Chapter 3 compiles the application of ground plane and strained engineering for high-performance silicon on insulator (SOI) FET devices. The first part introduces the concept of ground plane in SOI structures and the method of incorporation in fully depleted silicon on insulator (FDSOI) and fin field-effect transistor (FinFET) devices. The later part of the chapter discusses the strained-silicon concept to boost the device performance through higher carrier mobility and reduced source/drain resistance. The chapter discusses the basic concepts for introducing strain in CMOS transistor channels. The methods and structures for making strained FDSOI and strained FinFET device structure have been presented.

Improvement in the transistor structure is considered to be a novel way for sustaining performance improvement with aggressive device scaling. The benefits derived with various alternative device architectures can be seen as a possible alternative to the conventional planar MOSFET. Section II of the book presents such novel structures to improve device performance. It is organized with the aim of finding and building a successful novel device structure that can replace conventional MOSFET. The need for higher operating voltages and switching speeds has necessitated the use of alternative structures and materials in the power electronics domain. Basic principles of operation of the U-shaped gate trench MOSFET (UMOSFET) structure are discussed in Chapter 4. The superior on-resistance for this structure makes it a possible contender in high-frequency power circuits. Numerical analysis has been presented for evaluating the effectiveness of the device in comparison to vertical double-diffused MOSFET (VDMOSFET). The VDMOSFET dominates the medium voltage-level power electronic applications. Chapter 5 analyzes performance of the VDMOSFET structure including shielded accumulation-mode MOSFET (ACCUFET) structures for the development of monolithic power switches from silicon carbide. To accommodate future technology nodes, double-gate MOSFET (DGFET) has been shown to be more optimal due to the improved short-channel effects and reduced leakage current compared to conventional FET. The double-gate MOSFET (DG-MOSFET) has several advantages over conventional single-gate (SG) MOSFETs. Chapter 6 explores the scope of DGMOSFETs for use in future high-performance, low-power nanoscale device applications. Modeling and analysis of various configurations of DGMOSFETs, such as symmetric/asymmetric with gate and channel engineering, have been presented in the chapter.

Tunnel field-effect transistors (TFETs) are also considered to be a potential successor of MOSFETs by enabling the supply voltage (V_{DD}) scaling at nanometer dimensions due to the absence of a subthreshold swing limit of 60 mV/decade. The third section of the book would enhance the knowledge base of the reader in the area of tunnel FETs. Currently, it is of prime interest to develop compact analytical models, describing the device operation to meet the circuit and system design. Chapter 7 of the book deals with potentials and challenges of analog circuits based on unique characteristics of *TFET*. The chapter discusses parameters of TFET that are important for analog circuit operations. Different TFET architectures have been studied to overcome the challenges in conventional TFET. Doping-less structure has been proposed as a potential solution to overcome the need of creating abrupt junctions in TFETs. Chapter 8 deals with the performance analysis of a dual-metal double-gate tunnel FET device and its advantages over doping-less TFET.

The fourth section of the book focuses on nanocarbons, in particular carbon nanotubes and graphene, and their application in the area of electronics and plasmonics. The realization of these carbon-based materials has facilitated many interesting nanodevices for future technological revolutions. In recent years, there have been substantial efforts in filling the (*THz*) design gap. The recent advancement in photonics and nanofabrication technology opens up the development of plasmon-assisted devices. The prospects of graphene-based plasmonic devices for THz applications attract several research and application areas. Chapter 9 of the book provides an overview of optical and electronic properties of graphene and discusses graphene-based plasmonic waveguides for the terahertz range operation while targeting the multiplexing/demultiplexing and sensing applications. In nano-regime, carbon nanotube field-effect transistor (CNTFET) technology has been actively researched as an alternative to the conventional CMOS technology. CNTs have shown remarkable electronic properties that distinguish them as a good replacement for channel material in MOSFET also. CNTFET significantly outperforms the MOSFET and FinFET-based memory cells in the metrics of power, delays, and noise. Chapter 10 covers the important aspects of CNT for various electronic applications. The chapter presents the fundamental concepts of CNTFETs with a major focus on application of CNTFET for low-power static random access memory (SRAM) cell design. It is quite sensible to investigate newer circuit with CNFETs because of their unique structure and electrical characteristics. Ternary logic is a promising alternative to conventional binary logic, realizing energy-efficient circuits. Chapter 11 discusses the design of ternary logic circuits that are benefitted from the unique properties of carbon nano-tube field-effect transistors. The combination of ternary logic and CNFETs has the capability of achieving highly efficient digital systems.

The section V of the book presents recent approaches toward the modeling of emerging transistors employing non-silicon materials. The III-V compound semiconductors-based high electron mobility transistors (HEMTs) are rapidly emerging as front-runners in high-power millimeter, submillimeter, and microwave and THz applications. Chapter 12 provides readers with an overview of state-of-the-art HEMT devices. The chapter covers the working principle of the device followed by analysis of different structures of HEMTs. It develops a detailed theoretical framework of HEMT devices along with their application areas and future directions. Recently, the organic semiconductor (OSC)-based devices have been in focus due to low-cost fabrication techniques and their application in large-area electronics. Chapter 13 covers the emerging field of organic thin-film transistor (OTFT) while discussing the analytical methods for evaluation of device performance. The chapter deals with the device models that are developed using the underlying device physics and geometry-related parameters.

The last section of the book presents recent advances in nonvolatile memory and storage technology and provides an overview of future trends in the field. It covers a wide range of emerging memory device applications from state-of-the-art spintronic devices, magnetic tunnel junction to RRAM and memristor-based logic. Chapter 14 provides a comprehensive introduction to spintronics for an emerging class of high-speed, low-power memory and logic devices. Spintronic devices that utilize electron spin as a state variable for logic computation have recently emerged as one of the leading candidates for computing and data storage applications. The chapter focuses toward the future aspects of spintronics including spintronics-based energy-efficient on-chip memory and logic. Magnetic tunnel junctions (MTJs), are an important subset of spintronics devices. Magnetic storage technology has been under intensive study in recent years and is capable of overcoming the shortcomings of CMOS. Chapter 15 provides a platform to understand the fundamentals of MTJs, advanced device models, and their application in the area of field-programmable reconfigurable fabrics and logic-in-memory circuits. The past few years have seen an *increased interest* in resistive switching (RS) resistive random access memory (RRAM) for future data storage. Chapter 16 covers RRAM as one of the potential candidates for nonvolatile memories (NVMs) and their circuit applications. The chapter investigates the physics-based mathematical models and circuit models of the Ta_2O_5/TaO_x bi-layered RRAM. A SPICE model has been developed to study the behavior of devices to be used in circuit simulations. The memristor provides an opportunity for design of high-performance circuits and memories at a nanoscale level. Chapter 17 reports the latest advances in the area of memristor and its application in the area of digital and analog computing.

The book integrates the recent research and development in the area of nanoelectronics and provides a synergic platform for further advancement of devices and systems. This book shall prove to be a convenient and thorough reference for academicians and practitioners in the field of nanoelectronics.

Brajesh Kumar Kaushik
Roorkee, India

Acknowledgment

This book is the collective efforts of many researchers committed to the diversified fields of nanotechnology. I would like to acknowledge with gratitude all the authors for spending their precious time in creation of this edition.

I wish to express my heartfelt gratitude to one and all who directly or indirectly helped in the completion of this book.

Brajesh Kumar Kaushik

Editor

Brajesh Kumar Kaushik received doctorate of philosophy (PhD) in 2007 from Indian Institute of Technology, Roorkee, India. He joined Department of Electronics and Communication Engineering, Indian Institute of Technology, Roorkee, as an assistant professor in December 2009; and since April 2014 he has been an associate professor. He has served as General Chair, Technical Chair, and Keynote Speaker of many reputed international and national conferences. Dr. Kaushik is a senior member of IEEE and member of many expert committees constituted by government and non-government organizations. He is an editor of *IEEE Transactions on Electron Devices*; associate editor of *IET Circuits, Devices & Systems*; editor of *Microelectronics Journal*, Elsevier; editorial board member of *Journal of Engineering, Design and Technology*, Emerald; and editor of *Journal of Electrical and Electronics Engineering Research*, Academic Journals. He also holds the position of editor-in-chief of *International Journal of VLSI Design & Communication Systems*, and *SciFed Journal of Spintronics & Quantum Electronics*. He has received many awards and recognitions from the International Biographical Center (IBC), Cambridge. His name has been listed in Marquis Who's Who in Science and Engineering® and Marquis Who's Who in the World®. Dr. Kaushik has been conferred with Distinguished Lecturer award of IEEE Electron Devices Society (EDS) to offer EDS chapters with quality lectures in his research domain. His research interests are in the areas of high-speed interconnects, low-power VLSI design, memory design, carbon nanotube-based designs, organic electronics, FinFET device circuit co-design, electronic design automation (EDA), spintronics-based devices, circuits and computing, image processing, and optics- and photonics-based devices.

Contributors

Paula Ghedini Der Agopian
Telecommunication Engineering Department
Sao Paulo State University (UNESP)
Sao Joao da Boa Vista, Brazil

J. Ajayan
Department of Electronics and Communication Engineering
SNS College of Technology
Coimbatore, India

Haider A. F. Almurib
Department of Electrical and Electronic Engineering
University of Nottingham
Semenyih, Malaysia

Shashi Bala
VLSI Design Lab
Department of ECE
NIT Jalandhar
Jalandhar, India

Deepshikha Bharti
Department of Electronics and Communication Engineering
BIT, Mesra
Ranchi, India

W. Boukhili
Laboratoire de Physique des Matériaux: Structure et Propriétés
Groupe Physique des Composants et Dispositifs Nanométriques
Faculté des Sciences de Bizerte
Université de Carthage
Carthage, Tunisia

R. Bourguiga
Laboratoire de Physique des Matériaux: Structure et Propriétés
Groupe Physique des Composants et Dispositifs Nanométriques
Faculté des Sciences de Bizerte
Université de Carthage
Carthage, Tunisia

Rajeevan Chandel
Department of Electronics & Communication Engineering
National Institute of Technology
Hamirpur, India

Jyotirmoy Chatterjee
CEA, CNRS, Grenoble-INP, INAC-SPINTEC
Université Grenoble Alpes
Grenoble, France

Saurabh Chaudhury
Department of Electrical Engineering
NIT Silchar
Silchar, India

Wenchao Chen
Zhejiang Province Lab of Advanced Micro/Nano Electronic Devices and Smart Systems
State Key Lab of MOI
College of Information and Electronic Engineering
Innovative Institute of Electromagnetic Information and Electronic Integration
Zhejiang University
Hangzhou, China

and

ZJU-UIUC Institute
Zhejiang University
Haining, China

Cor Claeys
Electrical Engineering Department (ESAT)
KU Leuven
Leuven, Belgium

Ronald F. DeMara
Department of Electrical and Computer Engineering
College of Engineering and Computer Science
University of Central Florida
Orlando, Florida

Rohit Dhiman
Department of Electronics & Communication Engineering
National Institute of Technology
Hamirpur, India

Zhipeng Dong
Department of Electrical and Computer Engineering
University of Florida
Gainesville, Florida

Jing Guo
Department of Electrical and Computer Engineering
University of Florida
Gainesville, Florida

Firas Odai Hatem
Department of Electronics and Electrical Engineering
Liverpool John Moores University
Liverpool, United Kingdom

Aminul Islam
Department of Electronics and Communication Engineering
BIT, Mesra
Ranchi, India

Neetu Joshi
Department of Electronics and Communication Engineering
IIT-Roorkee
Roorkee, India

Ramandeep Kaur
Department of Electronics & Communication Engineering
National Institute of Technology
Hamirpur, India

Mamta Khosla
VLSI Design Lab
Department of ECE
NIT Jalandhar
Jalandhar, India

T. Nandha Kumar
Department of Electrical and Electronic Engineering
University of Nottingham
Semenyih, Malaysia

Joao Antonio Martino
Laboratory of Integrated Systems
Electronic System Department
University of Sao Paulo
Sao Paulo, Brazil

Marcio Dalla Valle Martino
Electronic System Department
University of Sao Paulo
Sao Paulo, Brazil

Chandrasekhar Murapaka
CEA, CNRS, Grenoble-INP, INAC-SPINTEC
Université Grenoble Alpes
Grenoble, France

D. Nirmal
Department of Electronics and Communication Engineering
School of Electrical Sciences
Karunya University
Coimbatore, India

Soumya Pandit
Institute of Radio Physics and Electronics
University of Calcutta
Kolkata, India

Nagendra P. Pathak
Department of Electronics and Communication Engineering
Indian Institute of Technology-Roorkee
Roorkee, India

Balwinder Raj
VLSI Design Lab
Department of ECE
NIT Jalandhar
Jalandhar, India

Arman Roohi
Department of Electrical Engineering and Computer Science
College of Engineering and Computer Science
University of Central Florida
Orlando, Florida

Sarmista Sengupta
Institute of Radio Physics and Electronics
University of Calcutta
Kolkata, India

Pankaj Sethi
CEA, CNRS, Grenoble-INP, INAC-SPINTEC
Université Grenoble Alpes
Grenoble, France

Eddy Simoen
Imec
Leuven, Belgium

Avtar Singh
Department of Electronics and Communication Engineering
Invertis University
Bareilly, India

Jeetendra Singh
VLSI Design Lab
Department of ECE
NIT Jalandhar
Jalandhar, India

M. B. Srinivas
School of Engineering and Technology
BML Munjal University
Gurgaon, India

Chetan Vudadha
Electrical and Electronics Engineering
BITS-Pilani
Hyderabad, India

Wen-Yan Yin
Zhejiang Province Lab of Advanced Micro/Nano Electronic Devices and Smart Systems
State Key Lab of MOI
College of Information and Electronic Engineering
Innovative Institute of Electromagnetic Information and Electronic Integration
Zhejiang University
Hangzhou, China

Ramtin Zand
Department of Electrical Engineering and Computer Science
College of Engineering and Computer Science
University of Central Florida
Orlando, Florida

Section I

Nanoscale Transistors

1

Simulation of Nanoscale Transistors from Quantum and Multiphysics Perspective

Zhipeng Dong, Wenchao Chen, Wen-Yan Yin, and Jing Guo

CONTENTS

1.1 Background and Introduction

The success of the semiconductor industry critically relies on collaborative efforts on fabrication, characterization, and design. It is estimated by the electronics industry that the use of computer-aided design saves the semiconductor development cost by 40%. Conventional technology computer-aided design (TCAD) simulation tools describe electrons in semiconductor devices as classical particles based on continuum theories. The physics base does not extend to nanoelectronic devices and sensors. As transistors scale down, multiphysics phenomena, quasi-ballistic transport, quantum effects, and even atomistic scale features of devices, interfaces, and surfaces become inevitably important for nanoscale devices and sensors.

This chapter describes our recent efforts in modeling nanoscale transistors, with an emphasis on modeling quantum transport and multiphysics phenomena. Section 1.2 discusses a simulation approach based on the non-equilibrium Green's function formalism and multiphysics approach. Section 1.3 gives examples of applying the simulation capability to assess transistors based on 2D semiconductor materials near the scaling limits, and Section 1.4 applies the multiphysics simulation capability to ultra-thin-body silicon on insulator (SOI) MOSFETs.

1.2 Overview on Quantum and Multiphysics Simulation of Nanoscale Transistors

1.2.1 Quantum Simulation Approach for Nanotransistors

Electrons in semiconductor devices are traditionally modeled as semiclassical particles based on a continuum theory, such as the drift-diffusion theory. For nanoelectronic devices, quantum effects and atomistic scale features become inevitably important. Recent research has led to the evolution of a unified and powerful quantum transport simulation framework based on the non-equilibrium Green's function (NEGF) formalism [1–6]. The NEGF formalism provides a general approach for quantum simulations of nanoscale transistors as diverse as nanoscale silicon FETs, nanowire FETs, CNTFETs, graphene and graphene nanoribbon FETs, molecular FETs, and TMDC FETs.

Figure 1.1 summarizes the procedure of applying the NEGF approach to a transistor. The transistor channel is connected to the source and drain contacts, and the gate modulates the conductance of the channel. One first identifies a suitable basis set and derives the Hamiltonian, $[H]$, for the isolated channel, which can be a molecule, CNT, graphene, or silicon. Then, the contact self-energy matrices, $[\Sigma_1]$ and $[\Sigma_2]$ are computed, which describe how the channel couples to the source and drain contacts. Next, the scattering self-energy $[\Sigma_S]$ is identified to describe the dissipative process. After that, the retarded Green's function is computed self-consistently with the potential matrix $[U]$. After self-consistency is achieved, the physical quantities of interest, such as the charge density and current, are computed from the Green's function. The NEGF formalism meets the needs of nanodevice modeling by providing (i) an atomistic description, (ii) treatment of open boundary and non-equilibrium transport, and (iii) treatment of inelastic scattering. The formalism, however, treats electron-electron interaction based on an independent-electron approximation. Many-body effects such as magnetism, single-electron charging, Kondo effect, and entanglement are left out [1–6].

Figure 1.2 shows our preliminary results on modeling a MoS_2 FET with a channel length of 20 nm by using the NEGF formalism. The device performance at the ballistic limit and in the presence of phonon scattering is presented. Phonon scattering plays an important role in establishing the intrinsic device performance limits and the impact of self-heating effects, which is carefully treated by using the self-consistent Born approximation with a

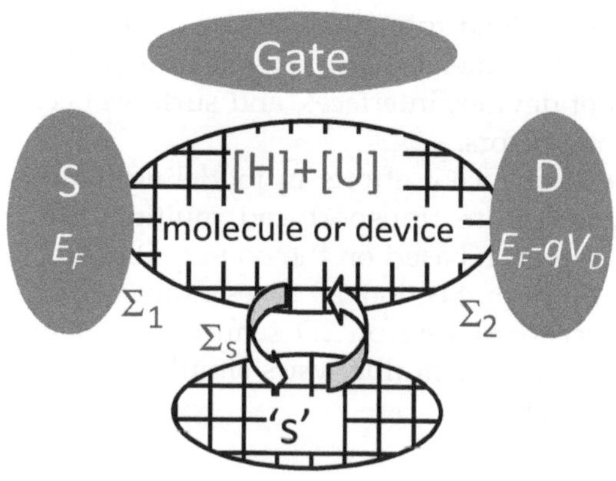

FIGURE 1.1
The NEGF applied to a generic nanodevice.

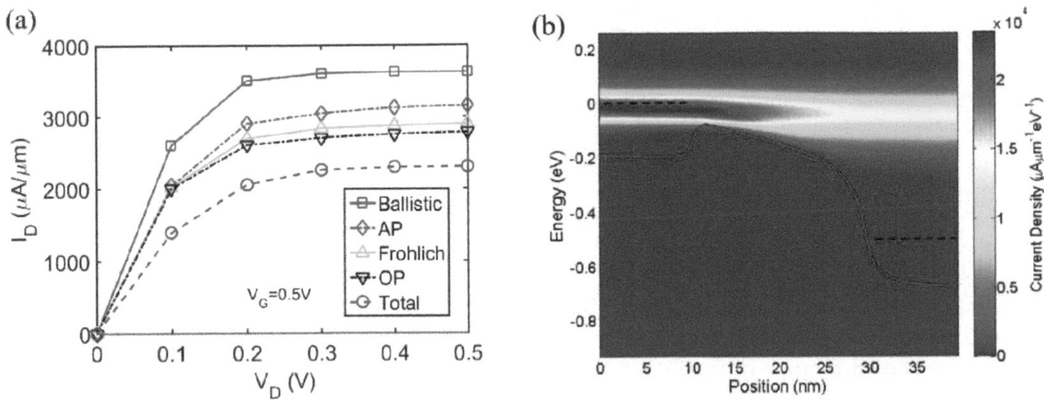

FIGURE 1.2
NEGF simulation of MoS_2 horizontal homojunction FETs with $L_G = 20$ nm. (a) I_D–V_D characteristics at the ballistic limit and in the presence of various phonon scattering mechanisms, including acoustic phonon (AP), optical phonon (OP), and Frohlich scattering. (b) The energy resolved current spectrum at on state with phonon scattering shows energy relaxation and back scattering. The red solid line is a potential profile, and the black dashed curve is the Fermi level at the source (drain).

first principle description of the phonon energy and electron-phonon coupling strength. The results in Figure 1.2 establish the ballistic performance limits as well as the practical intrinsic device performance limits with phonon scattering, and they provide a detailed and physical description of heat generation due to self-heating effects in the TMDC FET device.

Self-heating effects, which highlight the importance of modeling coupled with electron-thermal transport, can play an important role in nanoscale transistors. Take graphene transistor as an example: Figure 1.3 illustrates the corresponding thermal transport model of the simulated device. The top surface is assumed to be adiabatic. The interface thermal resistance between the graphene layer and the BN layer is equivalent to a 15-nm thick SiO_2 layer. g_S (g_D) is the source (drain) contact thermal conductance. Its value depends on the quality of the contact, and the quantum limit is about 1.5 $Wm^{-1}K^{-1}$ [7]. In simulation, we use the quantum limit value for the perfect thermal conductance contacts and a 10 times

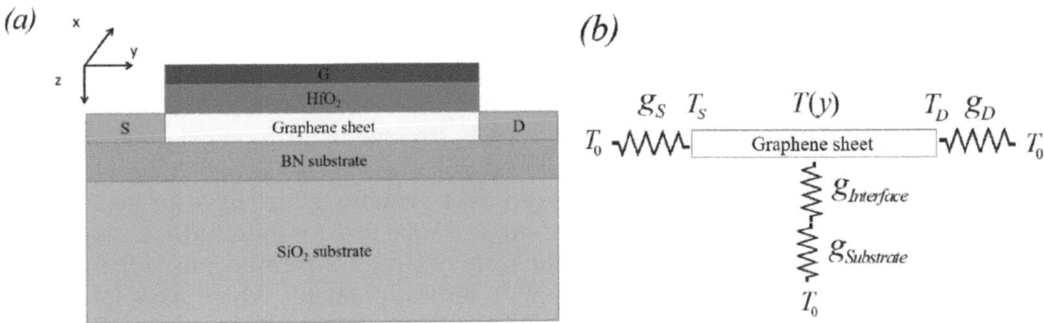

FIGURE 1.3
(a) Simulated device structure. HfO_2 is used for the top gate insulator with a thickness of 3 nm and dielectric constant of 25. Pd is used for the metal contacts, which have a Schottky barrier height of 0.2 eV for electrons. A BN layer deposited on the SiO_2 substrate is used as the improved substrate for the graphene sheet. The thickness of SiO_2 substrate is 300 nm. (b) The corresponding thermal boundary model of the device. The interface thermal resistance between the graphene sheet and BN layer is equivalent to a 15 nm-thick SiO_2.

smaller value for the realistic contacts. The steady state heat transport equation for 2D graphene along the transport direction is given by

$$A\frac{\partial}{\partial y}\left(k\frac{\partial T}{\partial y}\right)=-\left(p(y)-g(T-T_0)\right) \tag{1.1}$$

where A is the cross section of graphene, $k \approx 600\ \mathrm{Wm^{-1}K^{-1}}$ is the thermal conductivity of supported graphene [8], $p(y)=-\partial I_{\mathrm{heat}}/\partial y$ is the Joule heating rate where I_{heat} is the heat flux, $g=(g_{\mathrm{Interface}}^{-1}+g_{\mathrm{Substrate}}^{-1})^{-1}$ is the effective thermal conductance to the substrate where $g_{\mathrm{Interface}}$ is the thermal conductance between the graphene sheet and the BN layer, and $g_{\mathrm{Substrate}}$ is the thermal conductance of the SiO_2 substrate. T_0 is ambient temperature, which is fixed at 300K. The thermal boundary conditions at contacts are given by

$$\begin{cases}\text{Source}: g_S\left(T_0-T_S\right)=-k\left.\dfrac{\partial T}{\partial y}\right|_{y=0}\\[2ex] \text{Drain}: g_D\left(T_D-T_0\right)=-k\left.\dfrac{\partial T}{\partial y}\right|_{y=L}\end{cases} \tag{1.2}$$

The thermal transport equation is solved self-consistently by coupling it to the NEGF formalism with phonon scattering. The simulation results under both equilibrium and hot-phonon circumstances can be considered to study the self-heating effects in GFETs by this approach.

1.2.2 Approach of Multiphysics Simulation for Nanotransistors

Short-channel effect (SCE) limits further miniaturization of field effect transistors. To minimize the SCE, transistors based on new structures and new materials as discussed in previous sessions have been experimentally and theoretically investigated more recently. As the transistors keep shrinking, the integration density increases with the reduction of device size. Hence, self-heating effect (SHE) becomes more severe for smaller transistors [9–12], especially for silicon-on-insulator (SOI) MOSFETs with insertion of buried oxide (BOX) layer, which has low thermal conductivity. High temperature deteriorates hot carrier injection (HCI) and causes worse reliability for sub-100 nm Si-MOSFETs [13–15]. In order to capture the effects of HCI in MOSFETs stressed by different signals, transient electrothermal simulation needs to be employed [16].

The primary heat carriers in silicon are the acoustic phonons, which have group velocities around 5000 m/s for transverse modes and around 9000 m/s for longitudinal modes [12]. Scattering time of acoustic phonons is in the scale of tenths of picoseconds [17]. Electrons with energy above 50 meV scatter mainly with optical phonons, which have low group velocities around 1000 m/s [12]. Optical phonons decay into acoustic phonons in the scale of picoseconds [17]. The electrons gain energy from an electric field, and they need to travel several mean-free paths to release it to the lattice [12,18]. With an electron velocity of 10^7 cm/s, which is the saturation velocity in silicon [12], the transit time for electrons in the sub-100 nm channel MOSFET is in the scale of picoseconds. On the other hand, the transistors typically operate at GHz frequencies. Each operating pulse lasts for several hundreds of picoseconds, which is several orders higher than the heat generation process. Hence, the heat generation can be treated as an instantaneous process as the GHz pulse is applied to transistors.

The heat generation rate can be calculated by various methods, including drift diffusion, Monte Carlo, and NEGF, which treat carriers classically, semi-classically, and quantum mechanically, respectively. The heat generation rate in a quasi-ballistic transistor with a channel length of 20 nm is presented by using different methods as shown in [19]. The heat generation rate obtained by drift diffusion has larger magnitude and also large negative value, and the Monte Carlo simulation results indicate that heat dissipates far into the drain [19].

Thermal conductivity has temperature dependence. Specifically, the thermal conductivity of Si thin film depends not only on temperature but also its thickness [20–22]. As the thickness decreases, boundary scattering starts to play an important role in determining the overall thermal conductivity [20]. In particular, based on the modeling method given in [22], the temperature-dependent thermal conductivity of Si thin film with nanoscale thickness can be calculated as shown in Figure 1.4. For the thickness around 10 nm, its thermal conductivity shows linear dependence on temperature; it can be described by

$$\kappa_{\text{Si film}}(T) = \kappa_{\text{Si }0} + \gamma(T - 300) \tag{1.3}$$

where $\kappa_{\text{Si }0}$ is the thermal conductivity of Si thin film with certain thickness at 300K, and γ is the linear coefficient. The values of $\kappa_{\text{Si }0}$ and γ for Si thin film with different thicknesses are listed in Table 1.1.

Usually, the field effect transistor is buried inside several hundred micrometers thick Si substrate and several micrometers thick insulator. The heat will spread out through its surroundings. Hence, once the heat generation rate in the transistor is determined, heat transport on the circuit level can be described by the following heat conduction equation,

$$C(T)\partial T(\vec{r},t)/\partial t = \kappa(T)\nabla^2 T(\vec{r},t) + Q(t) \tag{1.4}$$

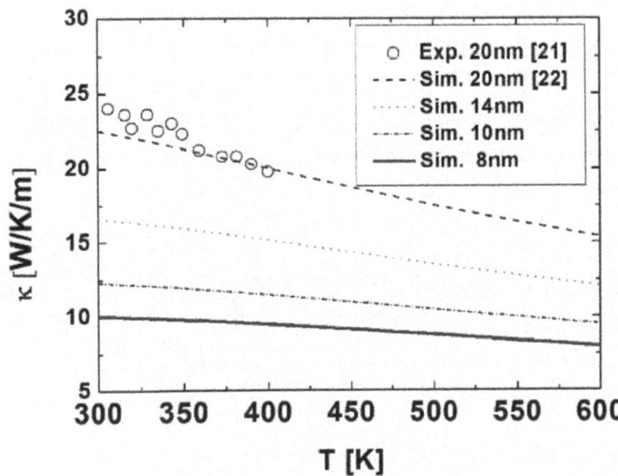

FIGURE 1.4
Experimental and simulated thermal conductivities of Si thin films with different nanoscale thicknesses. The simulated results are obtained by the method provided in Liu et al. (From Liu, W. et al., *IEEE Trans. Electron Devices*, 53, 1868–1876, 2006.)

TABLE 1.1

Thermal Conductivity of Silicon Thin Film with Nanoscale Thickness

Thickness of Si Film (nm)	$\kappa_{Si\,0}$ (W/K/m)	γ (W/K²/m)
12	14.60	-1.25×10^{-2}
11	13.52	-1.11×10^{-2}
10	12.41	-9.66×10^{-3}
9	11.29	-8.27×10^{-3}
8	10.14	-6.93×10^{-3}
7	8.96	-5.64×10^{-3}
6	7.76	-4.42×10^{-3}

where $C(T)$ is the heat capacity, $\kappa(T)$ is the thermal conductivity of materials involved, and $Q(t)$ is the heat generation. Equation (1.4) can be simplified into a static thermal conduction equation by setting the left-hand side to be zero,

$$\kappa(T)\nabla^2 T(\vec{r}) + Q = 0 \tag{1.5}$$

Following the standard procedure of time domain finite element method [23], the matrix form of Equation (1.4) can be obtained by doing backward difference with respect to time, i.e.,

$$\left([M] + \Delta t[K]\right)\{T\}^{t+\Delta t} = \Delta t[S]\{Q\}^{t+\Delta t} + [M]\{T\}^t + \{B\} \tag{1.6}$$

where $[M]$ is the time-dependent matrix, $[K]$ is the thermal conduction matrix, $[S]$ is the overlap matrix, $\{B\}$ is the boundary condition matrix, $\{Q\}$ is the heat generation vector, and Δt is the time step for time evolution. For steady case, the matrix form of Equation (1.5) can be written as

$$[K]\{T\} = [S]\{Q\} + \{B\} \tag{1.7}$$

The processes for solving time-dependent and static thermal conduction equations with temperature-dependent parameters are shown in Figure 1.5.

Different from conventional long-channel MOSFETs, the worst case of HCI effect happens at $V_g = V_d = V_{DD}$ for sub-100 nm Si-MOSFET [13]; it also gets worse as temperature increases [13–15]. Based on the empirical model of threshold voltage shift as a function of time [24], a temperature-dependent exponential coefficient α can be introduced,

$$\Delta V_{th} = \Delta V_{th0}(t/t_0)^{\alpha} \tag{1.8}$$

According to the measured threshold voltage shift given in [25], a linear model to characterize the temperature-dependent α can be given as

$$\alpha = \alpha_0 + \beta(T - T_0) \tag{1.9}$$

where nominal values $\alpha_0 = 0.1$, $T_0 = 300\text{K}$ and $\beta = 5\times10^{-4}/\text{K}$ are used in our simulation.

To numerically calculate the threshold voltage shift, we need to do a derivative of Equation (1.8) with respect to time,

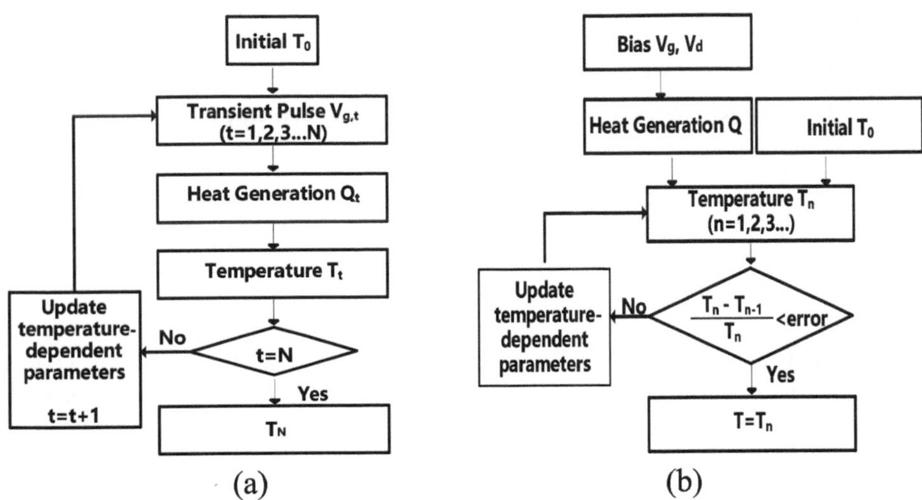

FIGURE 1.5
Flowchart for solving (a) time-dependent thermal conduction equation and (b) static thermal conduction equation with temperature-dependent parameters.

$$\frac{d\Delta V_{\text{th}}}{dt} = \frac{\alpha}{t_0}\Delta V_{\text{th0}}(t/t_0)^{\alpha-1} \tag{1.10}$$

At time step n, the threshold voltage shift is

$$\begin{aligned}\Delta V_{\text{th}}(n) &= \sum_{m=1}^{n}\frac{\alpha}{t_0}\Delta V_{\text{th0}}(t/t_0)^{\alpha_m-1}dt \\ &= \Delta V_{\text{th}}(n-1)+\frac{\alpha}{t_0}\Delta V_{\text{th0}}(t/t_0)^{\alpha_n-1}dt\end{aligned} \tag{1.11}$$

It should be noted that ΔV_{th} does not change as a function of time when the gate bias V_g is zero at time step n. Therefore, Equation (1.11) can be rewritten as

$$\Delta V_{\text{th}}(n) = \begin{cases} = \Delta V_{\text{th}}(n-1)+\dfrac{\alpha}{t_0}\Delta V_{\text{th0}}\left(\dfrac{t}{t_0}\right)^{\alpha_n-1} dt; & if\, V_{g,n}\neq 0 \\ = \Delta V_{\text{th}}(n-1)+0; & if\, V_{g,n}=0\end{cases} \tag{1.12}$$

where $V_{g,n}$ is the gate bias V_g at time step n.

It should be noted that the previous equations are derived under the condition that V_d is fixed at V_{DD}, and V_g is biased by step pulse or a sequence of AC pulses, which is a typical way in which hot-carrier injection is measured.

1.3 Assessment of Nanotransistors Near Scaling Limit

1.3.1 2D Material-Based MOSFET Characteristics with ~10 nm Gate Length

New emerging 2D semiconductor materials such as MoS_2, which belongs to the transition metal dichalcogenides (TMDCs) material family, have been extensively studied in research [26–28]. The ballistic performance limit of monolayer MoS_2 FET has been studied in literature [29,30]. Here, we investigate the effects of phonon scattering on the device performance as well as the scaling limit of MoS_2 transistors [31].

A double-gated MOSFET is studied here, as shown in Figure 1.6. Monolayer MoS_2 is used as the channel material with the source and drain heavily doped. By solving the Schrodinger equation coupled with Poisson equation, the $I_D - V_D$ characteristics of a monolayer MoS_2 FET are shown in Figure 1.7 with $L_G = 5$ nm. The electron-phonon interaction is treated by the self-consistent Born approximation within the NEGF formalism [32]. With scattering effect, the on-state current is degraded with $L_G = 5$ nm compared with the on-current in ballistic limit. This highlights the significance of phonon scattering on accurate estimation of performance limits of monolayer MoS_2 FETs even with a channel length

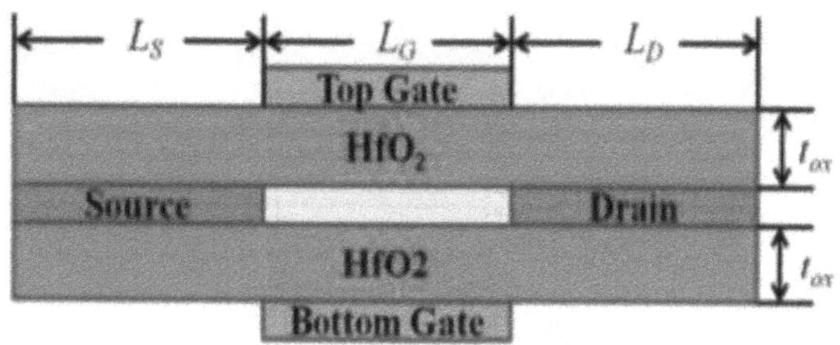

FIGURE 1.6
Cross section of studied MoS_2 MOSFET with L_G gate length. The source and drain are heavily doped.

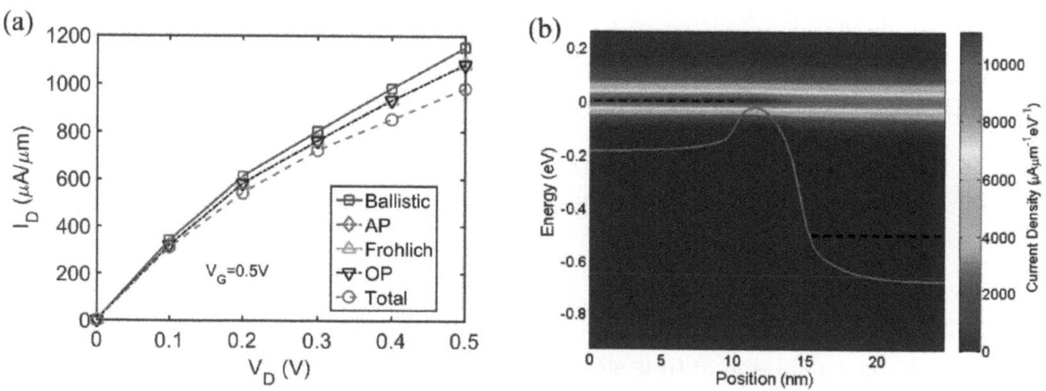

FIGURE 1.7
NEGF simulation of MoS_2 horizontal homojunction FETs with $L_G = 5$ nm. (a) I_D–V_D characteristics at the ballistic limit and in the presence of various phonon scattering mechanisms, including acoustic phonon (AP), optical phonon (OP), and Frohlich scattering. (b) The energy resolved current spectrum at on state with phonon scattering shows energy relaxation and back scattering. The solid line is a potential profile, and the black dashed curve is the Fermi level at the source (drain).

down to 5 nm. Because of the large direct source-drain tunneling, the FET with $L_G = 5$ nm does not have good $I_D - V_D$ saturation behavior as shown in Figure 1.7a. Figure 1.7b shows the on-state position-resolved current spectrums of monolayer MoS_2 FETs with $L_G = 5$ nm. It shows that when electrons travel across the top of a potential barrier, the energy of electrons is relaxed due to phonon scattering. Hence, the current spectrum is broadened, and the energy of peak current density is lowered after phonon scatterings. For a gate length with 5 nm, energy relaxation is not significant since electron transport is quasi-ballistic, compared with Figure 1.2b. In addition, direct source-to-drain tunneling is more significant since the energy barrier of a device with $L_G = 5$ nm is thinner.

We also compare the channel-length scaling behavior of drain-induced barrier lowering (DIBL) and subthreshold swing (SS) between monolayer MoS_2 FETs and ultra-thin-body Si FETs in Figure 1.8. The body thickness of Si film is 3 nm, and all the other device parameters are the same as that of MoS_2 FET. It shows that DIBL and SS are independent of phonon scattering since phonon scattering has little impact to the subthreshold region. Both DIBL and SS increase quickly when the channel length decreases due to large source-drain tunneling. For both monolayer MoS_2 FETs and ultra-thin-body Si FETs, the DIBL disappears and SS approaches the 60 mV/decade limit when the gate length is beyond 15 nm. In short-channel FETs, severe source-to-drain tunneling results in large degradation of DIBL and SS as shown in Figure 1.8b. When the channel length scales down to sub-10 nm, monolayer MoS_2 FETs show smaller DIBL and better SS compared to 3-nm-thick-body Si FETs. The reason is that MoS_2 has an atomic thin body that would provide better gate control and has higher effective mass, which reduces direct source-to-drain tunneling. If DIBL of ~100 mV/V is a practically acceptable value [33], Figure 1.8a suggests a minimum gate length of ~8 nm for monolayer MoS_2 FETs and 10 nm for ultra-thin-body Si FETs, respectively. Their corresponding SS of ~75 and 77 mV/decade, respectively, are also acceptable for practical applications [33]. In conclusion, the gate length of monolayer MoS_2 FETs can be scaled down to 8 nm, while the gate length scaling limit of 3-nm-thick-body Si FETs is 10 nm.

1.3.2 TMDC FET Performance at Sub-5 nm Gate Length Scale

Further scaling gate length to sub 5 nm is necessary since the ITRS requires a gate length of 5.9 nm in the year of 2026 [34]. Recently, S. B. Desai et al. have demonstrated a MoS_2 transistor

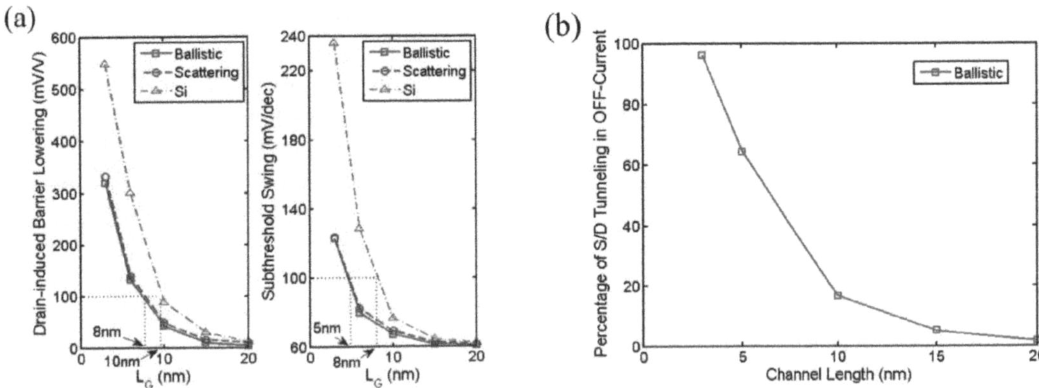

FIGURE 1.8
(a) DIBL and SS of monolayer MoS_2 FETs and ultra-thin-body Si FETs versus gate length. (b) Percentage of direct source-drain tunneling in off-state current in ballistic case. Different gate work functions are used to achieve the specific off-current (0.1 μA/μm).

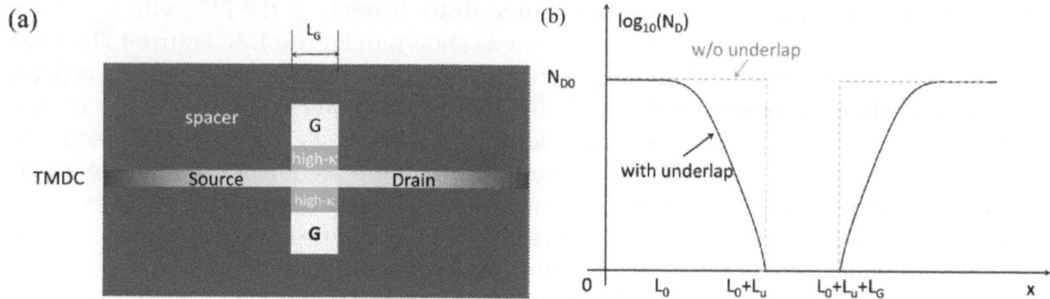

FIGURE 1.9
Schematic cross section of the modeled, double-gated MoS_2 FET. (a) The high-κ gate insulator, metal gate (*G*), and gate spacer are denoted. The dashed line shows the modeled region. (b) Sketch of the doping density as a function of the position with (solid line) and without (dashed line) the gate underlap doping. The modeled source extension length (L_0), gate underlap length (L_u), and metal gate length (L_G) are denoted. The doping density in the source extension is N_{D0}, and a Gaussian doping profile is used in the gate underlap region. The gate insulator dielectric constant is $\kappa_{ins} = 20$. Three FETs are simulated: (i) a baseline FET, which is without spacer (i.e., κ = 20 in the spacer region) and without underlap doping; (ii) a FET with spacer (i.e., $\kappa_{sp} = 4$ in the spacer region) but without underlap doping; (iii) an underlap-doped FET with spacer, which is with underlap doping and spacer (i.e., $\kappa_{sp} = 4$ in the spacer region).

with a 1 nm gate length [35], which shows the importance of a gate-fringing field to achieve ultimate channel-length scaling. By solving the non-equilibrium Green's function (NEGF) transport equation self-consistently with a Poisson equation, device scaling, design, and performance potentials of 2D TDMC FETs in the sub-5 nm gate length scale are studied [36].

Figure 1.9a is the schematic of the modeled double-gated monolayer TMDC FET with a gate length of L_G, high-κ insulator thickness of $t_{ins} = 3$ nm, and dielectric constant of $\kappa_{ins} = 20$. The monolayer MoS_2 material is used as a representative case with the effective mass as $m_x \approx m_y \approx 0.57m_0$. Three types of devices are proposed and simulated to study the effects of the gate spacer dielectric constant and the underlap doping: (i) a baseline FET, which is without a gate spacer (i.e., κ = 20 in the spacer region) and without the underlap doping, (ii) a FET with spacer, which is with a gate spacer of $\kappa_{sp} = 4$ and without underlap doping, and (iii) an underlap-doped FET with spacer, which is with a gate spacer of $\kappa_{sp} = 4$ and underlap doping. For the modeled devices without underlap doping, the doping density changes abruptly from $N_{D0} = 10^{13}/cm^2$ in the source and drain regions to 0 in the intrinsic gated region, as noted in the dashed line in Figure 1.9b. For the modeled devices with underlap doping, there exists a gate underlap region with a length of L_u and a Gaussian profile of the doping density, as shown in the solid line in Figure 1.9b.

Figure 1.10a and b show the simulated I-V characteristics at $V_D = 0.05$V and $V_D = V_{DD} = 0.5$V for a baseline FET with $L_G = 1$ nm. The gate work function is tuned to produce an off-current of $I_{OFF} = 0.1$ μA/μm. A large transconductance of 5106 μS/μm, on-current of $I_{ON} = 820$ μA/μm, and an on-/off-current ratio of about 8200 are achieved even with a 1 nm gate length and lower power supply voltage 0.5V. The calculated SS and the DIBL are 84.6 mV/dec and 107 mV/V, respectively, which indicate relatively good subthreshold characteristics at a short gate length of $L_G = 1$ nm. The percentage of tunneling current in the off-current is also studied. Figure 1.10c shows the conduction band profile for the baseline FET at off state ($V_G = 0$V, $V_D = 0.5$V), and the energy-resolved current density is plotted in Figure 1.10d. The dashed line denotes the top of the barrier. The current density is from direct source-to-drain tunneling when the electron energy is below the top of the barrier. Although the relatively large effective mass of MoS_2 suppresses the

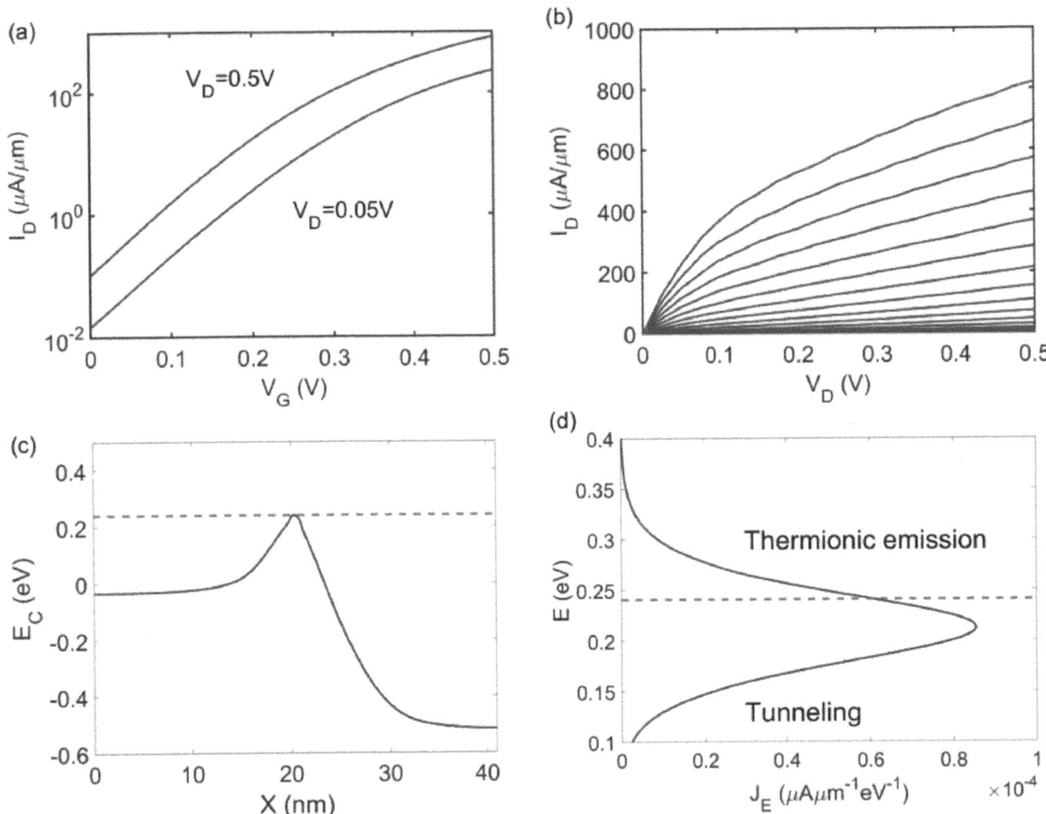

FIGURE 1.10
(a) I_D–V_G and (b) I_D–V_D characteristics of the baseline FET. The gate voltage ranges from 0 V to 0.5 V at 0.025 V per step. The modeled baseline FET is shown in Figure 1.9, with a gate length L_G = 1 nm. (c) The corresponding conduction band profile and (d) energy-resolved current density of the baseline FET is at the off state. The dashed line denotes the top of the barrier.

direct source-to-drain tunneling, source-to-drain tunneling is still dominant at off state for the baseline FET with L_G = 1 nm.

With the short gate length and the high-κ dielectrics in the gate spacer region in the baseline FET, a large gate-fringing capacitance is expected, which will lower the switching speed and frequency performance. Hence, we examine the effect of the spacer region by using dielectrics with κ_{sp} = 4. The simulated I-V characteristics for the FET with spacer are shown in Figure 1.11a and b. The figure shows that the design of the gate spacer has a considerable impact on the DC I-V characteristics of the device with L_G = 1 nm. At the same specified off-current and power supply voltage, the on-current reduces to 269 μA/μm, and the SS and DIBL increase to 101 mV/dec and 267 mV/V, respectively. Meanwhile, the output conductance increases, and the output I-V characteristics show lack of saturation. This indicates that the high-κ dielectrics in the gate spacer region can be used to improve the immunity to electrostatic SCEs at a short gate length by increasing the effective channel length through the gate-fringing field in the spacer region.

Next, we investigate the gate underlap doping effect on the device performance. Figure 1.11c and d show the I-V characteristics for the underlap-doped FET with spacer with L_G = 1 nm. A low SS of 81 mV/dec and a DIBL of 94 mV/V can be achieved, which are lower than both baseline FET and a FET with spacer. With the underlap doping,

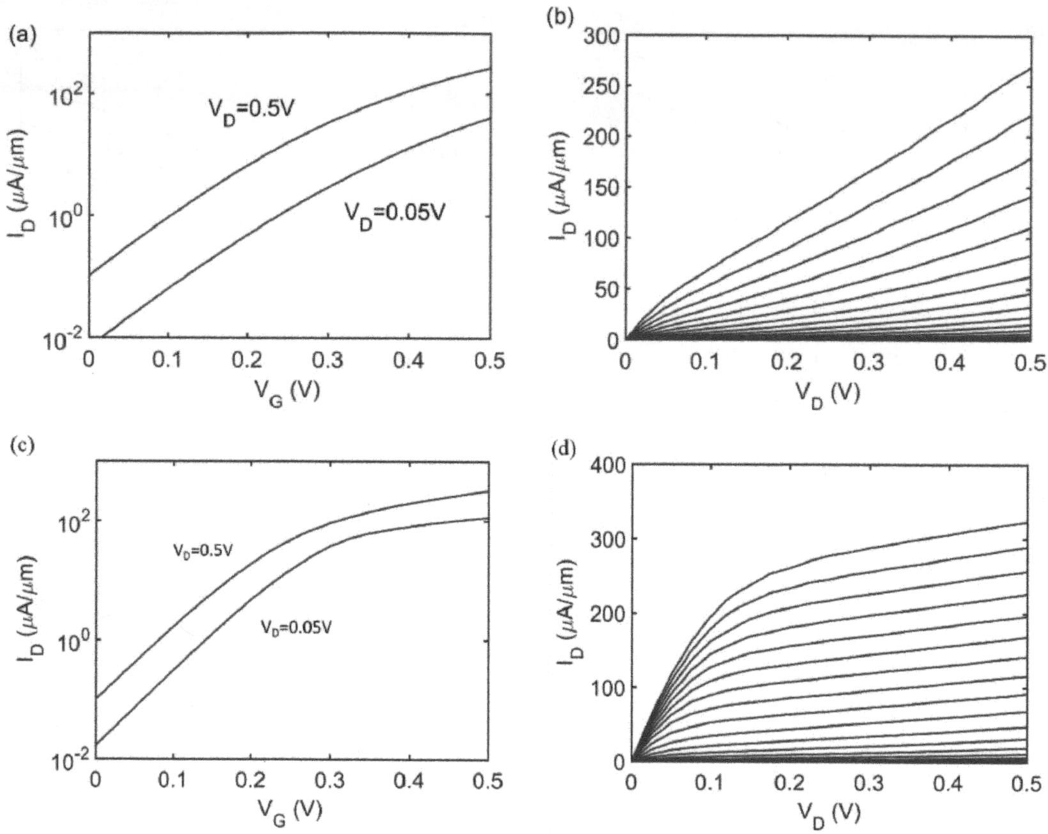

FIGURE 1.11
(a) I_D–V_G characteristics and (b) I_D–V_D characteristics of the FET with spacer and (c) I_D–V_G characteristics and (d) I_D–V_D characteristics of the underlap-doped FET with spacer at a gate length L_G = 1 nm. The gate voltage ranges from 0 V to 0.5 V at 0.025 V per step. The underlap doping length is L_u = 10 nm and the steepness is 1 nm/dec. The gate work function is tuned to achieve the specified off-current of I_{OFF} = 0.1 μA/μm.

the on-current value increases from 269 μA/μm for a FET with spacer to 323 μA/μm due to improved SS. And the output I-V characteristics saturate better with the output conductance value drops to 162 μS/μm. The underlap doping design can significantly improve the immunity to electrostatic SCEs.

Figure 1.12 presents the scaling behaviors with regard to the gate length L_G for both the baseline FET and the underlap-doped FET with spacer. When L_G scales from 5 nm down to 1 nm, the SS increases to a reasonable range for the underlap-doped FET with spacer and the baseline FET, respectively, as shown in Figure 1.12a. The scaling behaviors of the on-current, off-current, and on-/off-current ratios are shown Figure 1.12b. The figure shows that the on-current and off-current of underlap-doped FET with spacer are less sensitive to the gate length variation compared to the baseline FET, indicating that the FET designs with underlap doping are beneficial for reducing the device variability due to L_G variation.

1.3.3 Performance Limits of Device Applications

In addition, we access the performance limit of the 2D FET at sub-5 nm gate length for both high-frequency application and digital electronics application [36].

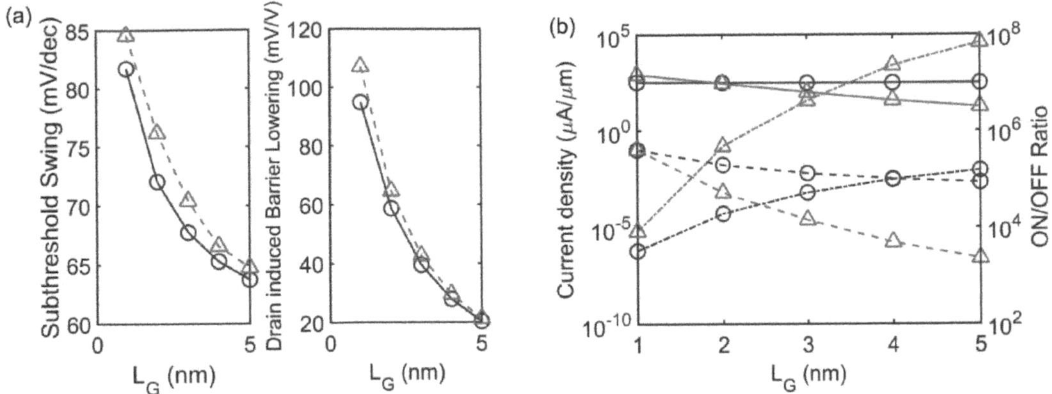

FIGURE 1.12
Scaling behaviors with regard to the gate length: (a) The SS and DIBL and (b) the on-current (solid), off-current (dashed), and on-/off-current ratio (dash-dot) as a function of the gate length L_G down to 1 nm. The lines with circles are for the underlap-doped FET with spacer, and the lines with triangles are for the baseline FET. For the underlap-doped FET with spacer, the underlap doping length is L_u = 10 nm and the steepness is 1 nm/dec.

A quasi-static treatment is used to study the high-frequency performance of the MoS_2 FETs. An equivalent circuit model is numerically extracted. The gate capacitance C_g and the transconductance g_m can be expressed as

$$c_g = \left.\frac{\partial Q_g}{\partial V_g}\right|_{V_d} \tag{1.13}$$

$$g_m = \left.\frac{\partial I_D}{\partial V_g}\right|_{V_d} \tag{1.14}$$

The intrinsic cut-off frequency f_T is computed as

$$f_T = \frac{1}{2\pi}\frac{g_m}{C_g} \tag{1.15}$$

The gate capacitance, transconductance, and cut-off frequency values are extracted with a DC bias of $V_{DD} = 0.5$V and a common specified DC current of $I_{DC} = 1000$ μA/μm for the three FET structures. The gate capacitance and transconductance as a function of the gate length L_G for the three devices are shown in Figure 1.13a. Due to the high-κ dielectrics in the gate-fringing region, which leads to large gate-fringing capacitance, the baseline FET has the largest gate capacitance. At L_G = 1 nm, the gate capacitance can be reduced by over a factor of 2 by introducing a gate spacer with a low dielectric constant. This also shows that the underlapped doping design can further lower the gate capacitance. Since the length scale of the gate length and insulator thickness are comparable, the gate capacitance values scale down with the gate length much slower than the proportional relation that is used for MOSFET with a long gate length. Due to the degraded gate control of the channel, the transconductance also decreases as the gate length scales down.

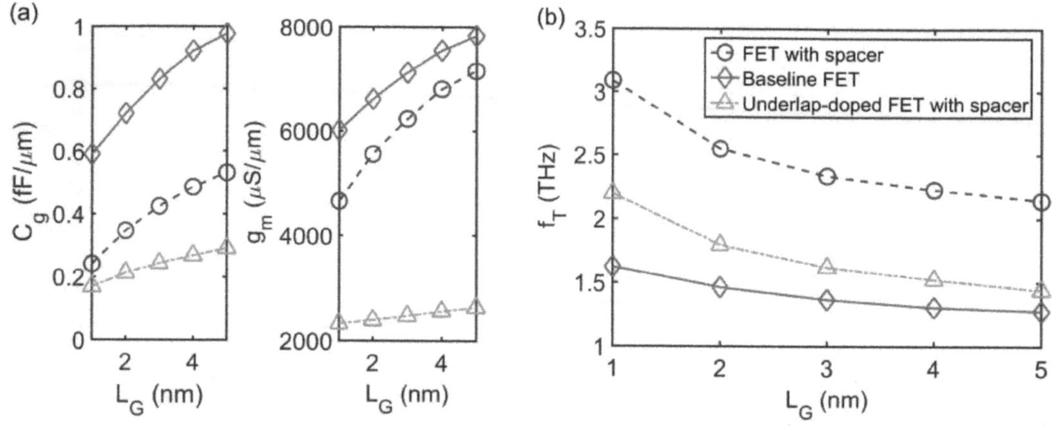

FIGURE 1.13
(a) Gate capacitance C_g and transconductance g_m and (b) cut-off frequency f_T as a function of the gate length L_G for the modeled MoS_2 FETs. The solid lines with diamonds are for the baseline FET. The dashed lines with circles are for the FET with spacer, and the dash-dot lines with triangles are for the underlap-doped FET with spacer.

Figure 1.13b shows that cut-off frequency increases as the gate length decreases at the common specified DC current. Both devices with a gate spacer have a larger cut-off frequency than the baseline FET without a gate spacer due to smaller gate-fringing capacitance. A lower dielectric constant value in the gate spacer region is favorable to improve the RF performance. At $L_G = 1$ nm, the FET with spacer has the largest cut-off frequency due to large transconductance and lower gate capacitance, while the baseline FET has the lowest cut-off frequency because of large gate-fringing capacitance. A good current saturation and small output conductance in the $I_D - V_D$ characteristics are preferred to achieve a good power gain frequency (f_{max}) performance. Among three modeled devices, the underlap-doped FET with spacer has the best saturation in the output I-V characteristics as well as relative large cut-off frequency, which are mostly preferred for improving the power gain frequency f_{max}. The results indicate that designs of the gate spacer and underlap doping are important for improving the RF high-frequency performance. TMDC FETs with a 1 nm gate length can have attractive RF performance with the cut-off frequency above 2 THz by proper design of the gate spacer and underlap doping.

The performance potential of studied devices in digital electronics is analyzed next. Figure 1.14 shows the intrinsic delay of the modeled underlap-doped FET with spacer as a function of the gate length. The intrinsic gate delay is computed as $\tau = \Delta Q / I_{ON}$, where $\Delta Q = (Q_{G,ON} - Q_{G,OFF})$ and $Q_{G,ON}$ and $Q_{G,OFF}$ are the gate charge computed at the on and off states, respectively, and I_{ON} is the on-current computed at a common specified off-current of $I_{OFF} = 0.1$ μA/μm. A slight increase of intrinsic delay is observed for the underlap-doped FET with spacer as the gate length scales down from 5 nm to 1 nm. The reason is that the decrease of the on-current outpaces the decrease of the gate charge change as L_G decreases. A similar trend is also observed for both the baseline FET and the FET with spacer, all of which have a gate insulator thickness of 3 nm. However, by scaling down the gate oxide thickness, we can reverse the trend. Figure 1.14b compares the underlap-doped FET with different gate insulator thickness values. A gate insulator with dielectric constant of 4 is used for the device with a gate insulator thickness of $t_{ins} = 1$ nm. It shows that for the FETs with $t_{ins} = 2$ nm and 1 nm, the intrinsic delay decreases as the gate length decreases. Meanwhile, a large decrease of the intrinsic gate delay is shown when the gate insulator thickness decreases from 3 nm to 1 nm. A thinner gate oxide improves the

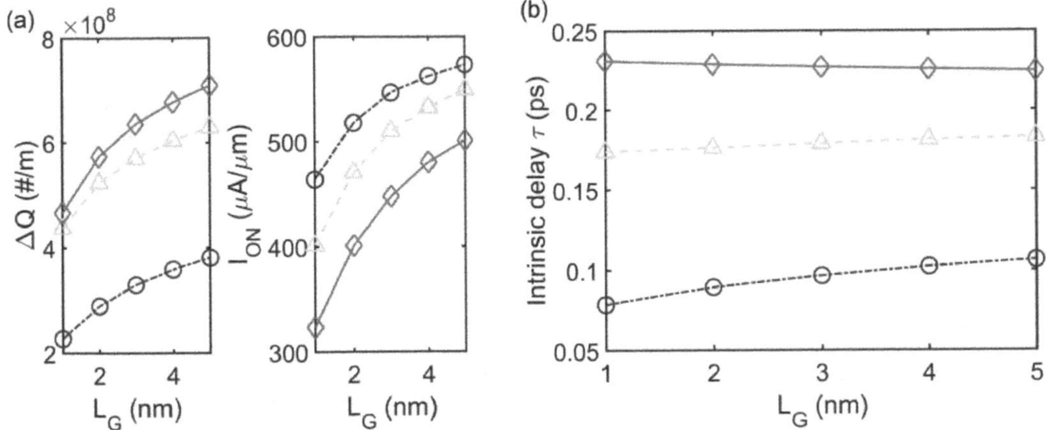

FIGURE 1.14
(a) Gate charge variation from the off to on state ΔQ and the on-current and (b) intrinsic delay as a function of L_G. Three FETs modeled all have the structure of the underlap-doped FET with spacer but with different gate insulator thickness and dielectric constant. The solid lines with diamonds have a gate insulator of $t_{ins} = 3$ nm and $\kappa = 20$. The dashed lines with triangles have a gate insulator of $t_{ins} = 2$ nm and $\kappa = 20$. The dash-dot lines with circles have a gate insulator of $t_{ins} = 1$ nm and $\kappa = 4$. A common off-current of $I_{OFF} = 0.1$ μA/μm is achieved by adjusting the gate work function.

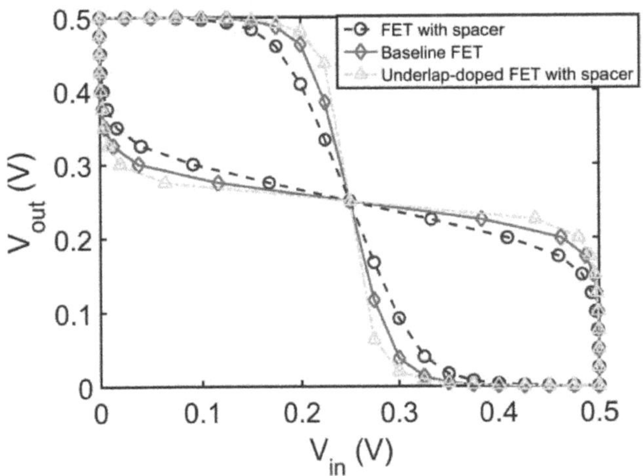

FIGURE 1.15
Simulation of inverter characteristics. The butterfly curves of the inverters based on the three modeled MoS_2 with $L_G = 1$ nm. The solid lines with diamonds are for the baseline FET, the dashed lines with circles are for FET with spacer, and the dash-dot lines with triangles are for the underlap-doped FET with spacer. Balanced *p*-type and *n*-type transistors are assumed, because the effective mass values for electrons and holes are close in monolayer MoS_2.

intrinsic gate delay because of a decreased ratio of the fringing capacitance to the total gate capacitance.

Finally, we simulate an inverter circuit based on the modeled monolayer MoS_2 FETs. Figure 1.15a shows the butterfly curves of the inverter for the three modeled devices with $L_G = 1$ nm. Balanced *p*-type and *n*-type transistors are assumed because the effective mass values for electrons and holes are close for monolayer MoS_2. The result shows that good inverter characteristics can be achieved for all three devices at a gate length of $L_G = 1$ nm. The underlap-doped FET with spacer shows the most preferred butterfly curve in terms of the largest voltage gain and static noise margin (SNM), whereas the FET with spacer shows the smallest noise margin due to lack of saturation in the output I-V characteristics as shown in Figure 1.11b. Our work demonstrates the good performance potential for 2D TMDC FET with carefully designed gate spacer and underlap doping in both RF and digital electronics applications.

1.4 Multiphysics Simulation of Nanoscale Transistors

A fully depleted silicon-on-insulator (SOI) MOSFET is taken as an example to study electrothermal effects on its hot-carrier reliability [16]. The channel length is 100 nm, and the thickness of the Si thin-film channel is 8 nm, as shown in Figure 1.16a. In order to get the heat generation rate in the channel region, various methods can be applied as mentioned previously. Here, a drift-diffusion method is applied since the channel length is in the scale of 100 nm, which is relatively long, and it can be a reasonable treatment. The simulated *J-V* curves agree well with experiments as shown in Figure 1.16b.

Typically, the SOI MOSFETs are fabricated with several hundred micrometers thick Si substrate and several micrometers thick Si_3N_4 insulator. To mimic the heat conduction process in a real chip and meanwhile save computational cost, two thin layers with $d = 50$ nm thickness are added on top and at bottom as shown in Figure 1.17a. The top thin layer has its thermal resistance R_{th} as that of $l_{Si_3N_4} = 2$ μm Si_3N_4, and the bottom one has R_{th} as that of $l_{Si} = 200$ μm Si. With the top layer as an example, its effective thermal conductivity can be written as

$$\kappa_{eff,\,top} = d/R_{th} = d/(l_{Si_3N_4}/\kappa_{Si_3N_4}) = \kappa_{Si_3N_4} d/l_{Si_3N_4} \tag{1.16}$$

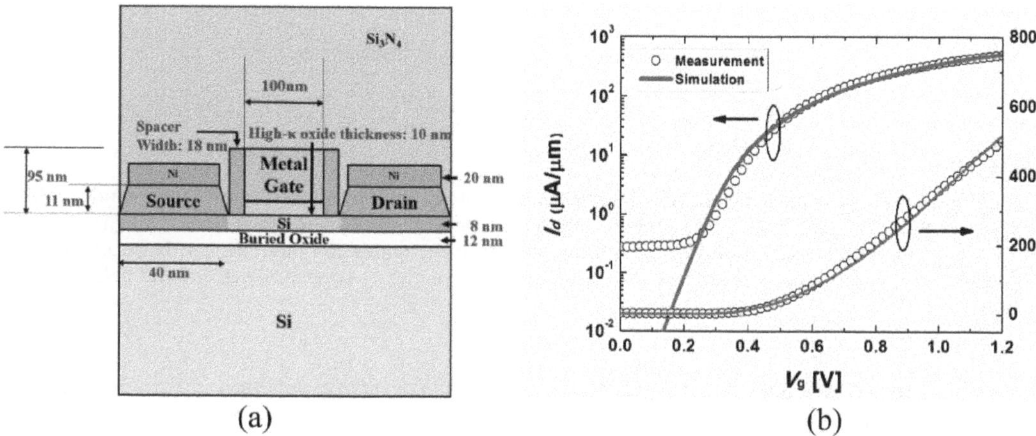

FIGURE 1.16
(a) Schematic of a fully depleted SOI MOSFET with its dimensions as noted. The channel length is 100 nm. (b) Comparison between the simulated and measured I_d–V_g curve in linear plot and logarithmic scale.

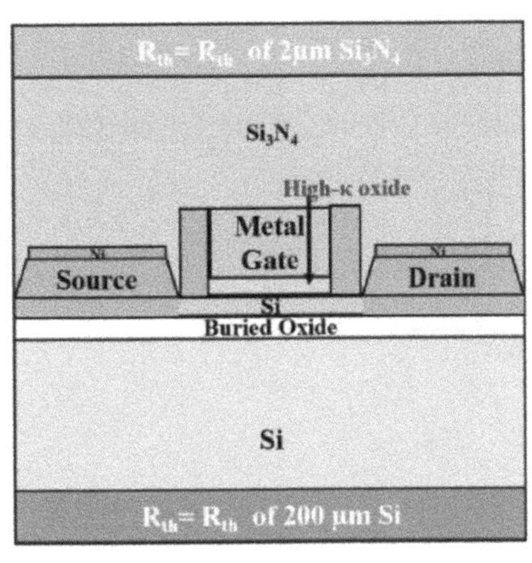

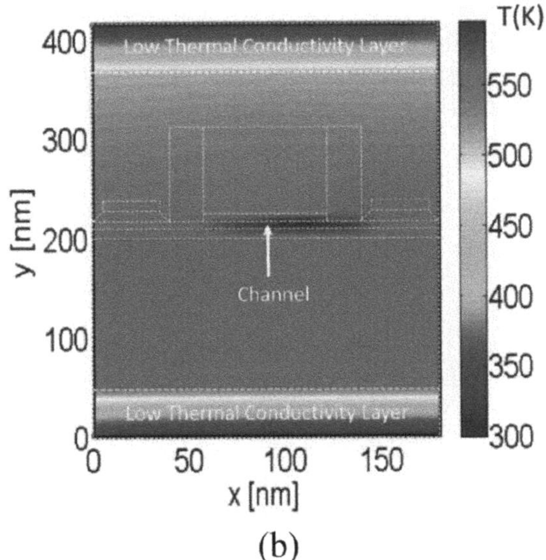

FIGURE 1.17
(a) The schematics for modeling of heat conduction. The top and bottom thin layers are with the same thermal resistance as 2 μm Si_3N_4 and 200 μm Si substrate to mimic the real environment around the active region of the transistor. (b) Temperature profile at steady state V_g = 1.2V and V_d = 1.2V. The top and bottom layers are with low thermal conductivity to mimic the thick insulator and substrate.

Similarly, the effective thermal conductivity of the bottom layer can be obtained.

By using the finite element method and following the simulation procedures illustrated in Figure 1.5, the steady temperature profile is obtained and plotted in Figure 1.17b. The white dotted line shows the skeleton of the modeling structure as in Figure 1.17a. The top and bottom low thermal conductivity layers are used to mimic the effects of 2 μm Si_3N_4 and 200 μm Si substrate. It is observed that the highest temperature is localized around the drain region and is about 600K.

In state-of-the-art digital ICs, transistors are typically operated at GHz frequencies. In traditional experiments and simulations, AC signals with MHz frequencies are usually used to characterize the HCI. To further capture the electrothermal effects on HCI behavior under transient signal stress, time-dependent thermal conduction needs to be further studied. The AC signal with frequency 2.5 GHz is shown in Figure 1.18. Both rising and falling times are 50 ps, and the period is 400 ps, with 1-bit endurance as 200 ps.

The temperature responses to the step pulse and AC signal with 1.2V magnitude are shown in linear and logarithmic timescale as plotted in Figure 1.19. The temperature is the highest temperature captured in the channel at each time step. The thermal response time is around 300ns, as shown in the temperature response to step pulse. The temperature for the step pulse is much higher than that of the AC pulse because the input power of the step pulse is higher.

Based on the temperature response in Figure 1.19 and Equation (1.12), the TVS evolution up to 10 years for the SOI MOSFET under different signal stresses can be obtained as shown in Figure 1.20. TVS for AC with time larger than 4 ms is extrapolation based on simulation for saving computational cost. The slope of extrapolation is obtained by linearly fitting the simulation results from 0.4 ms to 4 ms. The step pulse has the largest TVS due to these reasons: (i) the temperature response to step pulse has the largest magnitude

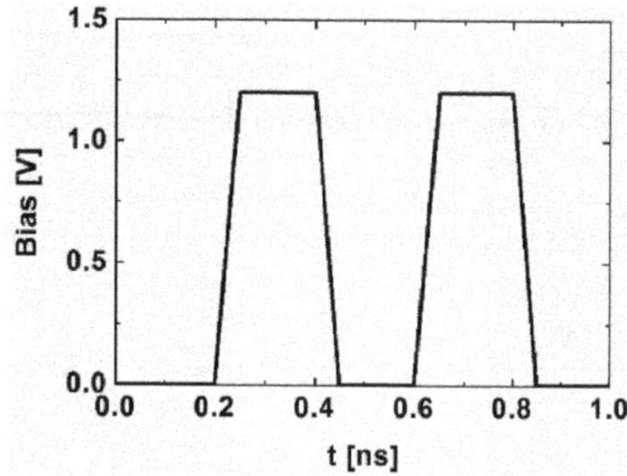

FIGURE 1.18
AC signal with both pulse rising and falling times as 50 ps. The period for the pulse is 400 ps and corresponds to frequency 2.5 GHz.

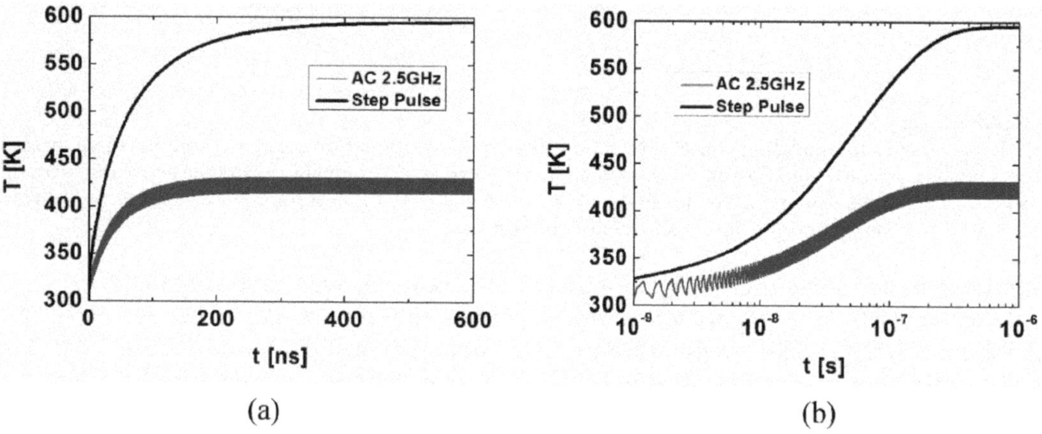

FIGURE 1.19
Temperature response in linear timescale (a) and logarithmic scale (b) to different signal stresses.

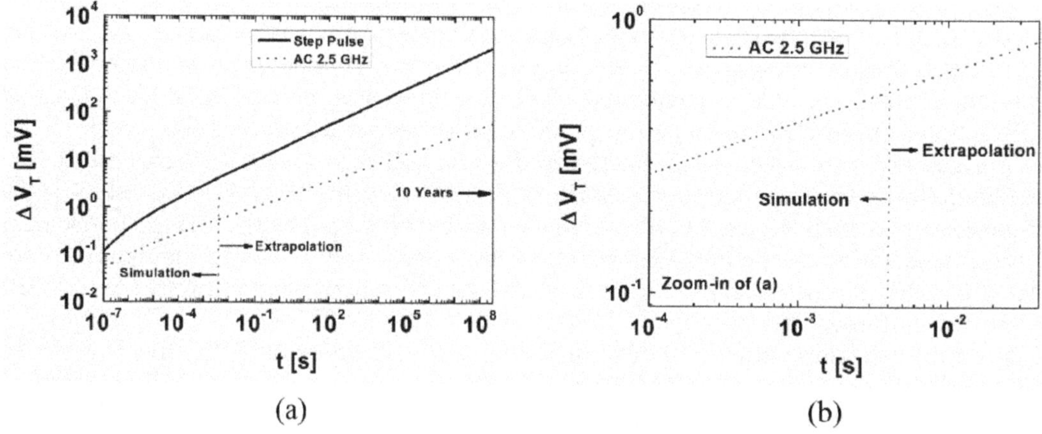

FIGURE 1.20
(a) Simulated threshold voltage shift ΔV_T as a function of time for different signal biases. Data for AC with time larger than 4 ms is extrapolation based on simulation for saving computational cost. (b) Zoom-in of (a). The slope of extrapolation is obtained by linearly fitting the simulation results from 0.4 ms to 4 ms.

and it leads to large temperature-dependent exponential α, and (ii) the SOI MOSFET is always in on state when it is stressed by the step pulse, while the device is half on and half off when it is stressed by the AC signals.

The temperature responses to AC signals with different frequencies are presented in Figure 1.21a. As the frequency decreases, the temperature oscillation magnitude increases. Since both on and off states last long for the low frequency signal, the transistor has a long, continuous time to absorb and dissipate the heat. As a result, long endurance of the on or off state leads to more time for the temperature to increase or fall.

Similarly, based on the temperature response to AC signals, the TVS evolution up to 10 years for the SOI MOSFET under AC signal stresses with different frequencies is shown in Figure 1.21b. As the frequency decreases, the slope of TVS evolution increases due to a large temperature magnitude during the on state when the device is stressed by a low frequency signal.

The values of TVS in 10 years for the SOI MOSFET under stress of various types of signals are summarized in Figure 1.22. It should be noted that there is a breakpoint in the y axis

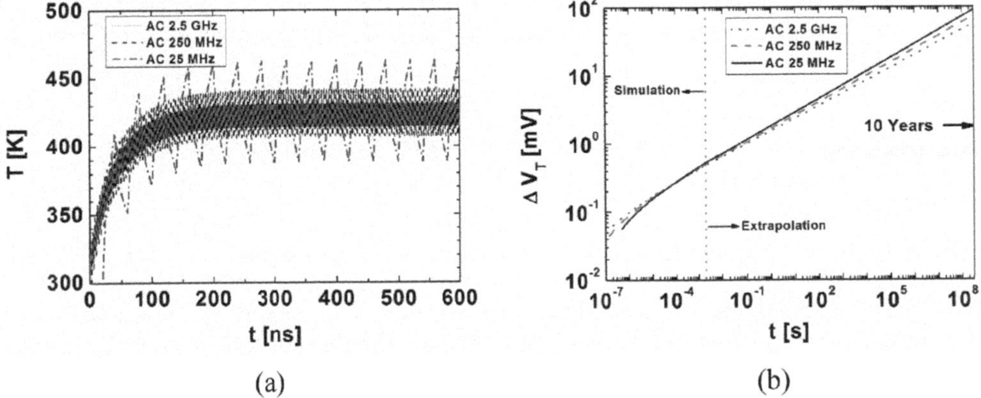

FIGURE 1.21
(a) Temperature response to the AC signals with different frequencies. The lower frequency results in large temperature oscillation. (b) Simulated threshold voltage shift ΔV_T as a function of time for the AC signals with different frequencies.

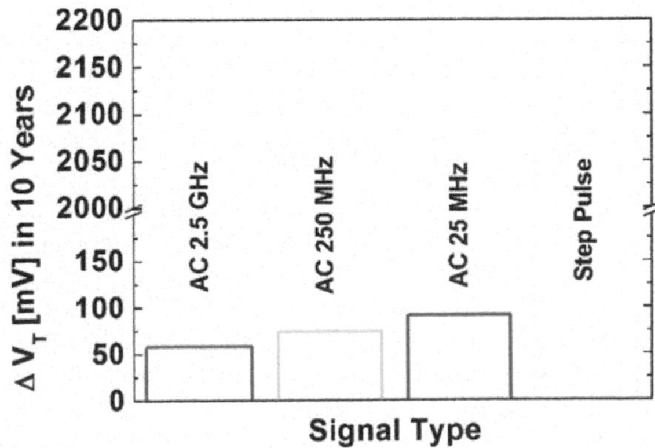

FIGURE 1.22
Prediction of the threshold voltage shift ΔV_T for the SOI MOSFET under 10-year stress of various signals.

in order to show the data more clearly. Although the TVS evolves differently for devices with different temperature exponentials as in Equation (1.9), the following trends should remain the same when the HCI is measured by applying AC or step pulse as gate bias with fixed $V_D = V_{DD}$. Firstly, for the sub-100 nm n-type SOI MOSFETs, the TVS under step pulse bias is worse than AC bias. Secondly, as the frequency of a measurement AC signal decreases from GHz to tens of MHz, the TVS increases.

Acknowledgment

Z. D. and J. G. were supported in part by the National Science Foundation under Awards #1618762, #1610387, and #1809770 . W. C. and W. Y. Y. were supported in part by the National Natural Science Foundation of China under grants 61431014, 61504121. The authors would also like to thank the supports from the ECE Department of University of Florida, the College of ISEE of Zhejiang University, and the ZJU-UIUC Institute of Zhejiang University. And this work was led by Principal Investigators: Jing Guo, Wen-Yan Yin, and Wenchao Chen.

References

1. L. V. Keldysh, "Diagram technique for non-equilibrium processes," *Sov. Phys. JETP*, vol. 20, p. 1018, 1965.
2. S. Datta, *Quantum Transport: Atom to Transistor*, Cambridge, UK: Cambridge University Press, 2005.
3. S. Datta, *Electronic Transport in Mesoscopic Systems*, Cambridge, UK: Cambridge University Press, 1995.
4. D. K. Ferry and S. M. Goodnick, *Transport in Nanostructures*, Cambridge, UK: Cambridge University Press, 1997.
5. A. Nitzan and M. A. Ratner, "Electron transport in molecular wire junctions," *Science*, vol. 300, pp. 1384–1389, 2003.
6. G. D. Mahan, *Many-Particle Physics*, 3rd ed., New York: Kluwer Academic/Plenum Publishers, 2000.
7. Y. Lu and J. Guo, "Thermal transport in grain boundary of graphene by non-equilibrium Green's function approach," *Appl. Phys. Lett.*, vol. 101, pp. 043112-1–043112-5, 2012.
8. J. H. Seol, I. Jo, A. L. Moore, L. Lindsay, Z. H. Aitken, M. T. Pettes, X. Li et al., "Two-dimensional phonon transport in supported graphene," *Science*, vol. 328, pp. 213–216, 2010.
9. C. Fiegna, Y. Yang, E. Sangiorgi, and A. G. O'Neill, "Analysis of self-heating effects in ultra-thin-body SOI MOSFETs by device simulation," *IEEE Trans. Electron Devices*, vol. 55, no. 1, pp. 233–244, 2008.
10. D. Vasileska, K. Raleva, and S. M. Goodnick, "Self-heating effects in nanoscale FD SOI devices: the role of the substrate, boundary conditions at various interfaces, and the dielectric material type for the BOX," *IEEE Trans. Electron Devices*, vol. 56, no. 12, pp. 3064–3071, 2009.
11. E. Pop, C. O. Chui, S. Sinha, R. Dutton, and K. Goodson, "Electro-thermal comparison and performance optimization of thin-body SOI and GOI MOSFETs," *IEEE IEDM Tech. Dig.*, 2004.
12. E. Pop, S. Sinha, and K. E. Goodson, "Heat generation and transport in nanometer-scale transistors," *Proc. IEEE*, vol. 94, no. 8, pp. 1587–1601, 2006.
13. E. X. Zhao, J. Chan, J. Zhang, A. Marathe, and K. Taylor, "Bias and temperature dependent hot-carrier characteristics of sub-100nm partially depleted SOI MOSFETs," *IRW Final Report*, 2002.

14. P. Aminzadeh, M. Alavi, and D. Scharfetter, "Temperature dependence of substrate current and hot carrier-induced degradation at low drain bias," *Symposium on VLSI Technology*, pp. 178–179, 1998.
15. N. Sano, M. Tomizawa, and A. Yoshii, "Temperature dependence of hot carrier effects in short-channel Si-MOSFET's," *IEEE Trans. Electron Devices*, vol. 42, no. 12, pp. 2211–2216, 1995.
16. W. Chen, R. Cheng, D. W. Wang, H. Song, X. Wang, H. Chen, E. Li, W. Y. Yin, and Y. Zhao, "Electrothermal effects on hot-carrier reliability in SOI MOSFETs-AC versus circuit-speed random stress," *IEEE Trans. Electron Devices*, vol. 63, no. 9, pp. 3669–3676, 2016.
17. D. K. Ferry, *Semiconductor Transport*, New York: Taylor & Francis Group, 2000.
18. M. Lundstrom, *Fundamentals of Carrier Transport*, 2nd ed., Cambridge, UK: Cambridge University Press, 2000.
19. E. Pop, J. Rowlette, R. W. Dutton, and K. E. Goodson, "Joule heating under quasi-ballistic transport conditions in bulk and strained silicon devices," in *Proceedings of the International Conference on Simulation of Semiconductor Processes and Devices* (*SISPAD*), pp. 307–310, 2005.
20. D. Vasileska, K. Raleva, and S. M. Goodnick, "Electrothermal studies of FD SOI Devices that utilize a new theoretical model for the temperature and thickness dependence of the thermal conductivity," *IEEE Trans. Electron Devices*, vol. 57, no. 3, pp. 726–728, 2010.
21. W. Liu and M. Asheghi, "Thermal conductivity of measurements of ultra-thin single crystal silicon layers," *J. Heat Transfer*, vol. 128, no. 1, pp. 75–83, 2006.
22. W. Liu, K. Etessam-Yazdani, R. Hussin, and M. Asheghi, "Modeling and data for thermal conductivity of ultrathin single-crystal SOI layers at high temperature," *IEEE Trans. Electron Devices*, vol. 53, no. 8, pp. 1868–1876, 2006.
23. Y. W. Kwon and H. Bang, *The Finite Element Method Using MATLAB*, 2nd ed., Boca Raton, FL: CRC Press, 2000.
24. E. Takeda and N. Suzuki, "An empirical model for device degradation due to hot-carrier injection," *IEEE Electron Device Lett.*, vol. 4, no. 4, pp. 111–113, 1983.
25. S. Poli, S. Reggiani, M. Denison, E. Gnani, A. Gnudi, and G. Baccarani, "Temperature dependence of the threshold voltage shift induced by carrier injection in integrated STI-based LDMOS transistors," *IEEE Electron Device Lett.*, vol. 32, no. 6, pp. 791–793, 2011.
26. B. Radisavljevic, A. Radenovic, J. Brivio, V. Giacometti, and A. Kis, "Single-layer MoS_2 transistors," *Nature Nanotechnol.*, vol. 6, no. 3, pp. 147–150, 2011.
27. H. Liu and P. D. Ye, "MoS_2 dual-gate MOSFET with atomic-layer deposited Al_2O_3 as top-gate dielectric," *IEEE Electron Device Lett.*, vol. 33, no. 4, pp. 546–548, 2012.
28. S. Das, H.-Y. Chen, A. V. Penumatcha, and J. Appenzeller, "High performance multilayer MoS_2 transistors with scandium contacts," *Nano Lett.*, vol. 13, pp. 100–105, 2012.
29. L. Liu, S. B. Kumar, Y. Ouyang, and J. Guo, "Performance limits of monolayer transition metal dichalcogenide transistors," *IEEE Trans. Electron Devices*, vol. 58, no. 9, pp. 3042–3047, 2011.
30. Y. Yoon, K. Ganapathi, and S. Salahuddin, "How good can monolayer MoS_2 transistors be?" *Nano Lett.*, vol. 11, no. 9, pp. 3768–3773, 2011.
31. L. Liu, Y. Lu and J. Guo, "On monolayer MoS_2 field-effect transistors at the scaling limit," *IEEE Trans. Electron Devices*, vol. 60, no. 12, pp. 4133–4139, 2013.
32. D. Nikonov, H. Pal, and G. Bourianoff, "Scattering in NEGF: Made Simple [Online]," November 2009. Available: https://nanohub.org/resources/7772.
33. D. J. Frank, R. H. Dennard, E. Nowak, P. M. Solomon, Y. Taur, and H.-S. P. Wong, "Device scaling limits of Si MOSFETs and their application dependencies," *Proc. IEEE*, vol. 89, no. 3, pp. 259–288, 2001.
34. The International Technology Roadmap for Semiconductors (ITRS). Available: http://www.itrs.net/.
35. S. B. Desai, S. R. Madhvapathy, A. B. Sachid, J. P. Llinas, Q. Wang, G. H. Ahn, G. Pitner et al., "MoS_2 transistors with 1-nanometer gate lengths," *Science*, vol. 354, pp. 99–102, 2016.
36. Z. Dong and J. Guo, "Assessment of 2D transition metal dichalcogenide FETs at sub-5nm gate length scale," *IEEE Trans. Electron Devices*, vol. 60, no. 12, pp. 4133–4139, 2013.

2

Variability in Nanoscale Technology and EδDC MOS Transistor

Sarmista Sengupta and Soumya Pandit

CONTENTS

2.1 Introduction

For the last 50 years, the driving force behind the semiconductor industry has been device scaling and the ability to manufacture devices with smaller geometries. Due to the exponential increase in the number of metal-oxide-semiconductor (MOS) devices on a chip, following Moore's law, the integrated circuit (IC) market dramatically evolved from small-scale integration (SSI) to large-scale integration (LSI) and then to very large-scale integration (VLSI) era [1]. The key factors behind this dramatic evolution are reduction in cost per chip and increase of functionality. Since the creation of the first microprocessors in the 1970s, both supply voltage and gate length have decreased from 15V and 10 μm, respectively, to 0.8 V and 15 nm in state-of-the-art technologies [2,3]. This dramatic downscaling of the MOS transistors in the deep sub-micrometer and nanometer regime makes the devices and circuits vulnerable to a number of new detrimental effects. These include effects associated with printing finer geometry features, increased atomic-scale effects, and increased on-chip power densities. These effects are manifested as variations in environmental and process parameters and device/circuit aging effect.

This chapter presents an overview of various sources of variability in nanoscale devices and circuits. The device-level process variability mitigation using planar MOS technology is discussed. Nominal substrate bias and mismatch characteristics for the device are presented. The performances of the device are analytically studied and verified by extensive TCAD simulations.

2.2 Sources of Variability in Nanoscale Technology

The effects of various variabilities in nano-dimension circuits cause significant deviation of the circuit performance from the prescribed specifications. This in turn eventually leads to circuit/chip failure, and the yield is severely reduced. The different sources of variations that critically affect the performance of an integrated circuit are broadly classified as follows:

- Environmental variation
- Aging or reliability
- Process variation

2.2.1 Environmental Variation

Environmental variation corresponds to the changes during the operation of a circuit. Variations in the supply voltage (V_{DD}) and the temperature come under this category.

2.2.1.1 Supply Voltage Variation

With the scaling of technology node, the supply voltage has also been scaled down, which in turn reduces the tolerance to voltage changes within power distribution networks in CMOS integrated circuits. Runtime fluctuations in the supply voltage levels in a chip cause significant variations in gate delay. The gate delay of a circuit can be mathematically represented as [4],

$$t_{pd} = \frac{k.C_L.V_{DD}}{(V_{DD} - V_T)^\alpha} \tag{2.1}$$

where k and α are two constants, V_{DD} is the supply voltage, and V_T is the threshold voltage of the MOS transistor. A small variation in supply voltage leads to significant variation of the gate delay, which in turn may even result in the logic failure of the whole circuit. In nanometer-scale technologies, the current densities have increased over previous generations, and spatial imbalances between the currents in various parts of a chip are accentuated—particularly with the advent of multi-core systems where some cores may switch on and off entirely.

2.2.1.2 Temperature Variation

The impact of temperature on the functioning of a chip is an important factor in inducing variation. It causes transistor threshold voltages to go down and carrier mobility to increase. Threshold voltage decrease tends to speed up a circuit, while mobility deterioration slows it down. Depending on which effect wins, a circuit may show either negative or positive or mixed temperature dependence if the trend is non-uniform. Leakage power also increases with temperature in cases where this increase is substantial; the increased power can raise the temperature further, causing a feedback cycle. This positive feedback can even cause thermal runaway, where the increase in the power goes to a point that cannot be supported by the heat sink, and the chip burns out.

2.2.2 Aging or Reliability

Reliability is a serious issue in VLSI circuits. The reliability of a design often degrades by various causes such as soft errors, electromigration, hot-carrier injection, negative bias temperature instability (NBTI), crosstalk, power supply noise, and variations in the physical design. With the continuous scaling down of circuit designs achievable by the advancement in technology, the sources of reliability degradation have more serious impact within the design. In this scenario, high-performance and energy-efficient circuit design becomes a crucial challenge for the chip designers. There are significant research contributions to reliable VLSI circuit design; however, such techniques are often computationally expensive or power intensive.

2.2.3 Process Variation

To overcome the scaling limitations of MOSFET devices, recent changes in device structures, processing materials, and processing conditions have enhanced the complexities of advanced CMOS technologies. Some of these technologies are strained-silicon channels, halo doping, ultrathin gate oxides, ultra-shallow source-drain junctions, and millisecond annealing [5–8]. The increasing processing complexities, along with atomistic and quantum-mechanical limitations, have introduced an increasing amount of process variability. Process variability is considered to be the most critical source of performance variation of nano-CMOS circuits. It can affect the performances of two transistors placed side by side in different wafers or fabricated at different times.

2.3 Sources of Process Variation

Process variation is broadly categorized into inter-die and intra-die, where inter-die variation is completely systematic in nature; intra-die variation can be both systematic and random. Inter-die variation is found globally; it can be between two lots of wafers, between two wafers of same lot, or even with a single wafer. In contrast, intra-die variations occur within the same die. As suggested by the name, systematic variations cause the performance mean values to shift systematically in one direction. Random variations are quite critical as they lead to the statistical distribution (spread) of the performances. In the subsequent sections we discuss various sources of process variations. Inter-die process variability causes a shift in the mean value of the design parameters such as transistor channel length (L), channel width (W), gate oxide thickness (t_{ox}), doping density (N_{ch}), etc. Sophisticated resolution enhancement techniques (RETs) such as optical proximity correction (OPC), phase shift mask (PSM), etc., are some of the sources of systematic process variations. Apart from these RETs, layout induced strain, well-proximity effect (WPE), chemical mechanical polishing (CMP), etc., also introduce systematic process variations. The systematic nature of inter-die variability enables us to address it through layout design and more controlled RETs [9]. However, for technology nodes below 90 nm, the impact of random intra-die variability on device/circuit performance is of critical concern. In the following subsections we focus on the sources of random process variations.

2.3.1 Random Process Variation

The random process variation, which occurs within a die (intra-die), is totally statistical in nature. The impact of random process variation is a serious bottleneck in nanoscale device/circuit design and hence deserves major attention to be paid. Different sources of process variation that fall under this category are random discrete dopant (RDD), line edge roughness/line width roughness (LER/LWR), and poly gate granularity/metal gate granularity (PGG/MGG). [10].

2.3.1.1 Random Discrete Dopant

RDD is the most important source of intra-die variation. It results from the discreteness of dopant atoms, mainly in the channel region. It is a common technique to control threshold voltage V_{th} of a MOS transistor by tuning doping concentration of the channel. For a transistor, if the channel concentration is N_a, the source/drain junction depth is x_j and the channel length and width are L and W, respectively, then the number of atoms enclosed in the channel is

$$N_{ch,tot} = N_a.\left(L.W.x_j\right) \tag{2.2}$$

With continuous scaling of x_j, L, and W in the nanometer regime, the number of dopant atoms drastically goes down as is evident from (2.2), even after the channel doping concentration increases following CMOS scaling rules [11,12]. Due to the comparable dimension of the dopant atoms and channel with downscaling of the device, continuous doping assumption no longer remains valid. The number of dopants in a transistor channel becomes a discrete statistical quantity with the probability to occupy any random location as shown in Figure 2.1.

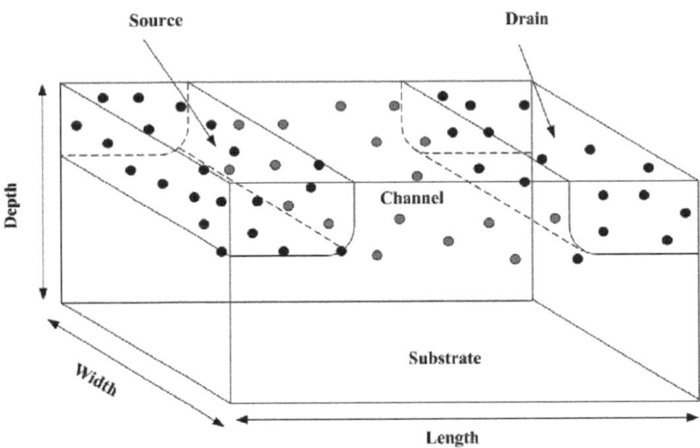

FIGURE 2.1
Random Discrete Dopant effect in both the channel and source/drain regions.

It is worth mentioning that this random nature of the dopant number and position is also observed in the source/drain regions as well. Thus, the electrical characteristics of two otherwise identical neighboring transistors in a circuit no longer remain identical. Following this physical phenomenon, threshold voltage variation (standard deviation) for a uniformly doped device is given by the Stoke's formula as [13],

$$\sigma_{V_T,\mathrm{RDD}} = \frac{q}{C_{\mathrm{ox}}}\sqrt{\frac{N_a W_{\mathrm{dm}}}{3WL}} \tag{2.3}$$

For a non-uniformly doped device, N_a has to be replaced by effective channel doping concentration N_{eff} given by [14],

$$N_{\mathrm{eff}} = 3\int_0^{W_{\mathrm{dm}}} N(x)\left(1-\frac{x}{W_{\mathrm{dm}}}\right)^2 \frac{dx}{W_{\mathrm{dm}}} \tag{2.4}$$

Here, $N(x)$ is the doping concentration along the depth of the device. Equation (2.3) shows that threshold voltage variation due to RDD continuously increases with a decrease in the channel area LW. However, it decreases with downscaling of the oxide thickness t_{ox}. It is reported that RDD contributes by 60% to the total V_T variation [10,15]. This V_T variability leads to the variation of overall performance of the circuit, such as drain current variation, delay variation, etc., as discussed earlier.

2.3.1.2 *Line Edge Roughness/Line Width Roughness (LER/LWR)*

The physical dimensions of a transistor—i.e., its length (L) and width (W)—are defined by the lithography process. Lithography is the process where a light-sensitive material, called photo-resist, is exposed to light for transferring the desired pattern from a mask or reticle to the wafer. There are various lithography techniques such as ultraviolet lithography, X-ray lithography, electron beam lithography, etc. However, each technique has its own merits and flaws. Roughness in the pattern is inherited from the lithography process itself. It may come from (i) the mismatch in the wavelength in light used with the dimension needed to be described or from (ii) intrinsic non-uniformities of the photoresist, which

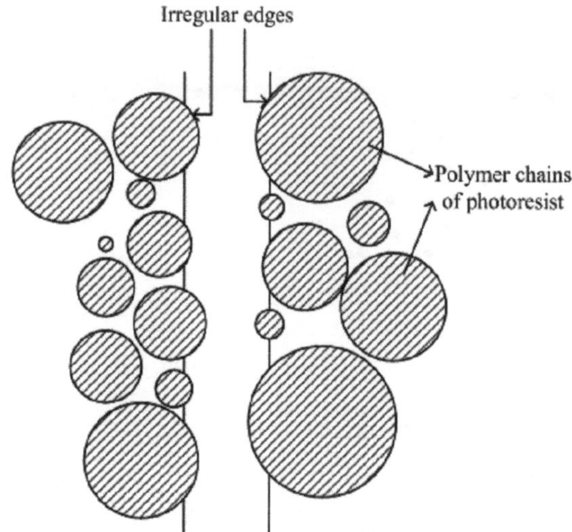

FIGURE 2.2
Polymer chain creating edge roughness.

occurs due to random dissolution and different sizes in the polymer chain as shown in Figure 2.2 [16]. This edge roughness of the pattern is termed as line edge roughness (LER). The LER is defined as the roughness of a single printed pattern edge. On the other hand, line width roughness (LWR) indicates the fluctuation in the physical distance between two printed pattern edges. The physical origin is the same for LER and LWR; these are mathematically related to each other.

Channel length, i.e., the distance between source and drain, becomes a random variable L_i as an impact of LER. In Figure 2.3, $<L>$ represents the average channel length. LER can be characterized by two quantities: (i) variability of an edge, σ_{LER} and (ii) correlation

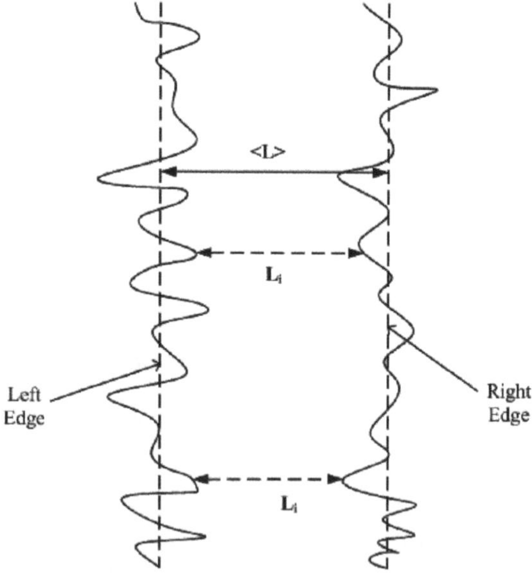

FIGURE 2.3
Line edge and line width roughness.

length, W_c. The correlation length denotes the distance beyond which the amplitudes of two points along an edge can be considered almost uncorrelated. Let the variability of left edge be σ_L and that of the right edge be σ_R. If σ_{LWR} denotes the variance of the line width, then it is given by [17],

$$\sigma_{LWR}^2 = \sigma_L^2 + \sigma_R^2 - 2\rho_X \sigma_L \sigma_R \tag{2.5}$$

Here, σ_L and σ_R are the left edge and right edge variances, respectively, of the channel; ρ_X is the cross-correlation coefficient between them. Under the assumption $\sigma_L = \sigma_R = \sigma_{LER}$,

$$\sigma_{LWR}^2 = 2\sigma_{LER}^2 (1 - \rho_X) \tag{2.6}$$

The variance of the average channel length is related to σ_{LWR} as follows [18].

$$\sigma_{<L>} = \frac{\sigma_{LWR}}{\sqrt{1 + 0.75\frac{W}{W_c}}} \tag{2.7}$$

As a random variable, the device channel length causes other device parameters such as threshold voltage V_T, drain current I_D, etc., to vary randomly.

2.3.1.3 Oxide Thickness Variation (OTV)

In CMOS technology, oxide thickness variation (OTV) occurs due to atomic-level interface roughness between the (i) semiconductor substrate and the gate dielectric material and between the (ii) gate material and gate dielectric. In MOSFETs below 65 nm, the value of the oxide thickness t_{ox} is equivalent to the length of a few Si atoms with a comparable interface roughness of about one to two atomic layers [19]. In present-day technology, where, t_{ox} is about 1 nm, its variation due to the comparable thickness of interface roughness introduces a gate current variation. This variation in turn produces a voltage drop in the gate and changes V_T significantly. If one assumes that the two interfaces are uncorrelated, the standard deviation of oxide thickness variation is expressed as [18]

$$\sigma_{t_{ox}} = \Delta H \frac{B\lambda}{\sqrt{2WL}} \tag{2.8}$$

Here, ΔH is the height of one atomic layer, λ denotes the correlation length of oxide surface roughness, and B is a fitting parameter. The corresponding variance in V_T comes out to be [18]

$$\sigma_{V_T} = C\frac{\sqrt{qN_a\varepsilon_{si}}}{\varepsilon_{ox}} \Delta H \frac{\lambda}{\sqrt{2WL}} \tag{2.9}$$

Here, C is a fitting parameter and the rest of the symbols have their usual significances.

2.3.1.4 Poly-Silicon/Metal Gate Granularity (PSG/MGG)

PSG causes enhanced gate dopant diffusion along the grain boundaries (GBs), leading to non-uniform polysilicon gate doping and potential localized penetration of the dopants through the gate oxide into the channel region. Fermi-level pinning at the boundaries between grains due to the high density of defect states is considered to be the most significant source of fluctuation within polysilicon. The uncertainty in doping concentration of polysilicon gates and transistor channels, as well as Fermi-level pinning, leads to parametric fluctuation comparable to that of RDD. However, the introduction of high-κ /metal gate technologies suppresses the issue of PSG. Still, the typically polycrystalline nature of the metal gate granularity (MGG) has become a new source of statistical variability [20]. In gate-first technologies, post-metallization annealing results in crystallization of the metal. Metal grains with different crystallographic orientations have different work functions at the metal/high-κ interface. Thus, the threshold voltage becomes a locally varying quantity from one metallic grain to the other. Hence, V_T becomes a random variable. However, MGG is technologically reduced to a significant extent in a gate-last process technology, where the metal gate does not suffer high-temperature thermal processing treatments, and hence poly-crystallization of the metal film can be avoided [21].

2.3.2 Device-Level Process Variability Mitigation Technique

Among the different process variability sources discussed previously, LER/LWR is limited by the lithography process used. In modern process technologies, the oxide thickness can be precisely controlled to have a few atomic layers leading to negligible oxide thickness variation (OTV) [22]. MGG can be reduced by implementing gate-last process technology. Only RDD is the effect that can be controlled with an appropriate design of the doping profile of the channel. Use of a low-doped channel can practically eliminate RDD. However, in practice, the channel is made low doped; this results in a highly reduced RDD effect. This technique is implemented in both silicon-on-insulator (SOI) and FinFET devices, which also have much improved electrostatic integrity compared to the conventional transistor structures [23]. Another approach is to use an epitaxial delta doped channel (EδDC) MOS transistor for controlled V_T variability [24–26]. A deeply depleted channel (DDC) MOS transistor, utilizing a layered channel profile, demonstrates the aggressive reduction of V_T variation [27,28].

2.4 Epitaxial Delta Doped Channel MOS Transistor

The novel transistor structures such as FinFETs and ultrathin body silicon-on-insulator (UTB-SOI) structures achieve low power consumption with superior electrostatic integrity. However, the fabrication process of these structures requires greater process complexity, so that increasing manufacturing cost is a concern. Hence, EδDC MOS transistor, being based on bulk CMOS technology, is a promising approach for extending the scalability of bulk metal oxide semiconductor (MOS) technology for low power system-on-chip applications.

2.4.1 Device Structure and Simulation

The schematic of an *n*-channel EδDC MOS transistor structure is shown in Figure 2.4. The doping concentration of the channel region is graded into two layers as shown in Figure 2.5. The first layer is very lightly doped and controls mismatch; this enhances carrier mobility, which reduces impurity scattering. This layer is usually formed by the epitaxy process and is termed an epitaxial layer. The second layer is very highly doped. The mobile charge carriers present in this high-doped layer terminate most of the electric

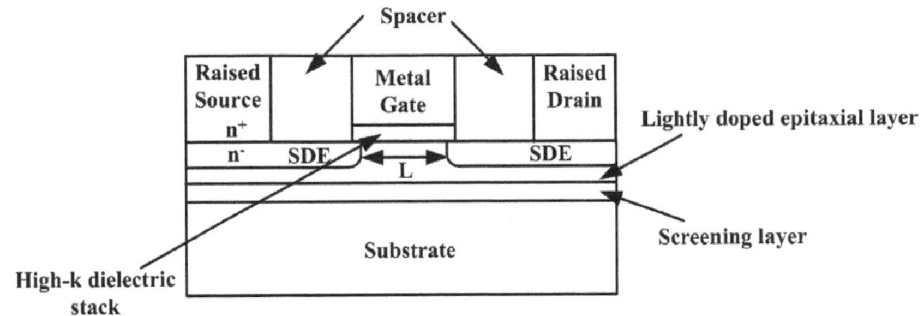

FIGURE 2.4
The schematic of an n-channel EδDC MOS transistor.

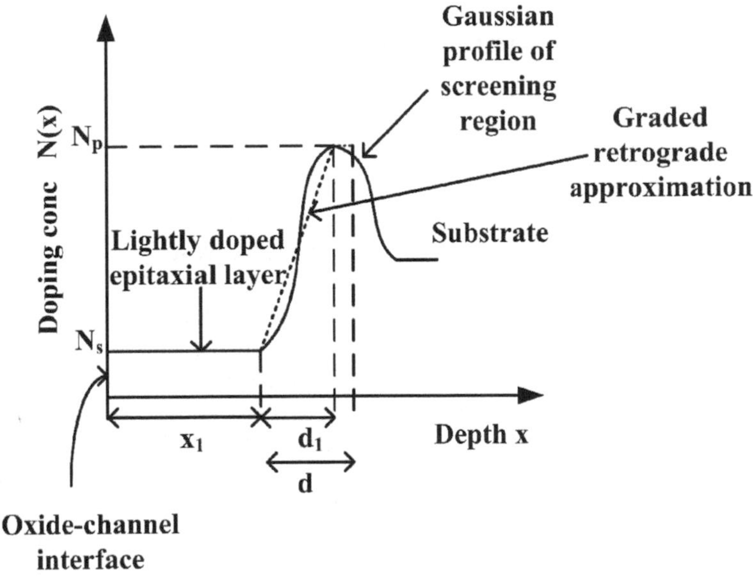

FIGURE 2.5
The doping profile in the channel region and its approximation.

field lines that originate from the drain and the source regions, thereby preventing them from penetrating into the channel region. This is referred to as screening phenomenon. The corresponding layer is termed a screening layer. Screening reduces the short-channel effects and any punch-through phenomenon between the source and drain. The source/drain extension (SDE) regions are shallower than the low-doped channel; the peak electric field associated with the source/drain-channel junction is less and hence band-to-band-tunneling (BTBT) leakage current is less compared to the conventional structures [13]. The supply voltage is considered to be $V_{DD} = 0.8$V. The lightly doped epitaxial layer has a concentration of $N_s = 5\times10^{16}/\text{cm}^3$. The screening layer has a Gaussian doping distribution with peak doping concentration of $N_p = 8\times10^{19}/\text{cm}^3$. The substrate doping concentration is $N_a = 4\times10^{18}/\text{cm}^3$. The epitaxial layer has thickness of $x_1 = 10$ nm. The depth of the peak doping concentration below the epitaxial layer is $d_1 = 5$ nm. The peak doping concentrations for the raised source/drain region and that for the SDE regions are taken to be $3.7\times10^{20}/\text{cm}^3$ and $3\times10^{19}/\text{cm}^3$, respectively. The transistor is implemented in High-κ Metal Gate (HKMG) technology. The equivalent oxide thickness is considered to be 0.8 nm. HfO_2

is used as the gate dielectric. The spacer material is Si_3N_4 and the length is 7 nm, for which acceptable current values are obtained.

In order to characterize the electrical performance of the device, we do device simulation using a Silvaco Atlas platform. Optimized meshing strategy is used for accurate results with less computational burden. The critical areas of the transistor such as the channel, gate oxide, and SDEs have dense meshing. The less critical regions such as deep inside the substrate have coarse meshing. A drift-diffusion (DD) model with density gradient quantum correction has been used to model the carrier transport. The Shockley–Read–Hall and Auger models take care of the carrier recombination process. For carrier mobility modeling, Arora, Lombardi CVT, and Caughey-Thomas models are incorporated. Kane's model takes care of the BTBT leakage mechanism. The model parameters and work function of the gate material are tuned to calibrate the simulator.

2.4.2 Study of Short-Channel Effect Characteristics

The simulation result of the variation of the magnitude of the lateral field along the channel-length direction is shown in Figure 2.6.

It is observed from Figure 2.6 that the magnitude of the lateral electric field near the middle of the channel is only 5 kV/cm for high drain bias where the channel length is as low as 16 nm. This is a significantly low field for a short-channel transistor compared to conventional transistors. The high-doped screening region restricts the p-n junction depletion depth associated with the source and the drain region. This in turn alleviates the impact of lateral field, and hence gate-controlled bulk depletion charge density becomes high. According to Poisson's equation, the source–drain controlled depletion charge density, therefore, becomes low. This implies that the contribution of the lateral electric field is low and the device becomes promising for controlled short-channel effect (SCE) and drain-induced barrier lowering (DIBL). A comparison of the DIBL coefficient of the EδDC transistor with reported values of FinFET and UTB-SOI is summarized in Table 2.1.

The variation of the maximum depletion depth W_{dm} of an EδDC transistor with various channel lengths is shown in Figure 2.7. With decreasing channel length, the depletion depth is found to remain constant for the device. This happens due to the presence of the screening region, which being highly doped restricts the gate depletion from penetrating

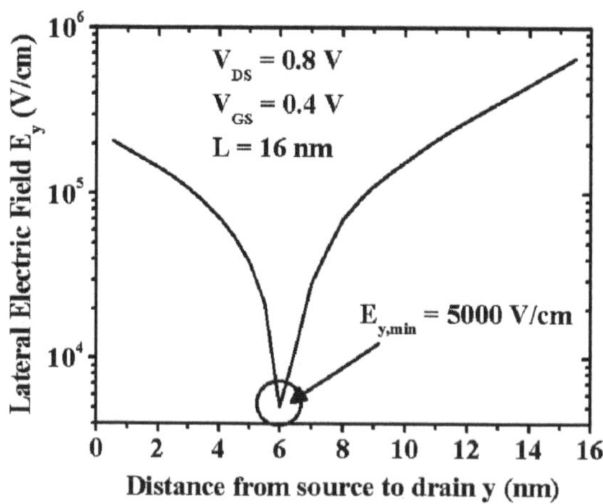

FIGURE 2.6
Variation of lateral electric field along the channel of an n-channel EδDC transistor.

TABLE 2.1
Comparison of DIBL Coefficient of EδDC Transistor with FinFET and UTB-SOI

Parameter	FinFET [29] $V_{DD} = 0.75V$ $L = 30$ nm	UTB-SOI [30] $V_{DD} = 0.9V$ $L = 20$ nm	EδDC $V_{DD} = 0.8V$ $L = 16$ nm
DIBL coefficient (mV/V)	42	80	63

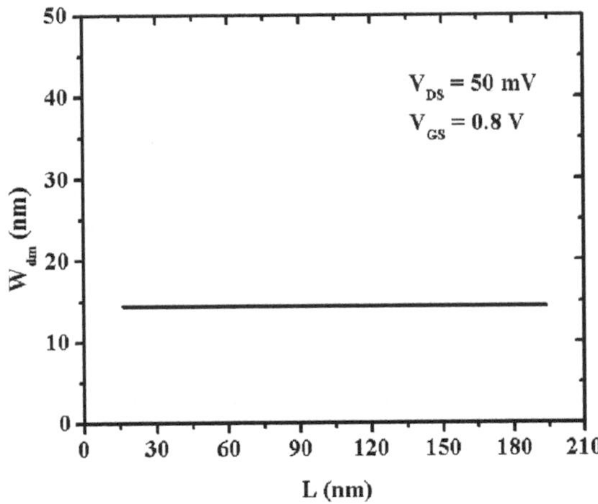

FIGURE 2.7
Variation of depletion depth with channel length for an *n*-channel EδDC transistor.

into the substrate. This is a remarkable improvement over the conventional devices where the gate depletion significantly increases removal of the gate's control over the channel and thereby enhances the short-channel effect. Figure 2.7 shows how the immunity is established from the SCE of the EδDC transistor.

The variation of the threshold voltage with applied drain bias for the EδDC transistor is shown in Figure 2.8. The corresponding DIBL coefficient is found to be 63 mV/V.

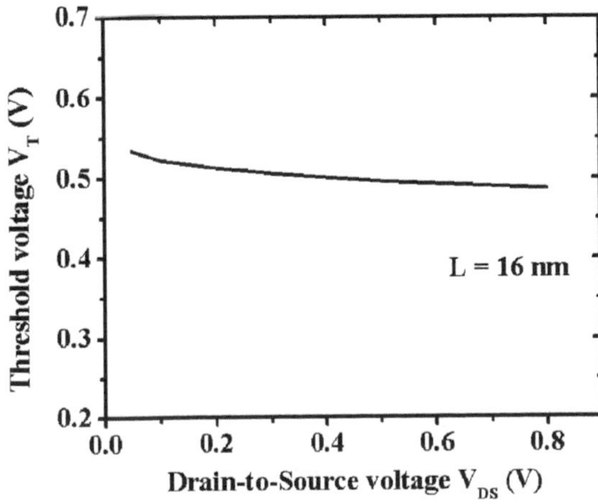

FIGURE 2.8
Variation of threshold voltage with applied drain bias.

This shows that the DIBL effect is much lower in the EδDC transistor and is comparable to state-of-the-art devices such as an ultra-thin-body silicon-on-insulator (UTB-SOI) transistor [31]. This is because the contribution of the lateral electric field is much lower for the EδDC MOS transistor compared to conventional transistor structures.

2.5 Substrate Bias Effect of EδDC MOS Transistor

Strong body effect makes adaptive substrate bias a useful strategy for dynamic V_T management to optimize circuit speed and power dissipation in SoC circuits [32,33]. However, body effect diminishes with transistor scaling in conventional nanoscale transistors. Channel engineering provides a good solution to boost the substrate bias effect. The non-uniform vertical doping of the EδDC transistor helps it to achieve a high substrate bias effect. The depletion width for an EδDC transistor very weakly depends upon the applied substrate bias, and with scaling down of channel length, the depletion width insignificantly widens. This way, substrate control over the channel becomes high, thereby terminating a significant amount of substrate depletion charge on the gate, instead of on the source and the drain. The substrate bias effect on the depletion depth and threshold voltage can be explicitly examined by an analytical model with suitable approximation of the nonlinear doping profile.

2.5.1 Substrate Bias Effect on Long-Channel Threshold Voltage

The channel doping profile given in Figure 2.5 can be mathematically approximated by a combination of abrupt retrograde and graded retrograde channel profiles [34] and is written as shown in the following equation:

$$N(x) = \begin{cases} N_s & x \le x_1 \\ N_s + \dfrac{N_p - N_s}{d_1}(x - x_1) & x_1 \le x \le x_1 + d_1 \\ N_p & x_1 + d_1 \le x \le x_1 + d \end{cases} \tag{2.10}$$

We start from a one-dimensional Poisson's equation, using (2.10), and the maximum value of the long-channel depletion layer width for the EδDC transistor is derived as shown in (2.11):

$$W_{\mathrm{dm}} = \sqrt{\frac{4\varepsilon_{\mathrm{si}}\Phi_F}{qN_p} + \left(1 - \frac{N_s}{N_p}\right)\frac{\left\{(x_1 + d_1)^2 + 2x_1^2 + d_1 x_1\right\}}{3}} \tag{2.11}$$

Here, Φ_F signifies the separation of the Fermi potential from the intrinsic potential. It may be noted from (2.11) that for the EδDC transistor, N_p is sufficiently high so that the depletion width very weakly depends on the potential term.

In order to simplify the computational task of using doping concentrations for low-doped and high-doped layers separately, we use Arora doping transformation [35] and compute an equivalent doping concentration of N_{eq}. The transformation assumes that the total charge under the gate is conserved, that is,

$$qN_{eq}W_{eq} = q\int_0^{W_d} N(x)dx \tag{2.12}$$

where

$$W_{eq} = \sqrt{\frac{2\varepsilon_{si}\psi_s}{qN_{eq}}} \tag{2.13}$$

When we substitute (2.13) in (2.12), the equivalent doping concentration is found to be

$$N_{eq} = \frac{q}{2\varepsilon_{si}\Psi_s}\left[N_s x_1 + d_1\left(\frac{N_s + N_p}{2}\right) + N_p\left\{W_{dm} - (x_1 + d_1)\right\}\right] \tag{2.14}$$

Under depletion approximation, the long-channel threshold voltage is

$$V_{Tl} = V_{FB} + 2\Phi_F + \frac{q}{C_{ox}}\int_0^{W_{dm}} N(x)dx \tag{2.15}$$

As we evaluate the integral in (2.12) by using the profile definition given in (2.10), the long-channel threshold voltage in the absence of a substrate bias is evaluated to be

$$V_{Tl0} = V_{FB} + 2\Phi_F + \frac{q}{C_{ox}}\left[N_s x_1 + d_1\left(\frac{N_s + N_p}{2}\right) + N_p\left\{W_{dm} - (x_1 + d_1)\right\}\right] \tag{2.16}$$

In the presence of a substrate bias, the long-channel depletion width and the long-channel threshold voltage, respectively, are written as

$$W_{dm} = \sqrt{\frac{2\varepsilon_{si}(2\phi_F + V_{SB})}{qN_p} + \left(1 - \frac{N_s}{N_p}\right)\frac{\left\{(x_1 + d_1)^2 + 2x_1^2 + d_1 x_1\right\}}{3}} \tag{2.17}$$

$$V_{Tl} = V_{Tl0} + \gamma_{eq}\left(\sqrt{2\Phi_F + V_{SB}} + \sqrt{2\Phi_F}\right) \tag{2.18}$$

Using first-order Taylor's series expansion of the square-root term of (2.18), we can write

$$V_{Tl} = V_{Tl0} + \frac{\varepsilon_{si}}{C_{ox}W_{eq}}V_{SB} = V_{Tl0} + \frac{C_{dm}}{C_{ox}}V_{SB} = V_{Tl0} + \alpha V_{SB} \tag{2.19}$$

Here, C_{dm} is the depletion capacitance per unit area; C_{ox} is the oxide capacitance per unit area; and α is the slope of V_{Tl} versus V_{SB}—that is, $\frac{dV_{Tl}}{dV_{SB}}$ is referred to as the substrate bias sensitivity.

The depletion width W_{dm} is almost independent of the substrate bias applied (evident by the high doping term in the first part of [2.17]); the associated capacitance, C_{dm}, is also independent of V_{SB}. Thus, we get the long-channel threshold voltage sensitivity α to be a constant. For high V_{SB}, however, the depletion width W_{dm} may vary slightly with V_{SB} so that the threshold voltage dependence on V_{SB} exhibits some classical behavior. This is incorporated through the following model of

$$V_{Tl} = V_{Tl0} + \alpha V_{SB} + k_1\left(\sqrt{\psi_s + V_{SB}} - \sqrt{\psi_s}\right) \tag{2.20}$$

where k_1 is a fitting parameter that takes into account the high substrate bias effect.

The variation of the depletion depth with substrate bias is shown in the Figure 2.9. As is already predicted from the analytical relation, simulation results also show a negligibly small increase in the depletion depth when substrate bias is increased. The highly doped screening region practically confines the channel depletion within the epitaxial layer. Therefore, as V_{SB} increases, the depletion layer width does not change significantly. The physical implication of this is as follows. The inversion channel layer forms a capacitor with the substrate, the dielectric being the depletion region. The channel-to-substrate capacitance is given by $C_{dm} = \varepsilon_{Si}/W_{dm}$. Thus, the lower value of W_{dm} signifies a larger C_{dm}, hence strong channel-to-substrate coupling, which in turn implies better substrate bias sensitivity. Therefore, we observe that with an increase of substrate bias, the channel-to-substrate coupling does not deteriorate in the EδDC transistor, as it happens for conventional transistors.

The variations of the threshold voltage with substrate bias for the EδDC transistor, as obtained from an analytical model (2.20), and simulation results are shown in Figure 2.10. A good match between the two indicates accuracy of the model. The variations of the substrate bias sensitivity with substrate bias for the EδDC transistor are also shown in it. It is observed that the substrate bias sensitivity of the EδDC transistor is nearly constant with the applied substrate bias. This validates the linear relationship (2.19). Physically, it is explained by the strong and invariant depletion capacitance with the applied substrate bias. The slight decrease of the substrate bias sensitivity is well taken care of by the parameter k_1 in (2.20) with its very small value.

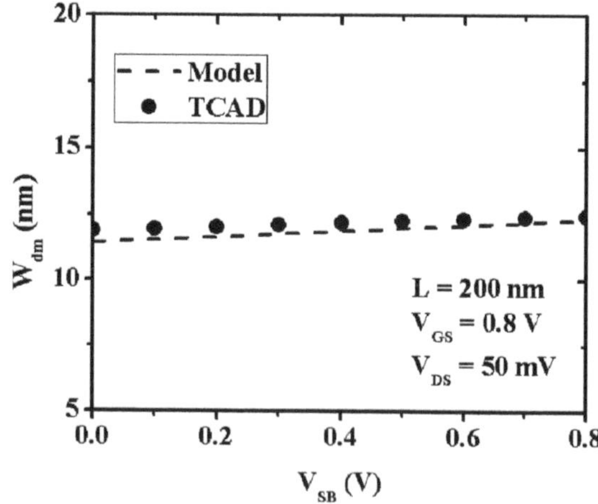

FIGURE 2.9
Variation of depletion depth with applied substrate bias.

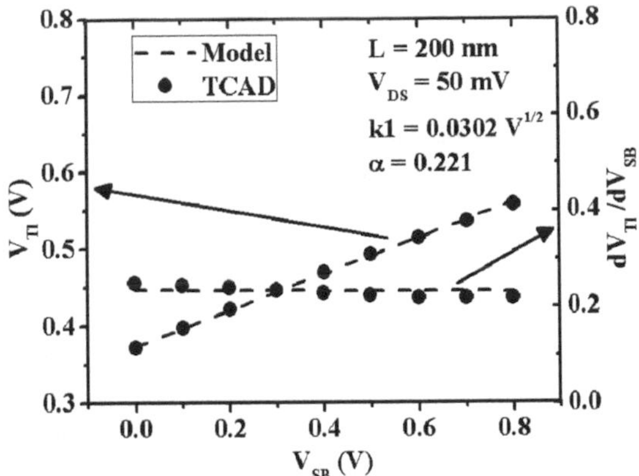

FIGURE 2.10
Variation of long-channel threshold voltage and its sensitivity with applied substrate bias.

2.5.2 Short-Channel Threshold Voltage Model

The threshold voltage model for an EδDC transistor is modeled [36] as follows:

$$V_T = V_{T1} + \Delta V_T\left(L, V_{SB}, V_{DS}\right) \tag{2.21}$$

where

$$\Delta V_T(L) = -\frac{\left[2\left(V_{bi} - \Psi_s\right) + V_{DS}\right]}{2\cosh\left(\dfrac{L}{2l_t}\right) - 2} \tag{2.22}$$

where $l_t = \sqrt{(3t_{ox}W_{dm}/\eta)}$ is the characteristic length. η is a fitting parameter and W_{dm} is given by (2.17). With the introduction of some fitting parameters for better accuracy, the short-channel effect becomes [36]

$$\Delta V_T\left(\text{SCE}\right) = -\frac{0.5\text{DVT0}}{\cosh\left(\text{DVT1}.\dfrac{L}{l_t}\right) - 1}\left(V_{bi} - \Psi_s\right) \tag{2.23}$$

With the application of substrate bias, the characteristic length becomes

$$l_t = l_{t0}\left(1 + \text{DVT2}.V_{SB}\right) \tag{2.24}$$

where l_{t0} is the zero substrate bias characteristic length.

Similarly, the threshold voltage shift due to drain-induced barrier lowering (DIBL) is modeled as

$$\Delta V_T\left(\text{DIBL}\right) = -\frac{0.5}{\cosh\left(\text{DSUB}.\dfrac{L}{l_{t0}}\right) - 1}\left(\text{ETA0} + \text{ETAB}.V_{SB}\right).V_{DS} \tag{2.25}$$

The DIBL coefficient, therefore, linearly increases with the substrate bias. The fitting parameters DVT0, DVT1, DVT2, ETA0, ETAB, and DSUB are extracted using optimization algorithm [37].

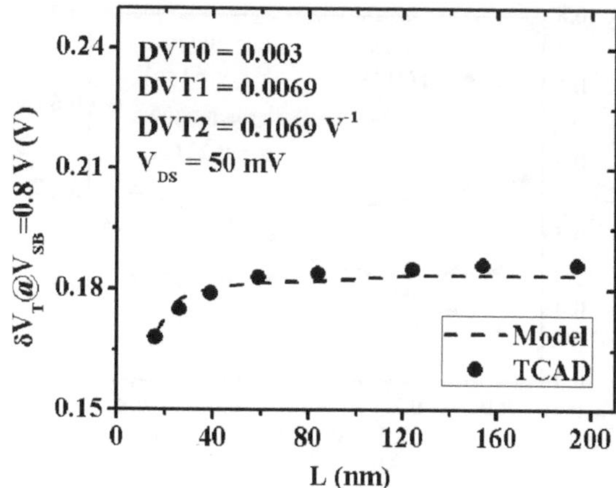

FIGURE 2.11
The variation of change in threshold voltage due to substrate bias with L.

The variations of the change in threshold voltage due to a substrate bias effect, that is, $\delta V_T = V_T|_{V_{SB}=0.8} - V_T|_{V_{SB}=0}$ with channel length L for the EδDC, are plotted in Figure 2.11. This quantity represents the amount of substrate effect. It reduces by only 10 percent when the channel length reduces to a mere 16 nm from 200 nm. Significant substrate bias effect of the EδDC transistor is attributed to the fact that the EδDC transistor is quite immune to the coupling of the source/drain lateral field on the channel due to the screening phenomenon. Consequently, the depletion width remains almost constant with scaling of the channel length (see Figure 2.9), and channel-to-substrate coupling remains significant. This explains the existence of a significant substrate bias effect in the EδDC transistor for short-channel lengths. The short-channel parameters noted are extracted using a parameter extraction algorithm and incorporated in the model. Once again, the close proximity of the model and TCAD data validates the model.

The variations of the threshold voltage roll off due to short-channel effect, that is, $\Delta V_T = V_{TI} - V_T$ with substrate is shown in Figure 2.12. However, ΔV_T it increases only by

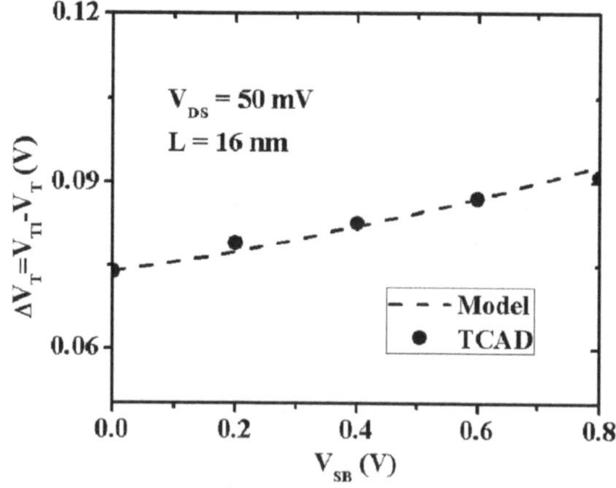

FIGURE 2.12
Variation of threshold voltage roll-off due to short-channel effect with substrate bias.

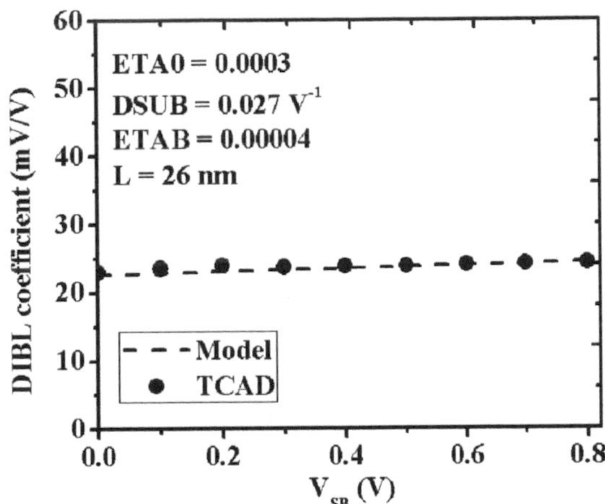

FIGURE 2.13
Variation of DIBL coefficient with applied substrate bias.

25% at a maximum V_{SB} with respect to zero substrate bias. This is another manifestation of a highly restricted short-channel effect in EδDC transistors.

In Section 2.4 we discussed the excellent DIBL performance of the EδDC transistor. Due to the depletion layer confinement even at a high substrate bias, the EδDC transistor gives good DIBL performance at a high substrate bias as well. There is an insignificant increase of the DIBL coefficient with increasing reverse substrate bias as shown in Figure 2.13. This effect is also captured very well by the model.

2.6 Channel Profile Design of an EδDC MOS Transistor for Low V_T Mismatch and High Intrinsic Gain

The motivation that drives us to design a device like an EδDC transistor is its variability-resistant device architecture. The two major concerns for analog circuit designers for designing analog circuits in nanoscale CMOS technologies are enhanced threshold voltage mismatch and the degradation of intrinsic gain of the transistors [38,39]. The major variability sources causing V_T mismatch are already outlined in Section 2.3. Different studies reveal RDD to be the most crucial source of mismatch in planar MOS technologies [10,15]. The primary reason behind the degradation of intrinsic gain of MOS transistors is the degradation of the output resistance, caused by the DIBL effect. As already discussed, device structures such as FinFET and UTB-SOI transistor are the two novel device structures with reduced RDD effect and improved electrostatic integrity [23]. However, enhanced manufacturing and/or design cost restricts their use for the low-cost system-on-a-chip (SoC) applications. Therefore, advanced planar bulk MOS transistor is still the cost-effective choice for SoC applications. A systematic methodology for the design of a channel profile of an EδDC MOS transistor based on planar bulk technology, for high gain and low V_T mismatch, is developed here. By design space exploration, the optimum values of the channel profile parameters are obtained.

2.6.1 Channel Profile Design

The intrinsic voltage gain of a MOS transistor is given by [38]

$$A_V = -\frac{g_m}{g_{\text{ds}}} \tag{2.26}$$

Here, g_m is the trans-conductance of the MOS transistor and g_{ds} is the conductance. In order to quantify the dependence of the output conductance on DIBL phenomenon, we express g_{ds} as follows:

$$g_{\text{ds}} = \frac{dI_D}{dV_{\text{DS}}} = \frac{dI_D}{dV_T} \cdot \frac{dV_T}{dV_{\text{DS}}} \tag{2.27}$$

Considering the fact that the drain current is a function of gate overdrive voltage, $(V_{\text{GS}} - V_T)$, we can write

$$g_m = \frac{dI_D}{dV_{\text{GS}}} = -\frac{dI_D}{dV_T} \tag{2.28}$$

Substituting (2.28) in (2.27), we can write

$$g_{\text{ds}} = -g_m \cdot \frac{dV_T}{dV_{\text{DS}}} \tag{2.29}$$

Comparing (2.26) and (2.29), we get

$$A_V = -\frac{g_m}{g_{\text{ds}}} = \frac{dV_{\text{DS}}}{dV_T} \tag{2.30}$$

Using (2.22) in (2.30), we get

$$A_V = 2\left[1 - \cosh\left(\frac{L}{2l_t}\right)\right] \tag{2.31}$$

The device-to-device V_T mismatch due to the RDD effect is quantified by computing the standard deviation of the threshold voltage difference between two closely spaced transistors. The device-to-device V_T variability for a non-uniformly doped channel bulk MOS transistor is given as [13],

$$\sigma_{V_T}{}^2 = \frac{q^2}{C_{\text{ox}}^2 WL} N_{\text{oi}} \tag{2.32}$$

Here, N_{oi} is an effective number of dopant atoms per unit area in the channel region and is given by

$$N_{\text{oi}} = \int_0^{W_{\text{dm}}} N(x)\left[1 - \frac{x}{W_{\text{dm}}}\right]^2 dx \tag{2.33}$$

With the help of (2.10) and (2.11), N_{oi} is found to be as follows:

$$N_{\text{oi}} = \frac{N_s W_{\text{dm}}}{3} + \frac{(N_p - N_s)}{3d_1} W_{\text{dm}}(W_{\text{dm}} - x_1)\left(1 - \frac{x_1}{W_{\text{dm}}}\right)^3 \tag{2.34}$$

It may be noted here that the formulation, (2.32 through 2.34), does not consider the random placement of the dopant atoms in the channel region. In order to find the device-to-device mismatch $(\sigma_{\Delta V_T})$, the device variability (σ_{V_T}) is to be multiplied by a factor of $\sqrt{2}$. Equations (2.31) and (2.32) are used to find the optimal values of the channel profile parameters by simultaneously studying the impact of their variations on intrinsic gain and V_T mismatch.

2.6.2 Design Space Exploration

The profile parameters to be optimized are x_1, d_1, N_s, and N_p. In order to restrict the channel depletion from spreading into the bulk of the device, N_p has to be considerably high. In that case, the epitaxial layer effectively represents the channel depletion layer. Practically, the channel depletion will be slightly higher due to the spread of the Gaussian profile of the screening region. For an ideal retrograde profile $(d_1 = 0)$, for $x_1 = 10$ nm (say), for $N_p \geq 8 \times 10^{19}\,\text{cm}^{-3}$, W_{dm} approximately equals to x_1. It can be found in Figure 2.14.

The plot of simultaneous variations of $\sigma_{\Delta V_T}$ and A_V with d_1 is shown in Figure 2.15. We can see that both the performances show detrimental effects with an increase in d_1. Also, $\sigma_{\Delta V_T}$ increases and A_V decreases with d_1. With an increase in d_1, it follows from (2.17) that the depletion depth W_{dm} increases, i.e., the effect of a screening phenomenon diminishes. This increases the characteristic length l_t, which in turn deteriorates the DIBL phenomenon, characterized through $\frac{dV_T}{dV_{DS}}$. It, therefore, follows from (2.30) that the intrinsic gain A_V reduces. On the other hand, with an increase in W_{dm}, the number of charges that take part in the RDD phenomenon increases, so that $\sigma_{\Delta V_T}$ increases. However, d_1 depends on the width of the Gaussian doping profile of the screening region. The width practically has to have a non-zero value. It can be assumed to have some minimum allowable value—for example, 5 nm [40].

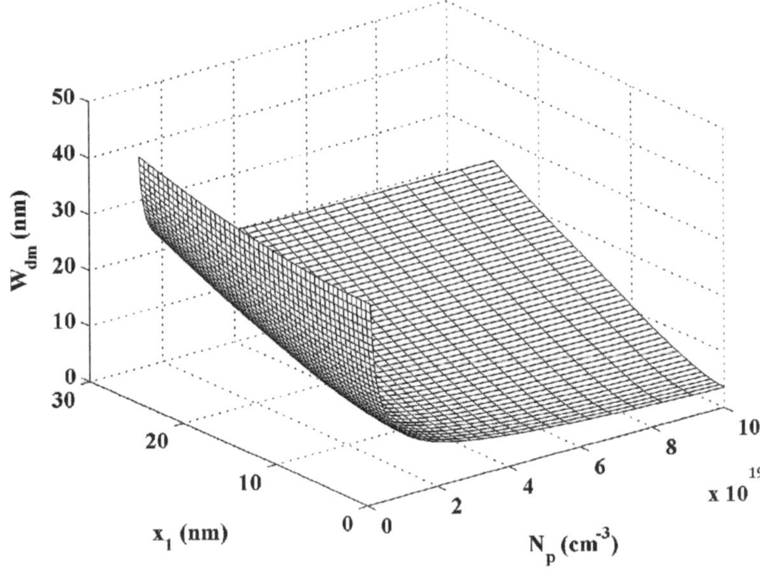

FIGURE 2.14
Variation of maximum depletion depth with simultaneous variations of epitaxial layer thickness and peak doping concentration using analytical equation.

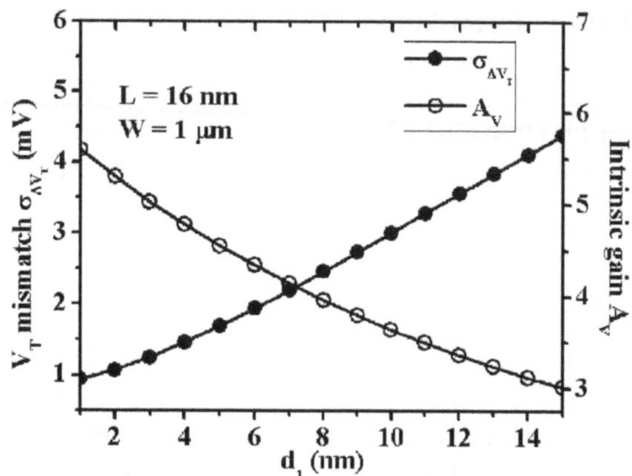

FIGURE 2.15
Simultaneous variations of V_T mismatch and intrinsic gain with depth of peak doping concentration below the epitaxial layer using analytical equations.

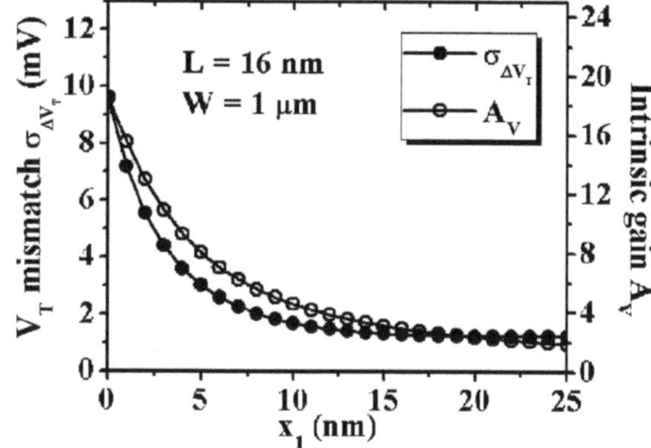

FIGURE 2.16
Simultaneous variations of V_T mismatch and intrinsic gain with epitaxial layer thickness using analytical equations.

The plot of simultaneous variations of $\sigma_{\Delta V_T}$ and A_V with x_1 is shown in Figure 2.16. With increasing x_1, the channel depletion depth increases according to the charge conservation rule. This is also analytically evident from (2.17). An increase in W_{dm} deteriorates intrinsic gain A_V as already explained. On the other hand, with the increase in x_1, the effective number of channel dopant atoms per unit area N_{oi} reduces, as evident from (2.34). Hence, $\sigma_{\Delta V_T}$, which is directly proportional to $\sqrt{N_{oi}}$, also decreases with increasing x_1. Selection of x_1 is not straightforward because of the conflicting requirement of high A_V and low N_{oi}. However, Figure 2.16 shows that a good solution may be found around $x_1 = 10$ nm.

Figure 2.17 shows the plot of simultaneous variations of $\sigma_{\Delta V_T}$ and A_V with N_s. Though an increase of N_s does not appreciably change the A_V value, it remarkably deteriorates the V_T mismatch. However large N_s becomes, it is orders less than N_p. In this limit, the depletion depth does not depend upon N_s and consequently A_V does not vary significantly. But an increase in the N_s level enhances the number of dopant atoms in the channel region, thereby detrimentally affecting RDD. Hence, $\sigma_{\Delta V_T}$ rapidly increases with an increase in N_s level.

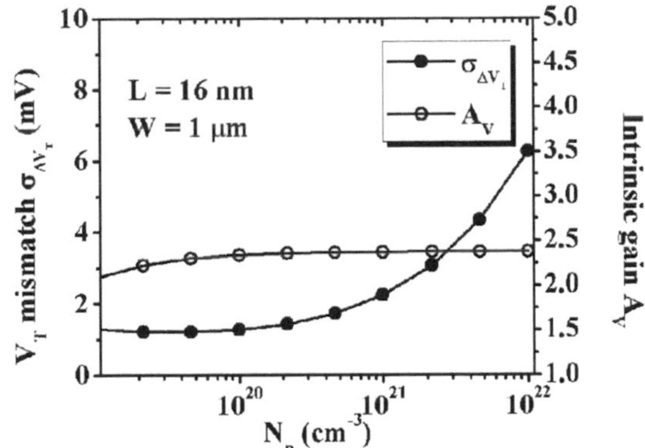

FIGURE 2.17
Simultaneous variations of V_T mismatch and intrinsic gain with epitaxial layer doping concentration using analytical equations.

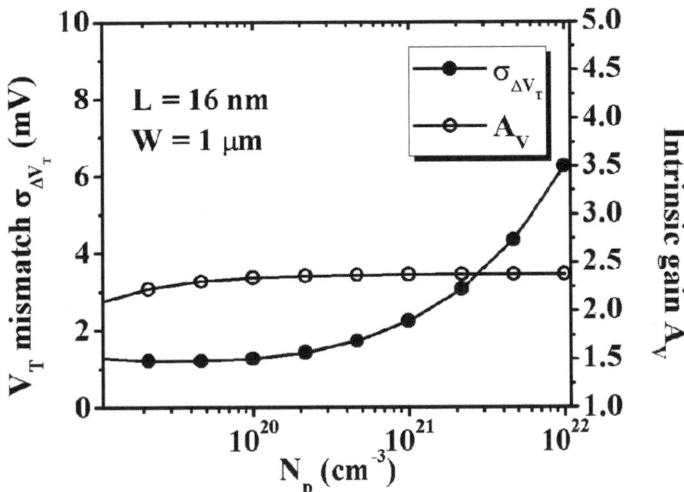

FIGURE 2.18
Simultaneous variations of V_T mismatch and intrinsic gain with peak doping concentration of screening layer using analytical equations.

Figure 2.18 represents the variations of $\sigma_{\Delta V_T}$ and A_V with N_p. We can see that A_V initially rises with the increase in N_p, then becomes almost independent of it. Due to charge conservation, W_{dm} is lower for higher N_p and vice versa. As a consequence, a higher N_p level results in better A_V. But for very high values of N_p, W_{dm} and hence A_V are independent of N_p as evident from (2.17) and (2.31). However, as was the case of N_s, a high N_p level also enhances the number of dopant atoms in the channel region, thereby increasing $\sigma_{\Delta V_T}$ rapidly.

2.6.3 Intrinsic Gain Characteristics

For the profile optimization purpose, we considered a simplified model of intrinsic gain and neglected its bias dependence. But for the validation purpose, we have to take into account the bias dependence of A_V. Consider that the bias dependence A_V can be written as

$$A_V = g_m.R_O = \frac{V_A.g_m}{I_{Dsat}} \tag{2.35}$$

where V_A is the early voltage parameter [38]. The early voltage due to channel-length modulation is calculated to be [38],

$$V_{ACLM} = \frac{\left(E_{sat}L + V_{gst}\right)\left(V_{DS} - V_{DSsat}\right)}{E_{sat}l} \tag{2.36}$$

where $E_{sat} = \frac{2v_{sat}}{\mu_s}$ and $l = \left(3t_{ox}x_j\right)^{1/2}$. Similarly, V_A due to the DIBL effect is modeled as [38],

$$V_{ADIBL} = \left(E_{sat}L + V_{gst}\right)\frac{0.5}{\cosh\left(\frac{L}{2lt}\right) - 1}\left(1 + 2E_{sat}L / V_{gst}\right) \tag{2.37}$$

When we consider the effective early voltage due to both channel-length modulation and DIBL effect, the intrinsic gain can be expressed as

$$A_V = \frac{g_m}{I_{Dsat}}\left(\frac{1}{V_{ACLM}} + \frac{1}{V_{ADIBL}}\right)^{-1} \tag{2.38}$$

The analytical results from (2.38) and results obtained from TCAD simulation are plotted in Figure 2.19. An excellent match between the two results proves the accuracy of the model. The presence of the screening region helps the device for better short-channel and DIBL performance w.r.t. the conventional transistor structures, as already discussed. Thus, V_{ACLM} and V_{ADIBL} are considerably higher for the EδDC transistor than the conventional one. This leads to better intrinsic gain of the EδDC transistor.

The V_T mismatch characteristics with the thickness of the low-doped region are shown in Figure 2.20. The nature of variation of the simulated and model predicted results with x_1

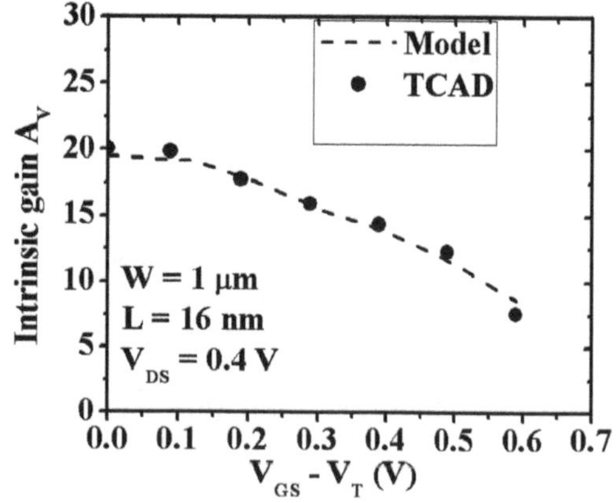

FIGURE 2.19
Comparison of model predicted and TCAD simulation results of variation of intrinsic gain with gate overdrive voltage.

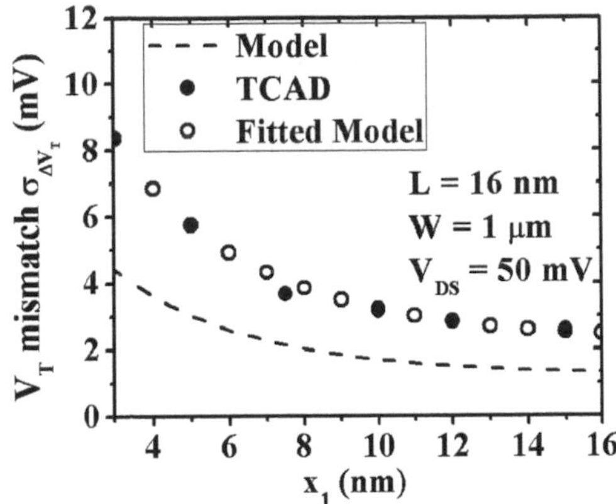

FIGURE 2.20
Comparison between V_T mismatch results obtained by model and TCAD simulation.

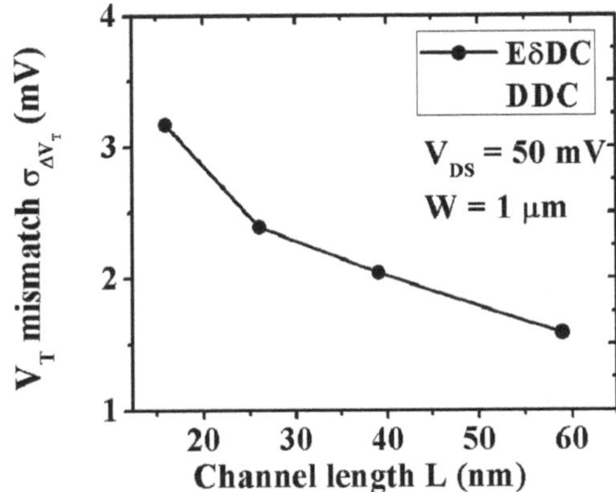

FIGURE 2.21
Channel-length dependence of V_T mismatch at a fixed channel width.

reveal that the TCAD simulated results are higher than predicted by the analytical model discussed earlier (2.32). This is due to the fact that the random placement of the dopant atoms in the channel region was not considered in the analytical model. However, we find from Figure 2.20 that a multiplication factor (~1.9), makes the results predicted by the analytical model a perfect match with the TCAD simulation results.

The channel-length dependence of $\sigma_{\Delta V_T}$ is depicted in Figure 2.21. We observe that $\sigma_{\Delta V_T}$ increases with decreasing channel length. This is quite obvious and in accordance with Pelgrom's law.

2.7 Summary and Conclusion

This chapter gives an overview of different sources of variability in nanoscale devices and circuits. RDD is reported to be the most crucial source of performance variability in bulk CMOS technologies. We prefer bulk CMOS technology as it is cost effective compared to various state-of-the-art technologies. However, RDD can be restricted with the help of a channel engineering technique. Following this route, we designed an epitaxial delta doped channel (EδDC) transistor. The beauty of the device lies in the screening property of the δ doped layer. This helps the device to be almost immune from short-channel effects and DIBL effects. In fact, the DIBL performance of the EδDC is found to be on par with a state-of-the-art device such as the UTB-SOI transistor. The highly doped screening region also helps to boost the substrate bias coefficient, which actually degraded in conventional transistor structures in scaled-down technologies. High substrate bias enables a device to be subjected to adaptive body biasing. Thus, significant leakage power reduction by applying reverse substrate bias is possible for the EδDC transistor. However, a careful selection of the thickness of the low-doped layer as well as concentration of the high-doped layer is required for the desired substrate bias effect of the transistor. The most important feature of the EδDC transistor, which makes it a deserving candidate for low-power SoC design, is its excellent mismatch characteristics. The channel being very low doped, RDD effect observed for the device is very low. Thus, an RDD induced V_T mismatch is highly reduced in these devices. However, the mismatch characteristics also depend on the selection of the different profile parameters. It is shown in the chapter that with a suitable choice of channel profile parameters, the device can give acceptable V_T mismatch characteristics with some compromise with the intrinsic gain.

References

1. G. E. Moore, "Cramming more components onto integrated circuits," *Electronics*, Vol. 38, no. 8, pp. 114–117, 1965.
2. International Technology Roadmap for Semiconductors: ITRS 2010 updates. http://www.itrs.net/Links/2010ITRS/Home2010.htm.
3. H. El-Aawar, "Increasing the transistor count by constructing a two-layer crystal square on a single chip," *International Journal of Computer Science & Information Technology (IJCSIT)*, Vol. 7, no. 3, pp. 97–105, 2015.
4. T. Sakurai and A. R. Newton, "Alpha-power law MOSFET model and its applications to CMOS inverter delay and other formulas," *JSSC*, Vol. 25, no. 2, pp. 584–594, 1990.
5. S. K. Saha, "Scaling considerations for high performance 25 nm metal-oxide-semiconductor field effect transistors," *Journal of Vacuum Science & Technology B: Microelectronics and Nanometer Structures*, Vol. 19, no. 6, pp. 2240–2246, 2001.
6. S. K. Saha, "Design considerations for 25 nm MOSFET devices," *Solid-State Electronics*, Vol. 45, no. 10, pp. 1851–1857, 2001.
7. K. Bernstein et al., "High-performance CMOS variability in the 65-nm regime and beyond," *IBM Journal of Research and Development*, Vol. 50, no. 4–5, pp. 433–449, 2006.
8. K. Kuhn et al., "Managing process variation in Intel's 45 nm CMOS technology," *Intel Technology Journal*, Vol. 12, no. 2, pp. 92–110, 2008.
9. N. B. Cobb, "Fast optical and process proximity correction algorithms for integrated circuit manufacturing," PhD dissertation, University of California, Berkeley, CA, 1998.
10. S. K. Saha, "Modeling process variability in scaled CMOS technology," *IEEE Design and Test of Computers*, Vol. 27, pp. 8–16, 2010.
11. K. J. Kuhn, "Considerations for ultimate CMOS scaling," *IEEE Transactions on Electron Devices*, Vol. 59, no. 7, pp. 1813–1828, 2012.
12. H. Iwai, "CMOS technology year 2010 and beyond," *IEEE Journal of Solid-State Circuits*, Vol. 34, no. 3, pp. 357–366, 1999.
13. Y. Taur and T. H. Ning, *Fundamentals of Modern VLSI Devices*, Cambridge, UK: Cambridge University Press, 1998.

14. K. Takeuchi, T. Tatsumi, and A. Furukawa, "Channel engineering for the reduction of random-dopany-placement-induced threshold voltage," *Proceedings of IEDM*, pp. 33.6.1–33.6.4, 1997.
15. S. K. Saha, "Compact MOSFET modeling for process variability-aware VLSI circuit design," *IEEE Access*, Vol. 2, pp. 104–115, 2014.
16. C. Shin, *Variation-Aware Advanced CMOS Devices and SRAM*, Springer Series in Advanced Microelectronics 56, Dordrecht, the Netherlands: Springer Science+Business Media, 2016.
17. K. Patel, T.-J. K. Liu, and C. J. Spanos, "Gate line edge roughness model for estimation of FinFET performance variability," *IEEE Transactions on Electron Devices*, Vol. 56, no. 12, pp. 3055–3063, 2009.
18. Y. Ye, S. Gummalla, C. C. Wang, C. Chakrabarti, and Y. Cao, "Random variability modeling and its impact on scaled CMOS circuits," *Journal of Computational Electronics*, Vol. 9, pp. 108–113, 2010.
19. A. Asenov, S. Kaya, and J. H. Davies, "Intrinsic threshold voltage fluctuations in decananometer MOSFETs due to local oxide thickness variations," *IEEE Transactions on Electron Devices*, Vol. 49, no. 1, pp. 112–119, 2002.
20. A. R. Brown, N. M. Idris, J. R. Watling, and A. Asenov, "Impact of metal gate granularity on threshold voltage variability: A full-scale three-dimensional statistical simulation study," *IEEE Electron Device Letters*, Vol. 31, no. 11, pp. 1199–1201.
21. K. Ohmori, T. Matsuki, D. Ishikawa, T. Morooka, T. Aminaka, Y. Sugita, T. Chikyow, K. Shiraishi, Y. Nara, and K. Yamada, "Impact of additional factors in threshold voltage variability of metal/high-k gate stacks and its reduction by controlling crystalline structure and grain size in the metal gates," in *IEDM Technical Digest*, 2008, pp. 409–412.
22. S. Markov, A. Zain, B. Cheng, and A. Asenov, "Statistical variability in scaled generations of n-channel UTB-FD-SOI MOSFETs under the influence of RDF, LER, OTF and MGG," In *SOI Conference*, 2012.
23. J. G. Fossum and V. P. Trivedi, *Ultra-Thin-Body MOSFETs and FinFETs*, Cambridge, UK: Cambridge University Press, 2013.
24. K. Noda, T. Tatsumi, T. Uchida, K. Nakajima, H. Miyamoto, and C. Hu, "A 0.1-μm delta-doped MOSFET fabricated with post-low- energy implanting selective epitaxy," *IEEE Transactions on Electron Devices*, Vol. 45, no. 4, pp. 809–814, 1998.
25. A. Asenov and S. Saini, "Suppression of random dopant-induced threshold voltage fluctuations in sub-0.1-μm MOSFET's with epitaxial and δ-doped channels," *IEEE Transactions on Electron Devices*, Vol. 46, no. 8, pp. 1718–1724, 1999.
26. J. Woo et al., "Improved device variability in scaled MOSFETs with deeply retrograde channel profile," *Microelectronics Reliability*, Vol. 4, no. 6–7, pp. 1090–1095, 2014.
27. K. Fujita et al., "Advanced channel engineering achieving aggressive reduction of VT variation for ultra-low-power applications," in *Proceedings of IEEE International Electron Device Meeting*, December 2011, pp. 32.3.1–32.3.4.
28. R. Rogenmoser and L. T. Clark, "Reducing transistor variability for high performance low power chips," *IEEE Micro*, Vol. 33, no. 2, pp. 18–26, 2013.
29. S.-Y. Wu et al., "A 16nm FinFET CMOS technology for mobile SoC and computing applications," in *Proceedings of IEEE International Electron Device Meeting*, December 2013, pp. 9.1.1–9.1.4.
30. Q. Liu et al., "High performance UTBB FDSOI devices featuring 20nm gate length for 14nm node and beyond," in *Proceedings of IEEE International Electron Device Meeting*, December 2013, pp. 9.2.1–9.2.4.
31. A. S. M. Zain, S. Markov, B. Cheng, and A. Asenov, "Comprehensive study of the statistical variability in a 22 nm fully depleted ultra-thin-body SOI MOSFET," *Solid State Electronics*, Vol. 90, pp. 51–55, 2013.
32. X. Wang, F. Adamu-Lema, B. Cheng, and A. Asenov, "Geometry, temperature and body bias dependence of statistical variability in 20-nm bulk CMOS technology: A comprehensive simulation analysis," *IEEE Transactions on Electron Devices*, Vol. 60, no. 5, 2013.
33. H. Koura, M. Takamiya, and T. Hiramoto, "Optimum conditions of body effect factor and substrate bias in variable threshold voltage MOSFETs," *Japanese Journal of Applied Physics*, 39(Part 1, No. 4B), pp. 2312–2317, 2000.

34. S. Saha, "Effects of inversion layer quantization on channel profile engineering for nMOSFETs with 0.1μm channel lengths," *Solid State Electronics*, Vol. 42, pp. 1985–1991, 1998.
35. N. Arora, "Semi-empirical model for the threshold voltage of a double implanted MOSFET and its temperature dependence," *Solid State Electronics*, Vol. 30, no. 5, pp. 559–569, 1987.
36. X. J. Xi, M. Dunga, J. He, W. Liu, K. M. Kao, X. Jin, J. J. Ou, M. Chan, A. M. Niknejad, and C. Hu, *BSIM4.6.0 MOSFET Model: User's Manual* [Computer software manual], Berkeley, CA: Department of Electrical Engineering and Computer Sciences, University of California.
37. S. Sengupta, S. Sikdar, and S. Pandit, "Substrate bias effect of epitaxial delta doped channel MOS transistor for low power applications," *International Journal of Electronics*, Vol. 104, no. 1, pp. 1–16, 2016.
38. S. Pandit, C. Mandal, and A. Patra, *Nano-Scale CMOS Analog Circuits: Models and CAD Techniques for High-Level Design*, Boca Raton, FL: CRC Press, 2014.
39. L. L. Lewyn, T. Ytterdal, C. Wulff, and K. Martin, "Analog circuit design in nanoscale CMOS technologies," *Proceedings of the IEEE*, Vol. 97, no. 10, pp. 1687–1714, 2009.
40. L. Shifren, P. Ranade, P. E. Gregory, S. R. Sonkusale, W. Zhang, and S. E. Thompson, "Advanced transistors with punch through suppression," U.S. Patent 8 421 162 B2, April 16, 2013.

3

Effect of Ground Plane and Strained Silicon on Nanoscale FET Devices

Saurabh Chaudhury and Avtar Singh

CONTENTS

3.1 Introduction

In this chapter, we will briefly discuss the effect of ground plane and strained silicon in different FET device structures along with their implementations issues. Section 3.2 introduces the concept of ground plane in SOI structures and the method of incorporation in FDSOI and FinFET devices. We will illustrate the concept of straining silicon and its advantages over conventional silicon devices in Section 3.3. We also mention different methods of introducing the strain in silicon. Finally, some emphasis is given on the structure of strained SOI and strained FinFET devices.

3.2 Concept of Ground Plane

The short-channel effects in nanoscale devices are largely due to the fringing electric field lines underneath the channel region (substrate), as shown in Figure 3.1. Potential applied at the drain causes field lines to emanate from the drain. The channel is safe from these field lines at the top because of the presence of a gate terminal, and the field lines penetrate laterally the part of the channel and through the body or substrate [1]. The field lines through the substrate connect through the drain to body potential and cause unwanted back-channel conduction. There occurs a peak in the potential at the body/channel interface, and a voltage rises in the substrate as well. The back channel is driven from the depletion region to a weak inversion region, and due to the interface coupling, the threshold voltage decreases. Further, due to the decrement in the subthreshold slope, the drain-induced barrier lowering (DIBL) [9] largely occurs. Due to short-channel effects such as DIBL, it has been observed that the back-channel conduction is controlled via a drain rather than the substrate.

It has been proved that ground plane is effective in suppressing the short-channel effects as shown in Figure 3.2. When a highly conducting surface (P^+ or N^+) is placed beneath the channel or buried oxide (in the case of SOI devices), or a single P^+ or N^+

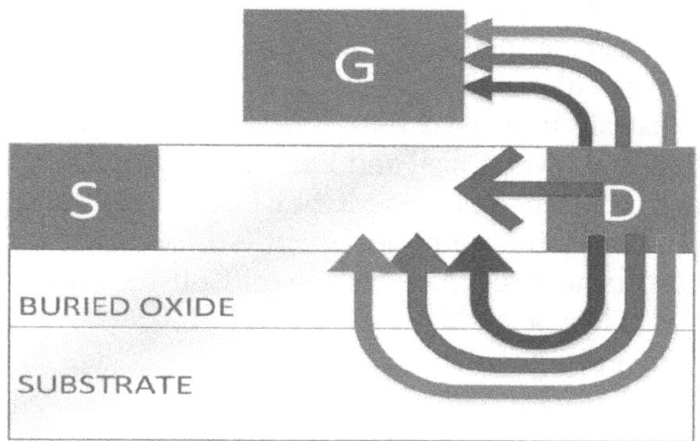

FIGURE 3.1
Schematic of fringing fields in SOI device. (From Ernst, T., and Cristoloveanu, S., The ground-plane concept for the reduction of short-channel effects in fully depleted SOI devices, *Electrochemical Society Proceedings*, pp. 329–334, 1999.)

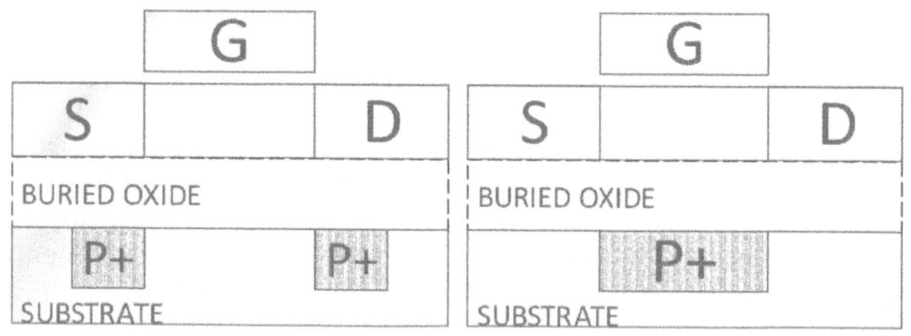

FIGURE 3.2
Ground-plane structures in SOI. (From Yanagi, S. et al., *IEEE Electron Device Lett.*, 22, 278–280, 2001.)

implant situated in the channel/buried oxide is aligned with the gate-oxide or two P^+ or N^+ implants located under the source and drain, this is termed a ground plane as shown in Figure 3.2. Doping of the ground plane layer is about $1 \times 10^{18}/cm^3$. A highly conducting surface acts as an electric field block and shields the moving fringing field lined from the drain coupling to the channel. When the substrate is lightly doped, the depletion region in the substrate is large and adds to the effective buried oxide thickness in the case of silicon on insulator (SOI). Deploying a ground plane reduces the body region depletion and stops the conduction via a back channel, which happens due to drain current.

Partial ground planes are also possible, but the fabrication steps and cost will increase in that case. Partial ground plane [8] is more effective in minimizing the short-channel effects in comparison with the continuous ground plane method.

3.2.1 Ground Plane in FDSOI

Buried oxide (BOX) is inserted in SOI between the channel and substrate as shown in Figure 3.3. Some advantages of SOI over conventional MOSFET are:

- Negligible drain-to-substrate capacitance
- Small body effects, fast stacked gates
- No latch-up
- Simple device isolation, smaller area
- Excellent radiation hardness
- Small junction leakage current
- Reduced short-channel effects

There are two ways to insert the ground plane in SOI structure:

- Ground Plane in Substrate (GPS)
- Ground Plane In BOX (GPB)

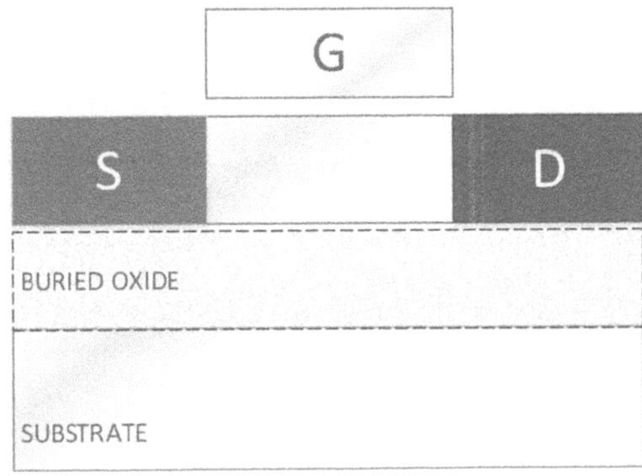

FIGURE 3.3
Silicon on insulator (SOI).

3.2.2 Ground Plane in Substrate (GPS)

When the ground plane is inserted between the body and buried oxide layer or after the BOX within the substrate, then it termed as the ground plane in substrate as shown in Figure 3.4.

The aforementioned structure is a continuous ground plane in substrate. For further reducing short-channel effects, there is a possibility to introduce the ground plane under the source/drain region or under the gate region [2,4].

Partial ground plane [3] is more effective in minimizing the short-channel effects. Figure 3.2 shows the partial ground plane design. Ground plane regions are formed by an ion-implantation technique after gate-electrode etching, whose edge locations can be self-aligned with the gate-side spacer whereas the widths can be well controlled by using a window-shaped photolithographic mask covering the gate electrode. Ion implantation is done in the resist window. L can be controlled by tilting an ion beam and by a slit size between edges of the gate electrode and the resist window. The SiO_2/poly-Si structure is used to prevent the gate electrode from being counter doped during the PGP implant. The partial ground plane (PGP) SOI MOSFET minimizes the short-channel effects (SCE) compared to the conventional single-gate SOI MOSFET because the gate-induced field in the SOI layer is held high by the PGP region. This results in lower standby leakage current [3]. The PGP SOI MOSFET shows much better switching performance because of its smaller parasitic capacitance as compared to the conventional GP device.

3.2.3 Ground Plane in BOX (GPB)

To realize the FD SOI MOSFETs in nano-regime, two device issues are most important. Firstly, we must have less drain-induced barrier lowering (DIBL); and secondly, the subthreshold slope must be very steep. In long-channel SOI MOSFET, the subthreshold slope can be improved by increasing the BOX thickness. In a short-channel SOI MOSFET, on the other hand, a thicker BOX causes a larger DIBL due to the electric field penetration through the BOX. Due to this, the subthreshold slope worsens with an increase in the BOX thickness. Ground plane (GP) concept is one of the techniques used to reduce the DIBL

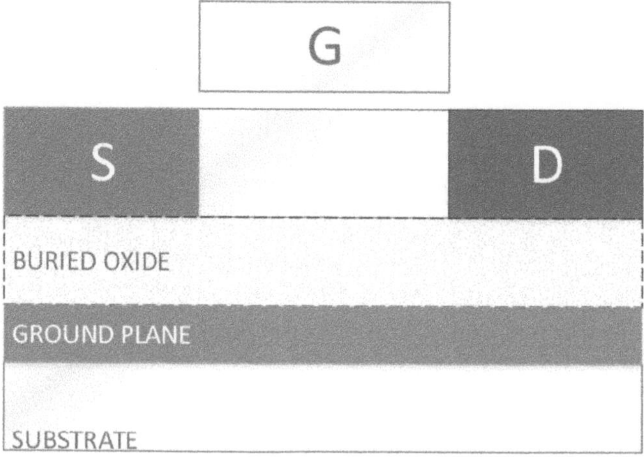

FIGURE 3.4
Ground plane in substrate.

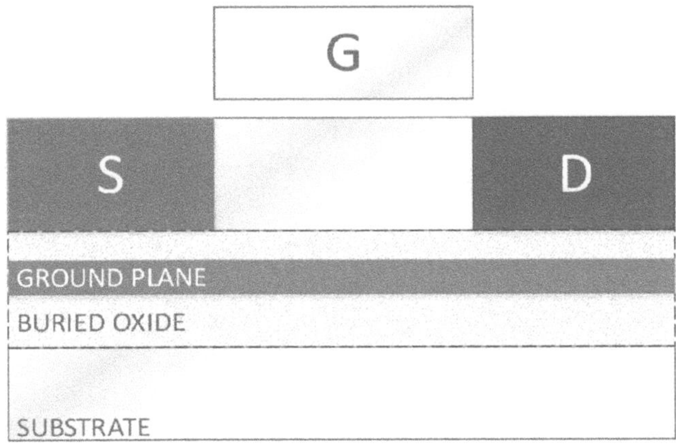

FIGURE 3.5
FD SOI with ground plane in BOX. (From Kumar, M. J. et al., *IEEE Trans. Device Mater. Rel.*, 7, 181–187, 2007.)

effect, and it is quite effective only when the distance between the ground plane and the drain is small compared to the channel length. Therefore, if the placing of the GP in the substrate (GPS) is to be done, then one should keep the BOX thickness as small as possible, but it will increase the subthreshold slope. Therefore, in nanoscale devices with a very short channel length, it is not possible to achieve both reduced DIBL effect and steep subthreshold slope. By using GP in the BOX, it may be possible to reduce the aforementioned shortcomings; thus, the requirement of reduced DIBL as well as an improved subthreshold slope can be obtained [5,7] (Figure 3.5).

Figure 3.6 shows the electric field contour of the FDSOI, FDSOI-GP, and FDSOI-GPB structures. In Figure 3.6a, conventional FDSOI have a large number of electric field lines passing through the channel-body, and hence the drain to substrate leakage is high in this structure. Whereas in FDSOI-GP and FDSOI-GPB, as in Figure 3.6b and c, fewer electric field lines pass through the channel, thereby reducing the leakage due to the addition of the ground plane in a conventional FDSOI structure. The GPB structure has the lowest leakage in all the aforementioned structures due to the reduced distance between the channel and the ground plane [5].

Figure 3.7 shows that the V_t roll-off effect—as we decrease the gate length, V_t decreases. The FDSOI devices have the lowest V_t as compared to the GP and GPB devices. The ground plane keeps the electric field lines from propagating into the channel region, which helps to improve the SCE. The blocked field leads to increases in the threshold voltage.

3.2.4 Ground Plane in FinFET

FinFET is the fin-shaped field effect transistor that has much higher immunity to short-channel effects and has higher charge mobility. Therefore, it is expected to have an increased saturation current. In recent years, different types of FinFETs have been proposed, fabricated, and deployed. Mostly, FinFETs are classified based on controlling the flow of charge carriers using different types of gates such as double gate, Pi- gate, surrounding gate, gate all around, and omega gate. FinFET devices are built on both types of wafers-bulk and SOI. SOI-type FinFET has relatively less leakage current, lower parasitic capacitances, higher saturation current, less sensitivity to body doping and better subthreshold slope as compared to bulk type. On the other hand, bulk FinFET is cheaper than SOI FinFET. Moreover, a self-heating problem is also not present in the bulk SOI [8].

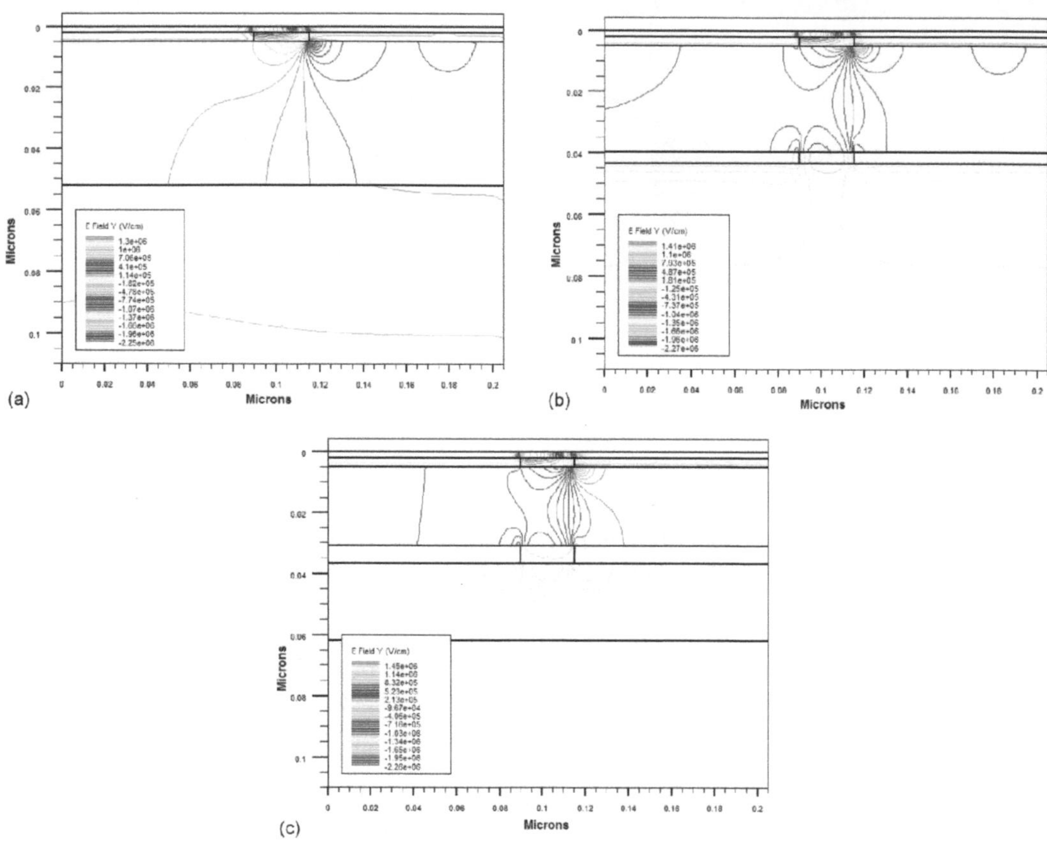

FIGURE 3.6
Electric field contour for FDSOI devices. (a) Conventional FDSOI (b) FDSOI with ground plane in substrate. (c) FDSOI with ground plane in substrate. (From Singh, A., and Chatterjee, A. K., Study of ground plane in FDSOI MOSFET, *Proceedings of the IEEE International Conference on ICONSET*, 2011.)

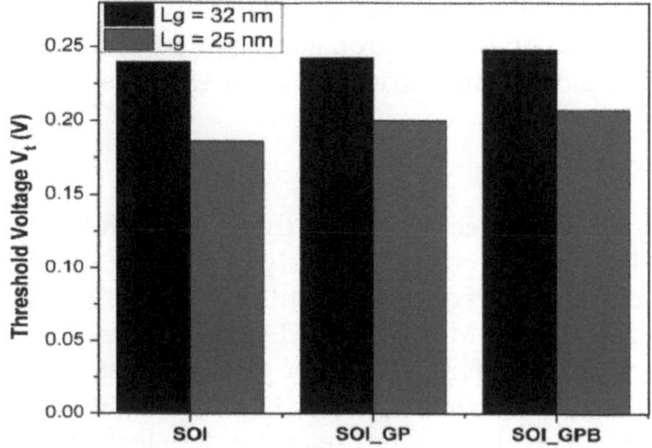

FIGURE 3.7
Threshold voltage for the FDSOI devices at gate lengths of 25 nm and 32 nm. (From Singh, A. et al., *Inform. MIDEM*, 45, 72–78, 2015.)

In SOI FinFET, the channel and substrate are isolated by a buried oxide layer while in bulk FinFET, the buried oxide layer covers almost everything except the channel region. The ground plane FinFET (GP-FinFET) structure is the same as the SOI FinFET structure except the two ground planes and polysilicon layer, which are inside the BOX layer. GP-FinFET is built by the SOIAS technology with the bonded SIMOX approach, and the ground planes are placed under the source and drain regions to control the electric field around the junction regions.

In the tri-gate FinFET structure, if the fin length is short, then the corner effect may influence the electrostatic potential profile between the gates and source/drain more than those of a SOI FET. To reduce the corner effects, tall spacers are used. Threshold voltage in these devices is controlled by the metal gate, which has the mid-gap work function. It has been proved that if the length of the ground plane is equal to the source/drain width, a better control over the channel can be achieved.

Figure 3.8 shows a 3D view of the FinFET with a ground plane. If the width of the ground plane is smaller than that of the source/drain, then the electrostatic potentials under the source/drain are not fully grounded, which will degrade the controlling of charge carriers in the channel.

By incorporating a ground plane in the FinFET structure, drain-induced barrier lowering (DIBL) can be minimized. A larger subthreshold slope can also be achieved in the ground plane FinFET structure as compared to other structures. Further, this leads to a considerable reduction in the leakage current. The on-/off-current ratio is also found to be larger in this structure, which is basically a parameter normally used to show a compromise between the speed and the leakage characteristics of a device. Subthreshold swing is reduced by increasing the channel doping with a consequent increase in control of the gate over the channel.

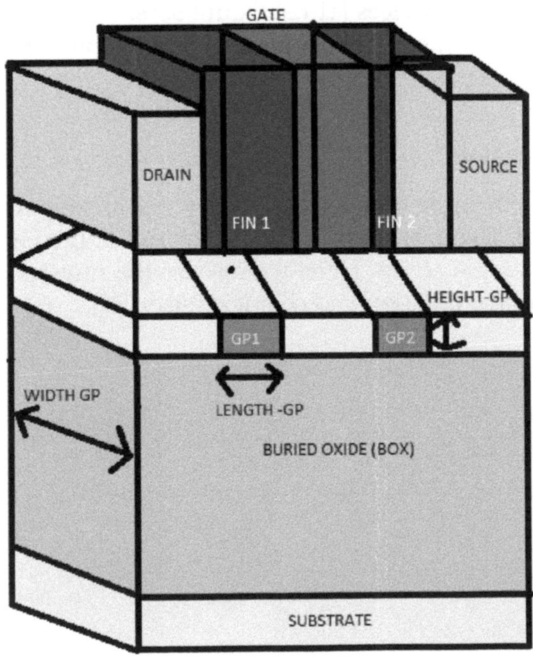

FIGURE 3.8
FinFET with ground plane.

The ground plane FinFET structure may be used in digital circuits in which the static power is of prime concern while the speed is of secondary issue. The structure may find applications in the circuits for mobile applications where the battery life is short.

3.3 Strained Silicon

As devices scale down to nanometer, alternative devices are emerging to minimize the short-channel effects. However, it can only be done at the cost of transistor sizes. When device dimensions are reduced, both the vertical as well as lateral dimensions are minimized. This causes channel length to reduce and thereby increases the associated electric field under the gate across the channel, which is further expected to increase as the scaling continues. The increase in the vertical electric field severely degrades silicon channel mobility. Mobility degradation leads to reduction in drive current [10].

One more issue with the short-channel devices is the requirement for a reduced silicon body thickness. For a 32 nm technology node, the silicon body needs to be as thin as 10 nm to minimize DIBL through it. A thinner body adds to the series resistance of the source and drain, which reduces the mobility and device performance. A reduction in ion translates to reduced operating speed. To increase the mobility or the drive capability of the nanodevices, new devices techniques should be implemented. Strained silicon is an attractive method by which one can increase the device performances. Strain in strained silicon causes increased carrier mobility in the channel region. This results in an increase in the number of ions on and speed, which is otherwise not possible because of a non-scalable subthreshold slope. Strain improves both electron and hole mobility. Most of the semiconductor industries such as Intel and Texas Instruments have already started working on strained silicon-based devices [11,12]. Figure 3.9 shows that after the application of strain in the channel, the charges attain larger mobility as compared to the channel that is not under strain.

3.3.1 Mobility Enhancement

There are a number of theories that were given by researchers and scientists about the straining of the channel. The most popular one is the use of the virtual substrate method, also known as biaxial tensile strain. Here, we discuss the physics of this method; a later part will discuss the implementation and characteristics [24].

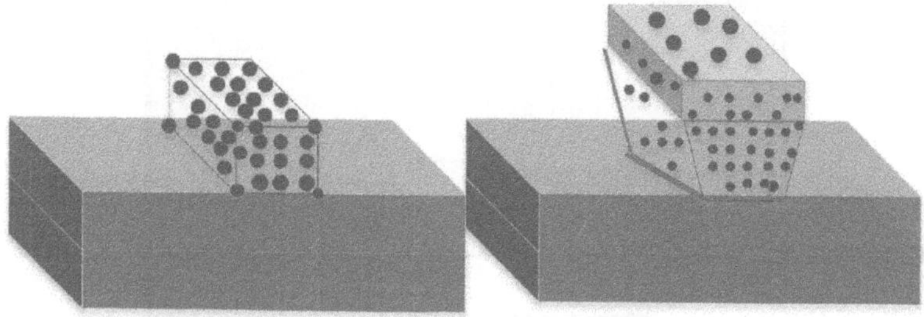

FIGURE 3.9
Mobility enhancement in strained silicon.

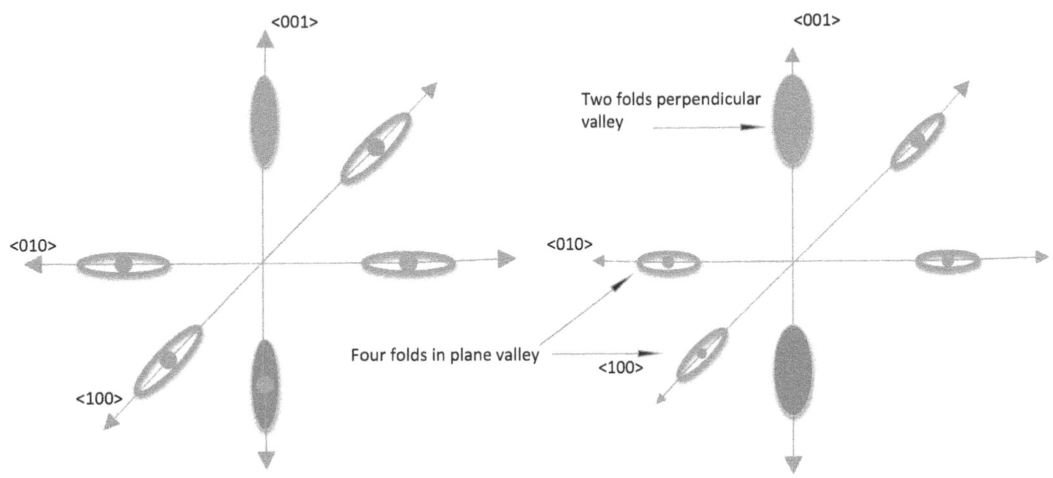

FIGURE 3.10
Strain-induced conduction band splitting in silicon.

Under the biaxial strain, six-fold degenerate valleys in silicon are split into two groups. The first group, which has the lower energy, is twofold degenerate and is also the primary contributor of the carrier transport at the low fields [22]. The effective mass of the electron within the plane is almost equal to the silicon transverse effective mass ($m_t^* = 0.19\ m_o$). On the other part, the effective mass perpendicular to the transport plane is equal to the longitudinal effective mass ($m_l^* = 0.19\ m_o$). We can see the schematic representation of energy ellipses in Figure 3.10. It has been shown that energy of the conduction band minima of the four valleys in the plane <100> axes rises with respect to the energy of the two valleys on the <100> axes perpendicular to the plane. Therefore, the electrons colonize the lower valleys with lighter effective mass, which helps in the reduction of the average conductivity effective mass.

The other mechanism of mobility enhancement for the hole was given by S.I.Takagi et.al., in which he discussed the abolishment of intervalley phonon scattering because of the energy splitting between the twofold and the fourfold valleys [23]. Strain uproots the degeneracy between the heavy-hole and light-hole bands, and the spin-orbit band shifts further down. This slows down the intervalley scattering of the holes.

Maxima of the valance band in unstrained materials is composed of three bands: the degenerate heavy-hole (HH) and light-hole (LH) bands at $k = 0$; and the split-off (SO) band, which is slightly lower in energy as shown in Figure 3.11 [25]. Further, the biaxial stress can be divided into a hydrostatic and a uniaxial stress component. The hydrostatic stress equally shifts all three valance bands, while the uniaxial stress lifts the degeneracy between light holes and heavy holes bands by raising the LH band higher than the HH. The SO band is at the lowest of the three bands. Stress application also changes the shape of the bands as shown in Figure 3.11. Thus, with the band deformation, the in-plane transport mass becomes smaller and the interband scattering is also quenched. In this way the hole mobility is improved.

Figure 3.11 Simplified hole valance band structure for longitudinal in plane direction [20].

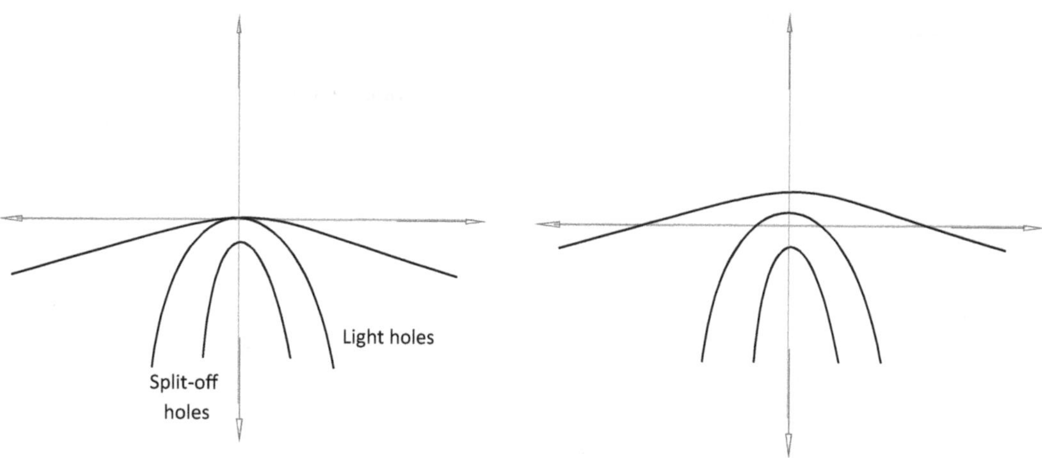

FIGURE 3.11
Simplified hole valance band structure for longitudinal in plane direction. (From Maikap, S. et al., *IEEE Electron Device Lett.*, 25, 40–42, 2004.)

3.3.2 Straining the Si Channel

There are the three ways by which we can introduce the strain in the channel. These are process-induced strain, substrate-induced strain, and external (mechanical), which is also known as package strain.

3.3.2.1 Process-Induced Strain

If strain is introduced at the time of processing of the device, then it is termed as process-induced strain [16]. When the conventional CMOS process flow is going on, the stress patterns can arise from various factors such as different processing temperatures, difference in thermal expansion coefficient, different growth conditions and mechanism, and dopant implantations. Due to the large number of processing steps, charge carriers distribution is non-uniform and increases the strain in some parts of the devices as compared to others. Process-induced strain is dependent on the device layout; therefore, it should be taken care of that the uniaxial stress components along different directions are not canceling each other. If this strain is deployed with some controlling action, then it helps to enhance the mobility of the device and increase the on-current of the device. It is also termed the local strain.

There are three ways we can introduce the process-induced strain in the devices: stress liners, STI stress, and heteroepitaxial strain.

3.3.2.1.1 Stress Liners

Strain is introduced into the FET devices using a capping layer. Capping layers are usually made up of Si nitride, which can be grown easily using a CVD process. The deposition conditions determine whether a strain produced is tensile or compressive in nature [14]. Both compressive and tensile stress are required for increasing the mobility of the transistor and the performance of the CMOS transistors. Generally, tensile stress is used in NMOS for enhancing the on-current, and similarly, compressive strain in PMOS is utilized for the same purpose. We can also use both compressive and tensile strain in a single architecture which is termed as dual stress linear technology. At first, a tensile silicon nitride layer is deposited over the entire wafer and then etched off the film over the PMOS part of the CMOS device [21]. Next, a compressive film is deposited over the whole device and then etched off it from the NMOS side. It is seen that the performance of both NMOS and PMOS device is improved simultaneously. It has also been found that both the linear and saturation current are also improved due to strained silicon.

3.3.2.1.2 Shallow Trench Isolation Stress

As we know, local oxidation of silicon (LOCOS), which was used for isolation, is obsolete and was replaced by shallow trench isolation (STI) for nanoscale gate lengths [16]. But due to the scaling factor, the distance between STI and the channel decreases, which induces stress into the channel [18]. Since it is a compressive stress, only PMOS-based devices benefit. It is reported that about 20% hole mobility is enhanced while STI stress is used, and at the same time the NMOS performance degrades by 20% (Figure 3.12).

3.3.2.1.3 Heteroepitaxial Strain

Compressive strain also originates in devices when there is a lattice mismatch between Si and SiGe channels. This is due to epitaxial growing of SiGe into the source and drain regions, then the Si channel introduces a compressive stress into the channel. With the help of this technique, we can improve the hole mobility enhancement factor that is dependent on the amount of strain coming from the Ge content into the SiGe. In addition, electron mobility also enhances with this technology, but tensile stress is needed that can be incorporated if a SiC material is used for making the epitaxial layer.

3.3.2.2 Substrate-Induced Strain

In substrate-induced strain [16], the most effective way to introduce high tensile strain to the channel is to grow epitaxial strained silicon on a relaxed silicon germanium (SiGe) layer. Bulk silicon and bulk germanium have different lattice constants, 5.43A and 5.65A, respectively. Si and Ge have a lattice mismatch of about 4.2% and can be combined to form a SiGe alloy, the lattice constant of which lies between those of Si and Ge. If a thin layer of Si is grown on a relaxed $\mathbf{Si_{(1-y)}Ge_y}$ substrate, and after that if a Si layer is grown, then it is

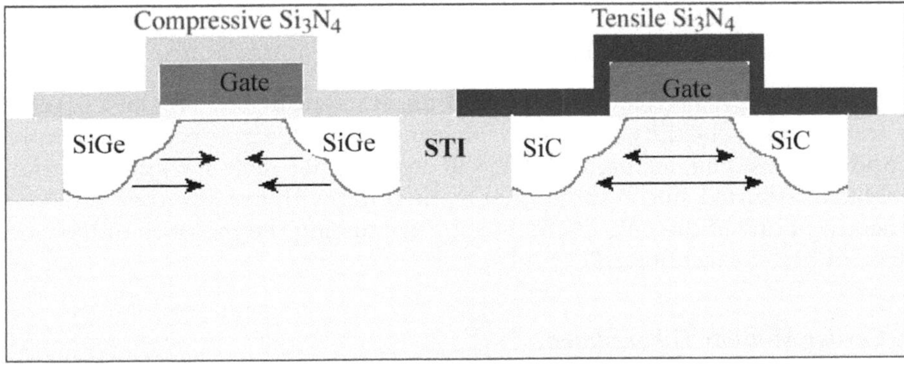

FIGURE 3.12
Different types of process-induced uniaxial strains including stress liners, stress from STI and hetero-epitaxial stress. (Courtesy of S. Dhar, Analytical Mobility Modeling for Strained Silicon-Based Devices.)

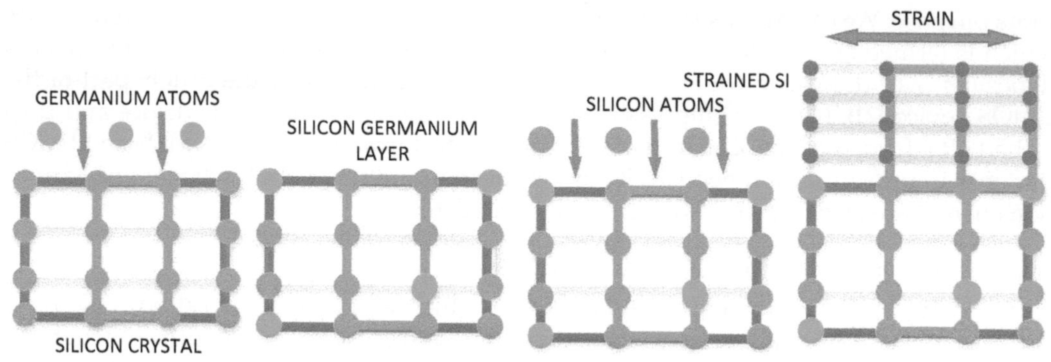

FIGURE 3.13
Formation of SiGe alloy and growth of strained Si on top. (From Chang, S.-T., *Jpn. J. Appl. Phys.*, 44, 5304–5308, 2005.)

said to be under biaxial tensile strain. In that case, strain is given by Ge content. The lattice constant in the alloy $Si_{1-x}Ge_x$ is between that of Si and Ge and varies with Ge concentration as $a(y) = a\text{Si} + (1 - y)a$ Ge [11] (Figure 3.13).

If a thin silicon film is grown on top of $Si_{1-x}Ge_x$, up to a critical thickness, the Si lattice follows the lattice of underlying substrate and gets stretched (or strained) in the plane of the interface. This results in biaxial tensile strain in the channel as shown in Figure 3.11 [13]. As a result of biaxial tensile strain, carriers experience a lower resistance in the strained layer and typically have 50%–70% higher mobility.

3.3.2.3 External Mechanical and Package Strain

In this type of strained silicon, the strain is either introduced through direct bending of the Si wafer or by bending a packaged substrate with a silicon chip [16].

The former technique is known as the mechanical strain, whereas the latter is known as the package strain. With the help of this technique, both short-channel and long channel-MOSFETs benefit. It is also very inexpensive as compared to all the other strain-inducing techniques. Uniaxial strain is produced using the one-end bending method, and the biaxial strain is generated when there is a displacement at the center of the wafer [19].

3.3.3 Advantages of Strained Silicon MOSFET

A few preferences have been found by presenting strained silicon in MOSFET [17]. It incorporates transporter versatility improvement, brings down the protection and power utilization, and the most vital feature is that it will defer the new gate stacking material needs. The benefits of strained silicon should not be tinkered with because they provide a kind of engineering done at the lattice structure for enhancing the mobility of the carriers in MOSFETs for better execution [17].

3.3.3.1 Carrier Mobility Enhancement

The variation of interatomic distance in silicon layers due to tensile or compressive force (strain) would eventually lead to an increase in the mobility of the device. By increasing mobility, the current and the speed requirement in a device can be maintained. The carriers can move almost 70% faster in strained silicon compared to intrinsic silicon. This will result in a device that will be 35% faster [17].

3.3.3.2 *Lower the Resistance and Power Consumption*

Since the interatomic distance between the atoms in the strained silicon has enhanced, it allows electrons to move faster, which means that it creates a lower resistance region. Fortunately, the power consumption in the device will be reduced as the resistance is lowered [17].

3.3.3.3 *New Gate Stack Material Needs Delay*

Strained silicon improves the MOSFET performances without further scaling of gate dielectric thickness, junction depth, or other dimensions of a transistor. The silicon germanium (SiGe) material could integrate well with silicon technology [17]. Further, the strained silicon products are expected to be less expensive than conventional silicon devices.

3.3.4 Strained Silicon-on-Insulator

Strained SOI MOSFET is formed using the biaxial strained silicon (strained silicon grown on relaxed Si_xGe_{1-x}) as already discussed in section 4.2 and shown in Figure 3.14. Strained SOI MOS can be made through a number of methods such as the etch-back and smart-cut method process [25], SIMOX technology [26], and the wafer-bonding method [28]. The presence of the SiGe layer in the strained silicon substrate gives rise to several challenges allied to materials and integration such as high density of defects in strained silicon on relaxed SiGe assured by the strain relaxation in SiGe and a meaningful difference in doping diffusion effects in SiGe. These effects can be minimized through more focus on the junction engineering to control SCEs and through the threshold voltage being set to the desired value. Self-heating is also found in strained SOI MOS devices because of the lower thermal conductivity in SiGe. Threshold voltage of the device can be adjusted by setting the SiGe profile in the channel, the channel and cap doping, and the gate oxide thickness. Gate materials are also the critical design parameter. The key design parameter of any

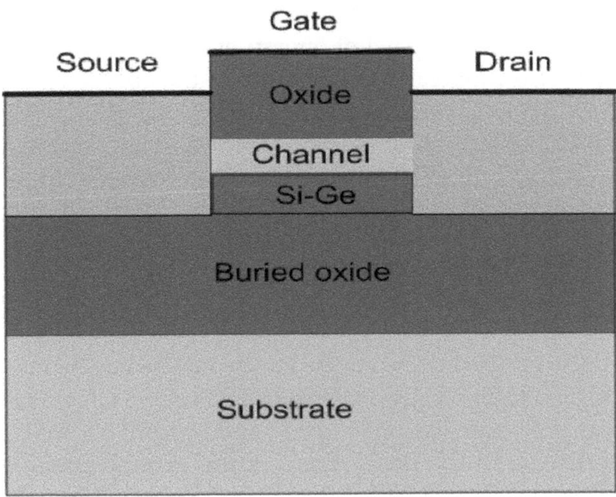

FIGURE 3.14
Strained FDSOI.

type of MOSFET is the device transconductance. It can be increased by maximizing the number of high mobility holes limited to the SiGe channel while minimizing the density low mobility holes, which flow at the Si/SiO_2 interface. The simulation results show that the thickness of the SiGe channel has very little effect on the hole densities in the SiGe and the surface channel. Decreasing the width of the SiGe layer improves the stability of the strained film, thus allowing higher Ge content to be used at the same stability [29].

Methods for fabricating both a P channel FET and N channel FET in a single SIGe channel region are proposed [30]. The fully integrated SiGe CMOS fabrication process solves the thermal budget and multi-deposition problems. Furthermore, this results in a planar structure at the expense of only one additional mask. In this process, a boron-doped layer would be grown with a typical concentration of 4×10^{17} cm^{-3} and a thickness of 350 ^{0}A. After that, an undoped SiGe channel is grown over it, then a thin intrinsic Si Cap is spread. This cap separates the gate oxide from the SiGe channel to minimize surface scattering.

Lastly, a quasi-uniform doping profile is obtained with a peak doping level of 3.5×10^{17} cm^{-3} behind the channel and a surface-junction depth of 400 ^{0}A. The doping concentration at the top of the channel (where most of the holes flow) is below 1×10^{16} cm^{-3}; hence, impurity scattering is minimized.

An additional mask is used to adjust the n-MOSFET threshold voltage. The final *N* channel MOSFet is a surface channel device, and the P channel is a subsurface channel device with holes confined in the SiGe channel.

3.3.5 Strained FinFETs

FinFET is the most innovative device for 14-nm technology node-based architectures. The main issue of any device is the controllability of the charge carriers passing through the channel with the number of gates and minimization of leakage current [27]. However, for ultra-low power (ULP), digital device applications are regularly demanding for high speeds, too. For the increment of the drive current, strained silicon is the most appropriate technique. Using a source/drain (S/D) stressor, we are able to enhance the mobility of charge carriers in the channel. SiGe [1] is used for compressive stress, and silicon carbide (SiC) [8] is generally employed for tensile stress. While SiGe has been used as a stressor for PMOS extensively, and SIC is used for NMOS, by changing the substrate, we are also able to control the driving current and speed of operation of the device [31].

The shape of the FIN also influences the mobility of the charge carriers. Due to this, the driving capability of the devices is strongly influenced by the triangular and rectangular fin shapes for strained FinFETs [finfet strained]. Symmetric and asymmetric profile is one of the major components in determining the compatibility of the device for high-speed applications.

3.4 Conclusion

In this chapter, we discussed the concept of ground plane in SOI device structures, the use of which can reduce the off-state leakage current and optimize it for the low-power applications. Further, we discussed the deployment of the ground plane in FDSOI MOSFET using both the methods of GPS and GPB. It has been found that the off-current is reduced significantly in the FD GPS SOI structure (ground plane behind the substrate) compared to conventional FD SOI MOSFET. Further, the current is reduced more in the case of the FD

SOI GPB structure (ground plane in between the substrate). The leakage current is less in GPB because in this structure the device region comes closer to the ground plane region. Hence, the drain-induced barrier lowering is reduced.

By incorporating the ground plane concept in the FinFET structure, DIBL can be minimized. It has also been found that a larger subthreshold slope is obtained in the ground plane FinFET structure as compared to other structures, which results in a remarkable reduction in the off-current. However, in these structures, the current drive capability of the transistors also reduces. To increase the on-current, a strained silicon channel has been used.

Besides the ground plane, we have studied how we can introduce the strain in the channel. Two types of strain technology are discussed. The physics of strain engineering through an E-K diagram are also discussed in this chapter. These are process-induced strain and substrate-induced strain. Further, their types are also discussed. It is seen that the on-current in the strained structure is greater compared to conventional FD SOI MOSFET structures. To decrease the leakage current, the ground plane concept—in accordance with the strained SOI structure—has been used. Lastly, we shed some light on incorporation of strain in FinFET and also discussed the different factors that affect the strain in FinFET such as fin shapes, etc.

To reduce the leakage concept of the ground plane, and to increase the drive capability, strain engineering should be used. Thus, by deploying the concepts of ground plane and strained silicon in a device, we are able to optimize SOI FET devices for low-power and high-speed applications.

References

1. T. Ernst and S. Cristoloveanu, "The ground-plane concept for the reduction of short-channel effects in fully depleted SOI devices," *Electrochemical Society Proceedings*, pp. 329–334, 1999.
2. T. Ernst and S. Cristoloveanu, "Buried oxide fringing capacitance: A new physical model and it implication on SOI device scaling and architecture," *IEEE International SOI Conference 1999*, pp. 38–39, 2010.
3. S. Yanagi, A. Nakaubo, and Y. Omura, "Proposal of a partial-ground-plane (PGP) silicon-on-insulator MOSFET for deep sub-0.1-μm channel regime," *IEEE Electron Device Letters*, vol. 22, no. 6, 278–280, 2001.
4. M. Saremi, B. Ebrahimi, A. A. Kusha, and M. Saremi, "Process variation study of ground plane," *2nd Asia Symposium on Quality Electronic Design*, 2010.
5. M. J. Kumar and M. Siva, "The ground plane in buried oxide for controlling short-channel effects in nanoscale SOI MOSFETs," *IEEE Transactions on Electron Devices*, vol. 55, no. 6, pp. 1554–1557, 2008.
6. A. Singh and A. K. Chatterjee, "Study of ground plane in FDSOI MOSFET," *Proceedings of the IEEE International Conference on ICONSET*, 2011.
7. A. Singh, S. Adak, H. Pardeshi, S. Sarkar, and C. K. Sarkar, "Comparative assesment of ground plane and strained based FDSOI MOSFET," *Journal of Microelectronics Component and Materials, Informacije MIDEM*, vol. 45, no. 1, pp. 72–78, 2015.
8. M. Saremi, A. Kusha, and S. Mohammadi, "Ground plane fin-shaped field effect transistor (GP-FINFET): A FINFET for low leakage power circuits," *Journal of Microelectronics Engineering*, vol. 95, pp. 74–82, 2012.
9. K. Roy, S. Mukhopadhyay, H. Mahmoodi et al., "Leakage current mechanisms and leakage reduction techniques in deep-submicrometer CMOS circuits," *Proceedings of the IEEE*, vol. 91, no. 2, pp. 305–327, 2003.

10. L. Ge, J. G. Fossum, and B. Liu, "Physical compact modeling and analyses of velocity overshoot in extremely scaled CMOS devices and circuits," *IEEE Transactions on Electron Devices*, vol. 48, pp. 2074–2080, 2001.
11. M. J. Kumar, V. Venkataraman, and S. Nawal, "Impact of strain or Ge content on the threshold voltage of nanoscale strained-Si/SiGe bulk MOSFETs," *IEEE Transactions on Device and Materials Reliability*, vol. 7, no. 1, pp. 181–187, 2007.
12. S. E. Thompson, M. Armstrong, C. Auth et al., "A logic nanotechnology featuring strained-silicon," *IEEE Electron Device Letters*, vol. 25, no. 4, pp. 191–193, 2004.
13. K. K. Rim, J. L. Hoyt, and J. F. Gibbons, "Fabrication and analysis of deep submicron strained-Si N-MOSFET's," *IEEE Transactions on Electron Devices*, vol. 47, pp. 1406–1415, 2000.
14. M. Reiche, O. Moutanabbir, J. Hoentschel, U. Gosele, S. Flachowsky, and M. Horstmann, "Strained silicon devices," *Solid State Phenomena*, vol. 156–158, pp. 61–68, 2010.
15. S.-T. Chang, "Nanoscale strained Si/SiGe heterojunction trigate field effect transistors," *Japanese Journal of Applied Physics*, vol. 44, no. 7A, pp. 5304–5308, 2005.
16. R. A. Bianchi, G. Bouche, and O. Roux-dit-Buisson, "Accurate modeling of trench isolation induced mechanical stress effects on MOSFET electrical performance," *IEDM Technology Digest*, pp. 117–120, 2003.
17. V. Moroz, X. Xu, D. Pramanik, F. Nouri, and Z. Krivokapic, "Analyzing strained-silicon options for stress-engineering transistors," *Solid State Technology*, vol. 47, no. 7, pp. 49–52, 2004.
18. A. S. Zoolfakar, N. I. Mohmad Tahiruddun, and L. N. Ismail, "Modeling of strain technology on 140 nm CMOS devices," *Journal of Electrical and Electronic Systems Research*, vol. 2, 2009.
19. A. Lochtefeld and D. A. Antoniadis, "Investigating the relationship between electron mobility and velocity in deeply scaled NMOS via mechanical stress," *IEEE Electron Device Letters*, vol. 22, no. 12, pp. 591–593, 2001.
20. S. Maikap, C.-Y. Yu, S.-R. Jan, M. H. Lee, and C. W. Liu, "Mechanically strained Si NMOSFETs," *IEEE Electron Device Letters*, vol. 25, no. 1, pp. 40–42, 2004.
21. S. E. Thompson, M. Armstrong, C. Auth et al., "A 90 nm logic technology featuring strained Si," *IEEE Transactions on Electron Devices*, vol. 51, no. 11, pp. 1790–1797, 2004.
22. R. People, "Physics and applications of GexSi1-X/Si strained-layer heterostructures," *IEEE Journal of Quantum Electronics*, vol. 22, pp. 1696–1710, 1986.
23. S. I. Takagi, J. L. Hoyt, J. J. Welser, and J. F. Gibbons, "Comparative study of phonon-limited mobility of two-dimensional electrons in strained and unstrained Si metal-oxide-semiconductor field-effect transistors," *Journal of Applied Physics*, vol. 80, pp. 1567–1577, 1996.
24. T. Vogelsang and K. R. Hofmann, "Electron-transport in strained Si layers on Si1-XGex substrates," *Applied Physics Letters*, vol. 63, pp. 186–188, 1993.
25. S. Takagi, N. Sugiyama, T. Mizuno, T. Tezuka, and A. Kurobe, "Device structure and electrical characteristics of strained-Si-on-insulator (strained-SOI) MOSFETs," *Materials Science and Engineering B-Solid State Materials for Advanced Technology*, vol. 89, pp. 426–434, 2002.
26. D. Hisamoto, T. Kaga, Y. Kawarnoto, and E. Takeda, "A fully depleted lean-channel transistor (DELTA)-a novel vertical ultra thin SOI MOSFET," *IEDM Technology Digest*, pp. 833–836, 1989.
27. X. Huang et al., "Sub-50 nm P-channel FinFET: PMOS," *IEEE Transactions on Electron Devices*, vol. 48, no. 5, pp. 880–886, 2001.
28. K. Rim et al., "Fabrication and mobility characteristics of ultrathin strained Si directly on insulator (SSDOI) MOSFETs," *IEDM Technology Digest*, p. 49, 2003.
29. G. Armstrong and C. Maiti, "Strained-Si channel heterojunction P-MOSFETs," *Solid-State Electron*, vol. 42, pp. 487–498, 1998.
30. S. H. Olsen, A. G. O'Neill, S. Chattopadhyay, L. S. Driscoll, K. S. K. Kwa, D. J. Norris, A. G. Cullis, and D. J. Paul, "Study of single- and dual-channel designs for high-performance strained-Si–SiGe n-MOSFETs," *IEEE Transactions on Electron Devices*, vol. 51, no. 7, 2004.
31. S. Dubey and P. N. Kondekar, "Performance comparison of conventional and strained FINFET inverters," *Microelectronics Journal*, vol. 55, pp. 108–115, 2016.

Section II

Novel MOSFET Structures

4

U-Shaped Gate Trench Metal Oxide Semiconductor Field Effect Transistor: Structures and Characteristics

Deepshikha Bharti and Aminul Islam

CONTENTS

4.1 Introduction

A Silicon Carbide (SiC) vertical power MOSFET is a next-generation switching device expected to replace conventional Silicon (Si) power-switching devices in many applications because it can operate with lower power loss at a higher switching frequency and at higher operating temperatures. In the past 10 years, a lot of effort has been devoted to develop SiC power MOSFETs, and great progress has been achieved.

It was found that the ability to reduce the specific on-resistance for silicon carbide power VDMOSFETs was constrained by the poor channel density and the JFET region resistance [1]. The trench-gate or UMOSFET structure enabled significant increase in the channel density and elimination of the JFET resistance contribution, resulting in a major enhancement in power MOSFET performance. The first SiC power MOSFET was demonstrated in 1994 in the form of a U-shaped Gate Trench MOSFET structure (UMOSFET) [2,3]. The UMOS structure is attractive in SiC because the base and source regions can be formed epitaxially, without the need for ion implantation and associated high-temperature annealing. In UMOSFETs, the MOS channel is formed on the sidewalls of trenches. The silicon carbide UMOSFET was developed by borrowing the trench technology originally developed for DRAMs.

However, problems with controlling the quality of the trench surface and oxide reliability problems needed to be solved before the introduction of commercial devices. Eventually, this technology overtook the planar VDMOSFET technology in the 1990s and has now taken a dominant position in the industry for serving portable appliances, such as laptops, PDAs, etc. In the silicon carbide power VDMOSFET structure, the channel and JFET resistances were found to become the dominant components when the breakdown voltage was reduced below 50 V because of the low resistance of the drift region. Due to the much lower specific on-resistance for the drift region in silicon carbide, the channel and JFET contributions become dominant even when the breakdown voltage approaches 5000 V, especially if the channel mobility is poor. This has motivated the development of the trench-gate architecture in silicon carbide. In addition, the P-well region for the trench-gate device could be fabricated using epitaxial growth, which was at a more advanced state than the ability to use ion implantation to create P-type layers in silicon carbide. But the performance of the trench-gate silicon carbide power MOSFET structure is severely compromised by the development of a high electric field in the gate oxide during the blocking mode of operation. This problem occurs because the trench spans the P-well region, exposing the gate oxide at the bottom of the trench to the high electric field in the silicon carbide drift region. The electric field in the gate oxide reaches its rupture strength well before the electric field in the semiconductor approaches its breakdown field strength. Consequently, in order to operate at any given blocking voltage, the drift region doping concentration has to be reduced and its thickness increased until the specific on-resistance becomes 25 times larger than that for the ideal drift region. This problem inhibited the performance of the first trench-gate 4H-SiC power MOSFETs. In order to suppress the development of high electric fields in the gate oxide, a UMOSFET structure has been proposed [4] with a shielding region incorporated at the bottom of the trench. In the blocking state, it guards the oxide from the high electric field and thus results in a higher breakdown voltage value. The conventional UMOSFET and shielded UMOSFET are inversion-channel devices, and their on-resistance is greatly affected by the resistance of the MOS channel. The SiC U-ACCUFET (accumulation layer UMOSFET) [5] is introduced as an encouraging way to overcome the problems associated with the low mobility of an inversion layer in the channels of the SiC UMOSFET structure. This leads to significant increase in research activity on SiC U-ACCUFETs. The N-type epilayer is created along the sidewalls of the gate groove, translates the device to an accumulation-mode MOSFET, or "ACCUFET," escalating the mobility of MOSFET and also dropping the on-resistance value.

This chapter reviews the basic principles of operation of the UMOSFET structure. The specific on-resistance for this structure is shown to be significantly lower than that for the VDMOSFET structure. However, the gate oxide in the UMOSFET structure is exposed to the very high electric field developed in the silicon carbide drift region during the blocking-mode. This is a major limitation to adopting the basic UMOSFET structure. Consequently, structures designed to reduce the electric field at the gate oxide by using a shielding region are essential to the realization of practical silicon carbide MOSFET structures. These structures are described and analyzed in further sections. The next section reviews the basic principle of accumulation mode UMOSFET utilizing the higher mobility of accumulation layer channels.

4.2 UMOSFET Structure

The UMOSFET is a rectangular (U-shaped) groove MOSFET with a high packing density and a reduced channel resistance. The basic structure of the UMOSFET is shown in Figure 4.1. In the UMOSFET (U-shaped gate trench MOSFET) structure, the channel as well as the drift region is vertically aligned. This permits reduced value of on-resistance and higher packing density than the Vertically diffused MOSFET (VDMOSFET), thus resulting in its enhanced popularity in modern power device technology. The structure can be fabricated by either the epitaxial growth of the P-well region over the drift region or by introducing the P-type dopants using ion-implantation. The device is fabricated by diffusing P-well and N^+-source regions into an N-epitaxial drift region of an N^+ wafer with the P-well region diffused deeper than the N^+-source. The first 4H-SiC UMOSFET structure was fabricated by the epitaxial growth of the P-well region [6]. However, this requires either removal of the P-type layer on the edges of the structure to form a mesa edge termination [6] or multiple trench-isolated guard rings [7]. Reactive ion etching is used to form rectangular grooves or trenches in the substrate, followed by a gate oxidation that was created by thermal oxidation, followed by refilling the trench with polysilicon, as done for silicon UMOSFET structures. A P^+-region is incorporated into the UMOSFET structure under the source region to suppress the extension of the depletion region in this portion of the cell structure. Also, the P-well region is short-circuited to the N^+-source region by the source metal through this P^+-region. This is also advantageous for reducing the cell pitch, which results in a larger channel density. A refractory gate electrode is required to allow diffusion of the dopants under the gate electrode at elevated temperatures. The N-drift region is designed to give the forward blocking capabilities. The forward blocking capability is achieved by the P-N junction between the P-well region and the N-drift region. The N-drift must be moderately doped so that the drain breakdown voltage is sufficiently large and the thickness of the N-drift region is made as thin as possible to minimize drain resistance. During the device operation, a fixed potential to the P-well region is established by connecting it to the source metal by the break in the N^+-source region. Without adding a gate bias, a high voltage can be supported in the UMOSFET structure when a positive bias

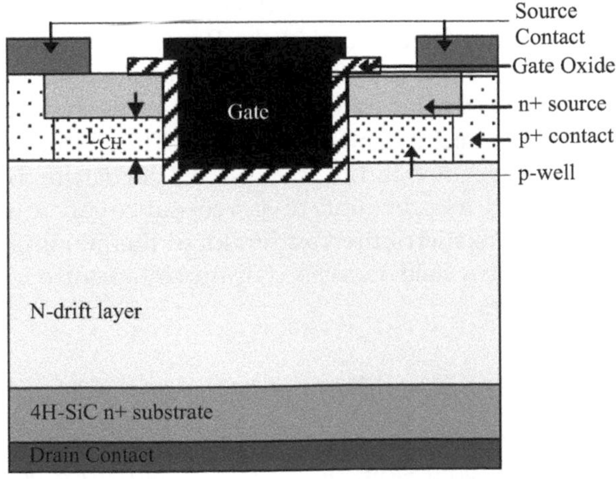

FIGURE 4.1
Cross section of the conventional UMOSFET structure.

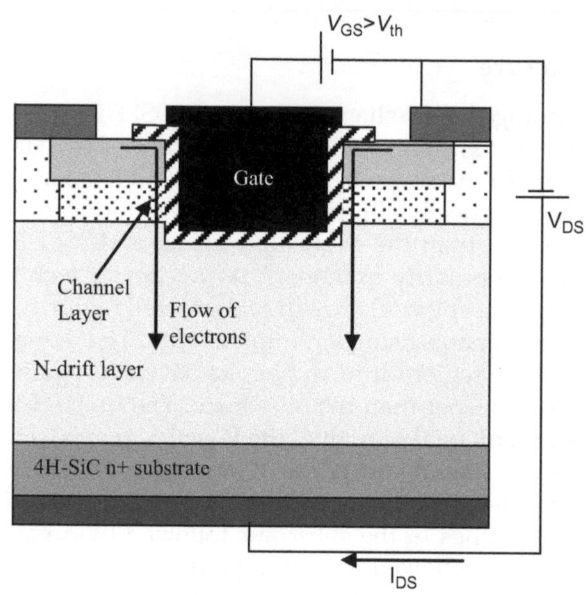

FIGURE 4.2
Principle of operation of UMOSFET.

is applied to the drain. In this situation—i.e., by short-circuiting the gate to the source and applying a positive bias to the drain—the junction formed between the P-well region and N-drift region becomes reversely biased, and this junction supports the drain voltage by the extension of a depletion layer on both sides. However, due to the higher doping level of the P-well layer, the depletion layer extends primarily into the N-drift region. The voltage is supported mainly within the thick lightly doped N-drift region. By adding a positive gate bias, drain current flow is introduced in the UMOSFET structure.

This produces an inversion layer in the P-well region alongside the gate electrode as shown in Figure 4.2. This inversion layer channel provides a path for transport of electrons from the source to the drain, thereby inducing the current flow (I_{DS}) from the drain to the source electrode when a positive drain voltage is applied. After transport from the source region through the channel in the P-well region, the electrons enter the N-drift region and then reach to the drain region. On the other hand, when the MOSFET is turned off, the gate voltage should be lower than the threshold voltage. In this way, the MOSFET switches to on-state and off-state. The JFET region, which was generated between the adjacent P-well regions in the VDMOSFET structure, is eliminated here in the UMOSFET structure. Thus, a lower value of on-resistance can be achieved. In the drift layer, the potential difference is distributed and reduces the maximal electric field in the transistor. Due to a lower electric field, increased drain voltages are allowed and the high-power performance improves.

4.2.1 UMOSFET Structure: Off-State Characteristics

When the UMOSFET operates in the forward blocking mode, the voltage is supported by a depletion region formed on both sides of the P-well/N-drift junction. The maximum blocking voltage can be determined by the electric field at this junction becoming equal to the critical electric field for breakdown if the parasitic $N^+/P/N$ bipolar transistors are completely suppressed. This suppression is accomplished by short-circuiting the N+-source and P-well regions using the source metalas shown on the upper left and right-hand side of the cross section. However, a large leakage current can occur when the depletion region in the P-well region reaches through to the N+-source region. The doping concentration and thickness of the P-well region must be designed to prevent the reach-through phenomenon from limiting the breakdown voltage.

The maximum blocking voltage capability of the UMOSFET structure is determined by the drift region doping concentration and thickness. The relation between maximum depletion width and breakdown voltage can be stated as

$$W_D = \left(\frac{2V_{BR}}{E_C}\right) \tag{4.1}$$

where V_{BR} is the breakdown voltage and E_C is the critical electric field. The doping concentration of the drift region (N_D) to achieve the required breakdown voltage V_{BR} is given as

$$N_D = \left(\frac{\varepsilon_S E_C^2}{2qV_{BR}}\right) \tag{4.2}$$

The blocking characteristics of the 4H-SiC UMOSFET proposed in [8] are presented in Figure 4.3. The blocking characteristics measurement was performed with a gate voltage of 0 V. The measured blocking voltage is 880 V.

In the UMOSFET structure, the gate extends into the drift region, exposing the gate oxide to the high electric field developed in the drift region under forward blocking conditions. For 4H-SiC, the electric field in the oxide reaches a value of 9 × 10⁶ V/cm when the field in the semiconductor reaches its breakdown strength. This value not only exceeds the reliability limit but also can cause rupture of the oxide, leading to catastrophic breakdown. The problem is exacerbated by electric field enhancement at the corners of the trenches in

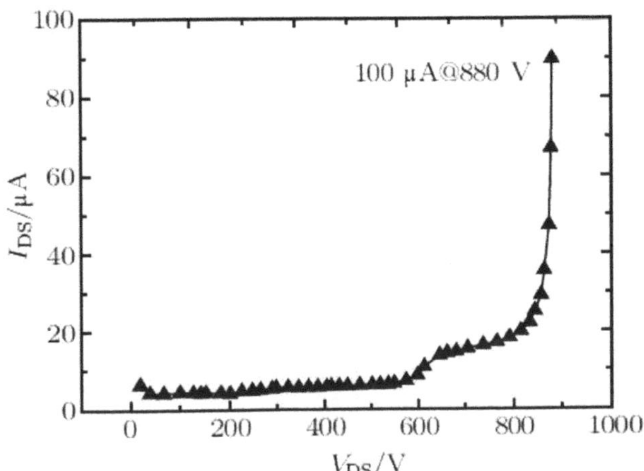

FIGURE 4.3
Blocking characteristics in a UMOSFET structure at $V_{GS} = 0$ V. (From Qing-Wen, S. et al., *Chin. Phys. B*, 22, 027302, 2013.)

the figure. A novel structure [4] that shields the gate oxide from a high electric field has been proposed and demonstrated to resolve this problem. This structure is discussed in the next section.

4.2.2 UMOSFET Structure: On-State Characteristics

Current flow between the drain and source can be induced by creating an inversion layer channel on the surface of the P-well region. The current path is illustrated in Figure 4.4 by the highlighted region. The current flows from the source region into the drift region through the inversion layer channel formed on the vertical sidewalls of the trench due to the applied gate bias. It then spreads into the N-drift region from the bottom of the trench at around a 45° angle and becomes uniform throughout the rest of the structure.

Figure 4.5 shows on-state source-drain (I_D – V_{DS}) characteristics at room temperature for the 4H-SiC UMOSFET with an active area of 9.88×10^4 cm² with the gate voltage (V_g)

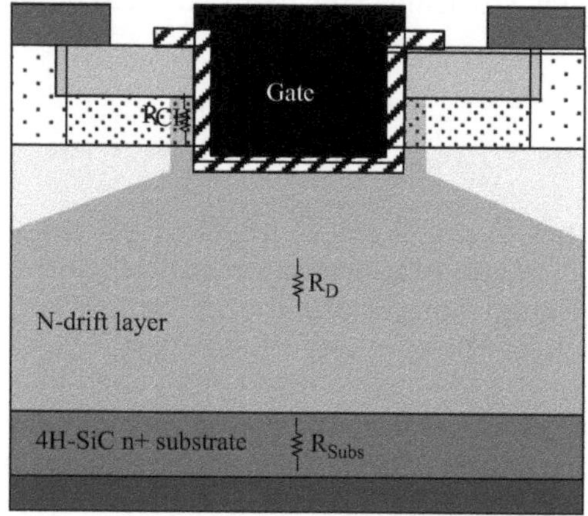

FIGURE 4.4
Current flow illustration during on-state in a UMOSFET structure.

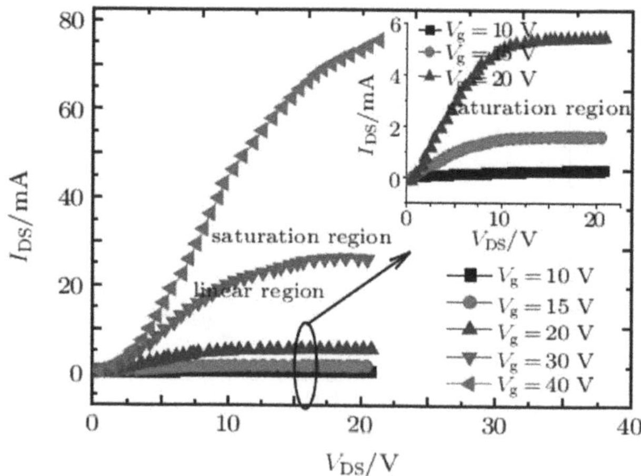

FIGURE 4.5
On-state characteristics of a UMOSFET structure. (From Qing-Wen, S. et al., *Chin. Phys. B*, 22, 027302, 2013.)

increased from 0 V to 40 V and the drain–source voltage (V_{DS}) increased from 0 V to 22 V. The device exhibits a drain current density of 65.8 A/cm² at V_g = 40 V and V_{DS} = 15 V. From the slope of the $I_D - V_{DS}$ characteristics in the linear region, the specific on-resistance ($R_{sp\text{-}on}$) can be calculated.

The total on-resistance for the UMOSFET power MOSFET structure is determined by the resistance of these components in the current path:

$$R_{on,sp} = R_{CH} + R_D + R_{Subs} \tag{4.3}$$

where R_{CH} is the channel resistance, R_D is the resistance of the drift region after taking into account current spreading from the channel, and R_{Subs} is the resistance of the N⁺-substrate. The various components of the on-resistance are shown in Figure 4.4.

The specific channel resistance is given by:

$$R_{CH} = \frac{(L_{CH}.p)}{2\mu_{inv}C_{ox}(V_G - V_T)} \tag{4.4}$$

where L_{CH} is the channel length determined by the width of the P-well region as shown in Figure 4.1, p is the cell pitch, μ_{inv} is the mobility for electrons in the inversion layer channel, C_{ox} is the specific capacitance of the gate oxide, V_G is the applied gate bias, and V_T is the threshold voltage. The specific capacitance can be obtained using:

$$C_{ox} = \frac{\varepsilon_{ox}}{t_{ox}} \tag{4.5}$$

where ε_{ox} is the dielectric constant for the gate oxide and t_{ox} is its thickness.

The drift region spreading resistance can be obtained by using:

$$R_D = \frac{\rho_D.p}{2}.\ln\left(\frac{p}{2W_T}\right) + \rho_D.(t - W_M) \tag{4.6}$$

where t is the thickness of the drift region below the P-well region and W_T, W_M are the widths of the trench and mesa regions, respectively.

The contribution to the resistance from the N⁺-substrate is given by:

$$R_{subs} = \rho_{subs}.t_{subs} \tag{4.7}$$

where ρ_{subs} and t_{subs} are the resistivity and thickness of the substrate, respectively.

Figure 4.6 shows the relationship between the specific on-resistance ($R_{sp\text{-}on}$) and the gate voltage (V_g), which indicates that the total specific on-resistance gradually reduces as the gate voltage increases. This is attributed to the reduced channel resistance that results from an increased charge in the inversion layer, and the drift region resistance gradually becomes the predominant one. The $R_{sp,on}$ of 181 mΩ.cm² can be obtained from the $I_D - V_{DS}$ characteristics at V_g = 40 V, as shown in Figure 4.6.

The specific on-resistance for this UMOSFET 4H-SiC MOSFET structure is plotted in Figure 4.7 as a function of the breakdown voltage using a channel mobility of 25 cm²/V-s.

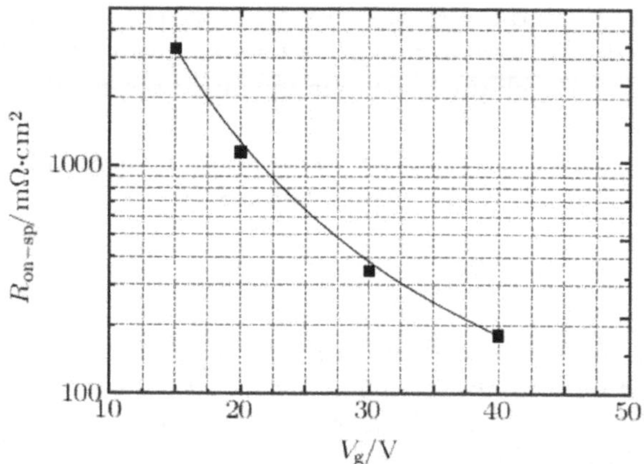

FIGURE 4.6
Relationship between $R_{sp\text{-}on}$ and V_g. (From Qing-Wen, S. et al., *Chin. Phys. B*, 22, 027302, 2013.)

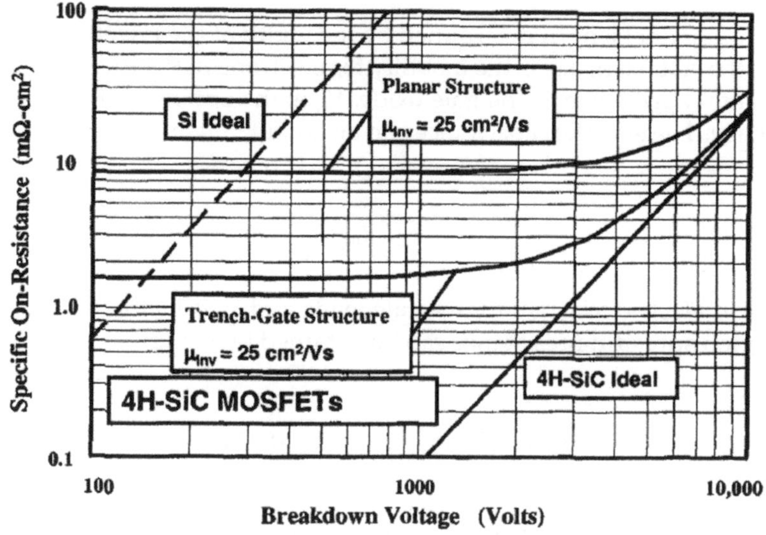

FIGURE 4.7
Specific on-resistance for 4H-SiC UMOSFETs. (From Baliga, J., *Silicon Carbide Power Devices*, World Scientific Publishing, Singapore, 2005.)

From Figure 4.7, it can be concluded that the specific on-resistance of 4H-SiC UMOSFETs (Trench-gate MOSFETs) approaches the ideal specific on-resistance when the breakdown voltage exceeds 5000 V, even with a channel mobility of 25 cm²/V-s. However, the channel resistance limits the performance of the UMOSFET 4H-SiC MOSFET structure when the breakdown voltage falls below 1000 V. For a better perspective, the performance of the planar 4H-SiC inversion-mode MOSFET (VDMOSFET) structure is also shown in the figure. The planar MOSFET analysis was based upon using the same channel length, gate oxide thickness, and channel mobility of 25 cm²/V-s, same as the trench-gate UMOSFET structure. A cell-pitch of 3 μ was used for the planar structure based upon the same 0.5 μ design rules. It can be seen that the UMOSFET structure offers a substantial (five-fold) reduction of the specific-on-resistance due to the higher channel density. These results highlight the need for development of aggressive processing techniques, such as those routinely used for low voltage silicon power UMOSFET structures, to achieve a small cell pitch in the UMOSFET 4H-SiC MOSFET structure. This model assumes that all the applied drain bias is supported within the N-drift region. For devices with lower breakdown voltages, the doping concentration in the N-drift layer becomes comparable to that for the P-well region. Consequently, a substantial fraction of the applied drain bias is supported within the P-well region as well. A model for the specific on-resistance that takes this into consideration indicates further reduction of the specific on-resistance [10]. However, the specific on-resistance of the UMOSFET structure is still limited by the channel inversion layer mobility for devices designed to support 1000 V. To address this issue, an accumulation layer UMOSFET (U-ACCUFET) structure is utilized, as discussed in further sections.

4.2.3 UMOSFET Structure: Transfer Characteristics

The transfer characteristics of the MOSFETs were obtained by measuring the drain current as the gate-source voltage was swept for a fixed drain-source voltage.

Figure 4.8 shows the room-temperature transfer characteristics of the 4H-SiC UMOSFET measured at a drain voltage of 100 mV. A threshold voltage of 5.5 V is obtained by linear extrapolation from the linear region of the transfer characteristics. From the semi-log plot of the transfer characteristics, it can be seen that the device is completely pinched off with a drain current of as low as ~3.0×10^{-10} A at $V_g = 1.3$ V and $V_{DS} = 100$ mV.

The threshold voltage of the UMOSFET structure is determined by the doping concentration of the P-well region, along the sidewalls of the trench region. A minimum threshold voltage must be maintained at above 1 V for most system applications to provide immunity against inadvertent turn-on due to voltage spikes arising from noise. At the same time, a high threshold voltage is not desirable because the voltage available for creating the charge in the channel inversion layer is determined by $(V_G - V_T)$, where V_G is the applied gate bias voltage and V_T is the threshold voltage.

The threshold voltage can be modeled by defining it as the gate bias at which onset of strong inversion begins to occur in the channel. This voltage can be determined using [1]:

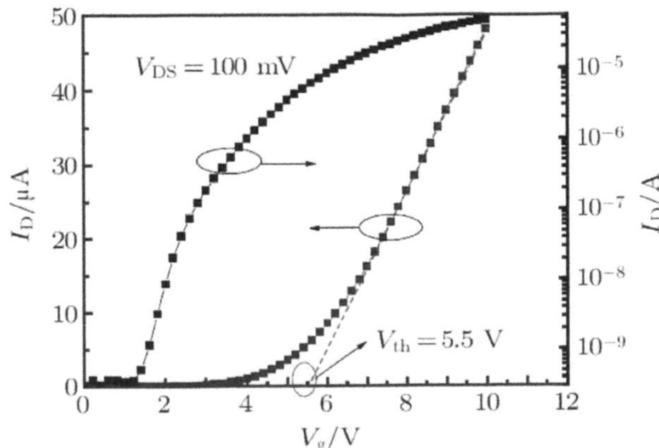

FIGURE 4.8
Transfer characteristics of an UMOSFET structure. (From Qing-Wen, S. et al., *Chin. Phys. B*, 22, 027302, 2013.)

$$V_t = \frac{\sqrt{4\varepsilon_s kTN_A \ln\left(\frac{N_A}{n_i}\right)}}{C_{OX}} + \frac{2kT}{q}\ln\left(\frac{N_A}{n_i}\right) \tag{4.8}$$

where N_A is the doping concentration of the P-well region, k is the Boltzmann's constant, and T is the absolute temperature. The presence of positive fixed oxide charge shifts the threshold voltage in the negative direction by:

$$\Delta V_{TH} = \frac{Q_F}{C_{OX}} \tag{4.9}$$

A further shift of the threshold voltage in the negative direction by 1 V can be achieved by using heavily doped N-type polysilicon as the gate electrode as is routinely done for silicon power MOSFETs.

The much larger threshold voltage for silicon carbide is physically related to its larger bandgap as well as the higher P-well doping concentration required to suppress reach-through breakdown. This indicates a fundamental problem for achieving reasonable levels of threshold voltage in silicon carbide power MOSFETs if the conventional silicon structure is utilized. Innovative structures that shield the P-well region can overcome this limitation as discussed in the next section.

4.3 Shielded UMOSFET Structure

The basic structure of the shielded power UMOSFET is shown in Figure 4.9. The shielding region consists of a heavily doped P-type (P^+) region located at the bottom of the trench. The P^+-shielding region is connected to the source electrode at a location orthogonal to

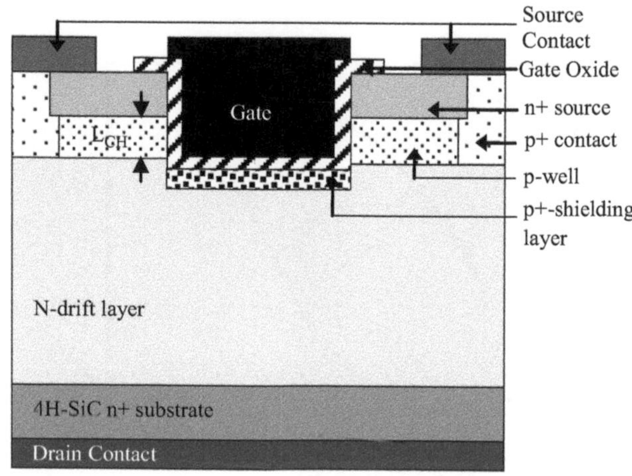

FIGURE 4.9
Cross section of the shielded UMOSFET structure.

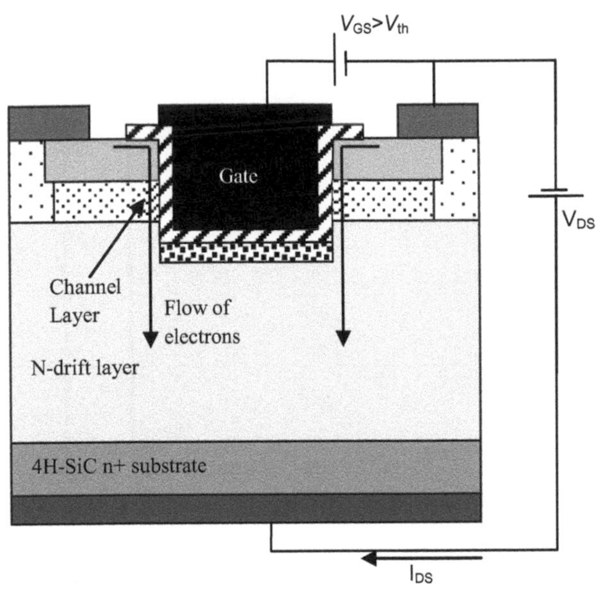

FIGURE 4.10
Principle of operation of shielded UMOSFET.

the device cross section shown above. The shielded power UMOSFET structure can be fabricated by using the same process for the conventional UMOSFET device structure discussed in a previous section, with the addition of an ion implantation step to form the P+-shielding region. The ion implant used to form the P+-shielding region must be performed after etching the trenches. The shielding is effective as long as the gate oxide is buffered from the high electric field in the N-drift region.

Application of positive gate bias higher than threshold voltage produces an inversion layer in the P-well region alongside the gate electrode as shown in Figure 4.10. This inversion layer channel provides a path for transport of electrons from the source to drain, thereby inducing the current flow (I_{DS}) from the drain to the source electrode when a positive drain voltage is applied.

4.3.1 Shielded UMOSFET Structure: Off-State Characteristics

The shielded UMOSFET power MOSFET operates in the forward blocking mode when the gate electrode is shorted to the source by the external gate drive circuit. At low drain bias voltages, the voltage is supported by a depletion region formed on both sides of the P-well/N-drift junction. Consequently, the drain potential appears across the MOSFET located at the top of the structure. This produces a positive potential at a location below the P-well region, so that the junction between the P+-shielding region and the N-drift region becomes reverse bias because the P+-shielding region is held at zero volts. The depletion region that extends from the P+/N junction pinches off the JFET region, producing a potential barrier at that location. The potential barrier tends to isolate the P-well region from any additional bias applied to the drain electrode. Consequently, a high electric field can develop in the N-drift region below the P+-shielding region while the electric field at the P-well region remains low. This has the beneficial effects of mitigating the reach-through of the depletion region within the P-well region and in keeping the electric field in the gate oxide low at a location where it is exposed to the N-drift region. The maximum blocking

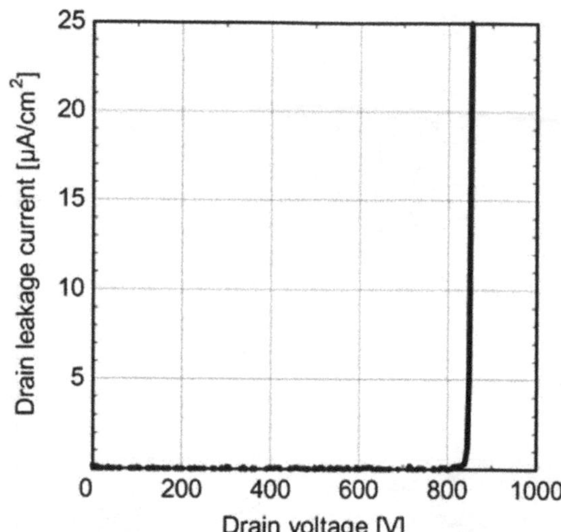

FIGURE 4.11
Blocking characteristics in a shielded UMOSFET structure at $V_{GS} = 0$ V. (From Furuhashi, M. et al., *Semicond. Sci. Technol.*, 31, 034003, 2016.)

voltage of the shielded UMOSFET is determined by the properties of the drift region. This allows reduction of the drift region resistance close to that of the ideal case. In addition, the reduction of the electric field in the vicinity of the P-well region allows reduction of the channel length as well as the gate oxide thickness. This is beneficial for further reduction of the device on-state resistance. It is worth pointing out that the P^+-shielding region must be adequately short-circuited to the source terminal in order for the shielding to be fully effective. The location of the P^+-region at the bottom of the trench implies that contact to it must be provided at selected locations orthogonal to the cross section of the device. Since the sheet resistance of the ion implanted P^+-region can be quite high, it is important to provide the contact to the P^+-region frequently in the orthogonal direction during chip design. This must be accomplished without significant loss of channel density if low specific on-resistance is to be realized.

Figure 4.11 shows the breakdown characteristics of the shielded UMOSFET in [11], having the value of breakdown voltage of 850 V.

4.3.2 Shielded UMOSFET Structure: On-State Characteristics

In the shielded UMOSFET structure, current flow between the drain and source can be induced by creating an inversion layer channel on the surface of the P-well region. The current flows from the source region into the drift region through the inversion layer channel formed on the vertical sidewalls of the trench due to the applied gate bias. The current then spreads into the N-drift region at around a 45° angle and becomes uniform throughout the rest of the structure.

The current path is illustrated in Figure 4.12 by the highlighted area. It can be seen that the addition of the P^+-shielding region introduces two JFET regions into the basic UMOSFET structure. The first one is formed between the P-well region and the P^+ shielding region with the current constricted by their zero-bias depletion boundaries. The second JFET region is formed between the P^+-shielding regions. Since the cross section for current

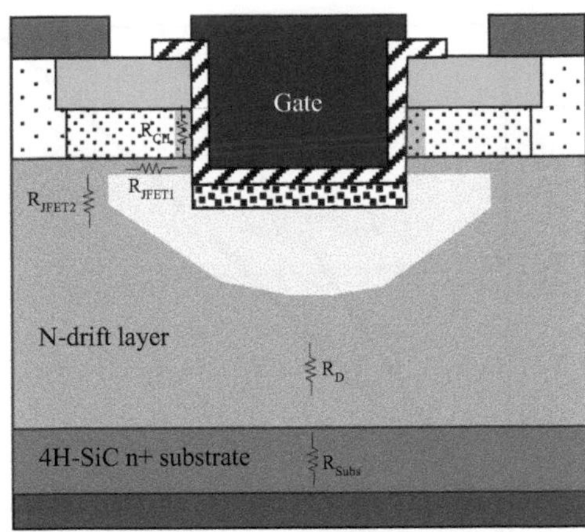

FIGURE 4.12
Current flow illustration during on-state in the shielded UMOSFET structure.

flow through this region is constricted by the zero-bias depletion width of the P+/N junction, it is again advantageous to increase the doping concentration in the JFET region to avoid having to enlarge the mesa width. A smaller mesa width allows for a smaller cell pitch to be maintained, which reduces the specific on-resistance due to a larger channel density. The total on-resistance for the shielded UMOSFET structure is determined by the resistance of all the components in the current path:

$$R_{on,sp} = R_{CH} + R_{JFET1} + R_{JFET2} + R_D + R_{Subs} \tag{4.10}$$

where R_{CH} is the channel resistance, R_{JFET1} and R_{JFET2} are the resistances of the two JFET regions, R_D is the resistance of the drift region after taking into account current spreading from the channel, and R_{Subs} is the resistance of the N+-substrate. These resistances can be analytically modeled by using the current flow pattern indicated by the shaded regions in Figure 4.12.

The specific resistance of the first JFET region can be calculated using:

$$R_{JFET1} = \frac{\rho_{JFET} \cdot p}{2}\left(\frac{x_{P+} + W_p}{t_B - 2W_P}\right) \tag{4.11}$$

where ρ_{JFET} is the resistivity of the JFET region, x_{P+} is the junction depth of the P+-shielding region, and W_P is the zero-bias depletion width in the JFET region. The resistivity and zero-bias depletion width used in this equation must be computed using the enhanced doping concentration of the JFET region.

The specific resistance of the second JFET region can be calculated using:

$$R_{JFET2} = \frac{\rho_{JFET} \cdot p}{2}\left(\frac{t_{P+} + 2W_p}{W_M - x_{P+} - W_P}\right) \tag{4.12}$$

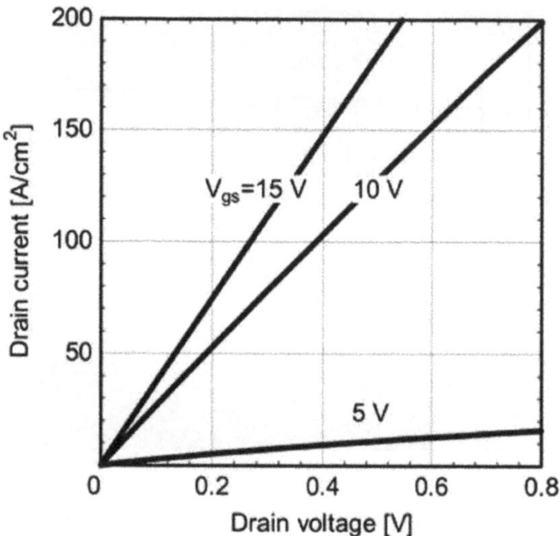

FIGURE 4.13
On-state characteristics of a shielded UMOSFET structure. (From Furuhashi, M. et al., *Semicond. Sci. Technol.*, 31, 034003, 2016.)

The drift region spreading resistance can be obtained by using:

$$R_D = \frac{\rho_D.p}{2}.\ln\left(\frac{p}{W_M - x_{P+} - W_P}\right) + \rho_D.(t - W_T - x_{P+} - W_P) \quad (4.13)$$

where t is the thickness of the drift region below the P+-shielding region and W_T, W_M are the widths of the trench and mesa regions, respectively.

Due to the shielding of the P-well region, the channel length for the shielded UMOSFET structure can be reduced when compared with that for the unshielded conventional UMOSFET structure.

Figure 4.13 shows $I_{DS} - V_{DS}$ characteristics of the SiC UMOSFET structure proposed in [11]. The $R_{on,sp}$ of the UMOSFET is 2.8 mΩ.cm².

In all the inversion layer mobility cases, the specific on-resistances of the shielded UMOSFET structure with a channel length of 0.4 μ are found to be very close to those for the conventional trench-gate structure with a channel length of 1 μ. This demonstrates the ability to obtain low specific on-resistance with the shielded UMOSFET structure while resolving the problems of P-well reach-through and high electric field in the gate oxide observed for the conventional UMOSFET structure. The behavior of the shielded structure with respect to other blocking voltages is similar to that provided for the conventional UMOSFET structure in the previous section.

4.3.3 Shielded UMOSFET Structure: Transfer Characteristics

The threshold voltage of the UMOSFET structure is determined by the doping concentration of the P-well region along the sidewalls of the trench region. As discussed in the previous sections, the reach-through in the P-well region of the shielded UMOSFET structure is mitigated by the reduced potential under the P-well/N-drift junction. This allows the P-well doping concentration to be decreased in order to reduce the threshold voltage. In addition, a P+-region

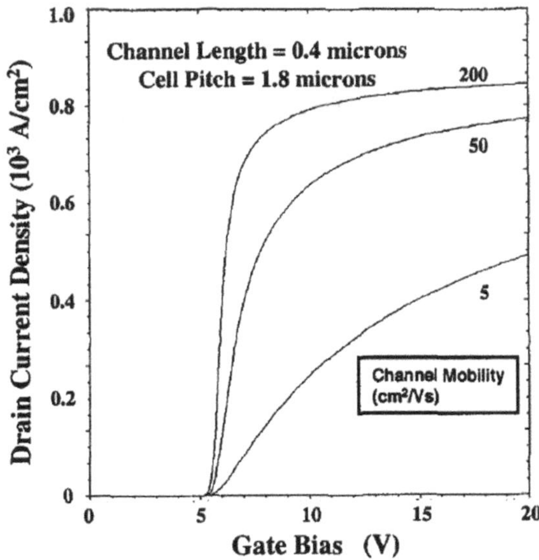

FIGURE 4.14
Transfer characteristics of the shielded UMOSFET structure. (From Qing-Wen, S. et al., *Chin. Phys. B*, 22, 027302, 2013.)

is incorporated under the source region to suppress the extension of the depletion region in this portion of the cell structure as described in the case of conventional UMOSFET structure.

The transfer characteristics obtained with a drain bias of 1 V are shown in Figure 4.14 for the device with a cell pitch of 1.8 μ. From the transfer characteristics, it can be seen that the threshold voltage that determines the on-resistance is approximately 5 V (allowing operation of this device with a gate bias of 10 V). The impact of changing the channel inversion layer mobility is also shown in this figure. As expected, there is a reduction in the drain current as the channel mobility is reduced.

4.4 Accumulation Mode UMOSFET Structure

The aforementioned devices, both conventional UMOSFET and shielded UMOSFETs, are inversion-channel devices and their on-resistances are greatly affected by the resistance of the MOS channel. The SiC ACCUFET (accumulation layer MOSFET) [12] is introduced as an encouraging way to overcome the problems associated with the deprived mobility of an inversion layer in the channels of the SiC conventional and shielded UMOSFET structures, which leads to significant activity in accumulation-mode SiC MOSFETs [5,12]. The basic structure of the accumulation layer UMOSFET is shown in Figure 4.15 with an accumulation layer channel.

To turn down the electric field at the oxide-SiC boundary to a null value, at the bottom of the trench, a p-type region is used as in the case of shielded UMOSFET. It shields the oxide from high electric fields in the blocking state. Under the p-well region, the N-type epilayer is formed, which avoids pinch-off of the conducting channel in the on-state and assists lateral current dispersion into the drift region. The lightly doped N-type epilayer grown on the sidewalls of the gate trench translates the device into an accumulation-layer MOSFET,

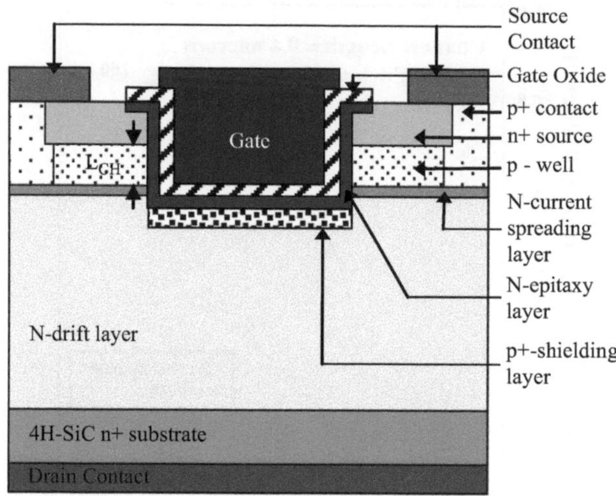

FIGURE 4.15
Cross section of the accumulation-mode UMOSFET structure.

or "ACCUFET," escalating the MOSFET mobility and further dropping on-resistance. When a positive bias is applied to the gate electrode, an accumulation layer channel is formed in the structure, enabling the conduction of drain current with a low specific on-resistance as shown in Figure 4.16.

The epitaxially grown *n*-type SiC trench sidewall layer that defines the channel region can be completely depleted by the built-in potential of the PN junction between the *p*-type SiC base layer and the *n*-type SiC trench sidewall epilayer. The potential created by the work function difference between the *n*-type SiC trench sidewall epilayer and the poly-Si gate electrode controls the channel condition during MOSFET operations.

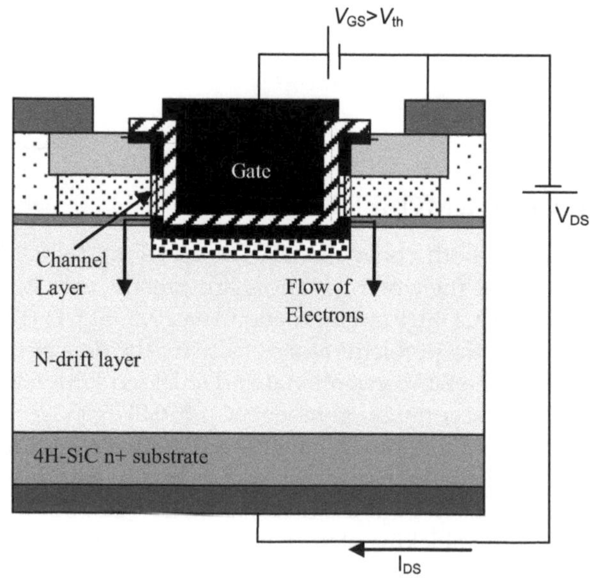

FIGURE 4.16
Principle of operation of accumulation mode UMOSFET.

4.4.1 Accumulation Mode UMOSFET Structure: Off-State Characteristics

The U-ACCUFET operates in the same forward blocking mode as the shielded UMOSFET—i.e., when the gate electrode is shorted to the source by the external gate drive circuit. In the accumulation-mode UMOSFET structure, the depletion of the trench sidewall *n*-epilayer region is accompanied by the formation of the potential barrier for the flow of electrons through the channel.

In the off-state, a depletion region is formed in the *n*-type SiC trench sidewall layer by the built-in fields of the SiC P-well and the poly-Si gate electrode. To obtain a sufficient barrier height to prevent conduction between the source and the drain, the trench sidewall epilayer that forms the channel must be thin (sub-μm). The maximum thickness of the epilayer that can be used depends on its impurity concentration, the SiO_2 film thickness, and the type of poly-Si used as the gate electrode. This enables normally-off operation of the accumulation mode UMOSFET with zero gate bias.

The N-epilayer is commonly designed with a much higher donor concentration than the N-drift layer. Since the N-layer doping between the P-well layer and the bottom trench P-implant layer is higher than that of the N-drift layer in the previously discussed UMOSFETs, the depletion region with the UMOSFET having this N-layer will be much shorter than the UMOSFETs without the N-epilayer. Also, it is well known that the breakdown field increases as doping increases. For typical UMOSFET doping parameters, the doping concentration of the P-well is much higher than that of the N-drift region. Therefore, the breakdown field of the P-well is higher than that of the N layer. If the P-well is adjacent to the N-layer, the breakdown field is determined by the lower doped N-layer. So it is possible to insert an N layer with the right thickness and right doping that makes the higher doped P-well layer and lower doped N-drift layer reach their respective breakdown fields at about the same point. This is another way of fully utilizing the high avalanche field property of SiC.

The blocking voltage of a proposed accumulation mode UMOSFET structure in [15] is 1400 V, and the device is robust in breakdown if the drain current is externally limited. The high-voltage portion of Figure 4.17 is swept four times, and all four measurements are shown in the figure, although the points overlay each other.

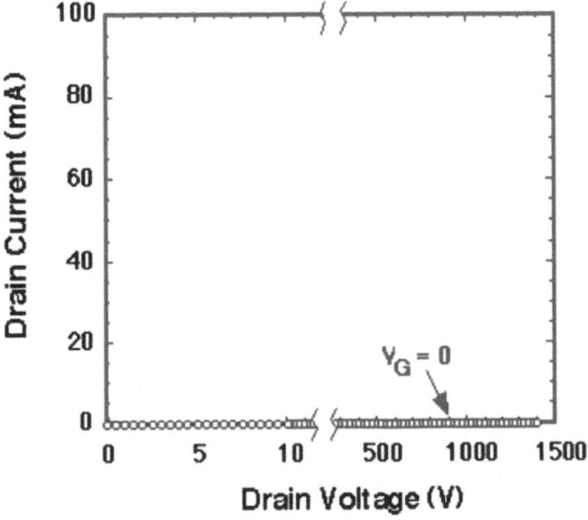

FIGURE 4.17
Blocking characteristics in a shielded UMOSFET structure at $V_{GS} = 0$ V. (From Tan, J. et al., *IEEE Electron Device Lett.*, 19, 487–489, 1998.)

4.4.2 Accumulation Mode UMOSFET Structure: On-State Characteristics

The device can be switched to the on-state by the application of a positive bias to the gate to create a surface accumulation layer at the interface between the SiO_2 and the *n*-type SiC surface epilayer extending from the N^+ source to the N/N-drift region. The electrons flow from the source region into the drift region through the accumulation layer channel formed on the vertical sidewalls of the trench in the N-epilayer region due to the applied gate bias. The current then spreads into the N-drift region and becomes uniform through the rest of the structure. The current path is illustrated in Figure 4.18 by the highlighted area.

The conventional UMOSFET has no problem of current flow, since the surface under the bottom trench is accumulated. For the shielded UMOSFET, the introduction of the bottom trench P-layer prevents the current from flowing directly downward from the bottom trench, and the current has to flow laterally from the sidewall first. Since this is usually a very small area and the N-epilayer along the sidewall of the gate trench is very lightly doped, the resistance here is usually very high, and for some cases this can totally cut out the current when pinch-off occurs. The introduction of a thin N-layer between the P-well and the bottom trench shielding the P-layer not only solves the pinch-off effect, but it also provides a highly conductive path for the current. The current actually first flows laterally in the N-layer, then flows downward to the drain. Since this N-layer is two orders of magnitude higher in doping than the trench sidewall N-epilayer, the resistance in the N-layer is very small and therefore not taken into account. It can be seen that utilizing the accumulation layer UMOSFET, one of the JFET regions formed between the P^+-shielding regions has been eliminated as compared to two JFET regions in the shielding UMOSFET, as discussed in the previous section. Although a JFET resistance still exists, the total drift region resistance is smaller than that of the conventional UMOSFET because of the elimination of the spreading resistance.

The total on-resistance for the shielded UMOSFET power MOSFET structure is determined by the resistance of all the components in the current path:

$$R_{\text{on,sp}} = R_{\text{CH}} + R_{\text{JFET2}} + R_D + R_{\text{Subs}} \tag{4.14}$$

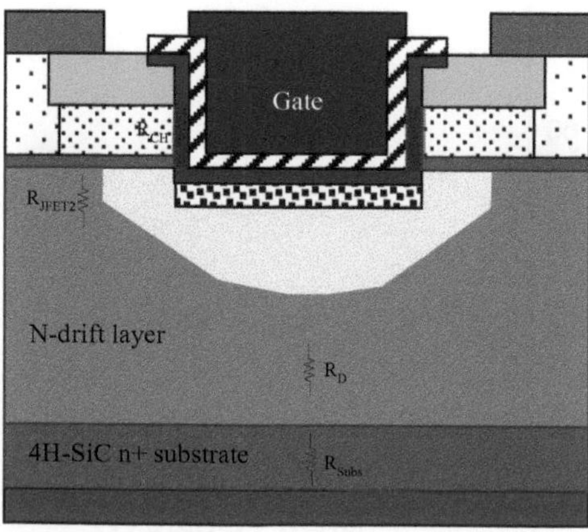

FIGURE 4.18
Current flow illustration during on-state in a U-ACCUFET structure.

where R_{CH} is the channel resistance, R_{JFET2} is the resistance of the JFET region, R_D is the resistance of the drift region after taking into account current spreading from the channel, and R_{Subs} is the resistance of the N$^+$ substrate. The resistance components are shown in Figure 4.18.

For the UMOSFET structure with the accumulation layer, the specific channel resistance is given by:

$$R_{CH} = \frac{(L_{CH} \cdot p)}{2\mu_a C_{ox}(V_G - V_T)} \tag{4.15}$$

where μ_a is the mobility for electrons in the accumulation layer channel.

The drain characteristics for the U-ACCUFET device are shown in Figure 4.19. The device is normally off (V_T = 1 V), since the accumulation channel is totally depleted by the built-in voltage. The specific on-resistance is 15.7 mΩ.cm^2 at a gate voltage of 40 V with oxide field of 3.1 MV/cm. TLM measurements indicate that the source has a specific contact resistance of 0.1–0.2 mΩ.cm^2 and a sheet resistance of 2–2.5 kΩ per square. These resistances contribute approximately 1 mΩ.cm^2 to the total on-resistance of the device. The combined resistance of the drift region and JFET region is difficult to determine from experimental measurements, but it is in the range of 6–12 mΩ.cm^2. The contribution of the MOSFET channel can be estimated by subtracting these values from the total resistance of 15.7 mΩ.cm^2.

Much larger accumulation layer mobility in silicon carbide allows reduction of the specific on-resistance. In addition, the threshold voltage for the accumulation mode is smaller than for the inversion mode, allowing further improvement in the channel resistance contribution. The rest of the resistance components in this UMOSFET structure can be modeled by using the same equations provided in a previous section.

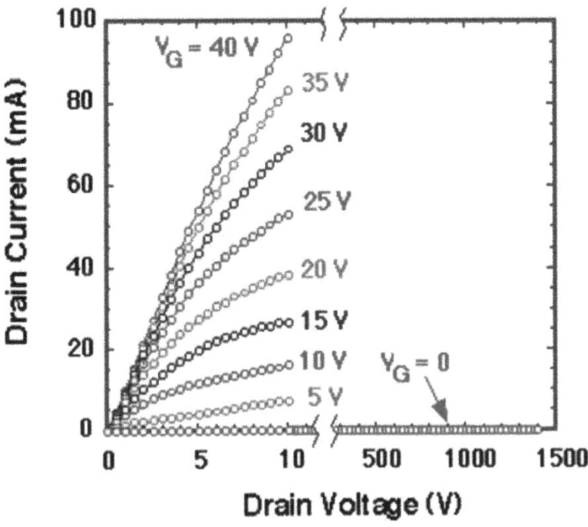

FIGURE 4.19
On-state characteristics of a UMOSFET structure. (From Tan, J. et al., *IEEE Electron Device Lett.*, 19, 487–489, 1998.)

4.4.3 Accumulation Mode UMOSFET Structure: Transfer Characteristics

The threshold voltage of the power MOSFET is an important design parameter from an application standpoint. The band bending required to create a channel in the accumulation-mode MOSFET is much smaller than required for the inversion mode device. This provides the opportunity to reduce the threshold voltage while obtaining the desired normally-off device behavior. A model for the threshold voltage of accumulation-mode MOSFETs has been developed using the electric field profile shown in Figure 4.20 when the gate is biased at the threshold voltage. In this figure, the electric fields in the semiconductor and oxide are given by:

$$E_1 = \frac{V_{bi}}{W_{NE}} - \frac{qN_D W_{NE}}{2\varepsilon_S} \tag{4.16}$$

$$E_2 = \frac{V_{bi}}{W_{NE}} + \frac{qN_D W_{NE}}{2\varepsilon_S} \tag{4.17}$$

$$E_{ox} = \frac{\varepsilon_S}{\varepsilon_{ox}} E_1 \tag{4.18}$$

Note that this model is based upon neglecting any voltage supported within the P-well region under the assumption that it is very heavily doped. Using these electric fields, the threshold voltage is found to be given by:

$$V_t = \Phi_{MS} + \left(\frac{\varepsilon_S V_{bi}}{\varepsilon_{ox} W_{NE}} - \frac{qN_D W_{NE}}{2\varepsilon_{ox}} \right) . t_{ox} \tag{4.19}$$

The first term in this equation accounts for the work function difference between the gate material and the lightly doped N-epilayer region. The second term represents the effect of the built-in potential of the P^+/N junction that depletes the N-epilayer region.

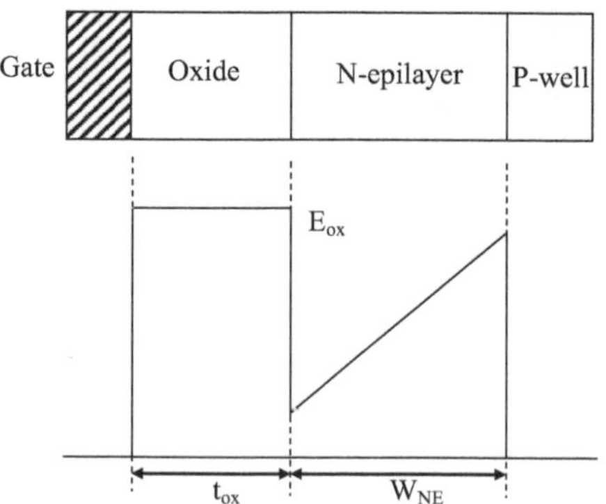

FIGURE 4.20
Electric field profile in the gate region for the accumulation-mode UMOSFET structure.

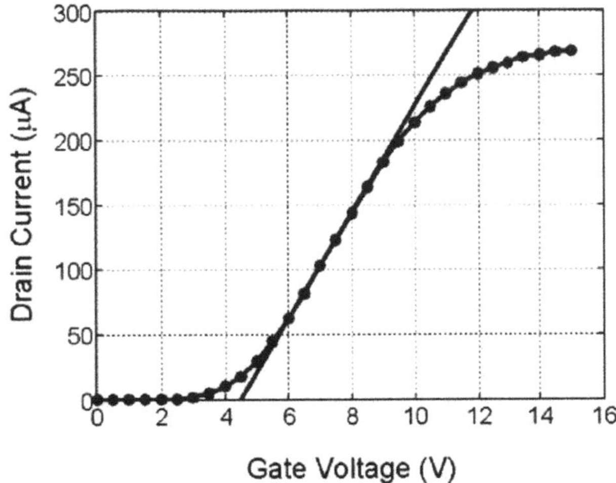

FIGURE 4.21
Transfer characteristics of an accumulation-mode UMOSFET structure. (From Sui, Y. et al., *IEEE Electron Device Lett.*, 26, 255–257, 2005.)

The room-temperature linear region versus characteristic of an accumulation mode UMOSFET is shown in Figure 4.21, measured at a drain voltage of 50 mV. A threshold voltage of 4.5 V is extracted from the linear portion of the curve. Of course, the most important benefit of the accumulation-mode is that lower threshold voltages can be achieved than in the inversion-mode structures.

4.5 Summary

The 4H-SiC vertical power UMOSFET structure was the first approach explored by the silicon carbide community because the P-well region could be epitaxially grown rather than formed by ion implantation, which was a less mature technology. Although devices with high breakdown voltages have been fabricated, their performance was severely limited by the onset of rupture of the gate oxide and the poor channel mobility. In order to overcome these problems, it is necessary to shield the gate oxide from the high electric field developed in the drift region. The shielded trench-gate power MOSFET structure was proposed for silicon carbide in order to shield the P-base region and the gate oxide from the high electric field generated in the drift region during the blocking mode. The shielding is provided by the addition of a P-type region located at the bottom of the trenches. This P-type shielding region must be connected to the source electrode. Its presence creates two JFET regions in the trench gate structure, which can increase the on-resistance unless the doping concentration in the vicinity of the trenches is enhanced. With the enhanced doping concentration, it has been found that the full blocking voltage capability of the drift region in 4H-SiC can be utilized with a low electric field at the P-base region and the gate oxide. Further, an accumulation mode UMOSFET has been discussed to overcome the low inversion layer mobility of the channel by incorporating an accumulation channel instead of the inversion channel. It also reduces the on-resistance as compared to the shielded UMOSFET.

References

1. B. J. Baliga, *Power Semiconductor Devices*, Chapter 7, pp. 335–425, Boston, MA, PWS Publishing Company, 1996.
2. D. Ueda, H. Takagi, and G. Kano, "A new vertical power MOSFET structure with extremely reduced on-resistance," *IEEE Transactions on Electron Devices*, vol. 32, pp. 2–6, 1985.
3. H.-R. Chang et al., "Ultra-low specific on-resistance UMOSFET," *IEEE International Electron Devices Meeting*, Abstract 28.3, pp. 642–645, 1986.
4. B. J. Baliga, "Silicon carbide switching device with rectifying gate," U. S. Patent 5,396,085, Issued March 7, 1995.
5. J. Tan, J. A. Cooper, and M. R. Melloch, "High-voltage accumulation-layer UMOSFET's in 4H-SiC," *IEEE Electron Device Letters*, vol. 19, no. 12, pp. 487–489, 1998.
6. J. W. Palmour et al., "4H-silicon carbide power switching devices," *Silicon Carbide and Related Materials - 1995, Institute of Physics Conference Series*, vol. 142, pp. 813–816, 1996.
7. B. J. Baliga, "Silicon carbide power MOSFET with floating field ring and floating field plate," U. S. Patent 5,233,215, Issued August 3, 1993.
8. S. Qing-Wen et al., "The fabrication and characterization of 4H SiC power UMOSFETs," *Chinese Physics B*, vol. 22, no. 2, p. 027302, 2013.
9. J. Baliga, *Silicon Carbide Power Devices*, Singapore, World Scientific Publishing, 2005.
10. P. M. Shenoy and B. J. Baliga, "The planar 6H-SiC ACCUFET," *IEEE Electron Device Letters*, vol. 18, pp. 589–591, 1997.
11. M. Furuhashi, S. Tomohisa, T. Kuroiwa, and S. Yamakawa, "Practical applications of SiC-MOSFETs and further developments," *Semiconductor Science and Technology*, vol. 31, no. 3, pp. 034003, 2016.
12. Y. Sui, T. Tsuji, and J. A. Cooper, "On-state characteristics of SiC power UMOSFETs on 115-μm drift layers," *IEEE Electron Device Letters*, vol. 26, no. 4, pp. 255–257, 2005.

5

Operational Characteristics of Vertically Diffused Metal Oxide Semiconductor Field Effect Transistor

Deepshikha Bharti and Aminul Islam

CONTENTS

5.1 Introduction

MOSFETs are the commonly fabricated unipolar semiconductor devices. The first power MOSFET structure commercially introduced by the power semiconductor industry was the vertical double-diffused or VDMOSFET structure [1]. The device fabrication process depends on the available planar gate technology used to manufacture CMOS integrated circuits. These devices initially found applications in power electronic circuits that are operated at low (<100 V) voltages. To reduce the area of the device, MOSFETs for high-power, high-voltage applications are normally designed with the source and drain at opposite sides of the wafer, the result of which is a vertical MOSFET. The operation of power MOSFET depends on the creation of a conductive channel at the surface of the semiconductor under the gate oxide layer. The channel is defined as the separation between the N^+/P-well junction and the P-well/N-drift junction under the gate electrode. Consequently, a reduction in channel length can be achieved toward sub-micron dimensions without the requirement for high-resolution lithography. This approach with VDMOSFETs served the industry from the 1970s into the 1990s and is still available for power electronic applications. VDMOS transistors are common in silicon power device technology where through a common mask opening, the P-well and N^+-source regions are created by impurity diffusion. However, impurity diffusion is impractical in Silicon Carbide (SiC) because of the very low diffusion coefficients at any temperature. The first VDMOS transistors in SiC were fabricated by the Purdue group using ion implantation for introducing dopants for the P-well and the N+-source. The implanted VDMOSFET requires separate masks to be used for defining the P-well and the N+-source. Thus, a vertical structure is constructed with a drift layer built on a highly conductive N+-substrate. The much lower resistance of the drift region in silicon carbide enables development of power MOSFETs with very high breakdown voltages. These devices offer not only fast switching speed but also a superior safe operating area when compared to high-voltage silicon IGBTs. This allows reduction of both the switching loss and conduction loss components in power circuits [2]. But the conventional VDMOSFET structure is not satisfactory for application in silicon carbide due to a reach-through problem, a high electric field developed in the gate oxide, a relatively high threshold voltage, and the relatively low inversion layer mobility in the channel. To address these issues in a satisfactory manner, shielding of the channel from this developed high electric field can be done. At PSRC in the early 1990s, the concept of shielding of the channel region was first proposed with a U.S. patent issued in 1996 [3]. The shielding was accomplished by formation of either a P-type region under the channel or by creating a conduction barrier region of high resistivity under the channel. An accumulation mode of operation addressed the issue of a low inversion layer channel [3].

This chapter begins with an analysis of the basic principles of operation of the VDMOSFET structure. Next, the DC characteristics of the structure such as off-characteristics, on-characteristics, and transfer characteristics are described in detail. The following section begins with the analysis of the basic principles of operation of the shielded VDMOSFET structure. Then, the DC characterization for the same is analyzed in detail. The shielded accumulation-mode MOSFET structures (named the ACCUFET) are discussed in this section and have the most promising characteristics for the development of monolithic power switches from silicon carbide.

5.2 VDMOSFET Structure

A vertical device such as VDMOSFET (vertically diffused MOSFET), which uses a double-diffusion process, is preferred to achieve a higher value of breakdown voltage. Figure 5.1 shows the cross section of the basic structure for the vertical DMOSFET. During the formation of this structure, an N-type drift layer is developed on a heavily doped N+-substrate. P-well and N+-source regions are also formed such that the P-well region is diffused deeper than the N+-source. The device channel is formed by the difference in lateral extension of the P-well region and N+-source region. A refractory gate electrode is desired to permit the dopant diffusion under the gate electrode at increased temperatures. The forward blocking capability is achieved by the P-N junction formed between the P-well region and N-drift region. The N-drift must be moderately doped so that the drain breakdown voltage is sufficiently large and the thickness of the N-drift region is made as thin as possible to minimize drain resistance. During the device operation, a fixed potential is established to the P-well region by connecting it to the source metal using a P+-contact.

Without adding a gate bias, a high voltage can be supported in the vertical DMOSFET structure when a positive bias is applied to the drain. In this situation, i.e., by short-circuiting the gate to the source and applying a positive bias to the drain, a junction is formed between the P-well region and N-drift region that becomes reversely biased. By the extension of a depletion layer on both sides, this junction supports the drain voltage.

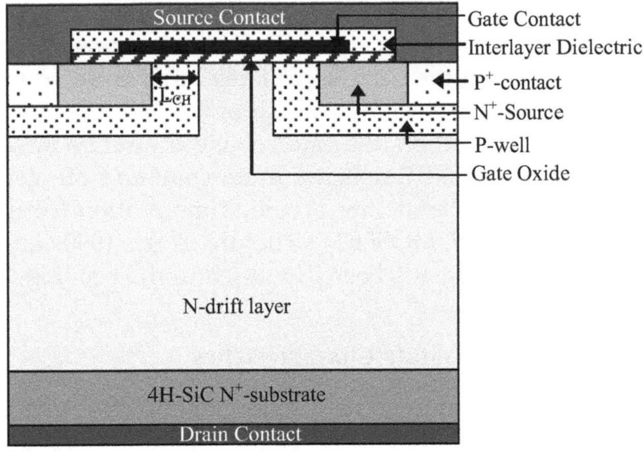

FIGURE 5.1
Cross section of the conventional VDMOSFET structure.

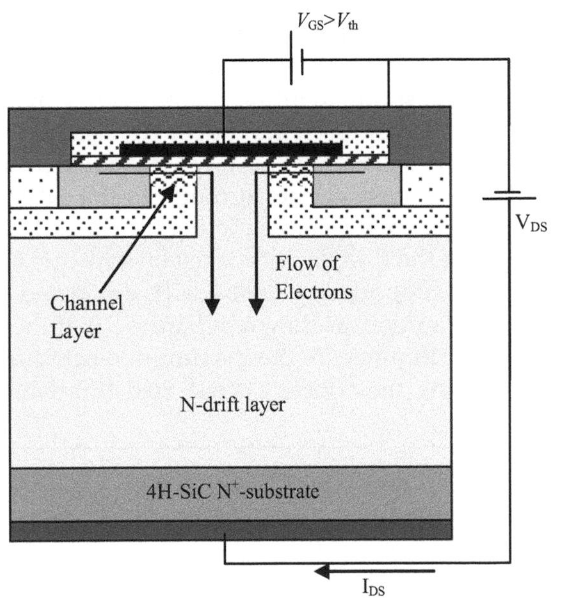

FIGURE 5.2
Principle of operation of VDMOSFET.

The depletion layer extends primarily into the N-drift region, due to the higher doping level of the P-well layer. By adding a positive gate bias, a drain current flow is introduced in the vertical DMOSFET structure. This produces an inversion layer at the surface of the P-well region under the gate electrode as shown in Figure 5.2. This inversion layer channel provides a path for the flow of electrons from the source to the drain electrode when a positive drain voltage is applied. After transport from the source region through the channel, at the upper surface of the device, the electrons enter the N-drift region. They are then transported through a relatively narrow JFET region, which is created between the adjacent P-well regions. Due to the restriction of current flow through the JFET region, the internal resistance would increase in the vertical DMOSFET structure. As a result, the gate width should be carefully chosen to minimize the internal resistance of the structure. After being transported through the narrow JFET region, the electrons enter the N-drift region and the current spreads to the entire width of the cell cross section. On the other hand, when the MOSFET is turned off, the gate voltage should be lower than the threshold voltage. In this way, the MOSFET switches to on-state and off-state. The large internal resistance of the VDMOSFET structure provided motivation for the development of the trench-gate power MOSFET (UMOSFET) structure in the 1990s and advanced power MOSFET structures, which has already been discussed in the previous chapter.

5.2.1 VDMOSFET Structure: Off-State Characteristics

In the drift layer, the potential difference is distributed and thus decreases the maximal electric field in the transistor. Due to a lower electric field, increased drain voltages are allowed and the high-power performance improves [6]. The thickness and doping concentration of the drift region determine the blocking capability of the MOSFET. On both sides of the P-well/N-drift junction, the depletion region is formed in the forward blocking mode of the VDMOSFET, which supports the voltage. When the electric field at this junction becomes equal to the critical electric field for breakdown, the maximum blocking voltage can be determined if the parasitic $N^+/P/N$ bipolar transistor is fully suppressed. This suppression is attained by short-circuiting the N^+-source and P-well regions using the source metal as shown on the upper left and upper right-hand side of the cross section of the structure shown in Figure 5.2. However, a large leakage current can occur when the depletion region in the P-well region reaches through to the N^+-source region. Thus, to avoid the reach-through phenomenon from restricting the breakdown voltage, the doping concentration and thickness of the P-well region must be strategic in a way.

The applied drain voltage is supported by the N-drift region and the P-well region with a triangular electric field distribution, as shown in Figure 5.3, if the doping is uniform on both sides. At the P-well/N-drift junction, the maximum electric field occurs. The maximum electric field (E_m) reaching the critical electric field (E_c) determines the maximum

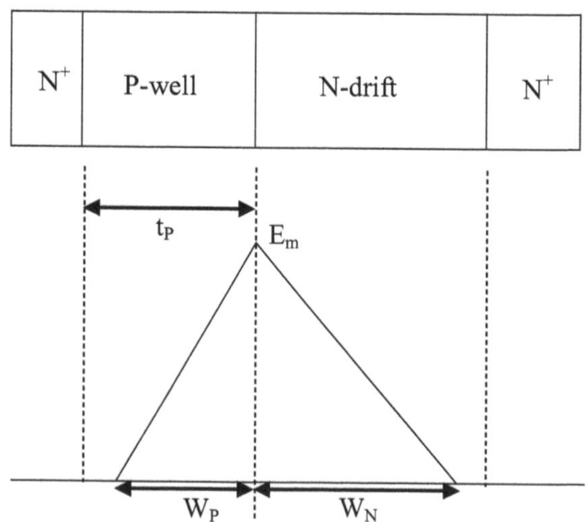

FIGURE 5.3
Reach-through in a power VDMOSFET structure.

voltage that can be supported by the drift region for breakdown for the semiconductor material. The critical electric field for breakdown and the doping concentration determine the maximum depletion width (W_P) on the P-well side given as:

$$W_P = \frac{\varepsilon_S E_m}{qN_A} \tag{5.1}$$

where N_A is the doping concentration in the P-well region and E_m is the maximum electric field located at the P-well/N-drift junction. If we assume that the maximum electric field at the P-well/N-drift junction reaches the critical electric field for breakdown when the P-well region is completely depleted, the minimum P-well thickness needed to avoid a reach-through restricted breakdown can be obtained as

$$t_P = \frac{\varepsilon_S E_c}{qN_A} \tag{5.2}$$

where E_c is the critical electric field for breakdown in the semiconductor. The minimum thickness of the P-well region required to avoid reach-through breakdown declines with increasing doping concentration.

The maximum blocking voltage capability of the power VDMOSFET structure is determined by the drift region doping concentration and thickness. The relation between maximum depletion width and breakdown voltage can be stated as

$$W_D = \left(\frac{2V_{BR}}{E_C}\right) \tag{5.3}$$

where V_{BR} is the breakdown voltage and E_C is the critical electric field. The doping concentration of the drift region (N_D) to achieve the required breakdown voltage V_{BR} is given as

$$N_D = \left(\frac{\varepsilon_S E_C^2}{2qV_{BR}}\right) \tag{5.4}$$

The previous equation states that the blocking voltage of the MOSFETs depends on the drift layer used. Devices fabricated on lighter doped drift layers achieve a higher value of blocking voltage as compared to those with heavier drift doping.

The blocking I-V characteristics of the VDMOSFET are shown in Figure 5.4 with $N_D = 1.4 \times 10^{16}$ cm^{-3}. The breakdown voltage is $V_{BR} = 990$ V with $V_{GS} = 0$ V ($I_{DS} = 10$ µA) [4]. However, under forward blocking conditions in the power MOSFET structure, a high electric field also appears in the gate oxide. The developed electric field in the oxide is associated with the electric field in the underlying semiconductor by Gauss's Law as

$$E_{Oxide} = \left(\frac{\varepsilon_{Semi}}{\varepsilon_{Oxide}}\right).E_{Semi} \tag{5.5}$$

where ε_{Semi} and ε_{Oxide} are the dielectric constants of the semiconductor and the oxide, and E_{Semi} is the electric field in the semiconductor. In the case of both silicon and silicon carbide, the electric field in the oxide is about three times larger than in the semiconductor. The electric field in the oxide does not exceed its reliability limit of about 3×10^6 V/cm because the maximum electric field in the silicon drift region remains below 3×10^5 V/cm. However, for 4H-SiC, the electric field in the oxide reaches a value of 9×10^6 V/cm when

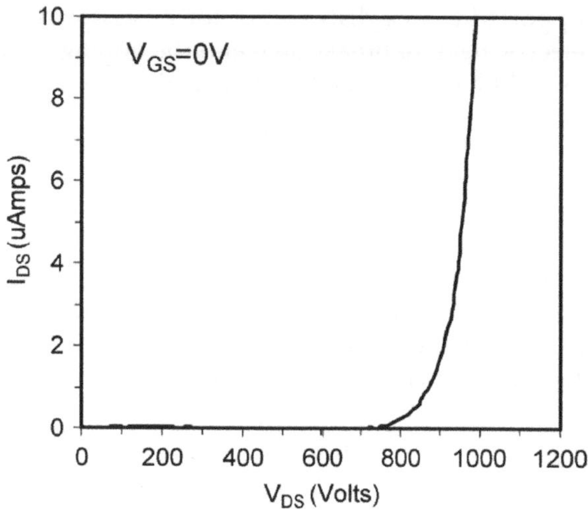

FIGURE 5.4
Blocking characteristics in a VDMOSFET structure at $V_{GS} = 0$ V. (From Losee, P. A. et al., *IEEE Trans. Electron Devices*, 55, 1824–1829, 2008.)

the field in the semiconductor reaches its breakdown strength. This value not only exceeds the reliability limit but also can cause rupture of the oxide, leading to a catastrophic breakdown. It is therefore important to monitor the electric field in the gate oxide when designing and modeling the silicon carbide MOSFET structures. Novel structures that shield the gate oxide from a high electric field have also been discussed to resolve this problem. This topic is addressed in the next section.

5.2.2 VDMOSFET Structure: On-State Characteristics

Current flow between the drain and source can be induced by creating an inversion layer channel on the surface of the P-well region. The channel in the VDMOSFET is aligned horizontally below the oxide; however, as it reached the *n*-region, the direction of the electron flow turns vertically in the drift layer. The current path is illustrated in Figure 5.5 by the highlighted area. The current drifts through the inversion layer channel (formed due to the applied gate bias) into the JFET region via the accumulation layer formed above it under the gate oxide. It then spreads into the N-drift region at about a 45° angle and becomes uniform throughout the rest of the structure. The total on-resistance for the VDMOSFET structure is determined by the resistance of these components in the current path given as

$$R_{\text{on,sp}} = R_{\text{CH}} + R_A + R_{\text{JFET}} + R_D + R_{\text{Subs}} \tag{5.6}$$

where R_{CH} is the channel resistance, R_A is the accumulation region resistance, R_{JFET} is the resistance of the JFET region, R_D is the resistance of the drift region after taking into account current dispersal from the JFET region, and R_{Subs} is the resistance of the N$^+$-substrate. The specific channel resistance is given by

$$R_{\text{CH}} = \frac{(L_{\text{CH}} \cdot p)}{\mu_{\text{inv}} C_{\text{ox}} (V_G - V_T)} \tag{5.7}$$

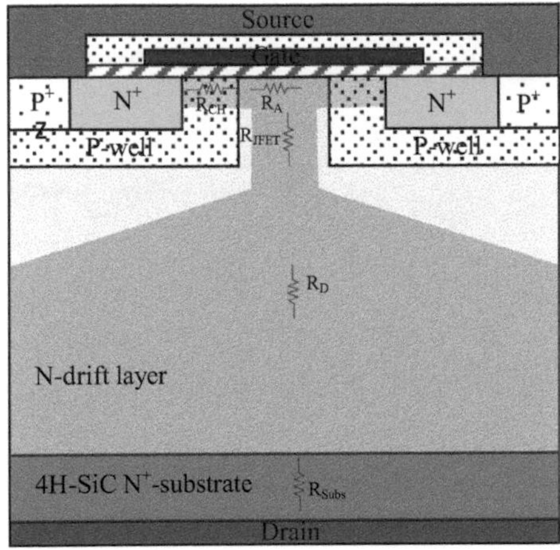

FIGURE 5.5
Electron flow illustration during on-state in a VDMOSFET structure.

where L_{CH} is the channel length, p is the cell pitch, μ_{inv} is the mobility for electrons in the inversion layer channel, C_{ox} is the specific capacitance of the gate oxide, V_G is the applied gate bias, and V_T is the threshold voltage. The specific capacitance can be obtained using:

$$C_{ox} = \frac{\varepsilon_{ox}}{t_{ox}} \tag{5.8}$$

where ε_{ox} is the dielectric constant for the gate oxide and t_{ox} is its thickness.

The specific resistance of the accumulation region is given by:

$$R_A = \frac{K(W_J - W_P)p}{\mu_a C_{ox}(V_G - V_T)} \tag{5.9}$$

where μ_a is the mobility for electrons in the accumulation layer, C_{ox} is the specific capacitance of the gate oxide, V_G is the applied gate bias, and V_T is the threshold voltage. The factor K is used to account for two-dimensional current spreading from the channel into the JFET region. In this equation, W_P is the zero-bias depletion width at the P-well/N-drift junction. It can be determined using

$$W_P = \sqrt{\frac{2\varepsilon_S V_{biP}}{qN_D}} \tag{5.10}$$

where the built-in potential V_{biP} for the P-N junction is typically 3.3 V for 4H-SiC.

The specific JFET region resistance is given by

$$R_{JFET} = \rho_D.t_P\left(\frac{p}{W_J - W_P}\right) \tag{5.11}$$

The drift region spreading resistance can be obtained by using

$$R_D = \rho_D.p.\ln\left(\frac{p}{W_J - W_P}\right) + \rho_D.(t - s - W_P) \tag{5.12}$$

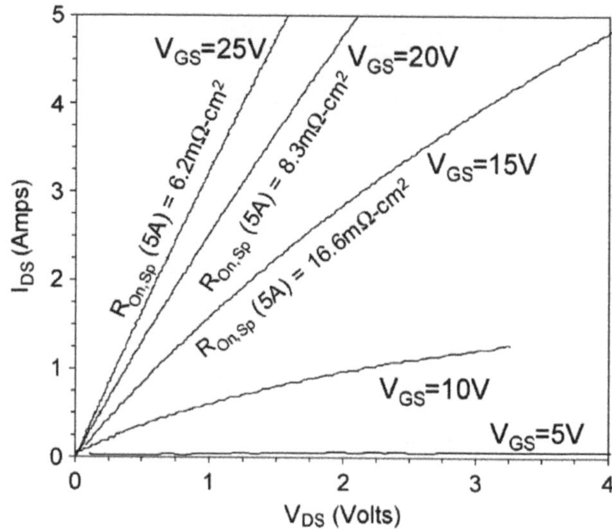

FIGURE 5.6
On-state characteristics of a VDMOSFET structure. (From Losee, P. A. et al., *IEEE Trans. Electron Devices*, 55, 1824–1829, 2008.)

where t is the thickness of the drift region below the P-well region and s is the width of the P-well region.

The contribution to the resistance from the N^+ substrate is given by

$$R_{subs} = \rho_{subs} . t_{subs} \tag{5.13}$$

where ρ_{subs} and t_{subs} are the resistivity and thickness of the substrate, respectively. The JFET width in the VDMOSFET structure must be optimized not only to obtain the lowest specific on-resistance but also to control the electric field at the gate oxide interface.

Figure 5.6 shows the drain I–V characteristics of a SiC VDMOSFET [4] with a 0.9-μm channel length. The device shown was designed with a cell pitch of 19 μm and device active area of 1.96×10^{-2} cm^2. Figure 5.6 shows linear I–V characteristics up to $I_{DS} = 5$ A with a modest gate drive ($V_{GS} = 15$ V). At a lower gate drive ($V_{GS} = 10$ V), the channel resistance dominates, and the drain current saturates at about $I_{DS} = 1$ A. With a gate bias of $V_{GS} = 15$ V (≈3 MV/cm), the on-resistance taken at $I_{DS} = 5$ A (current density J = 250 A/cm^2) is 850 mΩ ($R_{SP,ON} = 16.6$ mΩ·cm^2), reducing to 420 mΩ or $R_{SP,ON} = 8.3$ mΩ·cm^2 at $V_{GS} = 20$ V (≈4 MV/cm).

The inversion layer mobility in the channel of the VDMOSFET also influences the specific on-resistance of the device. The relative magnitude of the channel mobility on the performance of the VDMOSFET depends upon the drift region resistance, which is a function of the breakdown voltage of the device.

From Figure 5.7, it can be seen that when the breakdown voltage exceeds 5000 V, the specific on-resistance of 4H-SiC VDMOSFETs (also called planar MOSFETs) ranges toward the ideal specific on-resistance if a channel mobility of 100 cm^2/Vs is achieved. The channel resistance restricts the performance of the VDMOSFET when the breakdown voltage falls below 1000 V even at this relatively high inversion layer mobility for silicon carbide. For a breakdown voltage of 1000 V, the anticipated improvement in specific on-resistance over silicon devices is then about 100 times (as opposed to the 2000× improvement in the specific on-resistance of the drift region). When the inversion layer mobility is reduced to

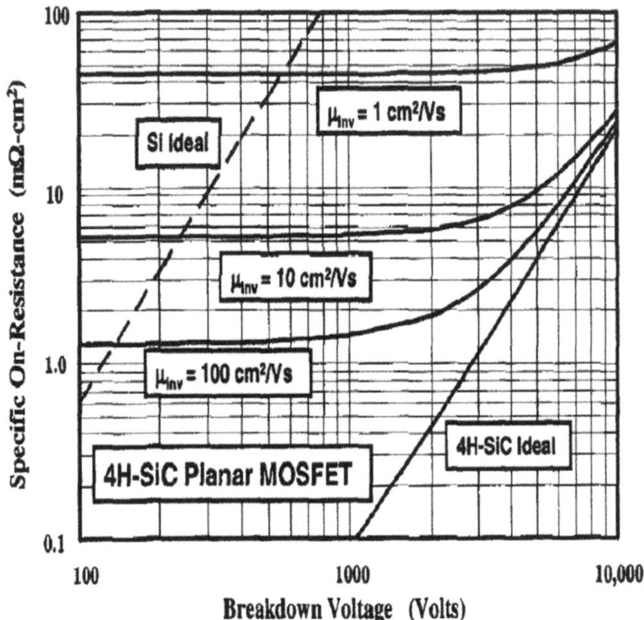

FIGURE 5.7
Specific on-resistance of a VDMOSFET structure. (From Baliga, J., *Silicon Carbide Power Devices*, World Scientific Publishing, Singapore, 2005.)

10 cm^2/Vs, the degradation in performance extends to much larger breakdown voltages. This highlights the importance of developing process technology to achieve high inversion layer mobility in 4H-SiC structures. In comparison with silicon, whose ideal specific on-resistance is shown by the dashed line in Figure 5.7, the 4H-SiC VDMOSFET can surpass the performance by an order of magnitude at breakdown voltages above 1000 V if a channel mobility of at least 10 cm^2/Vs is achieved [5].

5.2.3 VDMOSFET Structure: Transfer Characteristics

The transfer characteristic speaks about the response of the drain current (I_{DS}) to the input gate-source driving voltage (V_{GS}). Since the gate terminal is isolated electrically from the remaining terminals (source, drain and bulk), the gate current is effectively zero. Therefore, the gate current is not considered in most of the analysis.

From the transfer characteristic curve, we observe that at a certain gate voltage the device leaves the off-state and enters the on-state. This gate voltage is known as the threshold voltage (V_T) of the device. The transfer current–voltage (I–V) characteristics of a typical device [4] with L_{ch} = 0.9 μm are shown in Figure 5.8, exhibiting a threshold voltage of V_T = 5.7 V at room temperature.

The threshold voltage of the power MOSFET is an important design parameter from an application point of view. To offer immunity against unintended turn-on due to voltage spikes arising from noise, a minimum threshold voltage must be retained at above 1 V for most system applications. At the same time, a high threshold voltage is not desirable because the voltage available for building the charge in the channel inversion layer is determined by ($V_G - V_T$), where V_G is the applied gate bias voltage and V_T is the threshold voltage.

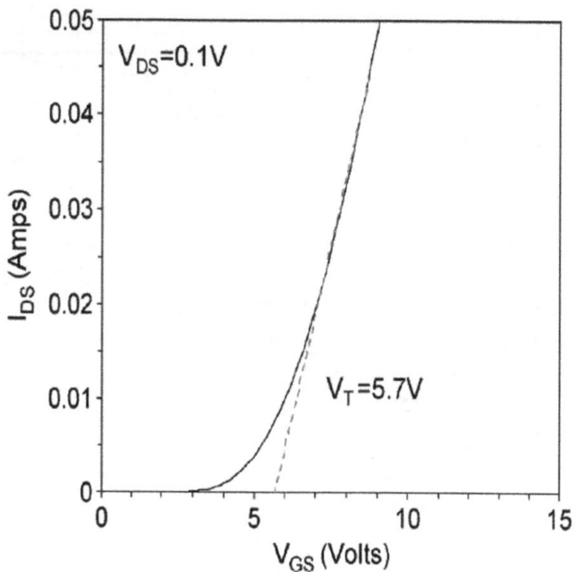

FIGURE 5.8
Transfer characteristics of a VDMOSFET structure. (From Losee, P. A. et al., *IEEE Trans. Electron Devices*, 55, 1824–1829, 2008.)

The gate bias at which onset of strong inversion begins to occur in the channel is defined as the threshold voltage. This voltage can be determined using [7]

$$V_T = \frac{\sqrt{4\varepsilon_s kTN_A \ln\left(\frac{N_A}{n_i}\right)}}{C_{OX}} + \frac{2kT}{q}\ln\left(\frac{N_A}{n_i}\right) \tag{5.14}$$

where N_A is the doping concentration of the P-well region, k is Boltzmann's constant, and T is the absolute temperature. The presence of a positive fixed oxide charge shifts the threshold voltage in the negative direction by

$$\Delta V_{TH} = \frac{Q_F}{C_{OX}} \tag{5.15}$$

A further shift of the threshold voltage in the negative direction by 1 V can be achieved by using heavily doped N-type polysilicon as the gate electrode as is routinely done for silicon power MOSFETs.

The much larger threshold voltage for silicon carbide is physically related to its larger bandgap as well as the higher P-well doping concentration required to suppress reach-through breakdown. This indicates a fundamental problem for achieving reasonable levels of threshold voltage in silicon carbide power MOSFETs if the conventional silicon structure is utilized. By decreasing the gate oxide thickness, the threshold voltage for a MOSFET can be reduced. However, operation of very high-voltage power MOSFETs with such thin gate oxides may create manufacturing and reliability issues when the high electric fields under the gate oxide in the semiconductor are taken into account. This limitation can be overcome by utilizing innovative structures that shield the P-well region, as discussed in the next section.

5.3 Shielded VDMOSFET Structure

The basic structure of the shielded VDMOSFET is shown in Figures 5.9 and 5.10 with an inversion layer channel or an accumulation layer channel, respectively. In the case of the structure with the inversion layer channel, the P+-shielding region extends under both the N+-source region as well as under the P-well region. It could also extend beyond the edge of the P-well region. In the case of the structure with the accumulation layer channel, the P+-shielding region extends under the N+-source region and the N-well region located under the gate. This N-well region can be formed using an uncompensated portion of the N-type drift region, or it can be created by adding N-type dopants with ion implantation to control its thickness and doping concentration. The gap between the P+-shielding regions is optimized to obtain a low specific on-resistance while simultaneously shielding the gate oxide interface from the high electric field in the drift region. In both the aforementioned

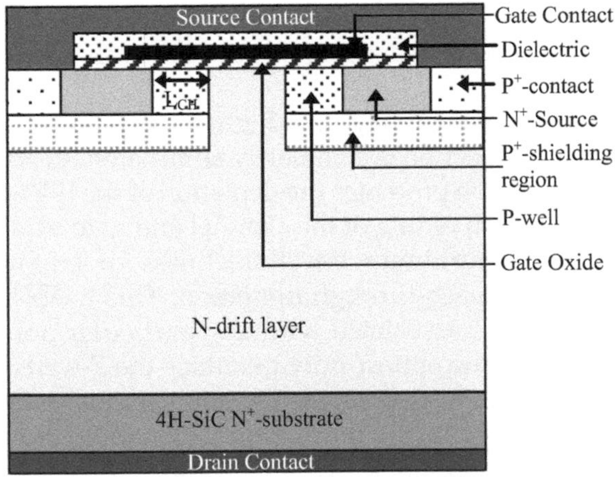

FIGURE 5.9
Cross section of the shielded inversion-mode VDMOSFET structure.

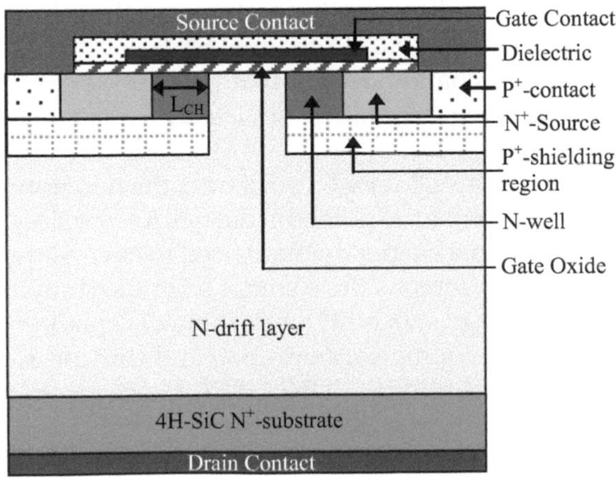

FIGURE 5.10
Cross section of the shielded accumulation-mode VDMOSFET structure.

structures, shielded VDMOSFET with an inversion layer channel and an accumulation layer channel, a potential barrier is formed at location after the JFET region becomes depleted by the applied drain bias in the blocking mode. This barrier prevents the electric field from becoming large at the gate oxide interface. When a positive bias is applied to the gate electrode, an inversion layer or accumulation layer channel is formed in the structures, enabling the conduction of the drain current with a low specific on-resistance.

5.3.1 Shielded VDMOSFET Structure: Off-State Characteristics

In the forward blocking mode of the shielded planar MOSFET structure, the voltage is supported by a depletion region formed on both sides of the P^+-region/N-drift junction. The maximum blocking voltage can be determined by the electric field at this junction, becoming equal to the critical electric field for breakdown if the parasitic N^+/P/N bipolar transistor is completely suppressed. This suppression is accomplished by short-circuiting the N^+-source and P^+-regions using the source metal as shown on the upper left and upper right-hand side of the cross section. This short circuit can be accomplished at a location orthogonal to the cell cross section, if desired, to reduce the cell pitch while optimizing the specific on-resistance.

If the doping concentration of the P^+-region is high, the reach-through breakdown problem discussed in the previous section is completely eliminated. In addition, the high doping concentration in the P^+-region promotes the depletion of the JFET region at lower drain voltages, providing enhanced shielding of the channel and gate oxide. With the shielding provided by the P^+-region, the minimum P-well thickness for 4H-SiC power MOSFETs is no longer constrained by the reach-through limitation. This enables the channel length to be reduced below the values associated with any particular doping concentration of the P-well region. In addition, the opportunity to reduce the P-well doping concentration enables the threshold voltage to be decreased. The smaller channel length and threshold voltage provide the benefits of reducing the channel resistance contribution.

In the case of the accumulation mode VDMOSFET structure, the presence of the sub-surface P^+-shielding region under the N-well region provides the potential required for completely depleting the N-well region if its doping concentration and thickness are appropriately chosen. This enables normally-off operation of the accumulation mode VDMOSFET with zero gate bias. It is worth pointing out that this mode of operation is fundamentally different from that of buried channel MOS devices. Buried channel devices contain an undepleted N-type channel region that provides a current path for the drain current at zero gate bias. This region must be depleted by a negative gate bias, creating a normally-on device structure.

In the accumulation-mode, the VDMOSFET structure, the depletion of the N-well region is accompanied by the formation of the potential barrier for the flow of electrons through the channel. The channel potential barrier does not need to have a large magnitude because the depletion of the JFET region screens the channel from the drain bias as well.

The maximum blocking voltage capability of the shielded planar MOSFET structure is determined by the drift region doping concentration and thickness, as already discussed in the previous section for conventional VDMOSFET. However, to fully utilize the high breakdown electric field strength available in silicon carbide, it is important to screen the gate oxide from the high field within the semiconductor. In the shielded VDMOSFET structure, this is achieved by the formation of a potential barrier by the depletion of the JFET region at a low drain bias voltage.

5.3.2 Shielded VDMOSFET Structure: On-State Characteristics

In the shielded VDMOSFET structure, current flow between the drain and source can be induced by creating an inversion layer channel on the surface of the P-well region or an accumulation layer channel on the surface of the N-well region. The current path is similar to that already shown in Figure 5.5 by the highlighted area. The current flows through the channel formed due to the applied gate bias into the JFET region via the accumulation layer formed above it under the gate oxide. It then spreads into the N-drift region at around a 45° angle and becomes uniform throughout the rest of the structure. The total on-resistance for the power VDMOSFET structure is determined by the resistance of these components in the current path

$$R_{on,sp} = R_{CH} + R_A + R_{JFET} + R_D + R_{Subs} \tag{5.16}$$

where R_{CH} is the channel resistance, R_A is the accumulation region resistance, R_{JFET} is the resistance of the JFET region, R_D is the resistance of the drift region after taking into account current spreading from the JFET region, and R_{Subs} is the resistance of the N+ substrate. These resistances can be analytically modeled by using the current flow pattern indicated by the highlighted regions in Figure 5.5.

For the shielded VDMOSFET structure with the P-well region, the specific channel resistance is given by

$$R_{CH} = \frac{(L_{CH} \cdot p)}{\mu_{inv} C_{ox} (V_G - V_T)} \tag{5.17}$$

where L_{CH} is the channel length, p is the cell pitch, μ_{inv} is the mobility for electrons in the inversion layer channel, C_{ox} is the specific capacitance of the gate oxide, V_G is the applied gate bias, and V_T is the threshold voltage. For the shielded VDMOSFET structure with the N-well region, the specific channel resistance is given by

$$R_{CH} = \frac{(L_{CH} \cdot p)}{\mu_a C_{ox} (V_G - V_T)} \tag{5.18}$$

where μ_a is the mobility of electrons in the accumulation layer channel.

Much larger accumulation layer mobility in silicon carbide allows reduction of the specific on-resistance. In addition, the threshold voltage for the accumulation mode is smaller than for the inversion mode, allowing further improvement in the channel resistance contribution. The rest of the resistance components in the shielded VDMOSFET structure can be modeled by using the same equations provided in the previous section.

The specific on-resistance for the shielded 4H-SiC VDMOSFETs is plotted in Figure 5.11 as a function of the breakdown voltage [5]. From Figure 5.11, it can be seen that specific on-resistance of 4H-SiC shielded VDMOSFET, also called planar MOSFETs, approaches the ideal specific on-resistance when the breakdown voltage exceeds 5000 V because the drift region resistance becomes dominant. However, when the breakdown voltage falls below 1000 V, the channel contribution becomes dominant. In this design regime, the advantage of using an accumulation channel becomes quite apparent. The accumulation-mode device has about two times lower specific on-resistance. When compared with the VDMOSFET structure without the shielded channel, the difference in specific on-resistance does not appear to be very significant. However, as discussed in the previous section, the

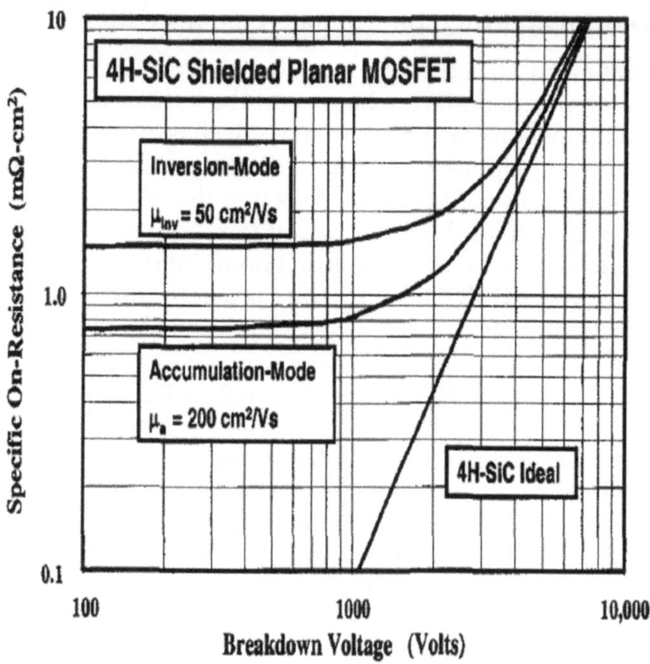

FIGURE 5.11
On-resistance for the shielded VDMOSFET structure. (From Baliga, J., *Silicon Carbide Power Devices*, World Scientific Publishing, Singapore, 2005.)

unshielded devices were able to support only about half the blocking voltage capability of the drift region due to the reach-through breakdown problem.

5.3.3 Shielded VDMOSFET Structure: Transfer Characteristics

As pointed out in the previous section, the threshold voltage of the power MOSFET is an important design parameter from an application standpoint. For the inversion-mode shielded planar MOSFET, the threshold voltage can be modeled by defining it as the gate bias at which onset of strong inversion begins to occur in the channel. This voltage can be determined using

$$V_T = \frac{\sqrt{4\varepsilon_s kTN_A \ln\left(\frac{N_A}{n_i}\right)}}{C_{OX}} + \frac{2kT}{q}\ln\left(\frac{N_A}{n_i}\right) \tag{5.19}$$

where N_A is the doping concentration of the P-well region, k is Boltzmann's constant, and T is the absolute temperature. The presence of a positive fixed oxide charge shifts the threshold voltage in the negative direction by

$$\Delta V_{TH} = \frac{Q_F}{C_{OX}} \tag{5.20}$$

A further shift of the threshold voltage in the negative direction by 1 V can be achieved by using heavily doped N-type polysilicon as the gate electrode, as is routinely done for silicon power MOSFETs.

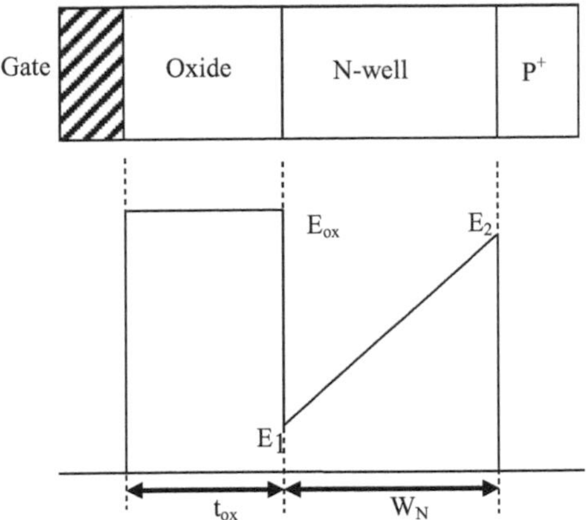

FIGURE 5.12
Electric field profile in the gate region for the accumulation-mode VDMOSFET structure. (From Baliga, J., *Silicon Carbide Power Devices*, World Scientific Publishing, Singapore, 2005.)

The band bending required to create a channel in the accumulation-mode planar MOSFET is much smaller than required for the inversion mode device. This provides the opportunity to reduce the threshold voltage while obtaining the desired normally-off device behavior. A model for the threshold voltage of accumulation-mode MOSFETs has been developed using the electric field profile shown in Figure 5.12 when the gate is biased at the threshold voltage. In this figure, the electric fields in the semiconductor and oxide are given by

$$E_1 = \frac{V_{bi}}{W_N} - \frac{qN_DW_N}{2\varepsilon_S} \quad (5.21)$$

$$E_2 = \frac{V_{bi}}{W_N} + \frac{qN_DW_N}{2\varepsilon_S} \quad (5.22)$$

$$E_{ox} = \frac{\varepsilon_S}{\varepsilon_{ox}} E_1 \quad (5.23)$$

Note that this model is based upon neglecting any voltage supported within the P+-region under the assumption that it is very heavily doped. Using these electric fields, the threshold voltage is found to be given by

$$V_t = \Phi_{MS} + \left(\frac{\varepsilon_S V_{bi}}{\varepsilon_{ox} W_{NE}} - \frac{qN_D W_{NE}}{2\varepsilon_{ox}} \right).t_{ox} \quad (5.24)$$

The first term in this equation accounts for the work function difference between the gate material and the lightly doped N-Well region. The second term represents the effect of the built-in potential of the P+/N junction that depletes the N-well region (Figures 5.13 and 5.14).

The analytically calculated threshold voltage for 4H-SiC accumulation-mode MOSFETs is provided in Figure 5.15 for the case of a gate oxide thickness of 0.05 μ, and N-well thickness of 0.2 μ as a function of the N-well doping concentration with the inclusion

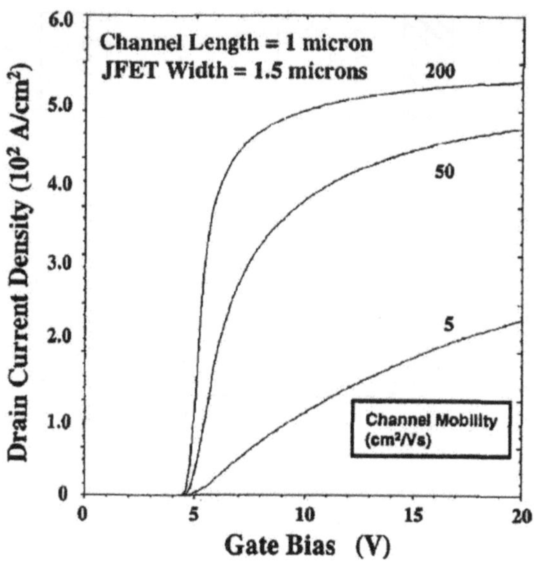

FIGURE 5.13
Transfer characteristics of an inversion-mode, shielded VDMOSFET structure. (From Baliga, J., *Silicon Carbide Power Devices,* World Scientific Publishing, Singapore, 2005.)

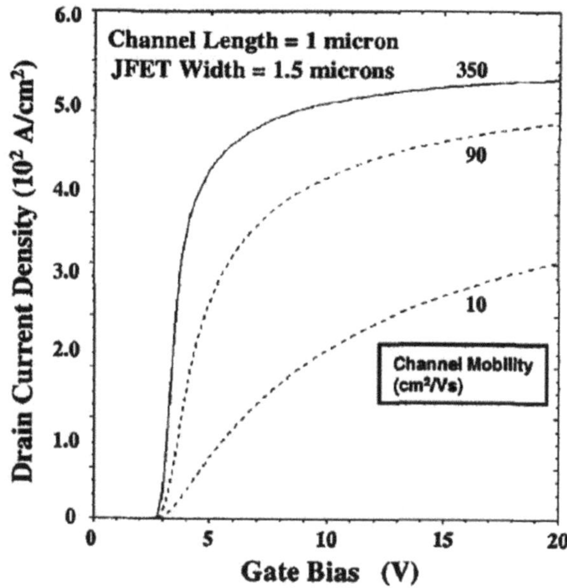

FIGURE 5.14
Transfer characteristics of an accumulation-mode VDMOSFET structure. (From Baliga, J., *Silicon Carbide Power Devices,* World Scientific Publishing, Singapore, 2005.)

of a metal-semiconductor work-function difference of 1 V. For comparison purposes, the threshold voltage for the inversion-mode 4H-SiC MOSFET is also given in this figure for the same gate oxide thickness. A strikingly obvious difference between the structures is a decrease in the threshold voltage for the accumulation-mode structure with increasing doping concentration in the N-well region. This occurs due to the declining influence of

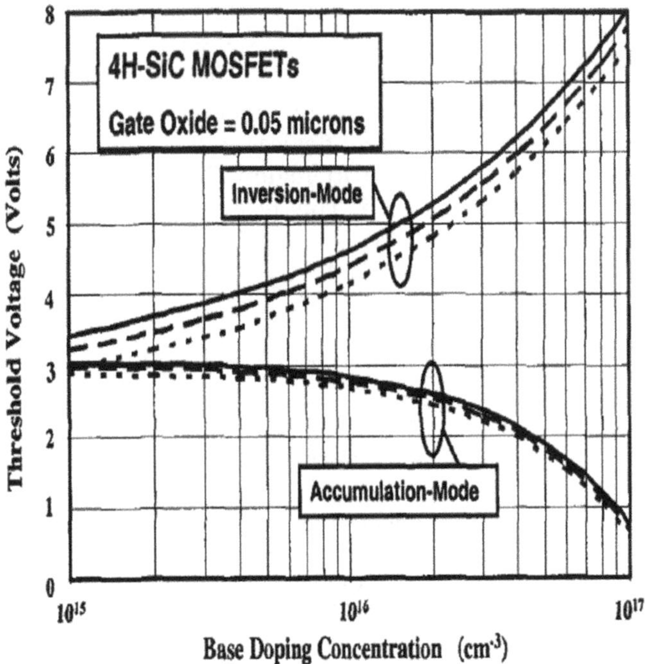

FIGURE 5.15
Threshold voltage of inversion mode and accumulation-mode VDMOSFET structures. (From Baliga, J., *Silicon Carbide Power Devices*, World Scientific Publishing, Singapore, 2005.)

the P^+/N junction at the gate oxide interface when the doping concentration of the N-well region is increased. It can also be noted that the temperature dependence of the threshold voltage is smaller for the accumulation-mode structure. Of course, the most important benefit of the accumulation-mode is that lower threshold voltages can be achieved than in the inversion-mode structures.

5.4 Summary

The silicon carbide-based power VDMOSFET structure has been critically examined in this chapter. It enables the device to achieve superior performance with silicon carbide as compared to silicon-based devices. Also, incorporation of a subsurface P+ region into the VDMOSFET structure enables the P-well region to be shielded from reach-through limited breakdown and prevents high electric fields from developing across the gate oxide during the blocking mode. Since relatively low inversion layer mobility has been reported for 4H-SiC MOSFETs, an accumulation-mode structure was proposed with an N-well region that is completely depleted by the built-in potential of the underlying P^+/N junction. This structure takes advantage of the much larger accumulation mobility observed in semiconductors. The operating principle of the shielded VDMOSFET structures has been reviewed in the chapter. It has been demonstrated that the JFET width is a critical parameter that controls the electric field at the gate oxide interface as well as the specific on-resistance.

Its optimization is important for obtaining high-performance devices. In the case of the accumulation-mode structure, the appropriate combination of the doping concentration and thickness of the N-well region must be chosen to ensure that it is completely depleted by the built-in potential of the underlying P^+/N junction. With adequate shielding of the well region, it is found that short-channel devices will support high blocking voltages, limited only by the properties of the drift region. These devices have excellent safe-operating-area and fast switching speed. This technology has the potential for use in systems operating at up to at least 5000 V.

References

1. D. A. Grant and J. Gowar, *Power MOSFETs: Theory and Applications*, John Wiley & Sons, New York, 1989.
2. B. J. Baliga, "Power semiconductor devices for variable frequency drives," *Proceedings of the IEEE*, vol. 82, pp. 1112–1122, 1994.
3. B. J. Baliga, "Silicon carbide semiconductor devices having buried silicon carbide conduction barrier layers therein," U. S. Patent 5,543,637, Issued August 6, 1996.
4. P. A. Losee et al., "DC and transient performance of 4H-SiC double-implant MOSFETs," *IEEE Transactions on Electron Devices*, vol. 55, no. 8, pp. 1824–1829, 2008.
5. J. Baliga, *Silicon Carbide Power Devices*, Singapore, World Scientific Publishing, 2005.
6. J. N. Shenoy, J. A. Cooper, and M. R. Melloch, "High voltage double implanted power MOSFETs in 6H-SiC," *IEEE Electron Device Letters*, vol. 18, pp. 93–95, 1997.
7. B. J. Baliga, *Power Semiconductor Devices*, Chapter 7, pp. 357–362, PWS Publishing Company, Boston, MA, 1996.

6

Modeling of Double-Gate MOSFETs

D. Nirmal and J. Ajayan

CONTENTS

6.1 Introduction

The world of electronics started with the invention of the vacuum tube (a two-terminal device—the terminals are anode and cathode) by the famous English physicist John Ambrose Fleming in 1895, based on the principle called "Edison Effect." In 1883, Thomas Alva Edison found that electrons could flow from one metal conductor to another through a vacuum. In other words, current flow is possible through a vacuum and this phenomenon is called the Edison effect. The main drawbacks of the first electronic device—a vacuum tube—were the low reliability and very high-power consumption; the device also required cooling mechanisms. Two years later, Lee De Forest from the United States discovered a similar type of device called a triode (a three-terminal device—the terminals are anode, cathode, and grid), which had the characteristics of an amplifier. These vacuum tube devices (diode, triode, tetrode, and pentode) were the main obstacles in the road of progress since the size of these devices could not be reduced too much. The search for new device architectures, initiated by the development of semiconductors, resulted in the discovery of the first solid-state device called a PN junction diode. The major breakthrough in the semiconductor industry was the invention of point contact transistors by Brattain and Bardeen at Bell Laboratories, USA, in 1947. At the same period of time, Shockley developed the first NPN transistor, which is widely used in radio frequency (RF) and analog circuits today. It has been more than five decades since the invention of the integrated circuit (IC) technology based on the idea given by Jack Kilby from Texas Instruments. Jack Kilby and Robert Noyce decided to throw away all the wires and tried to connect the resistors, capacitors, diodes, and transistors on the same piece of wafer internally. One of the most important aspects in the evolution of IC technology is the physical feature sizes of the transistors are reduced continually over time as the lithography technologies used to define these features become available. The tremendous and steady progress in IC technology is also propelled by the development of various etching and deposition techniques. The first metal oxide semiconductor field effect transistor (MOSFET) on a silicon substrate using SiO_2 as the gate oxide was fabricated in 1960 (Kahng and Atalla). Since then, complementary metal oxide semiconductor (CMOS) transistors have become the technology of choice for high-speed, low-power digital circuits, and bipolar junction transistors (BJTs) are used primarily in RF and analog circuits only.

The invention of MOSFETs led to the electronics revolution of the 1970s and 1980s, in which the microprocessor made possible powerful desktop computers, laptops, sophisticated handheld calculators, iPods, and other electronic systems. The MOSFET can be made very small, so high-density, very large-scale integration (VLSI) circuits and high-density memories are possible. In 1965, Intel co-founder Gordon Moore predicted that the number of transistors on a microchip would double approximately every two years. Over the course of time, his prediction proved accurate and has since been known as "Moore's Law." Advances in lithography, etching, and deposition techniques have enabled the semiconductor industry to scale down transistors in physical dimensions into the nanometer regime and to pack more transistors in the same chip area. Generally, by scaling down the geometrical dimensions of the device, the density of MOSFETs on the microchip is increased, power consumption per device is decreased due to the low operating voltages of the small devices, and the switching speed of the MOSFETs can also be increased due to reduced parasitics. However, in recent years the scaling has become increasingly challenging, due to the fundamental physical limits as well as the limits in existing fabrication technologies.

As the semiconductor industry moves toward even smaller devices, especially in the sub-100 nm regime, the greatest challenge is the transition from SiO_2 to high-k dielectrics. Below the 50 nm technology node, however, the major challenge is the transistor structure (ITERS 1999 and 2000). As the gate length (L_g) is scaled down, the source/drain (S/D) junction depth (x_j), depletion width (W_d), and oxide thickness (t_{ox}) has to be scaled as well, so that the gate maintains electrostatic control over the device. Reduction of oxide thickness is limited by gate leakage current due to tunneling, which has forced the semiconductor industry to change the gate oxide from SiO_2 to high-k oxides [1]. Reducing depletion width requires higher bulk doping, which reduces mobility of carriers and hence overall speed of the MOSFETs due to various scattering mechanisms such as interface roughness scattering, coulombic scattering, and phonon scattering. In order to overcome these scaling limitations, new device structures are being considered to replace the MOSFET when conventional scaling fails. Fully depleted (FD) structures with a thin Si body, such as ultra-thin-body (UTB), and multiple-gate (double-gate, tri-gate, and gate-all-around) MOSFETs have, in recent years, been the subject of intensive research [2–5]. It has been shown that the subthreshold swing and drain-induced barrier lowering (DIBL) increase with reduction in gate length and I_{on}/I_{off} also decreases with decreasing L_g, which indicates severe short-channel effects (SCEs) for low gate-length devices. The double-gate MOSFETs (DG-MOSFETs) are being considered as the most attractive candidate for the 10–50 nm gate length regime.

6.2 DG-MOSFET Structure

The schematic diagram of DG-MOSFET is shown in Figure 6.1. Double-gate MOSFETs (DG-MOSFETs) have the great potential for future CMOS applications as the gate length is scaled into the nanometer regime [1]. Device simulations performed by various research

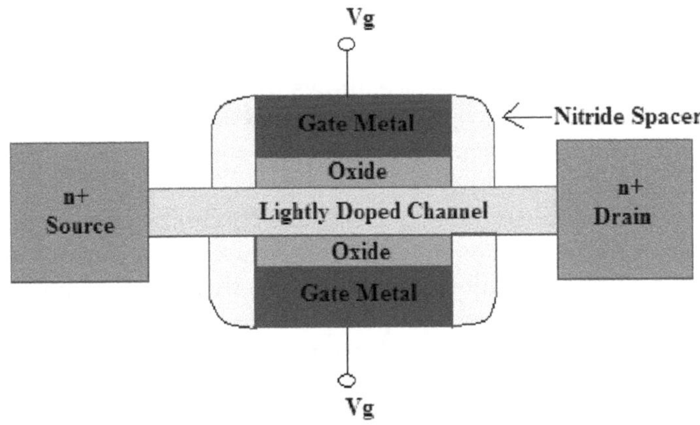

FIGURE 6.1
Structure of DG-MOSFET.

groups predicted that the DG-MOSFET is extendable to a gate length of 10 nm [2–10]. The DG-MOSFETs have several advantages over conventional single-gate (SG) MOSFETs. Some of the advantages are reduced short-channel effects, ideal subthreshold slope, and double the drain current. The ideal DG-MOSFETs should have a thin channel as well as thick source/drain regions for lower contact resistance. Because two gates are on either side of the lightly doped channel in the DG MOSFETs, both gates control the channel from both sides and provide additional gate-length scaling by a factor of 2. Due to better control on SCEs, DG MOSFET is superior to the conventional MOSFETs and it has higher current density, higher transconductance, higher subthreshold swings at low supply voltage, and reduced DIBL.

There are two different forms of DG MOSFETs, namely symmetric DG-MOSFET and asymmetric DG-MOSFET. The structure of symmetric as well as asymmetric DG-MOSFETs is shown in Figure 6.2. Nanoscale DG-MOSFETs with an undoped body, also called the intrinsic channel, are particularly attractive [11]. DG-MOSFETs with an intrinsic channel are very much effective in the suppression of SCEs by means of an undoped UTB instead of the conventional channel with high doping density. The absence of dopant atoms in the channel further enhances mobility by eliminating impurity scattering and avoiding random microscopic dopant fluctuations [12–14]. Figure 6.2a shows the schematic structure of symmetric DG n-type MOSFET. The symmetric DG-MOSFET has two gates (a front gate and a back gate) with the same oxide thickness. Figure 6.2b

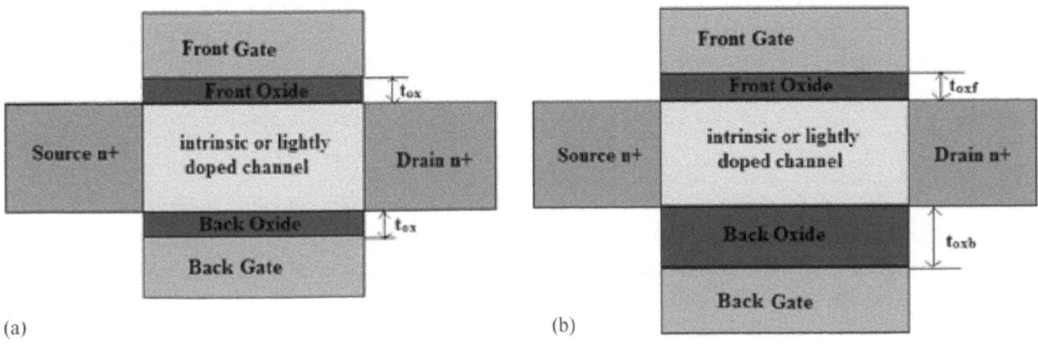

FIGURE 6.2
(a) Structure of symmetric DG-MOSFET and (b) structure of asymmetric DG-MOSFET.

shows the schematic structure of an asymmetric DG *n*-type MOSFET, and this device consists of two gates—a front gate and a back gate with two different oxide thicknesses. The back gate at the bottom of the device allows for full depletion of the channel. The symmetric device is a special case of the asymmetric device; that is, the symmetric device's two-oxide thickness is equal, the two gates have the same flat-band voltage, and the gates are connected together [14].

The threshold voltage of an asymmetric DG-MOSFET can be appropriately set by using n-type and p-type polysilicon gates in the DG structure; however, this will increase the transverse electric field since a built-in potential exists through the body due to the asymmetric gate work functions [15]. The appropriate threshold voltage of the DG MOSFET can also be set by providing proper channel doping; however, channel doping degrades the device performance in terms of carrier mobility and dopant fluctuations [16]. Increasing the doping density in the channel results in increased band-to-band tunneling leakage between the body and drain terminals [17]. Hence, the channel doping as well as an asymmetric DG MOSFET structure results in degraded device performance. This emphasizes the need for gate work function engineering as alternative solutions for nanometer devices. The threshold voltage of the DG-MOSFETs can be properly set by using work function engineering, which enables the device designers to maintain device performance in terms of drive current and reduced short-channel effects. Selection of the metal gate is also extremely important for achieving desired device performance. Molybdenum (Mo) is a potential candidate for a metal gate technology in nanoscale DG-MOSFETs due to the fact that it has a sufficiently high and stable work function. The work function of Mo can be significantly reduced by high-dose nitrogen implantation. Nitrogen implantation can be used to adjust the Mo gate work function in a controllable way without degrading device performance.

6.3 Effect of High-k Dielectrics on DG-MOSFETs

The semiconductor industry is working to introduce high-k gate oxides in DG-MOSFETs in order to meet the requirements for low off-leakage current while keeping power consumption under control. Many research groups have replaced SiO_2 with different high-k gate oxide materials in DG-MOSFETs. The International Technology Roadmap for Semiconductors (ITRS-2010) clearly predicted that high-k dielectric materials would soon replace the SiO_2 in the MOSFET. The ITRS-2003 clearly explained that the gate dielectric has emerged as one of the most difficult challenges for future nanoscale MOSFETs. Equivalent oxide thickness (EOT) for high-k materials can be mathematically calculated as

$$\mathrm{EOT} = \left(\frac{K_{\mathrm{SiO_2}}}{K_{\text{high-k}}}\right) t_{\text{high-k}} \tag{6.1}$$

Equation (6.1) clearly indicates that replacing SiO_2 with high-k gate dielectrics will increase the physical gate oxide thickness, which further increases the scalability of the device while retaining the same oxide capacitance (C_{ox}). This is evident from Equation (6.2).

$$C_{\mathrm{ox}} = \frac{\varepsilon_{\mathrm{ox}}}{t_{\mathrm{ox}}} \tag{6.2}$$

TABLE 6.1
High-k Dielectric Material Properties

Material	K	Eg (eV)	CBO (eV)	VBO (eV)	Breakdown Field (MV/cm)
SiO_2	3.9	9	3.2	4.7	10
Al_2O_3	8	8.8	2.8	4.9	10
TiO_2	80	3.5	0	2.4	3
ZrO_2	25	5.8	1.5	3.2	4–5
HfO_2	20	6	1.4	3.3	4–5
La_2O_3	30	6	2.3	2.6	3–4

Here, K_{SiO_2} is the dielectric constant of SiO_2, $K_{high\text{-}k}$ is the dielectric constant of high-k gate dielectric, $t_{high\text{-}k}$ is the thickness of high-k dielectrics, and C_{ox} is the oxide capacitance per unit area. Since a thicker gate dielectric layer is used for isolating the gate from the channel in DG-MOSFETs, the tunneling current can be significantly reduced in this nanoscale regime. The major requirements for high-k dielectrics are high thermal stability, smooth oxide-semiconductor interface, and high dielectric breakdown field. The properties of some of the high-k dielectric materials are given in Table 6.1. For the same EOT, the oxide capacitance remains the same for all the gate dielectrics due to the fact that the increase in dielectric constant is compensated by the large physical thickness.

6.4 Leakage Currents in DG-MOSFETs

Scaling down the transistors to smaller dimensions increases the speed of the transistors as well as minimizes the effective cost per transistor and also helps to increase the transistor density in the integrated circuit chips. In DG-MOSFETs, both the front gate as well as the back gate can control one-half of the device, and the operations are independent of each other. In DG-MOSFETs, there are two channels—one channel is under the front gate, and the second channel is under the back gate; therefore, the total drain current through the device is the sum of the current through the separate channels. In DG-MOSFET devices, the presence of two gates and ultra-thin body helps to reduce the short-channel effects, which drastically reduces the subthreshold leakage current. Reduced short-channel effects and higher drive currents allow the use of high-k gate oxides instead of SiO_2 in DG-MOSFET devices compared to conventional bulk CMOS devices. The large physical thickness of high-k dielectrics helps to reduce the gate leakage current by increasing the gate-channel separation.

The major leakage components in DG-MOSFETs are subthreshold leakage current, gate leakage current, and band-to-band tunnel leakage. Electrically coupled front and back gates and ultra-thin body reduce the short-channel effects in DG-MOSFET devices, resulting in a reduction of subthreshold leakage current. Moreover, for an equal ON current, DG-MOSFETs show lower subthreshold leakage current compared to the conventional single-gate MOSFETs, and this is because the surface electric field is lower in DG-MOSFETs due to low doping in the channel. Gate leakage in DG-MOSFETs is due to the tunneling of electrons from channel to gate through the gate oxide. Interface traps at the gate oxide-channel interface will also increase the leakage current. Therefore, a smooth oxide-semiconductor is required to minimize the trap density, thereby reducing the leakage current. Due to lower doping in the channel, the electric field across the oxide is lower in DG-MOSFETs compared to conventional single-gate (SG) MOSFETs. Hence, the DG-MOSFETs show lower gate leakage current compared to SG-MOSFETs. In the asymmetric DG-MOSFETs, due to the work function difference of front and back gates, the front surface field is very high whereas a back surface field is negligible. Hence, tunneling of electrons from channel to gate through the gate oxide takes place only through a front gate. The band-to-band tunneling (BTBT) across a reverse biased PN junction from the P-side valence band to the N-side conduction band is also a very important component of leakage in nanoscale DG-MOSFETs.

6.5 Double-Gate SOI MOSFET Fabrication Steps

Silicon-on-insulator (SOI) MOSFETs are emerging as one of the most promising next-generation technologies because of their inherent thin channel region, which can effectively suppress SCEs [18–24], and therefore, SOI-DG-MOSFETs are being considered as an ideal candidate to enable further dimension downscaling below the sub-50-nm regime [23,24]. Two applications showing evidence of the superiority of the SOI MOSFETs over bulk CMOS technology are SOI microprocessors with more than 20% percent speed improvement [25,26] and SOI RF power amplifiers with higher power efficiency [27]. SOI technology is now intensively studied because of its ability to overcome several inherent limitations of bulk silicon CMOS technology such as lateral isolation, radiation tolerance, lower parasitic capacitance and power, higher speed, and reduced short-channel effects [18,28]. SOI substrates are mainly formed by oxygen implantation (SIMOX), selective oxidation of porous silicon (FIPOS), or overgrowth and recrystallization techniques [28].

There are several methods have been proposed by various research groups to fabricate DG-MOSFETs. The direct wafer bonding and polishing technique is the most commonly used method to fabricate double-gate devices. However, this method involves crucial alignment of two wafers [29]. Another method used to fabricate DG-MOSFET is to etch a cavity on the buried oxide (BOX) for the bottom gate. However, it is very difficult to control the bottom gate length [30]. Selective epitaxy and epitaxial lateral overgrowth have been widely used to fabricate DG-MOSFETs [31,32]. However, these two methods are very complex and expensive. The lateral solid phase epitaxy (LSPE) is a very simple and compatible method used to fabricate DG-MOSFETs [33,34]. Haitao Liu et al. [35] have fabricated DG-MOSFET using the LSPE technique, and the major fabrication steps of the double-gate MOSFET are shown in Figure 6.3. The fabrication process begins with the commercial SIMOXSOI wafers. The first step is the etching of the complete SOI layer and part of the buried oxide layer to form a deep trench for the bottom gate, which is shown in Figure 6.3a; the smallest trench size is determined by the present lithography. The second step is the deposition of a thin layer of Si_3N_4 and a layer of low temperature oxide (LTO); the surface is polished using CMP, which is shown in Figure 6.3b. The oxide in the trench serves as the dummy bottom gate, and it will be removed later. The thin Si_3N_4 layers on the surface need to be removed in hot H_3PO_4 at 165°C. After being dipped in HF (1:50) solution, the wafers have to be loaded immediately into the LPCVD furnace for the deposition of a layer of amorphous silicon, which is shown in Figure 6.3c.

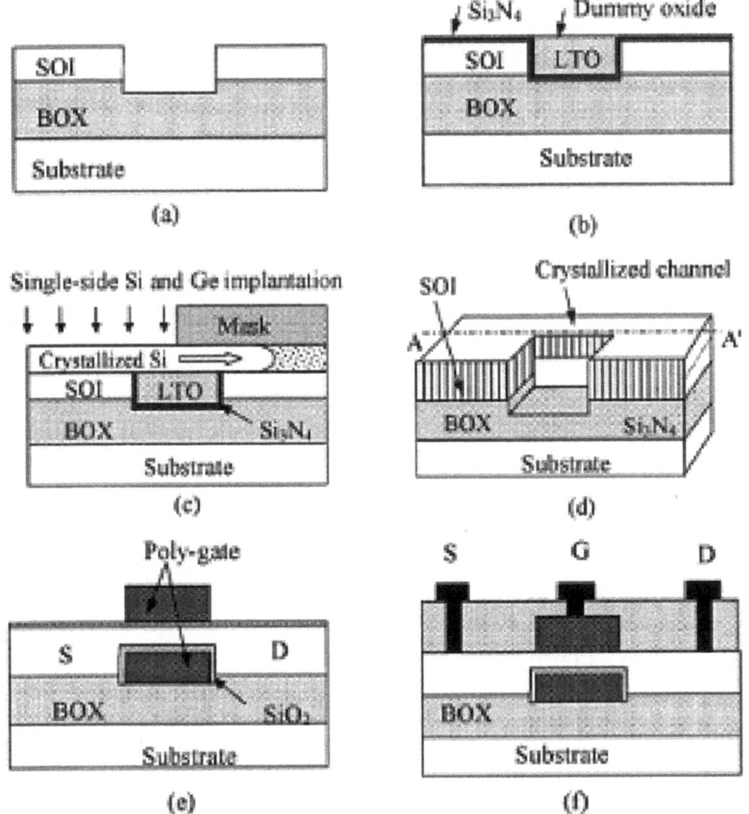

FIGURE 6.3
Major fabrication steps for the double-gate MOSFET using LSPE. (a) Etching of complete SOI layer and part of the BOX layer to form a deep trench for the bottom gate; (b) Deposition of a thin layer of Si3N4 and a layer of low temperature oxide (LTO); (c) Masked single-sided Si implantation is carried out to prevent the twin boundary in the middle of the channel and Ge implantation is used to enhance the lateral crystallization; (d) Removal of dummy oxide and Si_3N_4 under the channel region using BOE (1:6) and hot H_3PO_4; (e) Growth of a thin gate oxide and a phosphorus-doped poly-silicon; (f) Contact opening, metal sputtering, and metal patterning.

A masked single-sided Si implantation will be carried out to prevent the twin boundary in the middle of the channel, and Ge implantation is used to enhance the lateral crystallization [36]. This is because with the single-sided implantation, the crystallization will only start from either the source-side or drain-side [37]. It should be noted that the Si ion implantation energy and dose are very important to obtain a high-quality LSPE layer. After Si and Ge implantation, furnace annealing will be implemented to crystallize the amorphous silicon film, which is also shown in Figure 6.3c. The crystallized silicon will be thinned down to 100 nm using oxidation and is defined using an addition mask. Then, the dummy oxide and Si_3N_4 under the channel region needs to be removed in BOE (1:6) and hot H_3PO_4, which is shown in Figure 6.3d. The next step is the growth of a thin gate oxide and a phosphorus-doped polysilicon will be deposited and patterned, which is shown in Figure 6.3e. The top and bottom gates are inherently connected during polysilicon deposition. However, these two gates are not self-aligned because they are defined by different masks. After the polysilicon gate is patterned, source/drain implantation will be carried out. This is followed by the deposition of LTO and dopants activation. The rest of the processes include contact opening, metal sputtering, and metal patterning to complete the device fabrication, which is shown in Figure 6.3f.

6.6 Gate Engineering in DG-MOSFET

The control of the gate-to-source voltage (V_{GS}) on the threshold voltage (V_{th}) decreases as the channel length shrinks to a sub-50 nm regime because of the increased charge sharing from the source and drain. Therefore, the V_{th} reduction with decreasing channel lengths and DIBL are critical issues that need to be addressed while providing immunity against SCEs [4,18]. In order to improve the immunity against SCEs, a new structure called a dual-material gate (DMG) MOSFET was developed [38–41]. The schematic diagram of the gate-engineered DMG MOSFET is shown in Figure 6.4. The structure of DMG MOSFET consists of two metals in the gates M1 and M2 with different work functions. The dual metal gate structure reduces the SCEs and also improves transconductance due to a step in the surface-potential profile when compared with a single-gate MOSFET. In the DMG structure, the peak electric field at the drain end is reduced, which in turn increases the average electric field under the gate. This improves the reliability of the device, reduces the ability of the localized charges to increase drain resistance [42], and enables more control of the gate over the conductance of the channel so as to increase the gate transport efficiency. The step function profile of the surface potential ensures screening of the channel region under the material on the source side (M1) from drain-potential variations. After saturation, M2 absorbs any additional drain–source voltage, and the region under M1 is screened from drain-potential variations. However, the drive ability and transconductance of the DMG structure are not as good as that of the DG structure. To combine the advantages of both DG and DMG structures, G. Venkateshwar Reddy et al. [43] proposed a new structure, i.e., the dual-material double-gate (DMDG) SOI MOSFET, which is similar to that of an asymmetrical DG SOI MOSFET with the exception that the front gate of the DMDG structure consists of two materials (p^+ poly and n^+ poly). The DMDG-SOI MOSFET has a considerable reduction in the peak electric field near the drain end, high drain breakdown voltage, improved transconductance, reduced drain conductance, and a desirable threshold voltage "roll-up" even for channel lengths far below 100 nm [43].

The threshold voltage for the back gate (V_{thBG}) with n^+ -poly silicon gate for long channel devices is given by [44]

$$V_{thBG} = V_{FB,fp} + 2\phi_F + \frac{Q_{Si}}{2}\left(1 + V_T\frac{4C_{Si}}{Q_{Si}}\right)\left(\frac{1}{4C_{Si}} + \frac{1}{C_f}\right) + V_T \ln\left(V_T\frac{4C_{Si}}{Q_{Si}}\right) - \frac{\gamma t_f + t_{Si}}{\gamma t_f + \gamma t_b + t_{Si}}\Delta V_{FB} \quad (6.3)$$

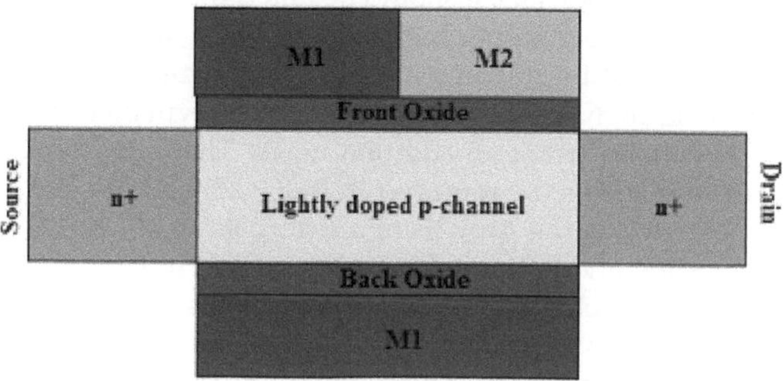

FIGURE 6.4
Gate-engineered DG-MOSFET.

The threshold voltage (V_{thFG}) for the front gate with p$^+$ poly and n$^+$ poly for long-channel devices is given by [44]

$$V_{thFG} = V_{FB,fp} + 2\phi_F + \frac{Q_{Si}}{2}\left(1 + V_T\frac{4C_{Si}}{Q_{Si}}\right)\left(\frac{1}{4C_{Si}} + \frac{1}{C_f}\right) + V_T\ln\left(V_T\frac{4C_{Si}}{Q_{Si}}\right) \tag{6.4}$$

where $V_{FB,fp}$ and $V_{FB,fn}$ are the front-channel flat-band voltages of p$^+$ poly and n$^+$ poly at the front gate, and they can be expressed as [43]

$$V_{FB,fp} = \phi_{MS1} = \phi_{M1} - \phi_{Si} \tag{6.5}$$

$$V_{FB,fn} = \phi_{MS2} = \phi_{M2} - \phi_{Si} \tag{6.6}$$

ϕ_{M1} and ϕ_{M2} are the work functions of M1 and M2 respectively, ϕ_{Si} is the work function of silicon, which is given by

$$\phi_{Si} = \chi_{Si} + \frac{E_g}{2q} + \phi_F \tag{6.7}$$

Here, E_g is the bandgap of the silicon at 300K, χ_{Si} is the electron affinity of silicon, $\phi_F = V_T \ln(N_A/n_i)$ is the Fermi potential, V_T is the thermal voltage, and n_i is the intrinsic carrier concentration.

$$\gamma = \frac{\varepsilon_{Si}}{\varepsilon_{ox}} \tag{6.8}$$

$$Q_{Si} = qN_At_{Si} \tag{6.9}$$

$$\Delta V_{FB} = V_{FB,fp} - V_{FB,bn} \tag{6.10}$$

$$C_{Si} = \frac{\varepsilon_{Si}}{t_{Si}} \tag{6.11}$$

$$C_f = \frac{\varepsilon_{ox}}{t_f} \tag{6.12}$$

$$C_b = \frac{\varepsilon_{ox}}{t_b} \tag{6.13}$$

Here, ε_{Si} is the permittivity of silicon, ε_{ox} is the permittivity of oxide, t_{Si} is the thickness of channel, t_f and t_b are the oxide thickness of front and back gates, respectively, and N_A is the body doping concentration.

Therefore, the expression for the threshold voltage of the DMDG SOI MOSFET is given by

$$V_{th} = V_{thL} - \Delta V_{th} \tag{6.14}$$

where V_{thL} can be either V_{thFG} or V_{thBG}.

The short-channel threshold voltage shift ΔV_{th} of the DMDG SOI MOSFET is given by [43]

$$\Delta V_{th} = 2\sqrt{\eta_s \eta_{L1}} e^{-\varsigma} \tag{6.15}$$

$$\eta_S = V_{bi} - V'_{GS,f1} + \frac{\Delta V_{FB}}{2\left(1+\frac{t_{Si}}{2\gamma t_f}\right)} \tag{6.16}$$

$$\eta_{L1} = \frac{1}{2}\left[\frac{\left(V_{bi}+V_{DS}-V'_{GS,f2}\right)\mathrm{Sinh}\left(\frac{L_1}{\lambda}\right)+\eta_S \mathrm{Sinh}\left(\frac{L_2}{\lambda}\right)}{\mathrm{Cosh}\left(\frac{L_1}{\lambda}\right)\mathrm{Sinh}\left(\frac{L_2}{\lambda}\right)+\mathrm{Sinh}\left(\frac{L_1}{\lambda}\right)\mathrm{Cosh}\left(\frac{L_2}{\lambda}\right)}\right] \tag{6.17}$$

$$\varsigma = \frac{L_1}{\sqrt{2\gamma t_{Si} t_f}} \tag{6.18}$$

$$V'_{GS,f1} = V_{GS} - V_{FB,fp} \tag{6.19}$$

$$V'_{GS,f2} = V_{GS} - V_{FB,fn} \tag{6.20}$$

$$V_{bi} = V_T \ln\left(\frac{N_A N_D}{n_i^2}\right) \tag{6.21}$$

V_{bi} is the built-in potential across the body-source junction, and N_A and N_D are the body and source/drain doping concentrations, respectively.

The channel current for DMDG SOI MOSFET is given by [44]

$$I_{ch,lin} = \sum_{i=1,2}\left(\frac{W\mu_{neff}C_{ox}}{L\left(1+\frac{V_{DS}}{LE_C}\right)}\right)\left[\left(V_{GS}-V_{thi}\right)V_{DS}-\frac{V_{DS}^2}{2}\right] \tag{6.22}$$

$$I_{ch,sat} = \sum_{i=1,2}\left(\frac{W\mu_{neff}C_{ox}}{L\left(1+\frac{V_{DS,sati}}{LE_C}\right)}\right)\left[\left(V_{GS}-V_{thi}\right)V_{DS,sati}-\frac{V_{DS,sati}^2}{2}\right] \tag{6.23}$$

Here, $V_{th1} = V_{th,BG}$ and $V_{th2} = V_{th,FC}$, and E_C is the critical electric field at which the electron velocity (v_{ns}) saturates and $V_{DS,sati}$ is the saturation voltage.

$$E_C = \frac{2v_{ns}}{\mu_{neff}} \tag{6.24}$$

$$V_{DS,sati} = \frac{V_{GS} - V_{thi}}{1 + \dfrac{V_{GS} - V_{thi}}{LE_C}} \tag{6.25}$$

The effective mobility (μ_{neff}) of the inversion layer electrons is given by [44]

$$\frac{1}{\mu_{neff}} = \frac{1}{\mu_{ph}} + \frac{1}{\mu_{sr}} \tag{6.26}$$

where μ_{ph} is the mobility associated with the phonon scattering and μ_{sr} is the mobility associated with the surface roughness scattering.

6.7 Channel Engineering in DG-MOSFETs

The continuous downscaling of complementary metal oxide semiconductor (CMOS) transistors into sub-100 nm dimensions seems to have reached its limits with SiO_2 as the gate oxide material [45]. As a result, the ever-growing semiconductor industry is focusing on high-k gate dielectrics in transistor manufacturing processes to meet the need for higher switching speed transistors while keeping power consumption under control [45–48]. DG-MOSFETs with channel lengths below 100 nm exhibit considerable threshold voltage roll-off and DIBL effects. The decrease in threshold voltage with reduction in channel length is widely used as an indicator of the short-channel effect (SCE) in evaluating CMOS cutting-edge technologies. This threshold voltage roll-off will determine the minimum acceptable channel length for future CMOS technology.

The threshold voltage roll-off can be reduced by locally raising the channel doping next to the drain or drain/source junctions [49]. This method is called lateral channel engineering, e.g., halo or pocket implants, and the engineered channel is known as a lateral asymmetric channel or graded channel (GC) [50]. Halo-implanted devices show excellent output characteristics with low DIBL, higher drive currents, flatter saturation characteristics, and slightly higher breakdown voltages compared to the conventional MOSFETs [51]. The use of polysilicon gates in MOS devices can significantly increase the capacitance equivalent thickness, resulting in high sheet resistance, and cause dopant diffusion through the high-k layer. One viable way to solve these challenges is to use a metal for the gate electrodes. Metal gates eliminate polydepletion effects, resulting in the reduction of inversion capacitance. Molybdenum is taken as the gate material so that the gate work function can be fixed at 4.577 eV to obtain the threshold voltage of 0.3 V at a drain voltage of 0.1 V [52,53]. The structure of channel-engineered DG-MOSFET is shown in Figure 6.5.

The effects of high-k dielectrics on I_D *Vs* V_{GS} characteristics of 45 nm DG-MOSFET are shown in Figure 6.6 which indicates that the drain current of DG-MOSFETs increases with an increase in dielectric constant. The DG-MOSFETs with high-k gate oxides improve the g_m/I_D ratio. This is mainly due to the improved transconductance of DG-MOSFETs with high-k gate oxides, which is shown in Figure 6.7. The effects

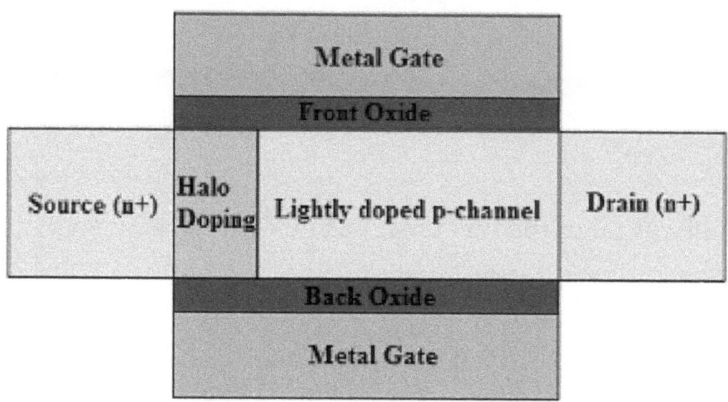

FIGURE 6.5
Channel-engineered DG-MOSFET.

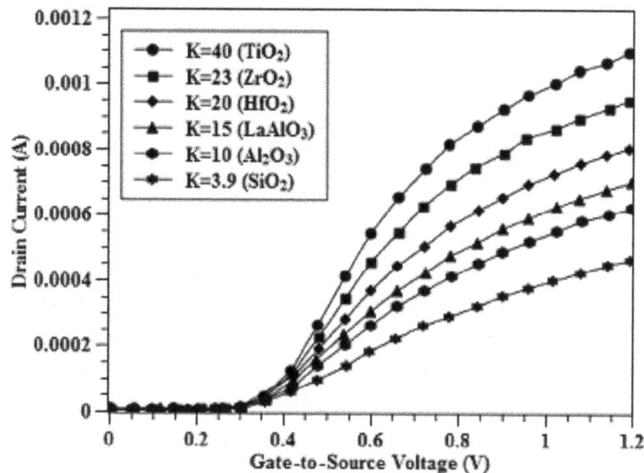

FIGURE 6.6
Effects of high-k dielectrics on the performance of DG-MOSFET.

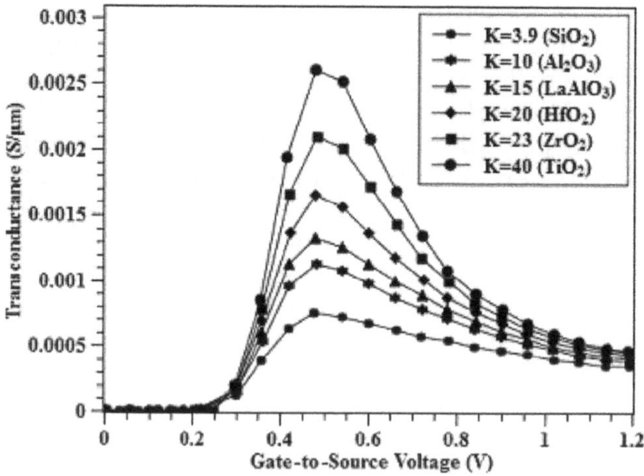

FIGURE 6.7
Effects of high-k dielectrics on transconductance of DG-MOSFET.

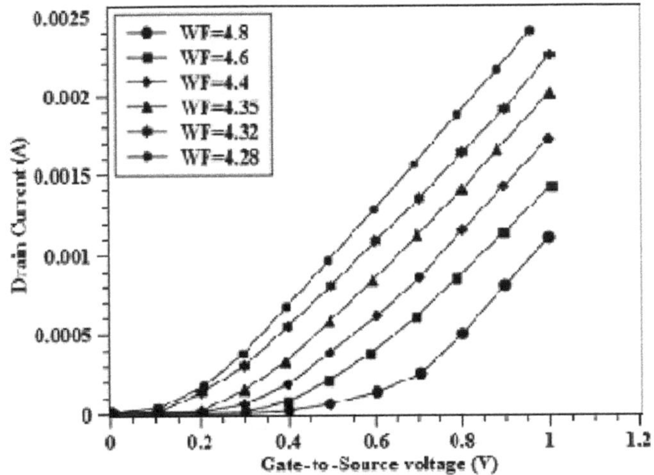

FIGURE 6.8
Effects of work function (WF in eV) on I_D V_S V_{GS} characteristics of DG-MOSFET.

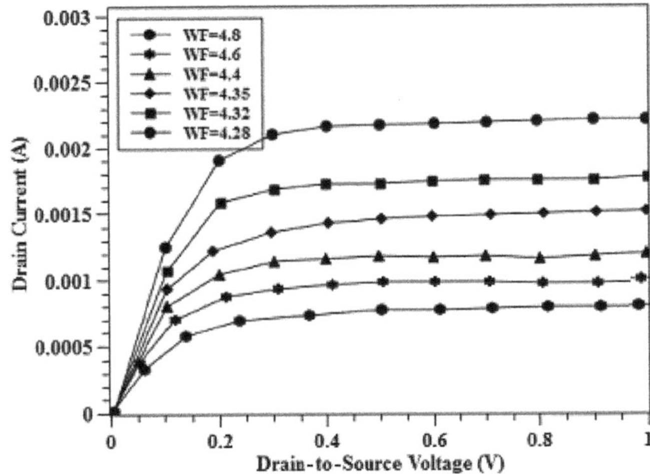

FIGURE 6.9
Effects of (WF in eV) on I_D V_S V_{DS} characteristics of DG-MOSFET.

of work function on I_D *Vs* V_{GS} and I_D *Vs* V_{DS} characteristics of 45 nm DG-MOSFET are shown in Figures 6.8 and 6.9, respectively. The metal gates with lower work function improve the drain current of DG-MOSFETs.

References

1. Y. Taur, D. A. Buchanan, W. Chen, D. J. Frank, K. E. Ismail, S. H. Lo, G. A. Halasz et al., "CMOS scaling into the nanometer regime," *Proc. IEEE*, vol. 85, pp. 486–503, 1997.
2. K. Kim and J. G. Fossum, "Double-gate CMOS: Symmetrical-versus asymmetrical-gate devices," *IEEE Trans. Electron Devices*, vol. 48, pp. 294–299, 2001.

3. M. Ieong, H. S. Wong, Y. Taur, P. Lodiges, and D. J. Frank, "DC and AC performance analysis of 25 nm symmetric/asymmetric double-gate, back-gate and bulk CMOS," in *Proceedings of the IEEE Simulation of Semiconductor Processes and Devices Conference*, Seattle, WA, 2000, pp. 147–150.
4. D. J. Frank, S. E. Laux, and M. V. Fischetti, "Monte Carlo simulation of a 30 nm dual-gate MOSFET: How short can Si go?" in *Proceedings of the IEDM Technology Digest*, San Francisco, CA, December 1992, pp. 553–556.
5. F. G. Pikus and K. K. Likharev, "Nanoscale field-effect transistors: An ultimate size analysis," *Appl. Phys. Lett.*, vol. 71, no. 25, pp. 3661–3663, 1997.
6. Z. Ren, R. Venugopal, S. Datta, M. Lundstrom, D. Jovanovic, and J. G. Fossum, "The ballistic nanotransistor: A simulation study," in *IEDM Technology Digest*, December 2000, pp. 715–718.
7. M. Ieong, H. S. P. Wong, E. Nowak, J. Kedzierski, and E. C. Jones, "High performance double-gate device technology challenges and opportunities," *Proceedings of the International Symposium on Quality Electronic Design*, 2002, pp. 492–495.
8. H. K. Jung and S. Dimitrijev, "Analysis of subthreshold carrier transport for ultimate DGMOSFET," *IEEE Trans. Electron Devices*, vol. 53, no. 4, pp. 685–691, 2006.
9. H.-S. P. Wong, K. K. Chan, and Y. Taur, "Self-align (top and bottom) double-gate MOSFET with a 25 nm thick silicon channel," in *IEDM Technology Digest*, 1997, pp. 427–430.
10. S. Tang, L. Chang, N. Lindert, Y.-K. Choi, W.-C. Lee, X. Huang, V. Subramanian, J. Bokor, T.-J. King, and C. Hu, "FinFET: A quasi planar double-gate MOSFET," in *IEEE International Solid-State Circuits Conference, Digest of Technical Papers*, 2001, pp. 118–119.
11. Q. Chen, K. A. Bowman, E. M. Harrell, and J. D. Meindl, "Double jeopardy in the nanoscale court," *IEEE Circuits Devices Mag.*, vol. 19, no. 1, pp. 28–34, 2003.
12. J. P. Colinge, "Multiple-gate SOI MOSFETs," *Solid State Electron.*, vol. 48, no. 6, pp. 897–905, 2004.
13. D. Jiménez, B. Íñiguez, J. Suñé, L. F. Marsal, J. Pallarés, J. Roig, and D. Flores, "Continuous analytic I–V model for surrounding-gate MOSFETs," *IEEE Electron Device Lett.*, vol. 25, no. 8, pp. 571–573, 2004.
14. A. Ortiz-Conde, F. J. García-Sánchez, J. Muci, S. Malobabic, and J. J. Liou, "A review of core compact models for undoped double-gate SOI MOSFETs," *IEEE Trans. Electron Devices*, vol. 54, no. 1, p. 131, 2007.
15. H. Lu, W.-Y. Lu, and Y. Taur, "Effect of body doping on double-gate MOSFET characteristics," *Semicond. Sci. Technol.*, vol. 23, no. 1, 015006–015011, 2008.
16. Y. Taur, C. H. Wann, and D. J. Frank, "25 nm CMOS design considerations," in *International Electron Devices Meeting Technical Digest*, 1998, pp. 789–792.
17. L. Chang, S. Tang, T.-J. King, J. Bokor, and C. Hu, "Gate-length scaling and threshold voltage control of double-gate MOSFETs," in *International Electron Devices Meeting Technical Digest*, 2000, pp. 719–722.
18. F. Balestra, S. Cristoloveanu, M. Benachir, J. Brini, and T. Elewa, "Double-gate silicon-on-insulator transistor with volume inversion: A new device with greatly enhanced performance," *IEEE Electron Device Lett.*, vol. EDL-8, p. 410, 1987.
19. D. J. Frank, R. H. Dennard, E. Nowak, P. M. Solomon, Y. Taur, and H.-S. P. Wong, "Device scaling limits of Si MOSFETs and their application dependencies," *Proc. IEEE*, vol. 89, no. 3, pp. 259–288, 2001.
20. B. Doyle, R. Arghavani, D. Barlage, S. Datta, M. Doczy, J. Kavalieros, A. Murthy, and R. Chau, "Transistor elements for 30 nm physical gate lengths and beyond," *Intel Technol. J.*, vol. 6, no. 2, pp. 42–54, 2002.
21. J. G. Fossum, V. P. Trivedi, and K. Wu, "Extremely scaled fully depleted SOI CMOS," in *Proceedings of the IEEE International SOI Conference*, October 2002, pp. 135–136.
22. J.-P. Colinge, "Novel gate concepts for MOS devices," in *Proceedings of the 34th ESSDERC*, September 2004, pp. 45–49.
23. Z. J. Lemnios, D. J. Radack, and J. C. Zolper, "The future of silicon-on insulator (SOI) technology in microelectronic systems," in *Proceedings of the International SOI Conference*, October 2004, pp. 9–13.

24. S. Cristoloveanu and V. Ferlet-Cavrois, "Introduction to SOI MOSFETs: Context, radiation effects, and future trends," *Int. J. High Speed Electron. Syst.*, vol. 14, no. 2, pp. 465–487, 2004.
25. G. G. Shahidi, "SOI technology for the GHz era," *IBM J. Res. Develop.*, vol. 46, no. 2/3, pp. 121–131, 2002. http://researchweb.watson.ibm.com/journal/.
26. S. B. Park, Y. W. Kim, Y. G. Ko, K. I. Kim, I. K. Kim, H.-S. Kang, J. O. Yu, and K. P. Suh, "A 0.25-μm, 600-MHz, 1.5-V, fully depleted SOI CMOS 64-bit microprocessor," *IEEE J. Solid-State Circuits*, vol. 34, no. 11, pp. 1436–1445, 1999.
27. J. G. Fiorenza and J. A. del Alamo, "Experimental comparison of RF power LDMOSFETs on thin-film SOI and bulk silicon," *IEEE Trans. Electron Devices*, vol. 49, no. 4, pp. 687–692, 2002.
28. T. Sekigawa and Y. Hayashi, "Calculated threshold-voltage characteristics of an XMOS transistor having an additional bottom gate," *Solid-State Electron.*, vol. 27, p. 827, 1984.
29. T. Tanaka, K. Suzuki, H. Horie, and T. Sugii, "Ultrafast operation of V_{th}-adjusted p^+ -n^+ double-gate SOI MOSFETs," *IEEE Electron Device Lett.*, vol. 15, pp. 386–388, 1994.
30. J. P. Colinge, M. H. Gao, A. R. Romano, H. Maes, and C. Claeys, "Silicon-on-insulator gate-all-around device," in *IEDM Technical Digest*, San Francisco, CA, 1990, pp. 595–598.
31. J. H. Lee, G. Taraschi, A. Wei, T. A. Langdo, E. A. Fitzgerald, and D. A. Antoniadis, "Super self-aligned double-gate (SSDG) MOSFETs utilizing oxidation rate difference and selective epitaxy," in *IEDM Technical Digest*, Washington, DC, 1999, pp. 71–74.
32. G. W. Neudeck, T.-C. Su, and J. P. Denton, "Novel silicon epitaxy for advanced MOSFET devices," in *IEDM Technical Digest*, 2000, pp. 169–173.
33. H. Liu, Z. Xiong, J. K. O. Sin, P. Xuan, and J. Bokor, "A high performance double-gate SOI MOSFET using lateral solid-phase epitaxy," in *Proceedings of the IEEE International SOI Conference*, Williamsburg, VA, 2002, pp. 28–29.
34. M. Kumar, H. Liu, J. K. O. Sin, J. K. O. Jun Wan, and K. L. Wang, "A 3-DBiCMOS technology using selective epitaxial growth (SEG) and lateral solid phase epitaxy (LSPE)," in *IEDM Technical Digest*, 2001, pp. 729–732.
35. H. Liu, Z. Xiong, and J. K. O. Sin, "Implementation and characterization of the double-gate MOSFET using lateral solid-phase epitaxy," *IEEE Trans. Electron Devices*, vol. 50, no. 6, pp. 1552–1554, 2003.
36. J. H. Oh, C. J. Kim, and H. Ishiwara, "Enhanced growth mechanism in lateral solid phase epitaxy of Si films simultaneously doped with P and Ge atoms," *Jpn. J. Appl. Phys.*, vol. 35, no. 3, pp. 1605–1610, 1995.
37. Y. C. Yeo, V. Subranmanian, J. Kedzierski, P. Xuan, T. J. King, J. Bokor, and C. Hu, "Nanosacle ultra-thin-body silicon-on-insulator P-MOSFET with a SiGe/Si heterostructure channel," *IEEE Electron Device Lett.*, vol. 21, pp. 161–163, 2000.
38. W. Long, H. Ou, J.-M. Kuo, and K. K. Chin, "Dual-material gate (DMG) field effect transistor," *IEEE Trans. Electron Devices*, vol. 46, no. 5, pp. 865–870, 1999.
39. X. Zhou and W. Long, "A novel hetero-material gate (HMG) MOSFET for deep-submicron ULSI technology," *IEEE Trans. Electron Devices*, vol. 45, no. 12, pp. 2546–2548, 1998.
40. A. Chaudhry and M. J. Kumar, "Controlling short-channel effects in deep submicron SOI MOSFET's for improved reliability: A review," *IEEE Trans. Device Mater. Rel.*, vol. 4, no. 1, pp. 99–109, 2004.
41. M. J. Kumar and A. Chaudhry, "Two-dimensional analytical modeling of fully depleted dual-material gate (DMG) SOI MOSFET and evidence for diminished short-channel effects," *IEEE Trans. Electron Devices*, vol. 15, no. 4, pp. 569–574, 2004.
42. U. K. Mishra, A. S. Brown, and S. E. Rosenbaum, "DC and RF performance of 0.1 μm gate length $In_{0.52}Al_{0.48}As{=}Ga_{0.47}In_{0.53}As$ pseudomorphic HEMT," in *International Electron Devices Meeting Technical Digest*, 1988, pp. 180–183.
43. G. Venkateshwar Reddy, "A new dual-material double-gate (DMDG) nanoscale SOI MOSFET—two-dimensional analytical modeling and simulation," *IEEE Trans. Nanotechnol.*, vol. 4, no. 2, pp. 260–268, 2005.
44. K. Suzuki and T. Sugii, "Analytical models for n^+ -p^+ double gate SOI MOSFET's," *IEEE Trans. Electron Devices*, vol. 42, no. 11, pp. 1940–1948, 1995.

45. D. Nirmal, P. Vijayakumar, K. Shruti, and N. Mohankumar, "Nanoscale channel engineered double gate MOSFET for mixed signal applications using high-k dielectric," *Int. J. Circ. Theor. Appl.*, vol. 41, pp. 608–618, 2013.
46. D. Nirmal and P. VijayaKumar, "Fin field effect transistors performance in analog and RF for high-k dielectrics," *Def. Sci. J.*, vol. 61, no. 3, pp. 235–240, 2011.
47. M. H. Chowdhury, M. A. Mannan, and S. A. Mahmood, "High-k dielectrics for submicron MOSFET," *Int. J. Emerg. Technol. Sci. Eng.*, vol. 2, no. 2, pp. 1–12, 2010.
48. D. Nirmal, K. Shruti, D. M. Thomas, P. C. Samuel, P. Vijaya Kumar, and N. Mohan Kumar, "Impact of channel engineering on FinFETs using high-k dielectrics," *Int. J. Micro Nano Electron, CircSyst*, vol. 3, no. 1, pp. 7–11, 2011.
49. N. Mohankumar, S. Binit, and S. C. Kumar, "Influenze of channel and gate engineering on the analog and RF performance of DG MOSFETs," *IEEE Trans. Electron Devices*, vol. 57, no. 4, pp. 820–826, 2010.
50. N. Mohankumar, S. Binit, and C. K. Sarkar, "Investigation of novel attributes of single halo dual-material double gate MOSFETs for analog/RF applications," *Microelectron. Rel.*, vol. 49, pp. 1491–1497, 2009.
51. R. G. Venkateshwar and K. M. Jagadesh, "Investigation of the novel attributes of a single-halo double gate SOIMOSFET: 2D simulation study," *Microelectron. J.*, vol. 35, pp. 761–765, 2004.
52. N. Mohankumar, S. Binit, and C. K. Sarkar, "Performance and optimization of dual material gate (DMG) short channel BULK MOSFETs for analog/mixed signal applications," *Int. J. Electron.*, vol. 96, no. 6, pp. 603–611, 2009.
53. H. Daewon, T. Hideki, C. Yang-Kyu, and K. Tsu-Jae, "Molybdenum gate technology for ultrathin-body MOSFETs and FinFETs," *IEEE Trans. Electron Devices*, vol. 51, no. 12, pp. 1989–1996, 2004.

Section III

Modeling of Tunnel FETs

7

TFETs for Analog Applications

Marcio Dalla Valle Martino, Paula Ghedini Der Agopian, Joao Antonio Martino, Eddy Simoen, and Cor Claeys

CONTENTS

7.1 TFETs: Operation Principle and Basic Characteristics

The drift-diffusion conduction mechanism governing transport in Metal-Oxide-Semiconductor Field-Effect Transistors (MOSFETs) physically limits the minimum sub-threshold swing (SS) to 60 mV/dec at room temperature, which is hampering low supply voltage operation without increasing exponentially the off-current (Figure 7.1a). Tunnel-FET (TFET) devices present a new class of devices with the conduction mechanism dominantly based on Band-to-Band Tunneling (BTBT) current, which can result in a SS lower

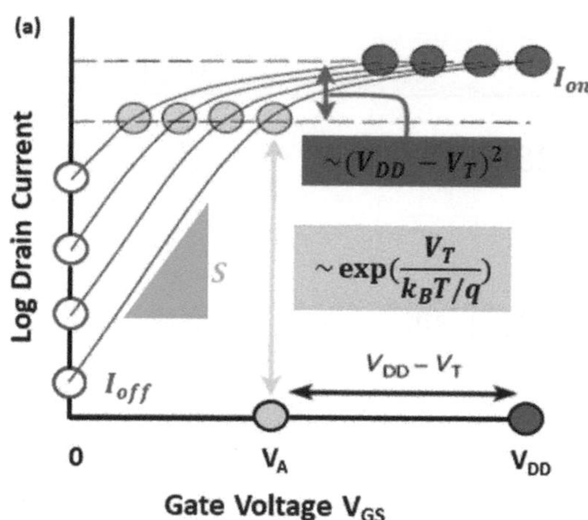

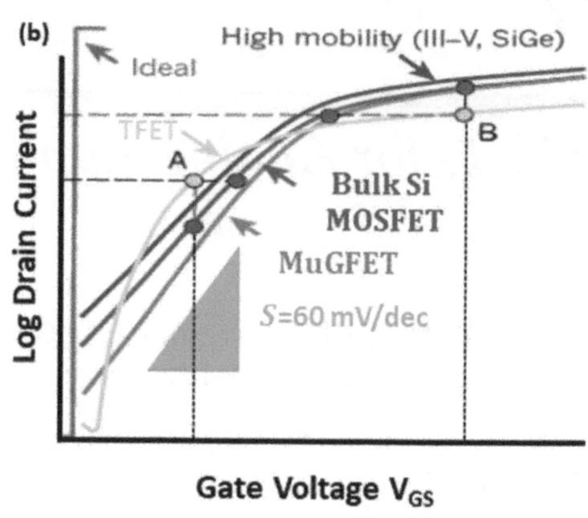

FIGURE 7.1
(a) MOSFET drain current as a function of gate voltage for different V_T; (b) Drain current as a function of gate voltage for different device architectures, i.e., bulk Si MOSFET, Multiple Gate FET (MuGFET), and TFET. (After Ionescu, A. M., and Riel, H., *Nature*, 479, 329–337, 2011.)

than 60 mV/dec. This means that it is possible to adopt a lower supply voltage (which requires a low threshold voltage V_T in order to keep the performance) without increasing the leakage or off-state current (I_{OFF}) approaching the ideal switch behavior, compared to the other device technologies (Figure 7.1b).

A TFET cross section and its band diagram at off-state and on-state conditions are shown in Figure 7.2. The Si- and Ge-source TFETs are in the off-state (Figure 7.2b) when the PIN diode (Source/Channel/Drain) is reverse biased and the gate voltage (V_{GS}) is zero. When V_{GS} increases, the channel band diagram shifts down, and when the source valence band becomes higher than the channel conduction band, a BTBT current starts and the TFET is turned to the on-state (Figure 7.2c). The choice of the source material is crucial for the magnitude of the tunneling current. For Si TFET devices I_{ON} is much lower than for a MOSFET with the same dimensions. In order to increase I_{ON}, it is necessary to use a source material with a lower bandgap (i.e., SiGe, Ge, or III-V materials) for improving the tunneling current. Figure 7.2b and c, illustrating the band diagram for a Si and Ge source, clearly point out that for the Ge source, the tunneling path is higher than for the Si one.

If V_{GS} is negative enough, it is possible to have an undesirable tunneling current between drain and channel, called the ambipolar current. In order to avoid the ambipolar current, the gate-to-drain cannot be self-aligned as shown in Figure 7.2a. Simulations point out that reduction of the total gate length with 20 nm diminishes the off-current with three to five orders of magnitude [2]. Another important remark is that in spite of the desirable predominant BTBT current between source and channel, other conduction mechanisms can also take place in TFETs for lower V_{GS}, such as Trap-Assisted Tunneling (TAT) and Shockley-Read-Hall (SRH) generation.

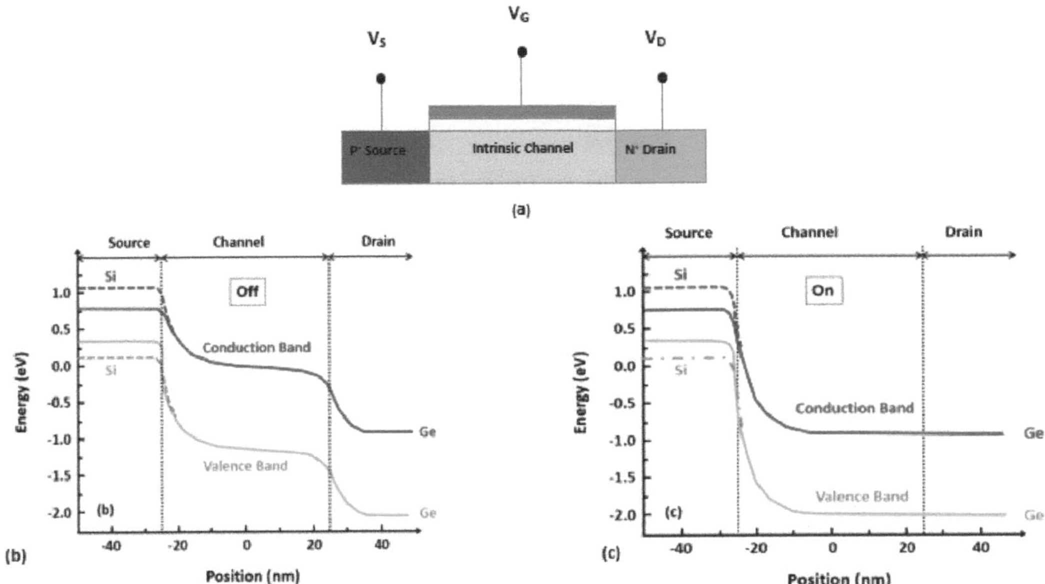

FIGURE 7.2
TFET cross section (a) and band-diagram for TFET in off-state (b) and in on-state (c) for silicon (dashed lines) and germanium (full lines) source materials.

Figure 7.3 shows a nanowire TFET (with gate-to-drain underlap and a small gate to source overlap – Figure 7.3a) for different V_{GS} conditions, showing the drain current and the band diagram for three different predominant conduction mechanisms: SRH, TAT, and BTBT in Figure 7.3b–g, respectively [3]. J_{SRH}, J_{TAT}, and J_t are the current density due to SRH, TAT, and BTBT, respectively. In Figure 7.3, k_B is Boltzmann's constant, T the absolute temperature, and ΔE_t the energy difference of the SRH generation level with respect to the intrinsic level in the semiconductor. One can see that based on the first-order model, SRH and TAT currents are thermally assisted and the relative influence of the temperature on it is much higher than for BTBT, which depends on the small narrowing of the bandgap (E_g) with higher temperature. Figure 7.4a shows the drain current as a function of gate voltage from 100 K to 400 K, where it is possible to see that for higher V_{GS} the I_{DS} changes due to increasing temperature are relatively lower (when the BTBT is the predominant current) than for low V_{GS}, where TAT and SRH are dominant. The dominant conduction mechanism for a determined bias condition can be obtained through the analysis of the activation energy (E_A), which has to be lower than 0.1 eV (for silicon devices) in order to have BTBT dominant (Figure 7.4b).

In spite of the silicon, TFETs do not present a SS lower than 60 mV/dec as expected (since the TAT contribution is not negligible and it degrades the subthreshold region, and I_{ON} is low compared to the MOSFET/FinFET on-current); it is observed that the output conductance is much better (lower) compared to the case of a MOSFET. As a result, although the digital behavior is still not good enough, the main analog parameters such as the early voltage and intrinsic voltage gain are improved compared to MOSFET/FinFET devices, as discussed in the next section.

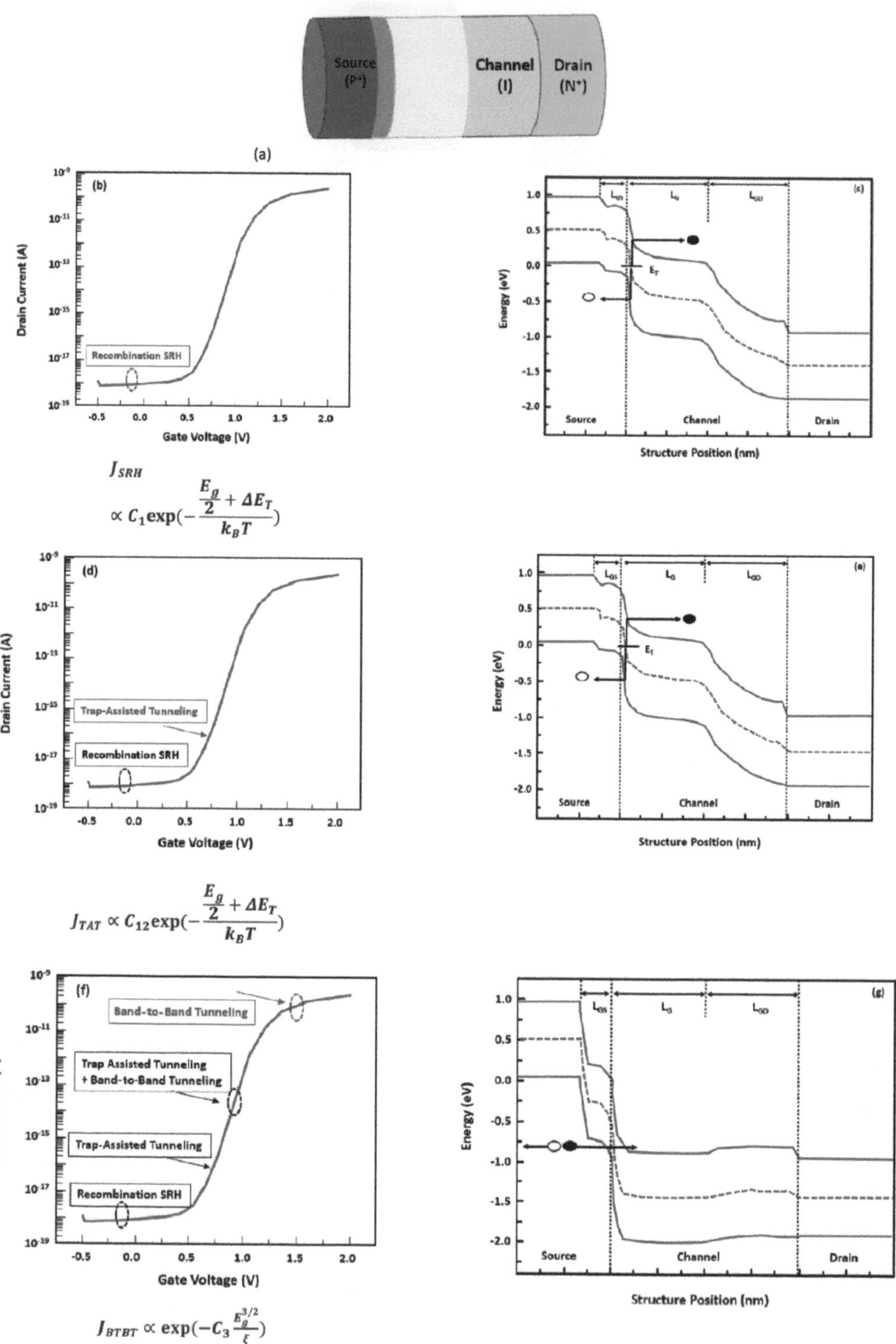

FIGURE 7.3
(a) Nanowire TFET cross section; drain current and band-diagram at SRH (b and c), TAT (d and e) and BTBT (f and g) conduction mechanisms. (After Martino, M.D.V. et al., *J. Integr. Circuits Syst.*, 8, 110–115, 2013.)

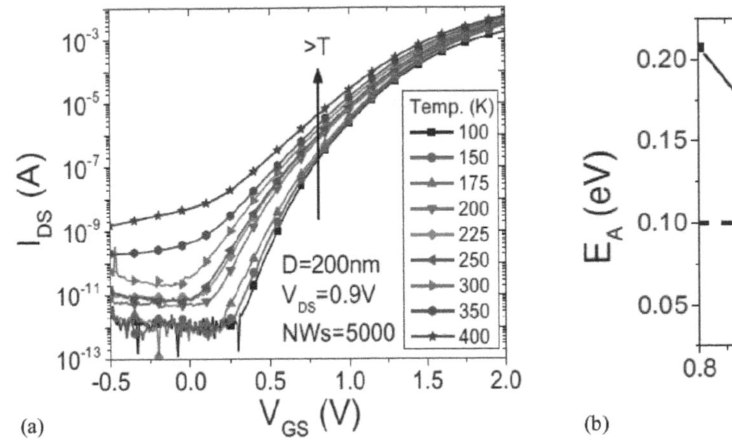

(a)

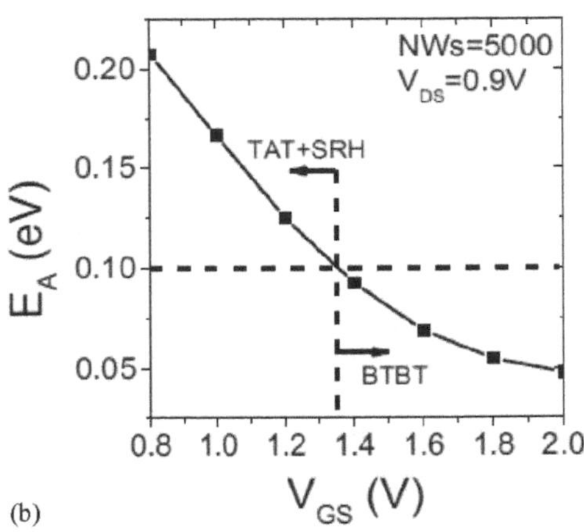

(b)

FIGURE 7.4
(a) Drain current as a function of the gate voltage for different temperatures; (b) Activation energy as a function of the gate voltage for V_{DS} = 0.9 V. (After Neves, F.S. et al., Transconductance hump in vertical gate-all-around tunnel-FETs, in *IEEE SOI-3D-Subthreshold Microelectronics Technology Unified Conference* (S3S), IEEE Explore, pp. 1–3, 2015.)

7.2 Analog Operation of TFETs

Although the tunnel-FET devices were developed for digital applications, some researchers have demonstrated their potential for analog circuit design. In order to discuss this topic, two important figures of merit for analog application were chosen: the intrinsic voltage gain and the unit gain frequency. The intrinsic voltage gain (A_V) depends on the good control of the drain current by the gate voltage and the low influence of the drain voltage on the output characteristics. The A_V can be obtained by:

$$|A_V| = \frac{g_m}{g_d} = \frac{g_m}{I_{DS}} \cdot V_{EA} \tag{7.1}$$

where I_{DS} is the drain current, g_m is the transconductance, g_d is the output conductance, and V_{EA} is the early voltage.

The unit gain frequency is an important parameter to define the operating frequency limit and can be represented by

$$f_T = \frac{g_m}{2\pi C_{gg}} \tag{7.2}$$

where C_{gg} is the total gate capacitance.

The analog performance of TFETs can be evaluated by investigating the A_V and f_T parameters at room and high temperatures. Comparing simulations, performed to better understand the obtained results, with experimental data is a very useful approach. Data has been reported on the analog performance of full silicon devices, silicon-based nanowire TFETs with different source compositions ($Si_{1-x}Ge_x$) and line-TFETs. Interesting results are also published on the performance of the latest generation of devices, fabricated with III-V materials.

7.2.1 Analog Performance of Full Silicon TFETs

Considering that FinFETs are one of the adopted structures in the semiconductor industry for the current technological nodes, the first analog evaluation will be performed by the comparison between FinFETs and TFETs fabricated in the same Silicon-on-Insulator (SOI) wafer, with the same process conditions, changing only the source implantation. A schematic structure is shown in Figure 7.5. Technological details about the devices fabrication process can be found in [5,6].

Figure 7.6 presents the transconductance (g_m) and output conductance (g_d) (a) and the intrinsic voltage gain (b) for both TFETs and FinFETs as a function of temperature. An evaluation of g_m for FinFETs for different dimensions reveals that the shorter and wider device presents a higher transconductance, as expected, due to the well-known

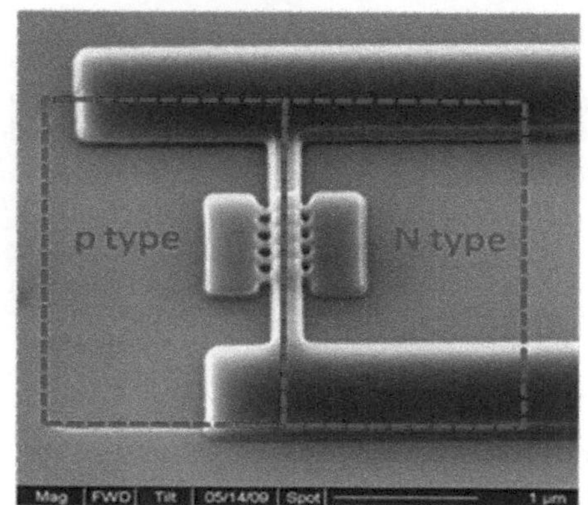

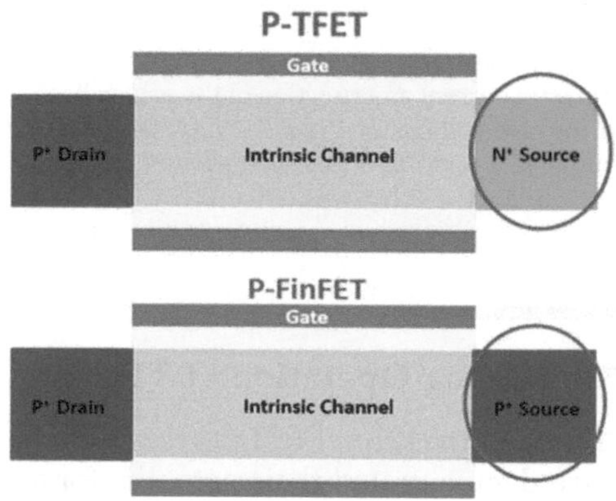

FIGURE 7.5
Scanning Electron Microscope (SEM) image of a FinFET structure with a TFET doping schema and TFET and FinFET 2D structures, highlighting the different doping in the source part.

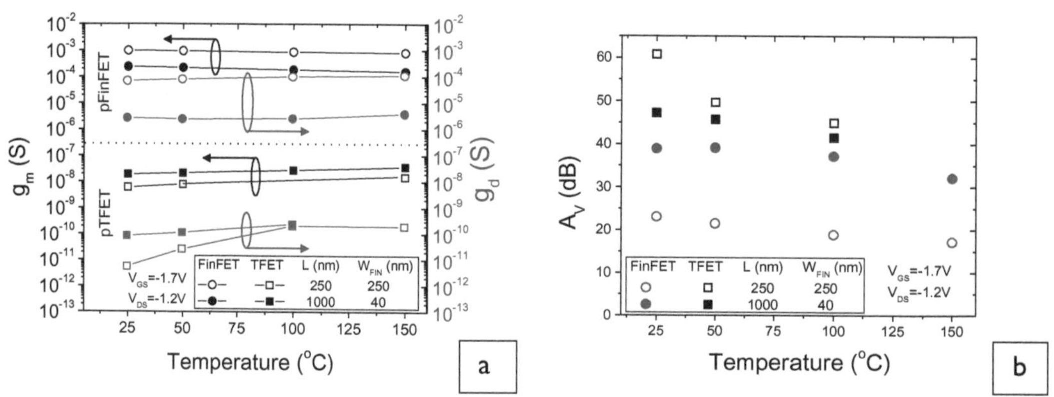

FIGURE 7.6
Experimental transconductance and output conductance (a) and the intrinsic voltage gain (b), for both TFET and FinFET transistors as a function of temperature. (After Agopian, P.G.D. et al., *IEEE Trans. Electron Devices*, 60, 2493–2497, 2013.)

W/L dependence. However, the small leakage current at the second interface degrades the g_m for shorter FinFETs, which appears due to the wide fin (weak coupling between lateral gates). As we focus on TFETs, we see that the better transconductance was obtained for narrow devices. Since for point tunneling transistors the influence of the channel length is negligible, only the fin width plays a role; therefore, for narrow devices the good coupling between the side gates ensures that the electric field is stronger along the entire source/channel junction and the tunneling current is enhanced. When we compare both technologies, pFinFETs present a much higher transconductance (more than four orders of magnitude) than pTFETs [7]. The low g_m obtained for TFETs is a consequence of the wide silicon bandgap; that results in a low on-state current when compared with its MOSFET counterpart. Beside all these analyses, the opposite trend of g_m with the temperature increase also has to be taken into account. While for FinFETs the g_m is slightly reduced due to the mobility degradation at higher T, for TFETs a g_m improvement is observed, since the tunneling mechanism increases with temperature caused by bandgap narrowing.

When the output conductance is evaluated, better (lower) g_d values are obtained for pTFET devices, and the difference of g_d between both technologies is at least four orders of magnitude. The very low g_d obtained for TFETs can be explained by the weak dependence of the tunneling current on the drain voltage. When we analyze the temperature impact on g_d, it is possible to observe an unexpected slight increase in g_d values for the FinFET with equal channel length and width. The output conductance should decrease with increasing temperature due to reduced mobility, suffering a slight increase due to self-heating as reported in [7]. However, for pTFET devices, the bandgap is reduced with the temperature increase, which results in an enhancement of the tunneling current degrading (increasing) the g_d values.

As we analyze the experimental intrinsic voltage gain for different temperature values (Figure 7.6b), it is possible to observe that for pTFETs, independent on the device dimensions, the obtained A_V is higher than for pFinFETs. Although the FinFETs have a higher transconductance than the tunneling transistors, the output conductance obtained by the pTFETs is much lower (better) than for FinFETs. This excellent g_d behavior becomes predominant and results in a better intrinsic voltage gain.

The unit gain frequency (f_T) strongly depends on g_m and the total gate capacitance (C_{gg}) (Equation 7.2). The C_{gg} is also dependent on the applied bias as reported in [7]. When we consider the same bias condition for which the transconductance and intrinsic voltage gain were extracted ($V_{GS} = -1.7$ V and $V_{DS} = -1.2$ V), the total gate capacitance was estimated by simulation for the temperature range from 25°C to 150°C, as can be seen in the inset of Figure 7.7. Since the total gate capacitance for TFETs, when operating in the on-state, depends only on the gate-to-drain capacitance (C_{gd}) [8], C_{gg} for TFETs is approximately two times lower than for a pFinFET at the studied bias conditions. However, considering the transconductance impact on f_T, since g_m varies over four orders of magnitude when both technologies are compared, it becomes the predominant factor and results in a higher f_T for pFinFETs for all temperature ranges.

Considering the good intrinsic voltage gain, full silicon TFETs thus show good potential for analog applications if a high unit gain frequency is not required. However, the problem observed for low f_T, is the low transconductance that results from the low I_{ON}. Some strategies have been adopted to improve I_{ON}, as an example, the use of materials with a smaller bandgap at the source (heterojunction devices).

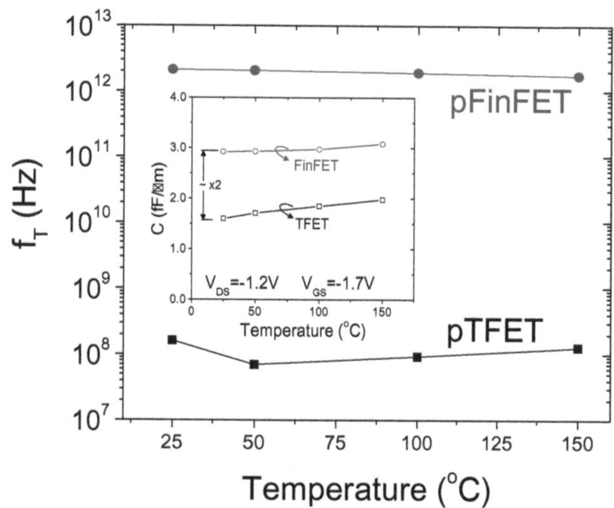

FIGURE 7.7
Experimental unit gain frequency for both pFinFETs and pTFETs as a function of temperature. The inset shows the calculated gate-to-drain capacitance as a function of T at $V_{DS} = -1.2$ V and $V_{GS} = -1.7$ V. (After Agopian, P.G.D. et al., *IEEE Trans. Electron Devices*, 60, 2493–2497, 2013.)

7.2.2 Analog Performance of $Si_{1-x}Ge_x$/Si Source to Channel Junction TFETs

With an aim to reach a higher on-state current, some important changes were considered in heterojunction devices. The first modification is the use of a gate-all-around (GAA) structure, which ensures a better coupling between gate and channel, increasing the tunneling rate. The second variation is to increase the germanium (Ge) amount in the source composition of GAA-TFETs. The third one is a change of the source geometry, resulting in a new type of tunneling device called line-TFETs. The structures of GAA (a) and line-TFET (b) are presented in Figure 7.8.

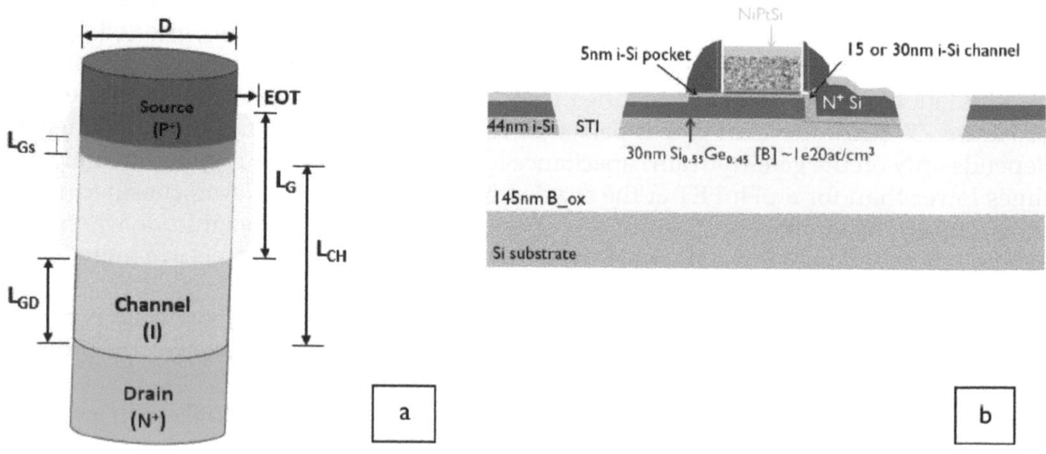

FIGURE 7.8
Gate-all-around structure (a) and a schematic cross section of a Line-TFET (b).

Although the GAA structure increases the tunneling rate, the improvement of I_{ON} of full silicon GAA transistors is not enough to achieve similar levels to those of the drain current supplied by a MOSFET. In order to obtain higher I_{ON} and higher g_m, the source material was changed to a SiGe alloy or to pure germanium.

Since germanium has a lower bandgap than silicon, the higher the germanium concentration in the source material, the higher the BTBT rate, which results in an increase of I_{ON} without degrading the off-state current (I_{OFF}). Besides the I_{ON} enhancement, it is still possible to observe in Figure 7.9a that by increasing I_{ON}, the switch velocity is also improved. The higher BTBT influence on total drain current as the Ge concentration increases can also be seen from the activation energy (E_A) graph (Figure 7.9b). It is possible to notice that the higher the Ge concentration, the smaller is E_A, i.e., the BTBT becomes the predominant mechanism ($E_A < 0.1$ eV) for lower V_{GS}.

To explain the analog behavior of each TFET structure based on the transport mechanism, it is important to keep in mind that since the tunneling mechanism occurs at the source/channel junction, the influence of the drain electric field is smaller than that shown by MOS transistors. Besides, when the tunneling transport mechanisms are compared, TAT is quite dependent on gate voltage and almost independent of the drain bias, while BTBT is more susceptible to the drain electric field.

Considering the I_{ON} improvement promoted by inserting the germanium at the source, it is reasonable to imagine that there is an increase in transconductance values as the Ge concentration is higher. The increase of g_m, in turn, takes the TFET to operate with a higher frequency of unit gain.

However, it is not only the transconductance value that increases with the increasing germanium percentage at the source; the output conductance also increases because the lower is the source bandgap, the more dominated by the BTBT mechanism is the device behavior and, therefore, it becomes more dependent on the drain voltage. When we compare the intrinsic voltage gain values for GAA-TFETs (Figure 7.10), it is possible to notice that there is a competition between the g_m improvement and the g_d degradation. The GAA-TFET with 27% of Ge at the source shows a better A_V because the g_m was improved

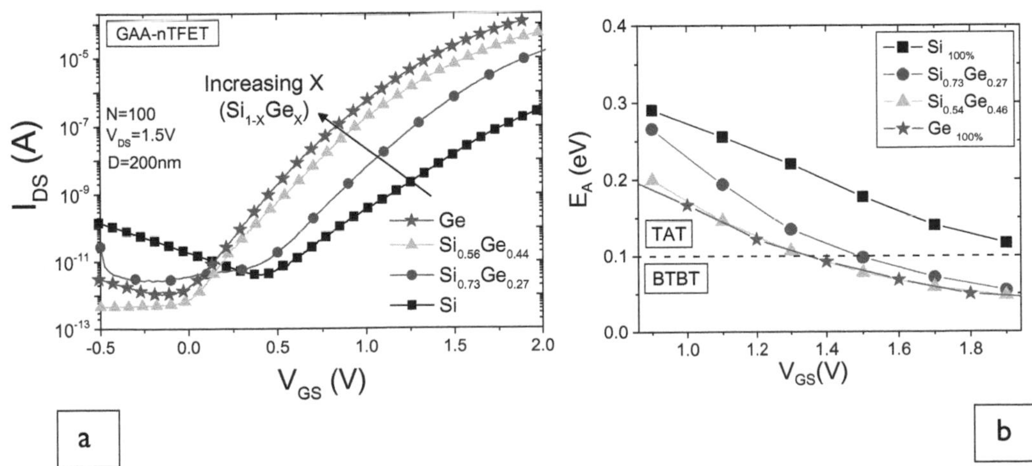

FIGURE 7.9
Experimental drain current (a) and activation energy (b) as a function of gate voltage for GAA-TFETs with different source compositions. (After Agopian, P.G.D. et al., *Solid-State Electron.*, 128, 43–47, 2017.)

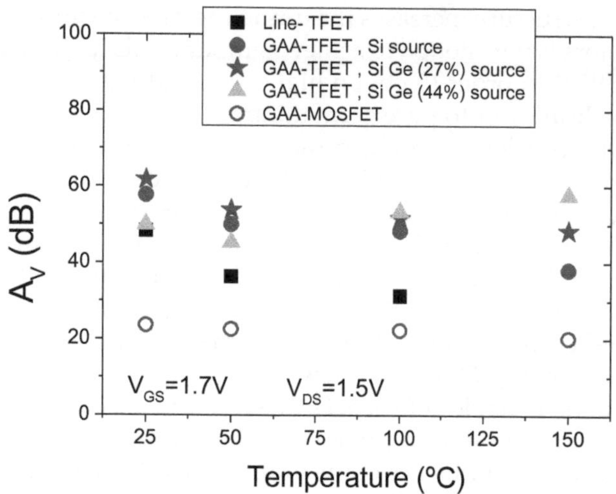

FIGURE 7.10
The A_V comparison as a function of temperature among Line-TFET, GAA-TFETs and GAA-MOSFET. (After Agopian, P.G.D. et al., Intrinsic voltage gain of line-TFETs and comparison with other TFET and MOSFET architectures, in *Joint International EUROSOI Workshop and International Conference on Ultimate Integration on Silicon* (*EUROSOI-ULIS*), IEEE Explore, pp. 13–15, 2016.)

due to the higher tunneling rate, but the BTBT mechanism is still not strong enough to degrade the g_d characteristic.

An extremely low g_d value is key for a very high A_V, and this is the reason why the full silicon device presents an intrinsic voltage gain at almost the same level as a TFET with a $Si_{0.73}Ge_{0.27}$ source.

The main variation in the line-TFET structure is the source extension under the gate and under the thin intrinsic silicon layer (channel). This new geometry allows an alignment between the tunneling and the electric field (as represented in Figure 7.11a). This alignment improves the tunneling rate, and extension of the source increases the tunneling area,

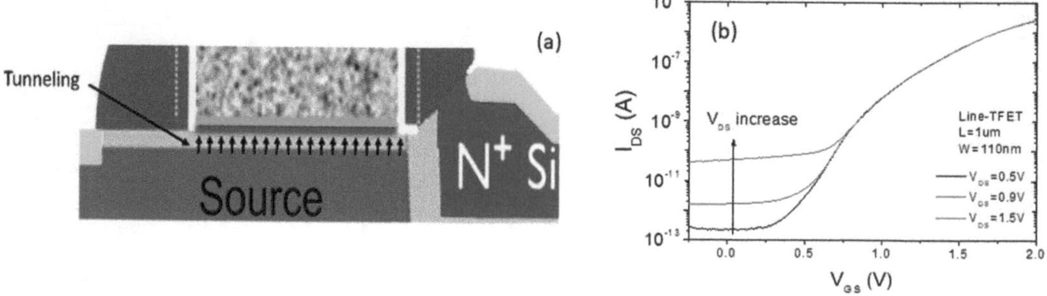

FIGURE 7.11
The schematic tunneling alignment (a) and drain current of Line-TFET as a function of gate bias for different drain voltages (b). (After Agopian, P.G.D. et al., Intrinsic voltage gain of line-TFETs and comparison with other TFET and MOSFET architectures, in *Joint International EUROSOI Workshop and International Conference on Ultimate Integration on Silicon* (*EUROSOI-ULIS*), IEEE Explore, pp. 13–15, 2016.)

improving the total drain current. The final I_{DS} for this TFET structure has the same level compared to conventional MOSFETs.

Although this I_{ON} improvement is very good for digital applications, when the A_V performance is required, the high g_d degrades A_V, resulting in the lowest A_V value among Group IV TFET devices, but it is still higher than for MOSFETs. Then, if a high I_{DS} level is required, the line-TFETs could be the best option for the application.

7.2.3 Analog Performance of III-V TFETs

III-V alloys usually have a direct bandgap compared with Si, SiGe alloys, or Ge, which offers another possibility to improve I_{ON} by using III-V TFETs. The presence of a direct gap increases the BTBT rate, which makes the transistor behavior almost independent on TAT.

The activation energy and intrinsic voltage are evaluated for $In_{0.53}Ga_{0.47}As$ TFETs, with TiN as the metal gate and a gate dielectric composed by 1 nm Al_2O_3 followed by two different gate HfO_2 thicknesses (3 and 2 nm) [11].

The activation energy of I_{ON} (Figure 7.12) drops from a 0.4 eV maximum at 0 V to a minimum plateau, independent of the gate oxide thickness. This abrupt transition to the BTBT regime confirms the small dependence of I_{ON} on TAT, which means a high switch velocity.

The III-V TFETs subthreshold slope reaches lower values than 60 mV/dec at room temperature [11], and this characteristic also contributes to a higher transistor efficiency (g_m/I_{DS}) and consequently high intrinsic voltage gain at a weak conduction regime. For strong conduction, the scenario changes a little. The strong conduction is not the optimum analog operation regime for this type of TFET, but it still reaches A_V~50 dB at room temperature (Figure 7.13), showing it to be better than MOSFET technologies.

It is important to highlight that when these III-V TFETs were analyzed at strong conduction, the gate and drain bias were equal to 0.6 V. It means that, besides these devices presenting a good performance at a weak inversion regime (useful for low-power/low-voltage applications), the necessary voltage to reach the strong conduction is not high, reducing the supply voltage for this technology.

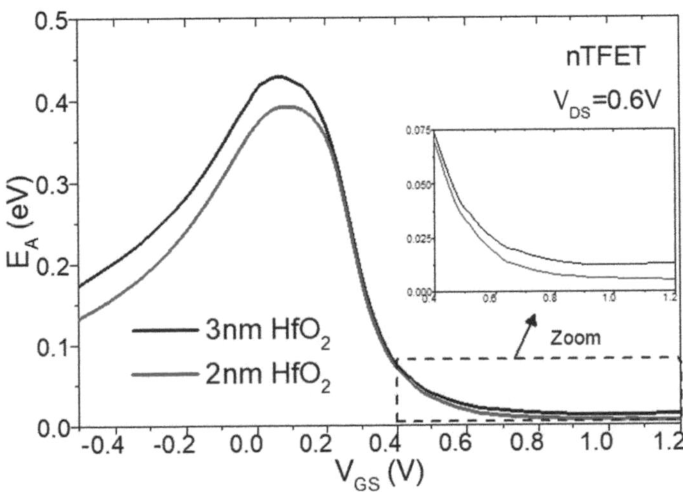

FIGURE 7.12
Activation energy as a function of V_{GS} for $In_{0.53}Ga_{0.47}As$ nTFETs, for two HfO_2 thicknesses, with a $V_{DS} = 0.6$ V.

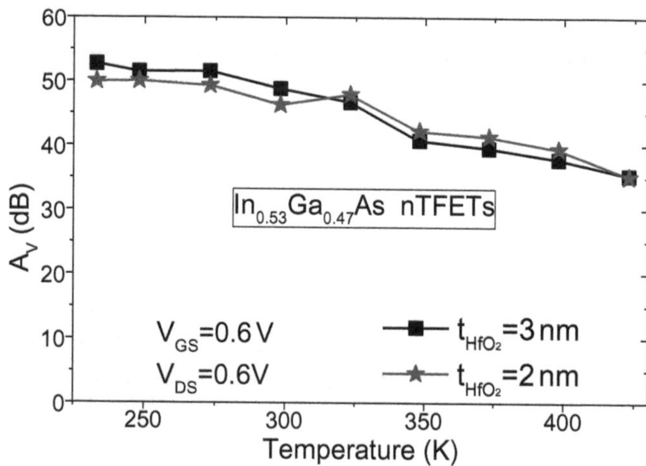

FIGURE 7.13
Experimental intrinsic voltage gain as a function of temperature for $In_{0.53}Ga_{0.47}As$ nTFETs, with a $V_{GS} = V_{DS} = 0.6$ V.

7.3 Low-Frequency Noise and Random Telegraph Signals

Low-frequency (LF) noise is an important parameter for the analog performance of solid-state devices and circuits. It sets a lower limit to the lowest detectable signal with respect to background fluctuations in the current. In addition, LF noise can be upconverted to high frequencies in the Radio Frequency (RF) domain by the device non-linearities, giving rise to phase noise in oscillator circuits. In this section, LF noise in TFETs will be outlined both from an experimental and a modeling viewpoint. First, the basic types of LF noise and their origin in semiconductor devices will be discussed. In a second part, the experimental LF noise behavior in TFETs, including RTS, will be described and compared with MOSFETs. This has led to dedicated analytical models. Finally, simulation results of the LF noise behavior of TFETs and simple TFET-based circuits will be summarized.

7.3.1 Noise Types and Origin

A general LF noise spectrum of a semiconductor device can be represented as follows [12]:

$$S_I = K_W + \frac{K_f}{f^\gamma} + \sum_{i=1}^{N} \frac{A_i}{1+(f/f_{0i})^2} \qquad \text{(in A}^2\text{/Hz)} \tag{7.3}$$

In Equation (7.3) S_I is the current noise power spectral density (PSD); K_W is the PSD of the white (frequency-independent) noise; K_f is the amplitude factor of the so-called $1/f$ or flicker noise with a frequency exponent γ close to 1 (f the frequency), and A_i and f_{0i} is the amplitude and the corner or characteristic frequency of Generation-Recombination (GR) noise, corresponding with a Lorentzian spectrum. Each of the noise types in Equation (7.3) can have different origins. For example, white noise has a fundamental origin, which is related to the transport in the device or across a junction (potential barrier). Thermal noise is driven by fluctuations in the thermal energy k_BT of the carriers, and for a resistor type of device (resistance R) the spectral density is [13]

$$S_I = 4k_BT/R \tag{7.4}$$

In case of a rectifying type of device (a Schottky barrier or a p-n junction), the white noise will be dominated by so-called shot noise, given by [13]

$$S_I = 2qI_D \tag{7.5}$$

for fully uncorrelated transitions (=full shot noise). For scaled transistors, thermal and shot noise will only dominate at rather high f, starting from a few hundred kHz to 1 MHz, up to the GHz range.

In the LF range from ~1 Hz to 100 kHz, non-fundamental excess noise determined by defects in the material or at the interfaces will govern the current fluctuations. Two types of $1/f$ noise sources have been proposed over the years, yielding two different models: mobility fluctuations often abbreviated as $\Delta\mu$ and number fluctuations also termed Δn. While the exact fundamental origin of $\Delta\mu$ noise is still a matter of debate, it has been described in terms of an empirical model, given by [14]

$$\frac{S_I}{I_D^2} = \frac{\alpha_H}{Nf} \tag{7.6}$$

with N the total number of fluctuators in the device, i.e., the number of carriers in a transistor channel and α_H the so-called Hooge parameter, which can depend on the bias conditions and has to be considered as a quality indicator for the noise of a device. A defective device has high values ($>10^{-3}$) while a good-quality device has a low α_H ($<<10^{-3}$). Usually, the α_H value is inversely correlated with the transport mobility.

While standard polysilicon/SiO_2 pMOSFETs can be described by the $\Delta\mu$ model, most of the nMOSFETs (and also scaled metal gate/high-κ pMOSFETs) are governed *by* Δn fluctuations. In other words, they are determined by carrier trapping and de-trapping by structural defects present in the gate oxide, in the vicinity of the interface with the semiconductor substrate [15]. These are also termed border traps, as they are able to communicate with the channel with time constants in the range of 1 ms to >1 s. Several models have been developed in the past to describe the corresponding $1/f$ noise. To date, the most popular description is based on the correlated mobility fluctuations Δn model, given by [16,17]

$$S_{V_G} = S_{V\text{fb}}(1 \pm \alpha_C \mu_{\text{eff}} C_{\text{ox}} V_{\text{GT}})^2 \tag{7.7}$$

with S_{V_G} as the input-referred voltage noise PSD ($=S_I/g_m{}^2$ in V^2/Hz) and $S_{V\text{fb}}$ as the value at flat-band voltage. Further, in Equation (7.7) α_C is the Coulomb scattering coefficient, μ_{eff} the effective mobility, C_{ox} the oxide capacitance density (F/cm^2), and V_{GT} the gate voltage overdrive. The sign in front of the second term between brackets depends on the nature of the oxide traps (acceptor or donor): if the trap becomes charged upon capture of a channel carrier, the Coulombic scattering associated with it adds up to the number fluctuations; the opposite holds in the other case. From the flat-band voltage spectral density, the oxide trap density N_{ot} can be derived, according to [16,17]:

$$S_{V_{\text{FB}}} = \frac{q^2 k_B T N_{\text{ot}}}{WLC_{\text{ox}}^2 \alpha_t f} \tag{7.8}$$

In Equation (7.8), α_t is the attenuation factor of the electron (hole) wave function in the gate dielectric, on the order of $10^8\ \text{cm}^{-1}$.

Finally, Generation-Recombination (GR) noise is due to subsequent carrier thermal generation and recombination through a deep level in the bandgap, according to the Shockley-Read-Hall (SRH) mechanism [18]. For a deep level in the semiconductor depletion region, maximum GR noise will be generated when the Fermi level crosses the trap level, which also pins the corner frequency f_{0i}, irrespective of the gate voltage [19,20]. The latter is related to the capture (τ_c) and emission time (τ_e) of the trap level through

$$2\pi f_0 = \frac{1}{\tau_c} + \frac{1}{\tau_e} \tag{7.9}$$

Therefore, the f_0 in function of the temperature T enables the so-called GR noise spectroscopy to be performed, yielding the trap activation energy and capture a cross section from an Arrhenius plot [21]. For scaled transistors, the number of active traps can be reduced to only a few. While the corresponding noise spectrum remains Lorentzian, the fluctuations in the time domain become discrete, as in Figure 7.14, switching between a high state and a low state. This phenomenon is called a Random Telegraph Signal (RTS) and originates from a single trap, which can reside on either side of the semiconductor/dielectric interface [22]. The high state usually corresponds with the trap empty (capture interval), while in the low state, the trap is occupied by a carrier (emission time). These time constants can depend strongly (exponentially) on the gate voltage, indicating that the trap is in the gate oxide [22,23]. A more detailed description of RTS (or RT Noise – RTN) can be found in Ref. [22].

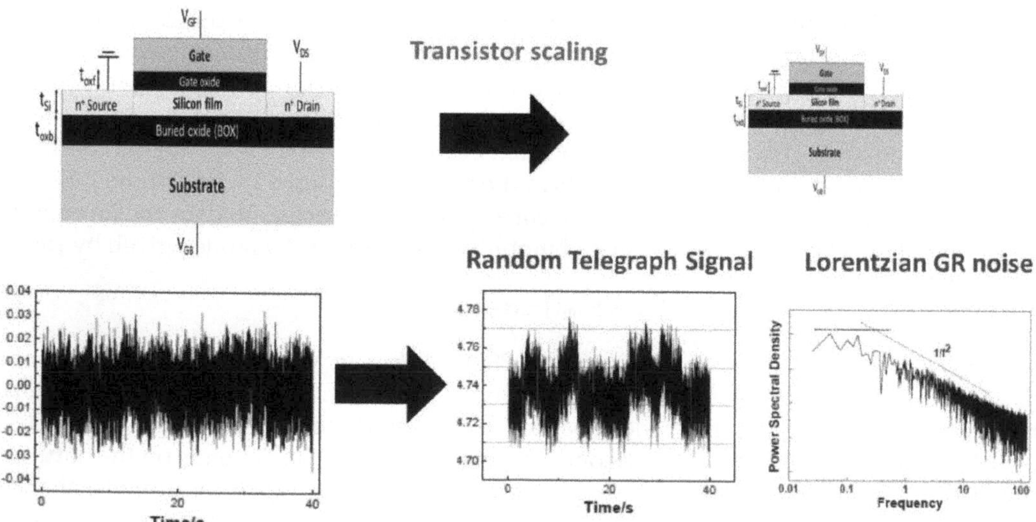

FIGURE 7.14
Impact of transistor scaling on the LF noise behavior. Reducing the transistor feature size results in a change in the time domain current fluctuations from structureless to RTS. The corresponding noise spectrum changes from predominantly $1/f^{\gamma}$-like to Lorentzian GR noise.

7.3.2 LF Noise in TFETs

Initial LF noise studies were performed on Si-based homojunction or hetero-junction TFETs [24–28]. A strong sensitivity to single-defect related RTS was reported [24–26], especially when the trap was close to the source tunneling junction. Other devices exhibited rather $1/f^{\gamma}$-like spectra [26–28], indicating a distribution of traps in the vicinity of the tunneling path contributing to the Δn fluctuations. Similar flicker noise dominance was found for $In_{0.7}Ga_{0.3}As$ homo-junction and $GaAs_{0.35}Sb_{0.65}/In_{0.7}Ga_{0.3}As$ hetero-junction TFETs [29] and in InAs-based nanowire TFETs [29]. Overall, the $1/f$ noise magnitude depends on the dominant transport mechanism, i.e., whether BTBT or defect-assisted mechanisms (TAT and SRH) are present [27,28]. An example is given in Figure 7.15, showing the input-referred voltage noise PSD versus gate voltage for Si-based TFETs with different types of source material (Ge or Si). It is evident that at low currents, where TAT dominates a higher S_{VG} is observed, while at higher currents, the noise magnitude saturates in the BTBT regime. Qualitatively, this behavior is opposite to what is typically found in a MOSFET, where S_{VG} is constant at low currents and starts to increase in strong inversion [13]. This is related to the different behavior of g_m: for a TFET, g_m continuous to increase for higher gate voltage V_{GS} (inset Figure 7.15), while for a MOSFET a peaked behavior *around* V_T is typically obtained. As the defect-assisted current mechanisms are usually thermally activated, low-temperature operation suppresses these in favor of BTBT. This is also found in the noise behavior [27,28], where compared with room temperature the excess flicker noise due to TAT is suppressed at 77 K, while the BTBT-related noise PSD remains constant [28].

As seen in Figure 7.15, heterojunction TFETs exhibit lower $1/f$ noise compared with homojunction devices, which can be explained, as seen later, by the more efficient BTBT in the former, yielding a shorter effective tunneling length L'. The higher the Ge content

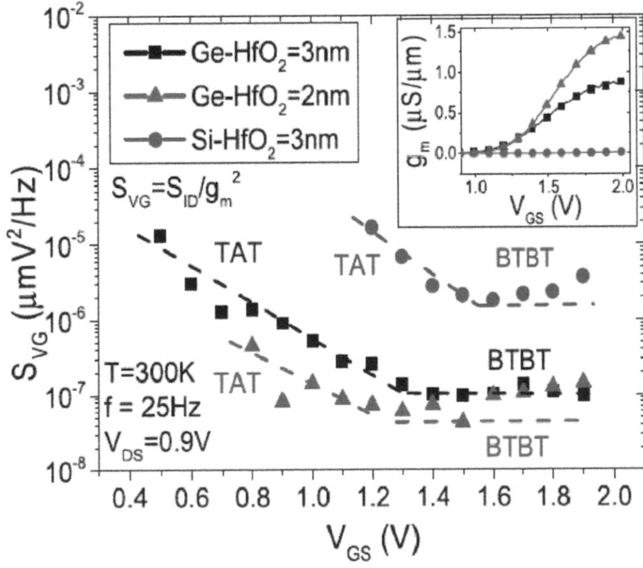

FIGURE 7.15
Input-referred voltage noise PSD versus V_{GS} at 25 Hz and V_{DS} = 0.9 V for a TFET with a Ge source and 3 nm HfO_2 gate dielectric (■), a Ge source and 2 nm HfO_2 (▲) and a silicon source and 3 nm HfO_2 (•). (After Neves, F.S. et al., *IEEE Trans. Electron Devices*, 63, 1658–1665, 2016.)

in the source, the smaller the $1/f$ noise PSD is, whereby the BTBT-related noise reduces for increasing Ge content x in the $Si_{1-x}Ge_x$ source [27,28]. A similar tendency has been reported for III-V TFETs [29]: higher $1/f$ noise PSD for homojunction versus heterojunction TFETs.

Reducing the gate oxide thickness also helps to improve the $1/f$ noise performance of Ge-source TFETs (Figure 7.15). This can be explained by the higher vertical electrical field at the same V_{GS}, which increases the BTBT generation rate. This should also reduce the tunneling length and increase the carrier density. The latter is important to screen the potential of a charged (interface) trap, contributing to the flicker noise.

In order to better understand the underlying noise mechanism(s) in TFETs, it can be helpful to investigate the RTS behavior as a function of the vertical and lateral trap position, for different device geometries (L and W) and for different bias conditions (V_{GS} and V_{DS}). Numerical simulation studies on RTS in nTFETs have shown that the trap amplitude is closely related to the lateral position along the channel: traps close to the tunneling path near the source junction will have the highest impact, while traps closer to the drain have smaller amplitudes ΔI_D [31]. The obtained RTS amplitude distribution is non-Gaussian, with large-amplitude tails. The up-and-down time constants follow an exponential distribution, as in MOSFETs [31]. The RTS amplitude can be estimated in the first instance by calculating the effective region of trap-induced potential perturbation. For a trap at the Si/high-κ interface, the maximum radius of the perturbed area is $0.8 \times t_{ox}$. This yields a theoretical maximum ΔI_D, which is smaller than $0.01 \times I_D$. The highest experimental amplitudes may amount, however, to several tens % of the channel current [31]. Such high amplitudes can only be explained by considering a non-uniform BTBT generation rate along the width of the device at the source-channel junction. These points of high generation rate can act as critical current paths, similar to the current filaments across the channel of a MOSFET [22]. An (interface) trap in close vicinity of this critical path will have a high impact on the BTBT current. Since there will be fewer critical paths in a narrow device, the impact of a single defect will become more pronounced [31]. As the BTBT current is determined by the maximum electric field (F_{max}), all variability sources impacting on this parameter will also influence the RTS amplitude, e.g., Line Edge Roughness (LER), Random Dopant Fluctuations (RDFs), and Work Function Variation (WFV). WFV will be important for metal-gate TFETs and less pronounced for polysilicon gates. LER will depend on the device architecture, i.e., planar versus multigate fin-type.

In InAs/GaSb-based heterojunction TFETs, pronounced RTS was found in the steep subthreshold regime [32] with maximum relative amplitude up to 50% of the DC current. This comes from the small diameter of the studied nanowires (~20 nm). The calculated step in gate voltage ΔV_{GS} corresponding with a single RTS was on the order of a few mV typically [30]. From the capture and emission time constants, a trap position in the gate oxide of 2.8 nm from the interface has been derived. It was also demonstrated that the collective action of a number of RTSs degrades the subthreshold swing [32].

An analytical model for the RTS amplitude in TFETs has been derived in Ref. [33], starting from the expression for the tunneling current:

$$I_D = AF^2 \exp\left(-\frac{B}{F}\right) \quad (7.10)$$

The RTS amplitude derived from that becomes:

$$\frac{\Delta I_D}{I_D} = \left(\frac{2}{F} + \frac{B}{F^2}\right)\Delta F \quad (7.11)$$

The change in the channel electric field near the tunnel junction by a charged trap can be approximated by [33]:

$$\Delta F = \frac{Q_{\text{eff}}}{\varepsilon_{\text{ch}} WL'} \tag{7.12}$$

with ε_{ch} the permittivity of the channel region and $Q_{eff} = \eta q$, where $\eta < 1$. This factor includes effects such as the screening of the trapped charge by the free carriers and the vertical position of the trap and the lateral distance from the source junction. From comparison with TCAD simulations, $\eta = 0.5$ appears to provide a good fit for a trap at the interface [33].

Extension of the RTS model to a number fluctuations model for the $1/f$ noise of a TFET is straightforward, yielding [29,33]

$$\frac{S_I}{I_D^2} = \left(\frac{2}{F} + \frac{B}{F^2}\right)^2 \frac{q^2 k_B T N_{\text{ot}}}{C_{\text{ox}}^2 WL' \alpha_t f} \tag{7.13}$$

This is rather similar to the Δn model for the $1/f$ noise of a MOSFET, with the exception of the field-dependent pre-factor and replacing the gate length by the effective tunneling distance. Based on the same approach, one can show that [28]

$$\frac{S_I}{I_D^2} = \frac{\beta^2}{SS^2} S_{V_G} \quad \text{with } \beta = k_B T \tag{7.14}$$

As shown in Figure 7.16, the relationship between the normalized noise PSD and the inverse square of the SS is followed in good approximation for the Ge-source TFETs.

7.3.3 Simulation of LF Noise in TFETs and Circuits

Further insights into the noise and RTN behavior of TFETs have been gained by TCAD simulations. Extensive efforts have been spent to investigate the dependence of the RTS amplitude ΔI_D (or ΔV_{GS}) on several device and trap parameters, for silicon [31,34,35] and

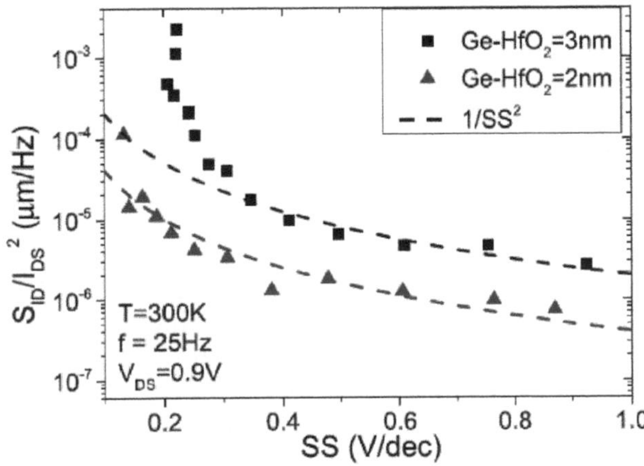

FIGURE 7.16
Normalized current noise PSD versus subthreshold swing at 25 Hz and $V_{DS} = 0.9$ V for Ge-source silicon nTFETs with 2 nm (■) or 3 nm (▲) HfO_2. (After Neves, F.S. et al., *IEEE Trans. Electron Devices*, 63, 1658–1665, 2016.)

III-V TFETs [32,33]. A general trend coming out of these simulations is that the highest amplitude for a single oxide trap occurs when the trap resides at the interface. In fact, the impact for a trap in the channel could be even higher [33]. Therefore, most studies considered a single defect right at the interface. A second important conclusion is that the maximum RTS impact is for a defect at or in close vicinity of the source-channel tunnel junction. In a TFET, the maximum RTS amplitude reduces with distance to the source and is determined by the distance of the trap with respect to the critical tunneling path [34]. It has been found by simulation that relative amplitudes larger than 100% can be obtained for a TFET, since the tunnel current can increase when the trap is occupied [34]. The effect of a trap on the tunnel current can be understood qualitatively by considering the reduction of the electric field on the source side of the channel, which reduces the BTBT generation rate. This is different for a FinFET, where the maximum impact of an interface trap is when it is located on the source side of the middle of the channel [33]. Other differences are that a single trap near the source increases the off-state current and degrades the SS [32,33]. In contrast, marginal impact on the off-state leakage is found for a FinFET. The role of the nature of the interface trap (donor or acceptor) on the RTS amplitude has also been studied [33,34]. An nTFET is shown to be more sensitive to donor traps, even away from the source [34], yielding a significant degradation of I_{OFF} and larger ΔI_D.

3D simulations revealed that the RTS amplitude increases for narrower TFETs [33], while at the same time, marginal gate length dependence is obtained. This is opposite to the case of a FinFET, where there is an increase in $\Delta I_D/I_D$ for reducing L. It is, however, well in line with the model of Equations (7.11) and (7.12) and indicates that only traps in the vicinity of the critical tunneling path with length L' have a pronounced impact on the current. Reducing the EOT gives rise to an increase in relative RTS amplitude, due to the improved SS [34].

An increase of the amplitude with V_{DS} has been found from TCAD, saturating when the current saturates. This can be understood by considering that for higher V_{DS}, the carrier density in the channel drops, leading to less screening of the trap potential and, hence, a stronger effect. The gate voltage dependence of the relative amplitude exhibits a pronounced increase below device turn-on, followed by a steep drop above V_T [33,34]. The latter results from the enhanced free carrier screening at higher currents. This stands in contrast to the V_{GS} dependence of the RTS amplitude of a FinFET, showing a plateau in weak inversion and a smoother reduction in strong inversion [33]. This follows from the fact that a FinFET turns on later (less steep SS) than a TFET.

Numerical simulations also allow for the investigation of the effect of other variability sources on the RTS amplitude. It has been shown that in the case of WFV, grains with a smaller WF result in a larger band bending (nTFET, acceptor interface trap at the source) and a smaller impact of the interface trap [34]. For a trap away from the source, the WFV can change the distance with the spot of maximum BTBT rate and, hence, the RTS amplitude. These results can be used to optimize the gate metal processing in view of LF noise and RTS reduction for analog/mixed-signal applications. The effect of RDF on the RTS amplitude has also been studied by TCAD, indicating that a higher source doping results in a more concentrated distribution and a reduced maximum value of the RTS amplitude [30]. The suppressed RTS generation may be originating from a reduced nonuniformity of the BTBT generation rate. This was verified experimentally by comparing the RTS amplitudes for silicon nTFETs with different source implantation doses [31]. Another way to suppress RTS is by using a two-step annealing of the source junction [31], which provides more diffusion of dopants and a more uniform electric field distribution at the junction at the expense of a less abrupt profile and a lower BTBT rate [31].

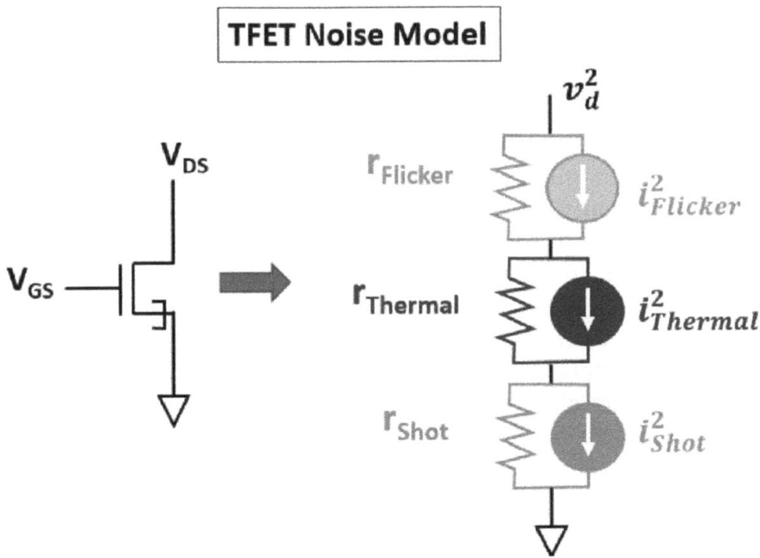

FIGURE 7.17
Representation of flicker, thermal, and shot noise models (noise current sources) at transistor level. (After Datta, S. et al., *Microelectron. Reliab.*, 54, 861–874, 2014.)

In order to perform circuit simulations, the noise model of Figure 7.17 has been implemented, including white and flicker noise sources in Verilog-A code [33,36]. The dominant white noise source in a TFET is the shot noise, which can be modeled by Equation (7.5), including a Fano factor $\Gamma = 2$, which represents the deviation of the shot noise magnitude from the nominal Poissonian value $2qI_D$.

The impact of RTS on circuit performance has been evaluated by simulations for SOI-based 8T SRAM cells and sense amplifiers [35] and for differential amplifiers and 6T and 10T III-V heterojunction TFET SRAM cells [33,36,37]. SRAMs in particular are vulnerable to RTS, as they are designed with minimum-sized transistors [37]. RTS endangers the cell stability and compromises the read/write noise margins. It has been shown that a 10T HTFET SRAM cell exhibits improved read noise margin at an operation voltage of 0.175 V, compared with a 6T Si-FinFET SRAM cell [37].

7.4 Analog Circuits Based on TFETs

7.4.1 Motivation

Taking into consideration the promising advantages of individual TFETs, basic circuits have also been analyzed in recent studies [38,39]. After all, interesting behaviors such as the lower susceptibility to the channel length and to the temperature could enhance the performance of basic circuits. Digital configurations for instance inverters, multiplexers and comparators have been analyzed based on simulations [40,41] and experimental data as well [42,43].

Better output conductance is also particularly interesting for analog applications. This way, widely used functional blocks of circuits such as operational amplifiers have been investigated recently [44–47].

This section selects two essential analog applications, namely current mirrors and differential pairs, in order to show the impact of designing analog circuits with TFET devices. Representative parameters for each circuit designed with TFETs are then compared to the ones extracted for FinFETs [48,49]. The results are obtained for different dimensions, bias conditions, and temperatures, so that a final summary highlights the advantages of each technology depending on the application requirements.

It is worth remembering that new concepts of TFET devices have also been studied in order to overcome low on-state current [50,51]. One of these new alternatives is the Line TFET, which is characterized by a source/gate overlap in a way that the direction of tunneling and the gate electric field become aligned [52,53]. The resulting boost in the drive current and increase in the susceptibility to the channel length affect circuits designed with this TFET alternative [54,55].

Therefore, while this section mainly highlights the results for Point TFETs (described in details in Section 7.4.2), the final summary includes comments on new TFET approaches as well [56].

7.4.2 Current Mirrors

A current mirror is widely used to provide a fixed bias current to a circuit, regardless of the loading, by copying the current from another active device. For this goal, basic performance parameters are the current transfer ratio (output and reference currents ratio) and the compliance voltage (maximum bias variation that still results in an acceptable output current).

It is key to evaluate the susceptibility of these key parameters to a variation in the temperature and to a mismatch in the transistors dimensions. A basic current mirror is schematically represented in Figure 7.18.

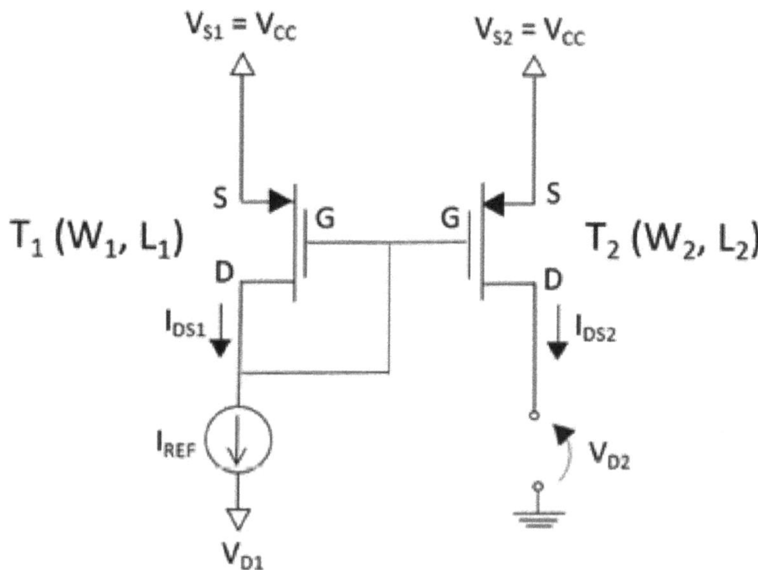

FIGURE 7.18
Representation of a current mirror basic circuit with P-type transistors.

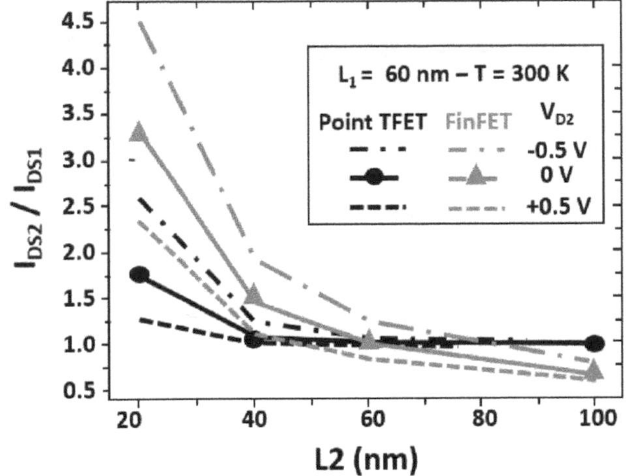

FIGURE 7.19
Current transfer ratio as a function of L_2 for current mirror circuits with Point TFETs and FinFETs, respectively, with $L_1 = 60$ nm. Solid lines represent the condition of $V_{D2} = 0$ V and dashed lines illustrate the results for a variation of 0.5 V.

First of all, it is relevant to highlight the impact of channel length mismatch and bias variation on the suitability of Point TFET and FinFET technologies for current mirror circuits. Keeping fixed values of $V_{D1} = 0$ V and $L_1 = 60$ nm, the current transfer ratio (I_{DS2}/I_{DS1}) has been studied for V_{D2} ranging from −0.5 V to +0.5 V and L_2 ranging from 20 nm to 100 nm.

Figure 7.19 sums up the results as a function of L_2. Solid lines represent $V_{D2} = 0$ V, when both transistors are under the same bias conditions, while the dashed curves refer to a 0.5 V bias variation. A perfect matching situation is represented by $L_2 = L_1 = 60$ nm and $V_{D2} = V_{D1} = 0$ V when the ratio is unitary.

It is possible to confirm that a Point TFET circuit is much less dependent on L_2 than the FinFET counterpart. In other words, current mirrors with Point TFETs present a very good mirroring behavior even when the transistor channel lengths do not exactly match. This is a direct result of the band-to-band tunneling happening quite close to the source/channel junction. The results also show that the drain current is affected by the output channel length only when it gets lower than 40 nm, which may be justified by the impact of DIBT on TFETs [54].

In contrast, a current mirror with FinFETs loses the ability to mirror the reference current when L_2 changes. When the output channel length changes from 40 nm to 100 nm, the I_{DS2} variation reaches 50% for FinFETs, while it varies less than 10% for Point TFETs. The worst-case scenario occurs for $L_2 = 20$ nm for both technologies. For this condition, the output drain current deviation reaches 80% for TFETs and 230% for FinFETs.

In terms of susceptibility to the output drain voltage, it is patent that current mirrors with Point TFETs are much less dependent, provided the DIBT does not play a relevant role. The dashed lines for TFETs are very close to the solid ones for L_2 higher than 40 nm, which can be connected to a higher compliance voltage observed for this technology when compared to FinFETs.

To evaluate the temperature impact on the current mirror output current, Figure 7.20 plots $I_{DS2}(T)/I_{DS2}$ (300 K) for perfect matched transistors ($L_1 = L_2 = 60$ nm). The curves reveal two main differences between Point TFETs and FinFETs. The first one is the opposite

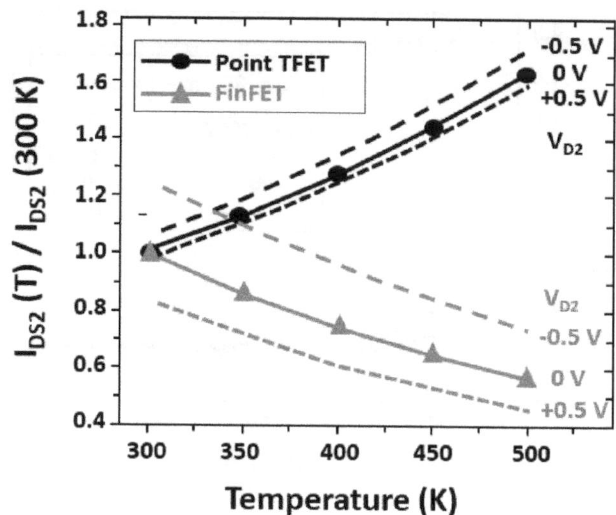

FIGURE 7.20
$I_{DS2}(T)/I_{DS1}(300\text{ K})$ as a function of the temperature for circuits with Point TFETs and FinFETs, respectively. Solid lines represent the condition of $V_{D2} = 0$ V, and dashed lines illustrate the results for a variation of 0.5 V.

trend, with an ascending curve only for the TFET case. The explanation comes from the differences in the prevailing transport mechanisms for each case. In Point TFET devices, higher temperatures slightly increase band-to-band tunneling, which is the most important component for high values of V_{DS}. Meanwhile, the mobility in FinFETs is degraded, causing a decrease in the drain current to almost half of its original value.

In addition to comparing solid lines to dashed ones, it is possible to analyze the susceptibility to the output drain voltage for current mirrors operating at higher temperatures. While for Point TFETs this variation causes an output current deviation of 6%, for FinFETs the deviation reaches almost 30%. In other words, circuits with TFETs tend to present a larger compliance voltage not only at room temperature (as previously shown in Figure 7.19) but also for higher temperatures. Therefore, Point TFETs present a better performance in terms of mirroring the input current even for output bias variations.

Regarding the most recent TFET approaches, Line TFET devices have also been studied as an option for current mirror circuits. Due to its previously explained working principle, it is possible to overcome the disadvantage of the lower on-state current observed in Point TFETs. On the other hand, the independence of the channel length is not achieved anymore, resulting in a current mirror that is more susceptible to transistor dimensions mismatch. Since the prevailing transport mechanism for Line TFETs is band-to-band tunneling, circuits with this technology keep the advantage observed for Point TFETs in terms of temperature and bias condition impact.

In comparing the suitability of these 3 technologies to design current mirror circuits, Table 7.1 summarizes the advantages presented by each of them. It can be stated that Point TFETs can provide a lower dependence on temperature, channel length mismatch, and bias condition simultaneously. Meanwhile, if the drive current magnitude is a strict requirement, Line TFETs are the best option, combining high on-state current, typical for FinFETs, and a low dependence on the temperature, a very important characteristic of tunneling devices.

TABLE 7.1

Features Presented by Current Mirrors Designed with FinFET, Point TFET and Line TFET Devices

	FinFET	Point TFET	Line TFET
High on-state current	✓		✓
Low susceptibility to the temperature		✓	✓
Low susceptibility to channel length mismatch		✓	
Low susceptibility to bias condition (high compliance voltage)		✓	✓

7.4.3 Differential Pairs

A differential pair circuit is obtained with two transistors connected to a common node and biased with a fixed current source. This way, transistors will transfer equal and opposite signals through source and drain, leading to a differential output linearly dependent on the input signals.

Fundamental parameters are the differential voltage gain (A_d) and the compliance voltage. Once more, it is relevant to analyze how these key indicators vary with temperature and transistors dimensions mismatch. Figure 7.21 illustrates two different configurations of differential pairs, namely passive load and active mode ones.

First of all, simulations have been performed to highlight the differential gain (A_d) for circuits with Point TFETs or FinFETs. By varying the input voltages, it was studied how the normalized drain current for each transistor changes. Figure 7.22 shows I_{DS1}/I_{SS} and I_{DS2}/I_{SS} for each case as a function of $v_{id} = v_{in1} - v_{in2}$, when $v_{in1} = -v_{in2}$.

In accordance to the expectations for perfect balanced differential pairs with matched transistors and drain resistors, drain currents are the same for both transistors when $v_{in1} = v_{in2} = 0$ V. When v_{in1} is positive and v_{in2} is negative, the current in T_1 increases and in T_2 decreases, with the opposite trend for negative v_{in1} and positive v_{in2}.

Two clear differences between the performance observed for Point TFETs and for FinFETs must be pointed out. Point TFETs present a higher normalized current susceptibility to the

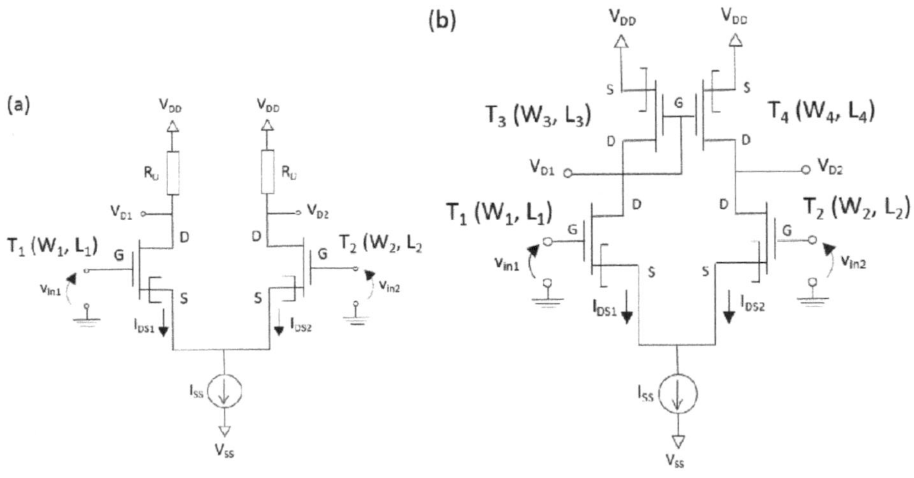

FIGURE 7.21
Representation of differential pair circuit with (a) passive load and (b) active load.

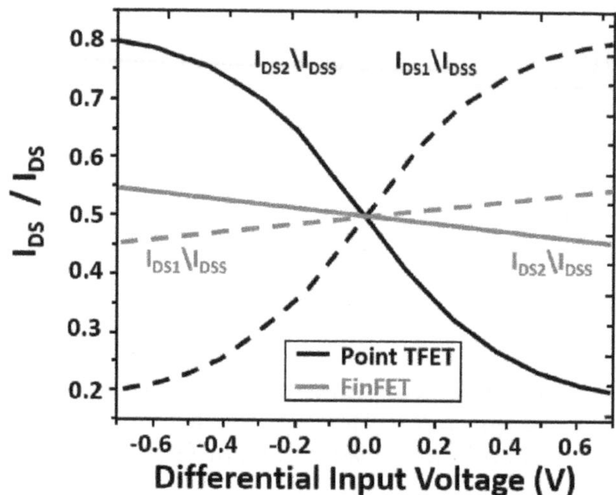

FIGURE 7.22
Normalized currents (I_{DS1}/I_{SS} and I_{DS2}/I_{SS}) as a function of differential input voltage for a balanced differential pair with Point TFETs and FinFETs, respectively.

input voltage. On the other hand, this variation is linear only when $|v_{in1}|$ is kept below 0.1 V, while the circuit with FinFETs presents a virtually linear behavior for the whole analyzed input range.

The calculated slope obtained with Point TFETs is about 12 times higher than the one for FinFETs when $|v_{in1}|$ <0.1 V. Meanwhile, FinFET differential pairs present a more constant behavior, varying less than 2% for the whole range, against a 38% variation for Point TFETs.

In terms of basic current mirror parameters, a bigger variation of the normalized current slope leads to a lower compliance voltage. Regarding the differential voltage gain, it is proportional to both the slope and the magnitude of the drive current. Therefore, for this, circuit FinFETs tend to provide a better compliance voltage and highest gain; note that the ratio of on-state current for FinFETs and Point TFETs is bigger than the 12 times the ratio observed for the slope.

The same procedure as used for current mirrors can be followed in order to focus on the impact of dimensions mismatching between T_1 and T_2 transistors. Once more, L_1 was kept fixed at 60 nm, while L_2 ranged from 20 to 100 nm.

Figure 7.23 illustrates the impact of L_2 on the differential gain. The results show the relative variation to the perfect matching condition ($L_1 = L_2 = 60$ nm). It is clear that for higher values of channel length, Point TFETs are less susceptible (around 4%) to length variation than FinFETs (around 9%). This statement is consistent with the Point TFET working principle, since it is known that tunneling happens quite close to the source/channel junction. On the other hand, Point TFETs are affected by DIBT, when the channel length is decreased to 40 nm or less [54].

These circuits have also been analyzed in terms of susceptibility to the temperature. Figure 7.24 shows the relative A_d variation for temperatures up to 450 K. It is clear that the values for FinFETs decreased by more than 50%, which may be explained by the mobility degradation for higher temperatures. Meanwhile, the TFET values remain nearly constant, since the bandgap narrowing is closely compensated by the output conductance increase. It is worth mentioning that a similar trend is observed for common-mode gain and for common-mode rejection ratio.

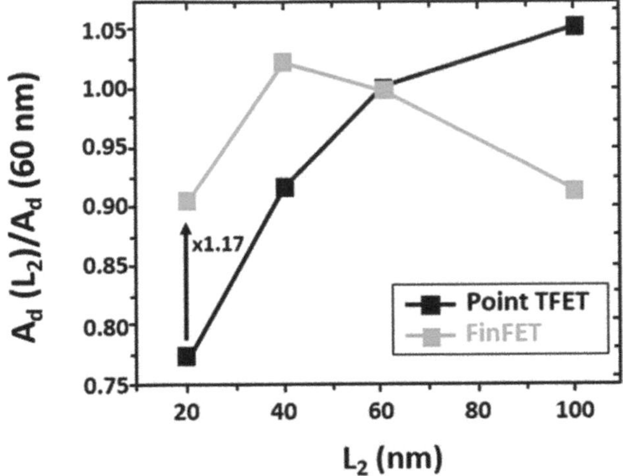

FIGURE 7.23
Differential gain variation ($A_d(L_2)/A_d$(60 nm)) as a function of T_2 transistor channel length (L_2) for a balanced differential pair with Point TFETs and FinFETs, respectively.

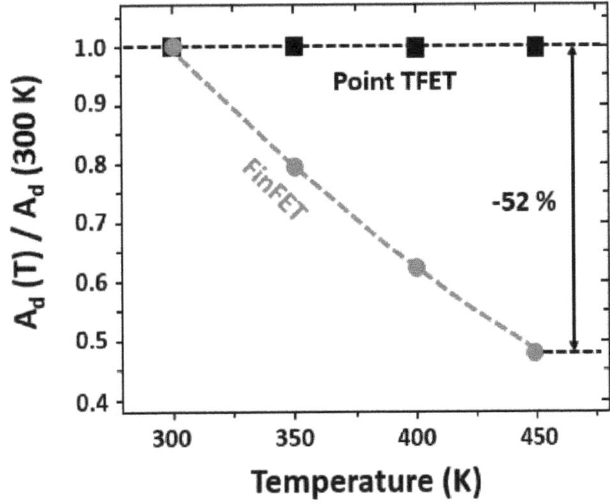

FIGURE 7.24
Differential gain variation (A_d/A_d(300 K)) as a function of temperature for a balanced differential pair with Point TFETs and FinFETs, respectively.

Following the same procedure described for current mirrors, the Line TFET alternative has also been studied in terms of suitability for differential pairs. This way, it was possible to obtain a differential pair with a reference current similar to FinFETs and a normalized current slope similar to Point TFETs. Temperature stability was also similar to the one extracted for Point TFETs, since band-to-band tunneling prevails for both technologies.

Therefore, it is possible to make a comparison between differential pairs designed with these three options in terms of compliance voltage, susceptibility to channel length mismatch, differential voltage gain, and susceptibility to the temperature. FinFETs would present a higher compliance voltage, due to lower values of normalized current slopes,

TABLE 7.2

Features Presented by Differential Pairs Designed with FinFET, Point TFET and Line TFET Devices

	FinFET	Point TFET	Line TFET
High differential voltage gain	✓		✓✓
High compliance voltage	✓		
Low susceptibility to channel length mismatch		✓	
Low susceptibility to the temperature		✓	✓

while Point TFETs would be the best for applications requiring a low susceptibility to channel length. Meanwhile, Line TFETs present the highest values of A_d, beating FinFETs (due to a much higher normalized current slope and a slightly lower I_{SS}) and Point TFETs (due to a similar slope and a much higher I_{SS}). Regarding the temperature impact, both Point TFET and Line TFET can take advantage of the lower temperature dependence of band-to-band tunneling and provide a less susceptible circuit in this point of view. Table 7.2 summarizes the advantages of each technology in terms of important parameters for differential pairs.

7.5 Conclusions

TFETs certainly have their space in the analog circuit landscape and, depending on the specific application needs, may offer a valuable substitute for standard CMOS, operating at a lower voltage. Developments over the years have established a rich portfolio of technology and design options to cover a broad span of existing and future needs. In many aspects, the poor digital performance of mainly silicon-based TFETs can be turned into advantageous analog performance parameters if the device can be operated in a BTBT-dominated mode. One analog parameter that remains challenging to control is the LF noise, which is in many cases (interface) trap dominated and forms an important source of device variability. Again, choosing the right technological option may help to reduce the sensitivity to noise and RTS and may even yield favorable signal-to-noise ratios compared with CMOS-based circuitry.

List of Acronyms

BOX	Buried Oxide
BTBT	Band-To-Band Tunneling
DIBT	Drain-Induced Barrier Thinning
EOT	Equivalent Oxide Thickness
FET	Field-Effect Transistor
FinFET	Fin Field-Effect Transistor

GAA	Gate-All-Around
GR	Generation-Recombination
HTFET	Heterojunction Tunnel Field Effect Transistor
LER	Line Edge Roughness
LF	Low-Frequency
MOSFET	Metal-Oxide-Semiconductor Field-Effect Transistor
MuGFET	Multiple Gate Field-Effect Transistor
PSD	Power Spectral Density
RDF	Random Doping Fluctuations
RF	Radio Frequency
RTN	Random Telegraph Noise
RTS	Random Telegraph Signal
SEM	Scanning Electron Microscope
SOI	Silicon-on-Insulator
SS	Subthreshold Swing
SRAM	Static Random Access Memory
SRH	Shockley-Read-Hall
TAT	Trap-Assisted Tunneling
TCAD	Technology Computer-Assisted Design
TFET	Tunnel Field-Effect Transistor
WF	Work function of the gate material
WFV	Work Function Variability
2D	Two-dimensional
3D	Three-dimensional
6T	Six-Transistor
8T	Eight-Transistor
10T	Ten-Transistor

List of Symbols

A	(Am^2/V^2)	Prefactor in the tunneling current
A_d	(V/V)	Differential voltage gain
A_i	(A^2/Hz)	Amplitude of ith GR noise component
B	(V/m)	Exponential coefficient in the tunneling current
C_{gd}	(F)	Gate-to-drain capacitance
C_{gg}	(F)	Total gate capacitance
C_{ox}	(F/m^2)	Oxide capacitance density
E_A	(eV)	Activation energy
E_g	(eV)	Bandgap
f	(Hz)	Frequency
F	(V/m)	Electric field
f_{0i}	(Hz)	Corner frequency of ith GR center
F_{max}	(V/m)	Maximum electric field
f_T	(Hz)	Unit gain frequency
g_d	(S)	Channel or output conductance
g_m	(S)	Transconductance

I_D	(A)	Drain current
I_G	(A)	Gate current
I_{OFF}	(A)	Off-state current
I_{ON}	(A)	On-state current
I_{REF}	(A)	Reference current for current mirror
I_{SS}	(A)	Bias current for differential pair
J_{BTBT}	(A/m^2)	Current density due to BTBT
J_{SRH}	(A/m^2)	Current density due to SRH
J_{TAT}	(A/m^2)	Current density due to TAT
k_B	J/K	Boltzmann constant
K_f	$(A^2Hz^{\gamma-1})$	$1/f$ noise strength factor
K_W	(A^2/Hz)	PSD of white noise
L	(m)	Transistor length
L′	(m)	Tunneling length
N		Total number of fluctuators
N_{ot}	$(cm^{-3}eV^{-1})$	Oxide trap density
q	(C)	Elementary charge (absolute value)
Q_{eff}	(C)	Effective charge of an RTS
R	(Ω)	Resistance
R_D	(Ω)	Drain resistance
S_{VFB}	(V^2/Hz)	Flat-band voltage noise PSD
S_I	(A^2/Hz)	Current noise power spectral density
SS	(mV/dec)	Subthreshold swing
S_{VG}	(V^2/Hz)	Input-referred voltage noise PSD
T	(K)	Absolute temperature
t_{ox}	(m)	Oxide thickness
V_{CC}	(V)	Supply voltage
V_{DD}	(V)	Supply voltage
V_{DS}	(V)	Drain voltage
V_{EA}	(V)	Early Voltage
V_{GS}	(V)	Gate voltage
V_{GT}	(V)	Gate voltage overdrive
v_{id}	(V)	Differential input voltage
v_{in1}	(V)	T1 input voltage
v_{in2}	(V)	T2 input voltage
V_{S1}	(V)	Source voltage for T1
V_{S2}	(V)	Source voltage for T2
V_T	(V)	Threshold voltage
W	(m)	Transistor width

List of Greek Symbols

α_C	(Vs/C)	Coulomb scattering coefficient
α_H		Hooge 1/f noise parameter
α_t	(cm^{-1})	Attenuation length of the electron wave function in the gate dielectric
ΔF	(V/m)	RTS amplitude in the electric field
ΔI_D	(A)	RTS amplitude in the current
$\Delta\mu$		Mobility fluctuations origin of 1/f noise
Δn		Number fluctuations origin of 1/f noise
ΔV_{GS}	(V)	RTS amplitude in gate voltage
ε_{ch}	F/m	Permittivity of the channel material
γ		Frequency exponent in $1/f^{\gamma}$ noise spectrum
Γ		Fano factor
κ		Dielectric constant
μ_{eff}	(cm^2/Vs)	Effective carrier mobility
τ_c	(s)	Capture time constant
τ_e	(s)	Emission time constant

References

1. A. M. Ionescu and H. Riel, "Tunnel field-effect transistors as energy-efficient electronic switches," *Nature*, vol. 479, pp. 329–337 (2011).
2. A. Verhulst, W. G. Vandenberghe, K. Maex and G. Groeseneken, "Tunnel-field-effect transistor without gate-drain overlap," *Appl. Phys. Lett.*, vol. 91, pp. 053102/1–3 (2007).
3. M. D. V. Martino, F. S. Neves, P. G. D. Agopian, J. A. Martino, R. Rooyackers and C. Claeys, "Nanowire tunnel field effect transistors at high temperature," *J. Integr. Circuits Syst.*, vol. 8, pp. 110–115 (2013).
4. F. S. Neves, P. G. D. Agopian, J. A. Martino, B. Cretu, A. Vandooren, R. Rooyackers, E. Simoen, A. Thean and C. Claeys, "Transconductance hump in vertical gate-all-around tunnel-FETs," in *IEEE SOI-3D-Subthreshold Microelectronics Technology Unified Conference* (S3S), IEEE Explore, pp. 1–3 (2015).
5. D. Leonelli, A. Vandooren, R. Rooyackers, S. De Gendt, M. M. Heyns and G. Groeseneken, "Drive current enhancement in p-tunnel FETs by optimization of the process conditions," *Solid-State Electron.*, vol. 65–66, pp. 28–32 (2011).
6. P. G. D. Agopian, M. D. V. Martino, S. G. dos Santos Filho, J. A. Martino, R. Rooyackers, D. Leonelli and C. Claeys, "Temperature impact on the tunnel FET off-state current components," *Solid-State Electron.*, vol. 78, pp. 141–146 (2012).
7. P. G. D. Agopian, J. A. Martino, R. Rooyackers, A. Vandooren, E. Simoen and C. Claeys, "Experimental comparison between trigate pTFET and pFinFET analog performance as a function of temperature," *IEEE Trans. Electron Devices*, vol. 60, pp. 2493–2497 (2013).
8. S. Mookerjea, R. Krishnan, S. Datta and V. Narayanan, "Effective capacitance and drive current for Tunnel FET (TFET) *CV/I* estimation," *IEEE Trans. Electron Devices*, vol. 56, pp. 2092–2098 (2009).
9. P. G. D. Agopian, J. A. Martino, A. Vandooren, R. Rooyackers, E. Simoen, A. Thean and C. Claeys, "Study of line-TFET analog performance comparing with other TFET and MOSFET architectures," *Solid-State Electron.*, vol. 128, pp. 43–47 (2017).

10. P. G. D. Agopian, J. A. Martino, R. Rooyackers, A. Vandooren, E. Simoen, A. Thean and C. Claeys, "Intrinsic voltage gain of line-TFETs and comparison with other TFET and MOSFET architectures," in *Joint International EUROSOI Workshop and International Conference on Ultimate Integration on Silicon (EUROSOI-ULIS)*, IEEE Explore, pp. 13–15 (2016).
11. C. Bordallo, J. A. Martino, P. G. D. Agopian, A. Alian, Y. Mols, R. Rooyackers, A. Vandooren et al., "Analog performance of III-V TFETs," *J. Nanoelectron. Devices.*
12. D. Boudier, B. Cretu, E. Simoen, R. Carin, A. Veloso, N. Collaert and A. Thean, "Low frequency noise assessment in n- and p-channel sub-10 nm triple-gate FinFETs: Part II: Measurements and results," *Solid-State Electron.*, vol. 128, pp. 109–114 (2017).
13. M. von Haartman and M. Östling, *Low-Frequency Noise in Advanced MOS Devices*, Springer, New York (2007).
14. F. N. Hooge, "1/f noise sources," *IEEE Trans. Electron Devices*, vol. 41, pp. 1926–1935 (1994).
15. E. Simoen, H.-C. Lin, A. Alian, G. Brammertz, C. Merckling, J. Mitard and C. Claeys, "Border traps in Ge/III-V channel devices: Analysis and reliability aspects," *IEEE Trans. Device Mater. Rel.*, vol. 13, pp. 444–455 (2014).
16. G. Ghibaudo, O. Roux, Ch. Nguyen-Duc, F. Balestra and J. Brini, "Improved analysis of low frequency noise in field-effect MOS transistors," *Phys. Status Solidi A*, vol. 174, pp. 571–581 (1991).
17. G. Ghibaudo and T. Boutchacha, "Electrical noise and RTS fluctuations in advanced CMOS devices," *Microelectron. Reliab.*, vol. 42, pp. 573–582 (2002).
18. W. Shockley and W. T. Read Jr., "Statistics of the recombination of holes and electrons," *Phys. Rev.*, vol. 87, pp. 835–842 (1952).
19. N. B. Lukyanchikova, *Noise and Fluctuations Control in Electronic Devices*, A. Balandin (Ed.), American Scientific, Riverside, CA, p. 201 (2002).
20. V. Grassi, C. F. Colombo and D. V. Camin, "Low frequency noise versus temperature spectroscopy of recently designed Ge JFETs," *IEEE Trans. Electron Devices*, vol. 48, pp. 2899–2906 (2001).
21. B. Cretu, D. Boudier, E. Simoen and N. Collaert, "Assessment of DC and low frequency noise performances of triple-gate FinFETs at cryogenic temperatures," *Semicond. Sci. Technol.*, vol. 31, p. 124006 (2016).
22. E. Simoen and C. Claeys, *Random Telegraph Signals in Semiconductor Devices*, The Institute of Physics, Bristol, UK (2016).
23. W. Fang, E. Simoen, M. Aoulaiche, J. Luo, C. Zhao and C. Claeys, "Distinction between silicon and oxide traps using single-trap spectroscopy," *Phys. Status Solidi A*, vol. 212, pp. 512–517 (2015).
24. J. Wan, C. Le Royer, A. Zaslavsky and S. Cristoloveanu, "Low-frequency noise behavior of tunneling field effect transistors," *Appl. Phys. Lett.*, vol. 97, no. 24, p. 243503/1–3 (2010).
25. J. Wan, C. Le Royer, A. Zaslavsky and S. Cristoloveanu, "Tunneling FETs on SOI: Suppression of ambipolar leakage, low-frequency noise behavior, and modeling," *Solid-State Electron.*, vol. 65–66, pp. 226–233 (2011).
26. Q. Huang, R. Huang, C. Chen, C. Wu, J. Wang, C. Wang and Y. Wang, "Deep insights into low frequency noise behavior of tunnel FETs with source junction engineering," in *2014 Symposium on VLSI Technology, Digest of Technical Papers*, IEEE Explore, pp. 88–89 (2014).
27. P. G. D. Agopian, M. D. V. Martino, S. D. dos Santos, F. S. Neves, J. A. Martino, R. Rooyackers, A. Vandooren, E. Simoen, A. V.-Y. Thean and C. Claeys, "Influence of the source composition on the analog performance parameters of vertical nanowire-TFETs," *IEEE Trans. Electron Devices*, vol. 62, pp. 16–22 (2015).
28. F. S. Neves, P. G. D. Agopian, J. A. Martino, B. Cretu, R. Rooyackers, A. Vandooren, E. Simoen, A. V.-Y. Thean and C. Claeys, "Low-frequency noise analysis and modeling in vertical tunnel FETs with Ge source," *IEEE Trans. Electron Devices*, vol. 63, pp. 1658–1665 (2016).
29. R. Bijesh, D. K. Mohata, H. Liu and S. Datta, "Flicker noise characterization and analytical modeling of homo and hetero-junction III-V tunnel FETs," in *Proceedings of the 70th Device Research Conference*, IEEE Explore, pp. 203–204 (2012).

30. M. Hellenbrand, E. Memišević, M. Berg, O.-P. Kilpi, J. Svensson and L.-E. Wernersson, "Low-frequency noise in III-V nanowire TFETs and MOSFETs," *IEEE Electron Device Lett.*, vol. 38, pp. 1520–1523 (2017).
31. C. Chen, Q. Huang, J. Zhu, Y. Zhao, L. Guo and R. Huang, "New understanding of random telegraph noise amplitude in tunnel FETs," *IEEE Trans. Electron Devices*, vol. 64, pp. 3324–3330 (2017).
32. E. Memišević, M. Hellenbrand, E. Lind, A. R. Persson, S. Sant, A. Schenk, J. Svensson, R. Wallenberg and L.-E. Wernersson, "Individual defects in InAs/InGaAsSb/GaSb nanowire tunnel field-effect transistors operating below 60 mV/decade," *Nano Lett.*, vol. 17, pp. 4373–4380 (2017).
33. R. Pandey, B. Rajamohanan, H. Liu, V. Narayanan and S. Datta, "Electrical noise in heterojunction interband tunnel FETs," *IEEE Trans. Electron Devices*, vol. 61, pp. 552–560 (2014).
34. M.-L. Fan, V. P.-H. Hu, Y.-N. Chen, P. Su and C.-T. Chuang, "Analysis of single-trap-induced random telegraph noise and its interaction with work function variation for tunnel FET," *IEEE Trans. Electron Devices*, vol. 60, pp. 2038–2044 (2013).
35. M.-L. Fan, V. P.-H. Hu, Y.-N. Chen, P. Su and C.-T. Chuang, "Investigation of single-trap-induced random telegraph noise for tunnel FET based devices, 8T SRAM cell, and sense amplifier," in *Proceedings of the International Reliability Physics Symposium 2013*, IEEE Explore, pp. CR1.1–CR1.5 (2013).
36. S. Datta, H. Liu and V. Narayanan, "Tunnel FET technology: A reliability perspective," *Microelectron. Reliab.*, vol. 54, pp. 861–874 (2014).
37. R. Pandey, V. Saripalli, J. P. Kulkarni, V. Narayanan and S. Datta, "Impact of single trap random telegraph noise on heterojunction TFET SRAM stability," *IEEE Electron Device Lett.*, vol. 35, pp. 393–395 (2014).
38. Q. Huang, R. Huang, C. Wu, H. Zhu, C. Chen, J. Wang, L. Guo, R. Wang, L. Ye and Y. Wang, "Comprehensive performance re-assessment of TFETs with a novel design by gate and source engineering from device/circuit perspective," *Technical Digest - International Electron Devices Meeting, IEDM*, IEEE Explore, pp. 13.3.1–13.3.4 (2014).
39. M. Alioto and D. Esseni, "Performance and impact of process variations in Tunnel-FET ultra-low voltage digital circuits," *Proceedings of the 27th Symposium on Integrated Circuits and Systems Design (SBCCI '14)*, IEEE Explore, pp. 1–6 (2014).
40. D. H. Morris, U. E. Avci, R. Rios and I. A. Young, "Design of low voltage tunneling-FET logic circuits considering asymmetric conduction characteristics," *IEEE J. Emerg. Sel. Top. Circuits Syst.*, vol. 4, pp. 380–388 (2014).
41. S. Shaik, K. Krishna and R. Vaddi, "Circuit and architectural co-design for reliable adder cells with steep slope tunnel transistors for energy efficient computing," in *Proceedings of the International Conference on VLSI Design*, IEEE Explore, pp. 306–311 (2016).
42. S. Richter, C. Schulte-Braucks, L. Knoll, G. V. Luong, A. Schäfer, S. Trellenkamp, Q. T. Zhao and S. Mantl, "Experimental demonstration of inverter and NAND operation in p-TFET logic at ultra-low supply voltages down to VDD = 0.15 V," in *72nd Device Research Conference*, IEEE Explore, pp. 23–24 (2014).
43. U. E. Avci, S. Hasan, D. E. Nikonov, R. Rios, K. Kuhn and I. A. Young, "Understanding the feasibility of scaled III–V TFET for logic by bridging atomistic simulations and experimental results," in *Symposium on VLSI Technology, Digest of Technical Papers*, IEEE Explore, pp. 183–184, June 2012.
44. B. Sedighi, X. S. Hu, H. Liu, J. J. Nahas and M. Niemier, "Analog circuit design using tunnel-FETs," *IEEE Trans. Circ. Syst. I: Regular Papers*, vol. 62, pp. 39–48 (2015).
45. G. Kaushal, K. Subramanyam, S. N. Rao, G. Vidya, R. Ramya, S. Shaik, H. Jeong, S. O. Jung and R. Vaddi, "Design and performance benchmarking of steep-slope tunnel transistors for low voltage digital and analog circuits enabling self-powered SOCs," in *International SoC Design Conference (ISOCC)*, IEEE Explore, pp. 32–33 (2014).

46. A. Biswas, G. V. Luong, M. F. Chowdhury, C. Alper, Q. T. Zhao, F. Udrea, S. Mantl and A. M. Ionescu, "Benchmarking of homojunction strained-Si NW Tunnel FETs for basic analog functions," *IEEE Trans. Electron Devices*, vol. 64, pp. 1441–1448 (2017).
47. F. Settino, M. Lanuzza, S. Strangio, F. Crupi, P. Palestri, D. Esseni and L. Selmi, "Understanding the potential and limitations of Tunnel FETs for low-voltage analog/mixed-signal circuits," *IEEE Trans. Electron Devices*, vol. 64, pp. 2736–2743 (2017).
48. M. D. V. Martino, J. A. Martino and P. G. D. Agopian, "Performance comparison between TFET and FinFET differential pair," in *30th Symposium on Microelectronics Technology and Devices (SBMicro)*, IEEE Explore, pp. 1–4 (2015).
49. M. D. V. Martino, J. A. Martino and P. G. D. Agopian, "Analysis of TFET and FinFET differential pairs with active load from 300 K to 450 K," in *Joint International EUROSOI Workshop and International Conference on Ultimate Integration on Silicon* (*EUROSOI-ULIS*), IEEE Explore, pp. 246–249 (2016).
50. D. Leonelli, A. Vandooren, R. Rooyackers, A. Verhulst, C. Huyghebaert, S. De Gendt, M. Heyns and G. Groeseneken, "Novel architecture to boost the vertical tunneling in tunnel field effect transistors," in *IEEE International SOI Conference*, IEEE Explore, pp. 1–2 (2011).
51. G. Zhou, R. Li, T. Vasen, M. Qi, S. Chae, Y. Lu, Q. Zhang et al., "Novel gate-recessed vertical InAs/GaSb TFETs with record high I_{ON} of 180 μA/μm at V_{DS} = 0.5 V," *International Electron Devices Meeting*, IEEE Explore, pp. 32.6.1–32.6.4 (2012).
52. H. Y. Chang, B. Adams, P. Y. Chien, J. Li and J. C. S. Woo, "Improved subthreshold and output characteristics of source-pocket Si Tunnel FET by the application of laser annealing," *IEEE Trans. Electron Devices*, vol. 60, pp. 92–96 (2013).
53. K. E. Moselund, H. Schmid, C. Bessire, M. T. Bjork, H. Ghoneim and H. Riel, "InAs–Si nanowire heterojunction tunnel FETs," *IEEE Electron Device Lett.*, vol. 33, no. 10, pp. 1453–1455 (2012).
54. M. D. V. Martino, J. A. Martino and P. G. D. Agopian, "Drain induced barrier thinning on TFETs with different source/drain engineering," in *29th Symposium on Microelectronics Technology and Devices (SBMicro)*, IEEE Explore, pp. 1–4 (2014).
55. D. K. Mohata, R. Bijesh, Y. Zhu, M. K. Hudait, R. Southwick, Z. Chbili, D. Gundlach et al., "Demonstration of improved heteroepitaxy, scaled gate stack and reduced interface states enabling heterojunction tunnel FETs with high drive current and high on-off ratio," in *Symposium on VLSI Technology* (*VLSIT*), pp. 53–54 (2012).
56. M. D. V. Martino, J. A. Martino, P. G. D. Agopian, R. Rooyackers, E. Simoen, N. Collaert and C. Claeys, "Experimental analysis of differential pairs designed with line tunnel FET devices," in *IEEE SOI-3D-Substhreshold Microelectronics Technology Unified Conference* (*S3S*), IEEE Explore, pp. 1–3 (2017).

8

Dual Metal–Double Gate Doping-Less TFET: Design and Investigations

Ramandeep Kaur, Rohit Dhiman, and Rajeevan Chandel

CONTENTS

8.1 Introduction

Ever since the invention of metal oxide semiconductor field effect transistor (MOSFET), this type of device has been continuously used in CMOS technology and has undergone significant advancements. MOSFETs are being continuously scaled down to obtain improved electrical characteristics [1,2]. Further scaling of MOSFETs has become a serious concern as it results in a reduced on–off switching rate at low bias voltage and increased leakage power dissipation. Tunnel Field-Effect Transistors (TFETs) have been investigated as a promising candidate for future low-power applications [3–7]. However, the fabrication of doped source and drain regions in TFET is challenging as well as expensive [8,9]. Effects such as random dopant fluctuation have become significant in highly scaled TFETs [10–13]. The other problem with TFETs is the low on-state current compared to the conventional MOSFETs. Different TFET architectures have been aggressively studied to improve their on-state current, subthreshold swing, and ambipolar behavior [14]. Recently, a doping-less structure has been proposed as a potential solution to overcome the need to create abrupt junctions in TFETs [15], which creates source and drain regions in intrinsic Si using a charge plasma concept [16–19]. Doping-less TFETs (DL TFETs) have low production cost but the on-current of these devices is even lower than the conventional TFET. This is mainly due to the presence of source-side and drain-side spacer resistances in a doping-less structure. Based on the previous reported results, it is seen that the electrical characteristics of TFET, such as on-current, threshold voltage, and subthreshold swing can be significantly improved by replacing single material gate with a dual metal gate. This happens to be the main focus of our research work. Consequently, in this work, we present a novel Dual Metal–Double Gate (DM–DG) doping-less TFET using the charge plasma concept and work function engineering, and analysis thereof. The proposed device consists of two metals of different work function as gate electrodes. Ideally, the size of tunneling metal gate should be narrow and its work function should be as small as possible to induce high electron concentration in the channel region [20].

The rest of the chapter is organized as follows. TFET basics and its operation are described in Section 8.2. The proposed device structure and design parameters are discussed in Section 8.3. The mathematical modeling of surface potential in the proposed device is presented in Section 8.4. The results and their implications are studied in Section 8.5. Finally, conclusions are drawn in Section 8.6.

8.2 TFET and Its Operation

The structure of TFET is quite similar to MOSFET except the doping of drain and source regions. The doping in the source and drain regions of TFET is of opposite polarity—i.e., it has *p–i–n* structure. TFET can be classified as *n*-typeor *p*-type depending on the type of majority carriers in the channel region when the device gets turned on. In an *n*-type TFET, the drain is of *n*-type and source is *p*-type, while in *p*-type TFET, the drain is *p*-type and the source is *n*-type. The source electrode is grounded in both TFETs, while positive bias voltages are applied to the drain and gate electrodes in an *n*-type TFET, and vice versa for *p*-type TFET. The channel region is either intrinsic or lightly *p* or *n* doped in comparison to the other two highly doped regions. The schematic of *n*-type TFET is shown in Figure 8.1.

According to the classical physics, a subatomic particle can cross a potential barrier only if it has sufficient energy to overcome the potential. However, according to the

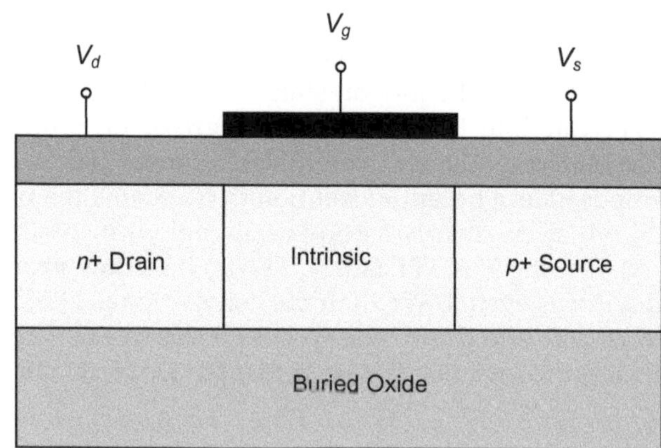

FIGURE 8.1
Cross-sectional view of *n*-type Silicon-on-Insulator TFET.

Schrodinger wave Equation theory, there exists a finite probability that a particle that does not have enough energy will tunnel to the other side of the potential barrier. This is made possible by quantum mechanical tunneling, which is the basic current conduction mechanism in a TFET. In sharp contrast, MOSFETs conduct using thermionic emission. Since the tunneling probability of a particle is inversely proportional to its mass and width of a potential barrier, the source region of TFET is heavily doped to get a narrower depletion width. Thus, Band–to–Band Tunneling (BTBT) of charge carriers in TFET takes place at a source–channel junction.

The energy band diagram representation of a conventional *n*-type TFET in the on and off states is shown in Figure 8.2. In the off-state, when gate voltage (V_g) is zero, the conduction band of the channel is not aligned with the valence band of the source, thereby inhibiting BTBT at the source–channel junction. When sufficient gate voltage is applied, the device gets turned on and band bending takes place at the source–channel interface. Energy bands in the channel get lower down such that the valence band of the source is above the conduction band of the channel. This results in tunneling of electrons from the source valence band to the channel conduction band.

The performance of TFET is measured by three parameters: On-current, ambipolar current, and subthreshold swing. On-current in a TFET is of the order of a few microamperes and off–current ranges in femtoamperes. The total drain current in a TFET is obtained by integrating the *Landauer* equation over energy range [21–24]. The drain current in a TFET is determined by the following expression [22,23]:

$$I_d = \frac{4|e|}{h}\sum_{k_{\text{trans}}} T_{\text{tot}}(k_{\text{trans}})\int_0^{\Delta\phi} dE\left(f_s(E) - f_d(E)\right) \tag{8.1}$$

Here, T_{tot} is the total transmission function, f_s and f_d are the source and drain Fermi functions (respectively), $\Delta\phi$ denotes the overlap of conduction band and valence band at the source channel interface, k_{trans} is the transverse wave vector, h is the Planck's constant, and e is the electronic charge in Coulombs. It may be noted that TFET is an ambipolar

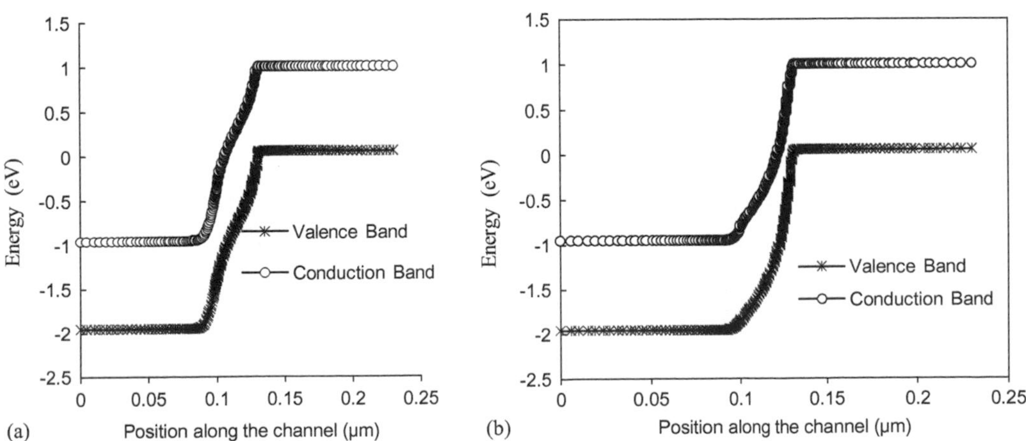

FIGURE 8.2
Valence and conduction band energy diagrams of conventional TFET in (a) off-state (b) on-state.

device—i.e., an n–type TFET can exhibit p-type behavior if negative voltage is applied at the gate, keeping the same drain bias. In that case, band bending and BTBT take place on the drain side. This is undesirable as it may lead to logic circuit failures. In order to suppress ambipolarity, the source region is heavily doped. Other ways to reduce ambipolarity is by making asymmetric structures such as hetro–TFETs or hetero–gate TFETs. Subthreshold swing is defined as the amount of gate voltage required to change the drain current by one order of magnitude. It is desirable to have low subthreshold swing as it signifies fast switching of the device. MOSFETs usually have subthreshold swing equal to 60 mV/decade at room temperature. Further lowering of subthreshold swing is not possible in MOSFETs because of their current conduction mechanism. Since TFET uses quantum tunneling for charge transport, it can overcome this problem and can have even lower subthreshold swing—i.e., nearly 30 mV/decade.

As TFETs are scaled down further, it becomes very difficult to maintain high concentration gradients and sharp doping profiles. Moreover, it is quite expensive to diffuse different regions in a TFET by using techniques such as ion–implantation and annealing. Recently studied DL TFETs provide a solution for these problems. Doping-less TFET does not have any metallurgical junctions, and the whole body is uniformly doped with intrinsic concentration. The charge–plasma concept is used to induce the source, drain, and channel regions in TFETs. It also overcomes the problem of random dopant fluctuation, which leads to significant high off-current in the device.

8.3 Proposed Device Structure

The cross-sectional view of DM–DG doping-less TFET is shown in Figure 8.3. The various design parameters used in the development of the proposed device are obtained using [15] and are as follows: Intrinsic Si doping, $N_i = 1 \times 10^{15}\ \text{cm}^{-3}$, Si film thickness, $T_{\text{Si}} = 10$ nm, Gate-oxide thickness, $T_{\text{ox}} = 3$ nm, Channel length = 50 nm. In DM–DG doping-less TFET, the top and bottom gates are composed of two materials of different work functions. The gate nearer to the source is called the tunneling gate and is of low work function while the gate nearer to the drain region is called the auxiliary gate

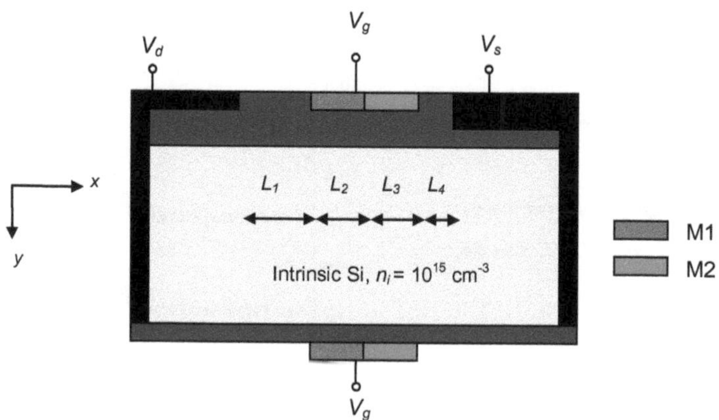

FIGURE 8.3
Cross-sectional view of DM–DG doping-less TFET.

and has a high work function. Optimized gate work functions can lead to improved electrical characteristics [25–29]. In the present research work, the work function of auxiliary and tunneling gates considered are 4.5 eV and 3.9 eV. The channel is divided into four tunneling regions—i.e., R_1, R_2, R_3, and R_4 include source depletion region, drain depletion region, and two channel regions under dual material gates, respectively. L_1, L_2, L_3, and L_4 denote drain depletion length, auxiliary gate length, tunneling gate length, and source depletion length, respectively. It may be noted that Si film thickness in our analysis is limited to 10 nm, since quantum confinement effect has not been considered. To avoid the possibility of silicide formation, a 0.5 nm thick layer of Silicon dioxide (SiO_2) is inserted between the source metal electrode and Si film. Isolations of 3 and 15 nm spacer oxide are provided between the source–gate and drain–gate electrodes, which are approximately equal to the depletion region width on the drain–channel side in a conventional doped TFET. The electrical characteristics of the device are simulated using nonlocal BTBT model [18]. Along with this, Shockley–Read–Hall recombination, Auger recombination, and concentration-dependent mobility models have also been enabled.

8.4 Modeling Surface Potential

The potential profile in the channel region R_i, where $\{i = 1, 2, 3, 4\}$ is given by 2-D Poisson equation as

$$\frac{\partial^2 \phi_i(x,y)}{\partial x^2} + \frac{\partial^2 \phi_i(x,y)}{\partial y^2} = \pm \frac{qN_i}{\varepsilon_{\text{Si}}} \tag{8.2}$$

where q is the electronic charge, i.e., 1.6×10^{-19} C; N_i is the doping concentration; and ε_{Si} is the dielectric constant of Si. The boundary conditions of DM–DG doping-less TFET for potential distribution in y direction are as follows:

The electric field at the center of TFET (at $y = T_{\text{Si}}/2$) is equal to zero:

$$\frac{\partial \varphi_i\left(x, \frac{T_{\text{Si}}}{2}\right)}{\partial y} = 0 \tag{8.3}$$

The potential at the Si–SiO_2 interface (i.e., $y = 0$) is equal to the surface potential ($\varphi_S(x)$), which gives

$$\varphi(x,0) = \varphi_S(x) \tag{8.4}$$

The electric field displacement is continuous across the Si–SiO_2 interface: therefore,

$$\frac{\partial \varphi_i(x,y)}{\partial y} = \frac{C_i\left(\varphi_{S,i}(x) - \varphi_{g,i}\right)}{\varepsilon_{\text{Si}}} \tag{8.5}$$

where C_i is the gate–body capacitance per unit area and is different for different regions, and $\varphi_{g,i}$ is the gate potential in region R_i and is expressed as a function of the gate–source voltage (V_{gs}) and flat–band voltage (V_{fb}). Thus, the gate potential is

$$\varphi_{g,i} = V_{gs} - V_{\text{fb},i} \tag{8.6}$$

The potential profile in vertical direction can be approximated using parabolic function [30] and is given as

$$\varphi_i(x,y) = \varphi_{S,i}(x) + C_{1,i}(x)y + C_{2,i}(x)y^2 \tag{8.7}$$

The parabolic approximation method and boundary conditions in (8.2) through (8.5) leads to

$$\frac{\partial^2 \varphi_{S,i}(x)}{\partial x^2} - \frac{2C_i \varphi_{Si}(x)}{T_{Si}\varepsilon_{Si}} = -\frac{qN_i}{\varepsilon_{Si}} - \frac{2C_i \varphi_{g,i}}{T_{Si}\varepsilon_{Si}} \tag{8.8}$$

The solution to previous equation is derived and is given as

$$\varphi_{S,i}(x) = A_i \exp\left(\frac{x - x_{i-1}}{L_{d,i}}\right) + B_i \exp\left(-\frac{x - x_{i-1}}{L_{d,i}}\right) + \frac{qN_i L_{d,i}^2}{\varepsilon_{Si}} + \varphi_{g,i} \tag{8.9}$$

In Equation (8.9), $L_{d,i}$ and x_{i-1} are expressed as

$$L_{d,i} = \sqrt{\frac{T_{Si}\varepsilon_{Si}}{C_i}} \tag{8.10}$$

$$x_{i-1} = \sum_{m=0}^{m=(i-1)} L_m, \mathrm{L}_0 = 0 \tag{8.11}$$

To determine A_i and B_i in (8.9), other surface boundary conditions must be considered. These boundary conditions are as follows.

The potential at the drain and source regions is given by (8.12) and (8.13) as

$$\varphi_{S,1}(0) = V_{\mathrm{bi},1} + V_{\mathrm{ds}} \tag{8.12}$$

$$\varphi_{S,4}\left(L_1 + L_2 + L_3 + L_4\right) = V_{bi,4} \tag{8.13}$$

Since the potential is continuous across the channel, so

$$\varphi_{S,1}\left(L_1\right) = \varphi_{S,2}\left(L_1\right) \tag{8.14}$$

$$\varphi_{S,2}\left(L_1 + L_2\right) = \varphi_{S,3}\left(L_1 + L_2\right) \tag{8.15}$$

$$\varphi_{S,3}\left(L_1 + L_2 + L_3\right) = \varphi_{S,4}\left(L_1 + L_2 + L_3\right) \tag{8.16}$$

The electric flux at the interface of regions is continuous, thus

$$\frac{\partial \varphi_{S(i)}(x)}{\partial x} = \frac{\partial \varphi_{S(i+1)}(x)}{\partial x} \tag{8.17}$$

Here, V_{bi} is the built-in potential, V_{ds} is the drain–source voltage, and i varies from 1 to 3. Substituting the boundary conditions (8.12) through (8.17) in (8.9) gives

$$A_i = -\frac{1}{2\sinh\left(\frac{L_i}{L_{d,i}}\right)}\left[\varphi_{i-1}\exp\left(-\frac{L_i}{L_{d,i}}\right) - \left(\frac{qN_iL_{d,i}^2}{\varepsilon_{Si}} + \varphi_{g,i}\right)\left(1+\exp\left(-\frac{L_i}{L_{d,i}}\right)\right) - \varphi_i\right] \tag{8.18}$$

$$B_i = \frac{1}{2\sinh\left(\frac{L_i}{L_{d,i}}\right)}\left[\varphi_{i-1}\exp\left(\frac{L_i}{L_{d,i}}\right) - \left(\frac{qN_iL_{d,i}^2}{\varepsilon_{Si}} + \varphi_{g,i}\right)\left(1+\exp\left(\frac{L_i}{L_{d,i}}\right)\right) - \varphi_i\right] \tag{8.19}$$

The vertical $E_{y,i}$ (x, y) and lateral $E_{x,i}$ (x, y) electric fields in region R_i can be obtained by differentiating the channel potential with respect to vertical and lateral axis:

$$E_{y,i}(x,y) = \frac{\partial\varphi(x,y)}{\partial y} = \frac{C_i}{\varepsilon_{Si}}\left(\varphi_{S,i}(x) - \varphi_{g,i}\right)(1-2y) \tag{8.20}$$

$$E_{x,i}(x,y) = \frac{\partial\varphi(x,y)}{\partial x} = \frac{1}{L_{d,i}}\left[A_i\exp\left(\frac{x-x_{i-1}}{L_{d,i}}\right) - B_i\exp\left(-\frac{x-x_{i-1}}{L_{d,i}}\right)\right]\left(1+\frac{C_i}{\varepsilon_{Si}}y - \frac{C_i}{\varepsilon_{Si}}y^2\right) \tag{8.21}$$

Using Kane's model for BTBT, the generation rate of carriers tunneling from the valence band of source to the conduction band of channel is expressed as [31]

$$G_{BTBT} = A_{kane}E^{\alpha}\exp\left(-\frac{B_{kane}}{E}\right) \tag{8.22}$$

where E is the local electric field and α is a material-dependent constant parameter. For direct bandgap materials (e.g., InAs), the value of α is 2; for indirect band gap materials (Si), it is equal to 2.5; A_{kane} and B_{kane} are the tunneling process–dependent parameters. The drain current (I_d) can be calculated numerically by integrating tunneling generation rate over the entire tunneling volume, given by (8.23).

$$I_d = q\iiint G_{BTBT}\, dV \tag{8.23}$$

8.5 Results and Discussion

The band diagram representation of DL TFET and DM–DG doping-less TFET in the off-state and on-state is shown in Figure 8.4a and b. It is observed that in the off-state, tunneling width between the conduction band of channel and valance band of source is too large for tunneling to occur. In the on-state (Figure 8.4b), application of $V_{gs} = 1\text{V}$ causes the conduction band of the channel to shift down and get aligned with the drain region conduction band. Due to the lower work function of the tunneling gate, energy band lowers at the source–channel interface, resulting in an abrupt change in tunneling width,

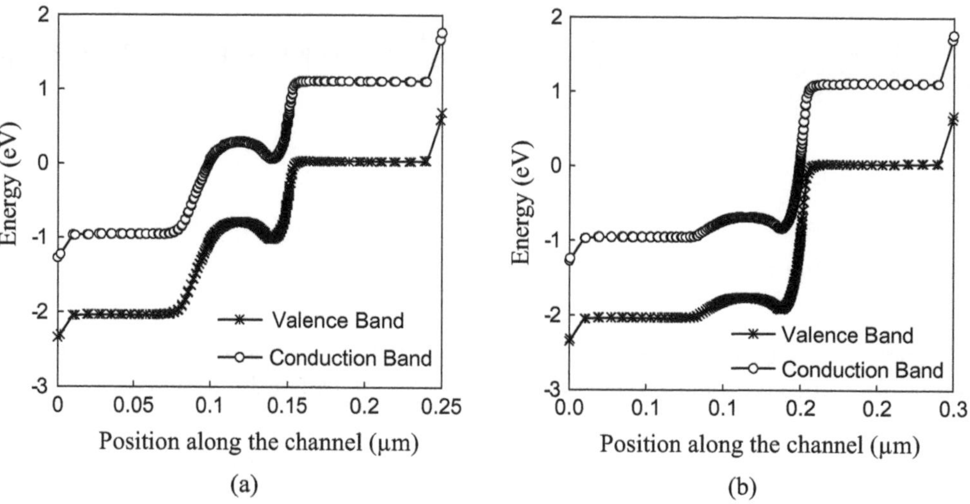

FIGURE 8.4
Valence and conduction band energy diagrams of DM–DG doping-less TFET at 1 nm below the Si–SiO_2 interface in (a) off-state ($V_{gs} = 0$, $V_{ds} = 1$ V) (b) on-state ($V_{gs} = V_{ds} = 1$ V).

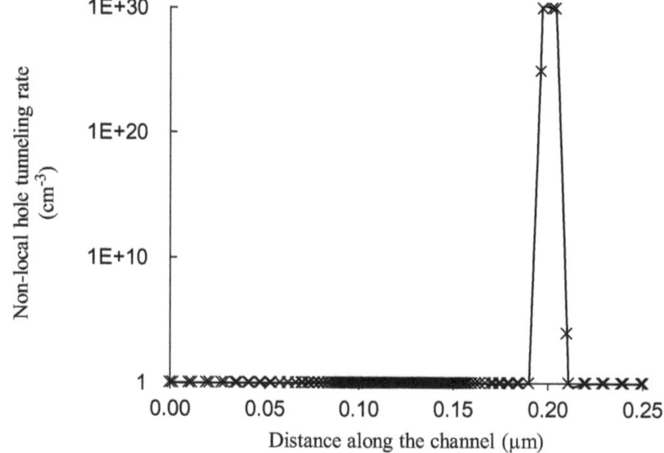

FIGURE 8.5
Tunneling rate at source–channel junction.

thereby increasing the probability of tunneling. The non-local hole tunneling rate at the source channel interface is shown in Figure 8.5. It can be seen that the maximum non-local hole tunneling rate is $1 \times 10^{30}\,cm^{-3}$.

The electron and hole concentrations of the proposed device in the off-state and on-state are shown in Figure 8.6a and b. It is observed that electron concentration in the channel region under the tunneling gate increases when gate voltage is applied as compared to the region under the auxiliary gate. This causes lowering of the tunneling barrier width, thereby resulting in a significant increase of the electric field under the tunneling gate at the source/channel interface, and it is shown in Figure 8.7. This phenomenon is quite similar to the charge plasma concept used to induce the $N+$ region to form drain in doping-less TFETs. Our proposed device DM–DG doping-less TFET is thus effective to create abruptness at the source/channel interface. The maximum value of the electric field at

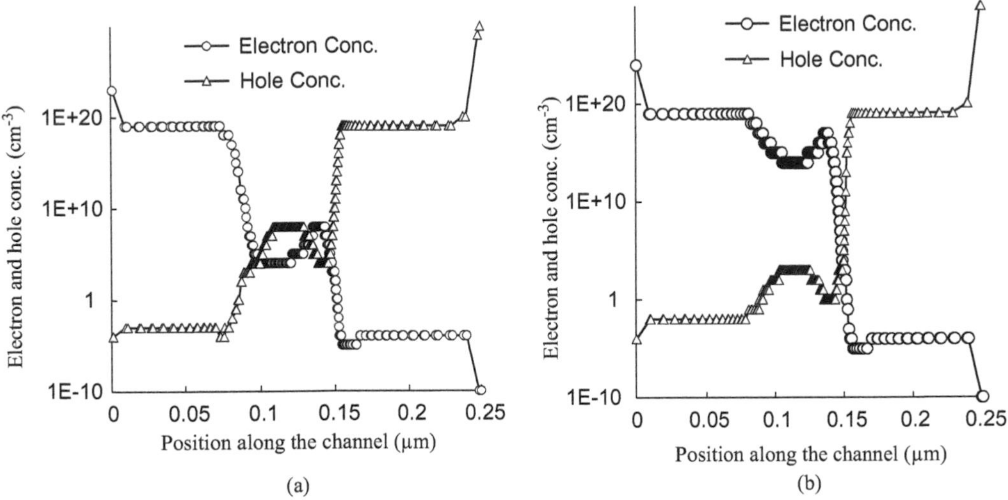

FIGURE 8.6
Electron and hole concentrations of DM–DG doping-less TFET at 1 nm below the Si–SiO_2 interface in (a) off-state ($V_{gs} = 0$, $V_{ds} = 1$ V) (b) on-state ($V_{gs} = V_{ds} = 1$ V).

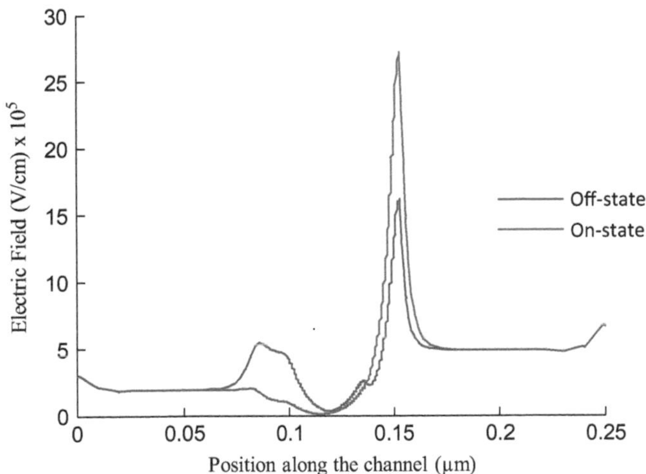

FIGURE 8.7
Variation of electric field in DM–DG doping-less TFET along the channel.

the interface between source and channel (at $x = 0.15$ μm) in the on-state is 2.72×10^6 V/cm while in the off-state, it is 1.51×10^6 V/cm. This large value of electric field under on-state condition is useful for achieving a better BTBT generation rate.

The variation of surface potential profile of the proposed device along the position of a channel in on-state is shown in Figure 8.8. It can be seen that the surface potential is constant up to 0.137 μm along the channel. At source–channel interface—i.e., at 0.15 μm—a peak is seen, which implies a sudden increase in the surface potential due to tunneling gate.

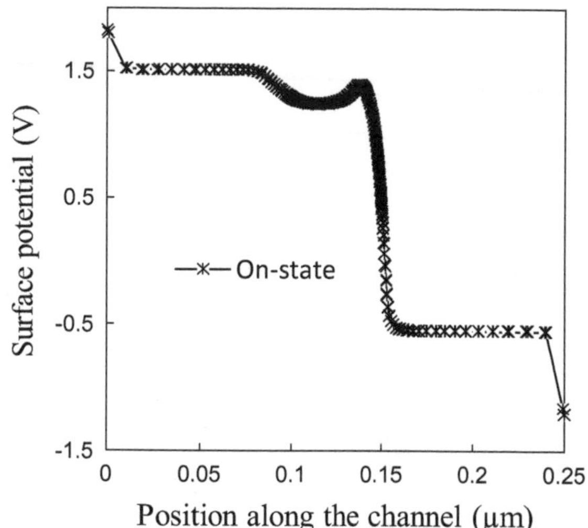

FIGURE 8.8
Surface potential profile of DM–DG doping-less TFET along the channel length in on and off states.

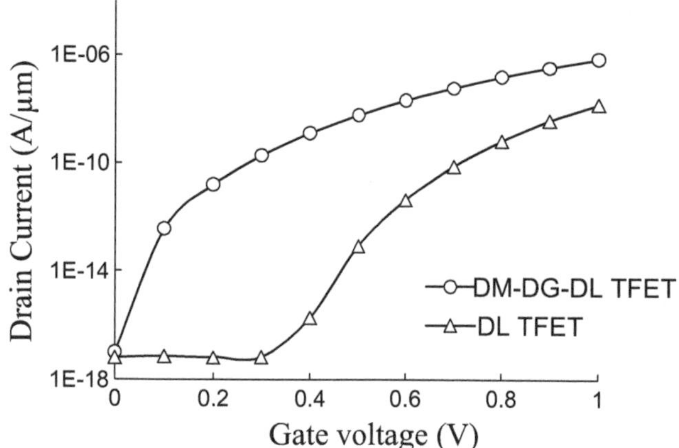

FIGURE 8.9
Transfer characteristics of DM–DG doping-less TFET and DL TFET.

The transfer characteristics of the proposed device and conventional DL TFET are presented in Figure 8.9. It is observed that due to the steeper characteristics of energy bands at the source/channel interface, abrupt switching of the device from off-state to on-state occurs. Thus, on-state current in DM–DG doping-less TFET is significantly higher than the conventional doping-less TFET. The value of on-current in conventional DL TFET is 10^{-8} A/μm, while for the proposed device, it is 10^{-6} A/μm. The improvement in subthreshold swing in the proposed device can also be seen in this figure.

The work function of the tunneling gate also plays a significant role in enhancing on-current of the device. The transfer characteristics of DM–DG doping-less TFET for different gate work functions are shown in Figure 8.10. It is evident from Figure 8.10 that with a decrease in gate work function, the device on-current increases and vice versa. This is attributed to the fact that low work function increases electron

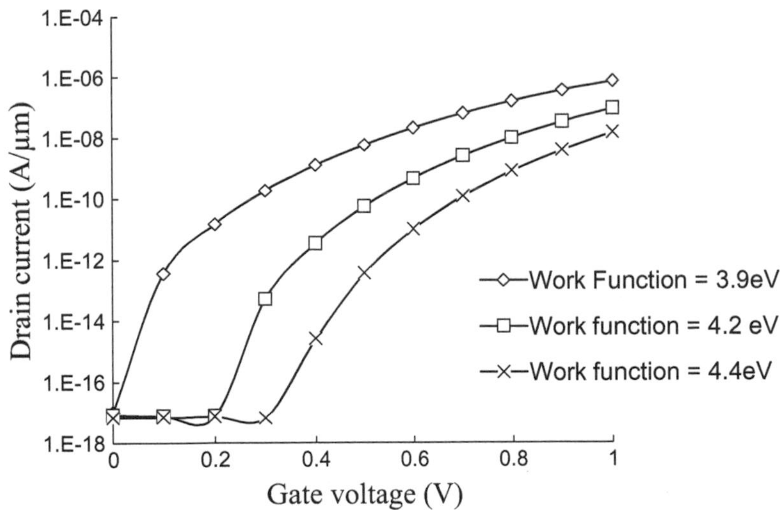

FIGURE 8.10
Transfer characteristics of DM–DG doping-less TFET with different gate work functions.

concentration in the region under the tunneling gate, resulting in greater alignment of energy bands, hence high on-current.

The output characteristic of the proposed device is shown in Figure 8.11. It is clear that at small drain voltages, the drain current increases exponentially and then gets saturated. The saturation of drain current shows that tunneling width does not change much and becomes constant after a certain drain voltage.

A comparative analysis of DL TFET and DM–DG doping-less TFET is provided in Table 8.1. Three key parameters, namely the ratio of on-current by off-current, subthreshold swing, and threshold voltage are considered.

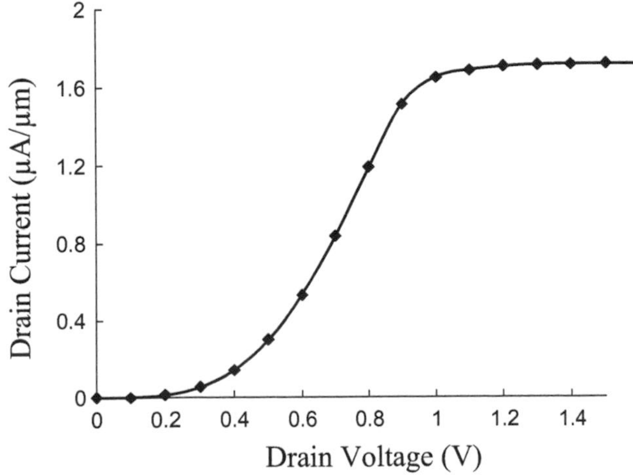

FIGURE 8.11
Output characteristics of DM–DG doping-less TFET at $V_{gs} = 1\,\text{V}$.

TABLE 8.1

Key Device Parameters for Two Types of TFETs

Parameters	DL TFET	DM–DG DL TFET
I_{on}/I_{off}	10^9	10^{11}
V_T	0.37 V	0.30 V
SS	37 mV/decade	22 mV/decade

The average subthreshold swing, SS, is computed using the relationship given in (8.24) as

$$SS = \frac{V_T - V_{off}}{\log I_{V_T} - \log I_{off}} \tag{8.24}$$

In Equation (8.24), the gate voltage at which the drain current becomes equal to 1×10^{-7} A/µm is taken as the threshold voltage V_T, V_{off} is the gate voltage at which the device is in the off-state, I_{VT} is drain current at threshold, and I_{off} is the current at zero gate voltage. The average sub-threshold swing for the DM–DG doping-less TFET is 22 mV/decade, which is better than the DL TFET (37 mV/decade). Table 8.1, also shows that the proposed device provides improved I_{on}/I_{off} ratio and low threshold voltage.

8.6 Conclusions

A detailed study of the DM–DG doping-less TFET using charge plasma concept and work function engineering is presented. The work function of auxiliary and tunneling gates considered are 4.5 eV and 3.9 eV, respectively. Analytical expressions governing surface potential in different regions across the channel are presented. It is observed that the maximum value of the electric field at the interface between source and channel in on-state is 2.72×10^6 V/cm while in the off-state, it is 1.51×10^6 V/cm. This large value of electric field under on-state condition is useful for achieving better BTBT generation rate. Moreover, optimization of gate work function leads to improved electrical characteristics of the proposed device. For example, low gate work function results in greater alignment of energy bands and hence high on-current. The DM–DG doping-less TFET shows significant improvements in on–off current ratio, threshold voltage, and subthreshold swing over conventional doping-less TFET. Thus, DM–DG doping-less TFET can be considered as a potential candidate for future VLSI applications due to its simple fabrication process, low thermal budget, less susceptibility to process variations, steeper subthreshold swing, and enhanced on-current.

Acknowledgment

The authors acknowledge with gratitude the technical and financial support received from the Science and Engineering Research Board, Department of Science and Technology (SERB–DST) through the Start-up Research Grant for Young Scientists (Ref. No.: YSS/2015/001122/ES).

References

1. D. Kahng, "A historical perspective on the development of MOS transistors and related devices," *IEEE Transactions on Electron Devices*, vol. 23, no. 7, pp. 655–657, 1976.
2. W. F. Brinkman, D.E. Haggan, and W.W. Troutman, "A history of the invention of the transistor and where it will lead us," *IEEE Journal of Solid–State Circuits*, vol. 32, no. 12, pp. 1858–1865, 1997.
3. A.C. Seabaugh and Q. Zhang, "Low-voltage tunnel transistors for beyond CMOS logic," in *Proceedings of the IEEE*, vol. 98, no. 12, pp. 2095–2110, 2010.
4. A.M. Ionescu and H. Riel, "Tunnel field-effect transistors as energy efficient electronic switches," *Nature*, vol. 479, no. 7373, pp. 329–337, 2011.
5. S. Saurabh and M.J. Kumar, "Novel attributes of a dual material gate nanoscale tunnel field-effect transistor," *IEEE Transactions on Electron Devices*, vol. 58, no. 2, pp. 404–410, 2011.
6. M.S. Ram and D.B. Abdi, "Single grain boundary tunnel field effect transistors on recrystallized polycrystalline silicon: Proposal and investigation," *IEEE Electron Device Letters*, vol. 35, no. 10, pp. 989–991, 2014.
7. D.B. Abdi and M.J. Kumar, "Controlling ambipolar current in tunneling FETs using overlapping gate-on-drain," *IEEE Journal of Electron Devices Society*, vol. 2, no. 6, pp. 187–190, 2014.
8. C. Le Royer and F. Mayer, "Exhaustive experimental study of tunnel field effect transistors (TFETs): From materials to architecture," in *Proceedings of the 10th International Conference on Ultimate Integration of Silicon*, March 2009.
9. D. Leonelli et al., "Optimization of tunnel FETs: Impact of gate oxide thickness, implantation and annealing conditions," in *Proceedings of the European Solid-State Device Research Conference*, pp. 170–173, 2010.
10. N. Damrongplasit et al., "Study of random dopant fluctuation effects in Germanium-source tunnel FETs," *IEEE Transactions on Electron Devices*, vol. 58, no. 10, pp. 3541–3548, 2011.
11. N. Damrongplasit et al., "Study of random dopant fluctuation induced variability in the raised–Ge source TFET," *IEEE Electron Device Letters*, vol. 34, no. 2, pp. 184–186, 2013.
12. G. Leung and C.O. Chui, "Stochastic variability in silicon double gate lateral tunnel field-effect transistors," *IEEE Transactions on Electron Devices*, vol. 60, no. 1, pp. 84–91, 2013.
13. K. Boucart, A.M. Ionescu, and W. Riess, "A simulation-based study of sensitivity to parameter fluctuations of silicon Tunnel FETs," in *Proceedings of the European Solid-State Device Research Conference*, 2010.
14. D. Verreck et al., "Quantum mechanical performance predictions of p-n-i-n versus pocketed line tunnel field-effect transistors," *IEEE Transactions on Electron Devices*, vol. 60, no. 7, pp. 2128–2134, 2013.
15. M.J. Kumar and S. Janardhanan, "Doping-less tunnel field effect transistor: Design and investigation," *IEEE Transactions on Electron Devices*, vol. 60, no. 10, pp. 3285–3290, 2013.
16. B. Rajasekharan et al., "Fabrication and characterization of the charge-plasma diode," *IEEE Electron Device Letters*, vol. 31, no. 6, pp. 528–530, 2010.
17. R.J.E. Hueting, B. Rajasekharan, C. Salm, and J. Schmitz, "Charge plasma P-N diode," *IEEE Electron Device Letters*, vol. 29, no. 12, pp. 1367–1368, 2008.
18. M.J. Kumar and K. Nadda, "Bipolar charge plasma transistor: A novel three terminal device," *IEEE Transactions on Electron Devices*, vol. 59, no. 4, pp. 962–967, 2012.
19. K. Nadda and M.J. Kumar, "Schottky Collector bipolar transistor without impurity doped emitter and base: Design and performance," *IEEE Transactions on Electron Devices*, vol. 60, no. 9, pp. 2956–2959, 2013.
20. D.B. Abdi and M.J. Kumar, "In-built N+ Pocket p-n-p-n tunnel field-effect transistor," *IEEE Electron Device Letters*, vol. 35, no. 12, pp. 1170–1172, 2014.

21. Y. Gao, T. Low, and M. Lundstrom, "Possibilities for VDD = 0.1 V logic using carbon-based tunneling field effect transistors," in *Proceedings of the IEEE International Symposium on VLSI Technology*, pp. 180–181, 2009.
22. J. Knoch and J. Appenzeller, "Tunneling phenomena in carbon nanotube field-effect transistors," *Physica Status Solidi (a)*, vol. 205, no. 4, pp. 679–694, 2008.
23. C. Sandow, J. Knoch, C. Urban, Q.-T. Zhao, and S. Mantl, "Impact of electrostatics and doping concentration on the performance of silicon tunnel field-effect transistors," *Solid-State Electronics*, vol. 53, no. 10, pp. 1126–1129, 2009.
24. B. Bhushan, K. Nayak, and V.R. Rao, "DC compact model for SOI tunnel field-effect transistors," *IEEE Transactions on Electron Devices*, vol. 59, no. 10, pp. 2635–2642, 2012.
25. W.Y. Choi and W. Lee, "Hetero-gate-dielectric tunneling field-effect transistors," *IEEE Transactions on Electron Devices*, vol. 57, no. 22, pp. 2317–2319, 2010.
26. N. Cui, R. Liang, and J. Xu, "Heteromaterial gate tunnel field effect transistor with lateral energy band profile modulation," *Applied Physics Letters*, vol. 98, no. 14, 2011.
27. C.Y. Hsu, C.Y. Chang, E.Y. Chang, and C. Hu, "Suppressing nonuniform tunneling in InAs/GaSb TFET with dual-metal gate," *IEEE Journal of the Electron Devices Society*, vol. 4, no. 4, pp. 60–65, 2016.
28. H.W. Kim, J.H. Lee, W. Kim, M.C. Sun, J.H. Kim, G. Kim, K.W. Kim, H. Kim, J.Y. Seo, and B.G. Park, "A tunneling field-effect transistor using side metal gate/high-κ material for low power application," in *Proceedings of the International Semiconductor Device Research Symposium*, pp. 1–2, 2011.
29. R. Liang, N. Cui, M. Zhao, J. Wang, and J. Xu, "High performance tunnel field effect transistor with a tri-material-gate structure," in *Proceedings of the International Semiconductor Device Research Symposium*, pp. 1–2, 2011.
30. S. Kumar, E. Goel, K. Singh, B. Singh, M. Kumar, and S. Jit, "A compact 2-D analytical model for electrical characteristics of double-gate tunnel field-effect transistors with a SiO2/High-k stacked gate-oxide structure," *IEEE Transactions on Electron Devices*, vol. 60, no. 8, pp. 3291–3299, 2016.
31. E.O. Kane, "Zener tunneling in semiconductors," *Journal of Physics and Chemistry of Solids*, vol. 12, no. 2, pp. 181–188, 1960.

Section IV

Graphene and Carbon Nanotube Transistors and Applications

9

Modeling of Graphene Plasmonic Terahertz Devices

Neetu Joshi and Nagendra P. Pathak

CONTENTS

9.1 Introduction

The terahertz frequency band 0.1–10 THz has enchanted a huge number of researchers across the globe for its vast potential in the trending era of mobile communications in order to provide wide spectrum and large capacity to the current high-speed wireless links scenario [1–7]. It will provide immensely high data rates above 10 gigabits per second (Gbps) to throw an impetus to the current technology toward a better future for the present generation of highly demanding wireless communication systems. The THz band will definitely arise as a boon to state-of-the-art wireless systems, resulting in large nanoscale applications in the field of communication, military equipment, and biomedical instrumentation. This band has been quite unexplored owing to lesser availability of sources and detectors in this regime as compared to the neighborhood microwave and IR frequency regions.

The THz frequencies can be used for data communication in terabits per second (Tbps) resorting to eye-blinking speeds as compared to the microwave frequency communication in the order of Gbps. The current digital modulation methods such as OFDM, ASK, PSK, and MIMO limit permissible data rates because of lesser spectral availabilities. For instance, 1 Gbps data rates are permitted in the present digital communication systems such as the 4-by-4 MIMO-based LTE-A system. The millimeter (mm) wave-based systems will allow data rates of 10 Gbps. However, it is still a long route to travel to the surge of Tbps data rates. Higher data rates will increase the complexity at the transmitter and receiver ends such as more power consumption and thermal losses management. In addition, the THz wireless communication systems are more prone to the atmospheric losses, resulting in development of ultra-small cells needing repeaters or regenerators at small distances in order to avoid mismatching between transmitter and receiver. Moreover, these effects are lesser in non-LOS wireless communication systems [8]. For wide-area networks, LOS communication at 10 Gbps [9] and FSO communication with 1 Gbps [10] and 1.28 Tbps [11] is available in the visible frequency range. Nevertheless, the communication equipment needed is comparatively larger, limiting this frequency range for wireless mobile communication.

The THz wave communication provides wideband and high-frequency data links that offer bandwidths higher than the present mm wave communications but lower than that of optical fiber links. Certain issues at the device, instrumentation, and communication levels need to be resolved in order to overcome the research gap, such as huge path loss, at these frequencies. Another issue is the frequency selective path loss, which requires the development of efficient ultra-broadband antennas. These issues are quite similar to those in microwave and mm wave wireless communication counterparts.

9.2 Material Properties of Graphene

Graphene, the nanoscale thick carbon material, provides quantized electronic vibrational resonances known as Dirac Fermions, with high speeds [12–14]. It behaves in an unusual manner with respect to applied magnetic and electric fields because of quantum Hall effect and Klein paradox. The graphene material is sp^2 hybridized with fully occupied s-orbitals that have valence bands, creating sigma bond between the C–C atoms. Thus, a honeycomb cross-sectional planar pattern has been developed. In addition, the p-orbitals have incomplete conduction bands creating pi bonds among the carbon-carbon atom. The graphene material has zero bandgap characteristics with infinitely small bandgap similar to metals. Next, we will describe the electronic, THz, and optical properties of graphene in the following subsections [13].

9.2.1 Electronic Properties

The shape of the lattice pattern of graphene resembles honeycomb, which is shown in Figure 9.1a. The unit cell basis vectors, presented in Figure 9.1b, can be written as:

$$\mathbf{a}_1 = \frac{\mathbf{a}}{2}\left(3, \sqrt{3}\right), \mathbf{a}_2 = \frac{\mathbf{a}}{2}\left(3, -\sqrt{3}\right) \tag{9.1}$$

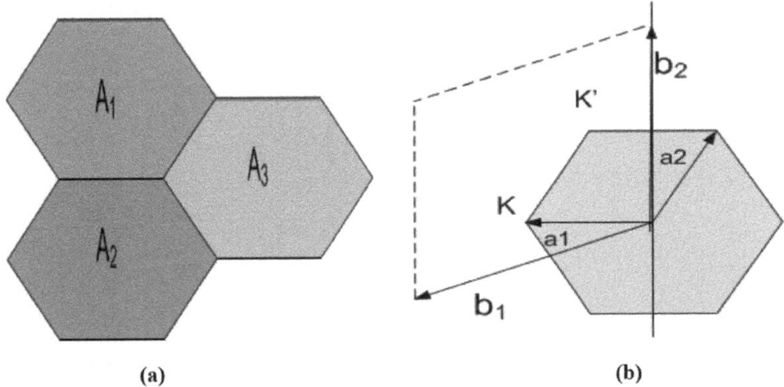

FIGURE 9.1
(a) Lattice pattern of graphene and (b) its first Brillouin zone.

Here, **a** is the radius of the first Brillouin zone, and the corresponding reciprocal lattice vectors represented by **b** are given as

$$\mathbf{b}_1 = \frac{2\pi}{3\mathbf{a}}\left(1,\sqrt{3}\right), \mathbf{b}_2 = \frac{2\pi}{3\mathbf{a}}\left(1,-\sqrt{3}\right) \tag{9.2}$$

In addition, the two Dirac points in the graphene Brillouin zone K and K' are given by

$$K = \left(\frac{2\pi}{3\mathbf{a}}, \frac{2\pi}{3\sqrt{3}\mathbf{a}}\right), K' = \left(\frac{2\pi}{3\mathbf{a}}, -\frac{2\pi}{3\sqrt{3}\mathbf{a}}\right) \tag{9.3}$$

According to the tight binding theory for graphene, there are two valence electrons in the valence band, but no electrons are found in the conduction band. At the Dirac points with $E_F = \hbar\nu_f$ where E_F is the Fermi energy level, ν_f the Fermi velocity, n is the concentration of charge carriers and $\hbar = h/2\pi$ (h is Planck constant), the dispersion curve remains nearly equal. It is rightly termed as a zero-bandgap semiconductor because of the zero values of Fermi velocity. The conductivity of graphene is a constant value given by $\sigma = e^2/4\hbar$ (where e is the charge of one electron), being dependent only on constant values. Moreover, 2.6% of the visible light is absorbed in monolayer graphene, which agrees reasonably well with the fine structure coefficient, which is $\pi\alpha$ where $\alpha = e^2/4\pi\varepsilon_0\hbar c$.

9.2.2 Terahertz Properties of Graphene

The material graphene provides excellent capabilities in the THz frequency regime because the plasmon polaritons developed at the surface have very high volume densities, leading to greater confinements and thus, localization. It provides a variety of applications such as in phase shifters, filters, and artificially engineered metamaterials [15–17]. Also, it has the potential to support a huge number of charge carriers, allowing tunabilities with applied chemical potential and relaxation times. It provides large values of surface resistance with the externally applied voltage. The surface conductivity (σ) of graphene has been defined by the Kubo formula in which the major contribution consists of interband transitions. It is given by Equation (9.4):

$$\sigma(\omega,\mu_c,\Gamma,T)=\frac{je^2(\omega-j2\Gamma)}{\pi\hbar^2}\left[\frac{1}{(\omega-j2\Gamma)^2}\int_0^\infty \varepsilon\left(\frac{\partial f_d(\varepsilon)}{\partial\varepsilon}-\frac{\partial f_d(-\varepsilon)}{\partial\varepsilon}\right)d\varepsilon -\int_0^\infty \frac{f_d(-\varepsilon)-f_d(\varepsilon)}{(\omega-j2\Gamma)^2-4\left(\frac{\varepsilon}{\hbar}\right)^2}d\varepsilon\right] \tag{9.4}$$

where e is the electronic charge, $\hbar=h/2\pi$ (h is Planck constant), k_B is the Boltzmann constant, ω is the frequency of the wave, Γ is scattering rate, μ is the chemical potential, ε is the electron energy, and T is the temperature. Here, $f_d(\varepsilon)=(e^{(\varepsilon-\mu_c)/k_BT+1})^{-1}$ gives the Fermi-Dirac distribution function. The desired characteristics can be obtained by change in the chemical doping, applied electrical voltage, or magnetic field via change in Fermi level, electrostatic field, or Hall effect. The condition to excite the SPPs in graphene is proper matching of the wave-vector of graphene plasmon polaritons (GPPs) with the incident electromagnetic wave. The quantized surface plasmonic resonances (SPRs) in graphene leads to greater confinements with lower losses in the THz frequency band. The sub-wavelength confinements of GPPs are considerably lower than the diffraction limits. The charge carriers induced through doping or applying electrical voltage bias provide effective tunability features as compared to inert metal counterparts where no such mechanism of tuning has been provided. The relaxation or switching time determines the values for damping coefficients in graphene. Graphene has compatibility with the prior silicon technologies and can be applied well in the field of nanophotonics.

9.2.3 Optical Properties

There are numerous factors such as temperature, carrier density, or frequency that affect the reflectance and transmittance properties of graphene. The surface conductivity (σ) given by the Kubo formula shown in Equation (9.4) consists of the major contribution of interband transitions.

9.3 Graphene-Based Plasmonic Waveguide Structures

The simplest graphene-based plasmonics waveguide structure supporting electromagnetic wave propagation is Graphene Parallel Plate Waveguide (GPPW), shown in Figure 9.2a and b. The characteristics of GPPWs have been explored rigorously in the last few years [18–21]. Hanson proved the propagation of quasi-TEM mode in GPPW [22]. The graphene layer possesses complex surface conductivities that depend on the chemical potential. Simultaneously, it may be tuned with the help of the doping concentration or by an externally applied electric field or a magnetic field via Hall effects, which can provide the isotropic and anisotropic values of the conductivity. There are various models for calculation of the conductivity in graphene such as the semi-classical model and relaxation

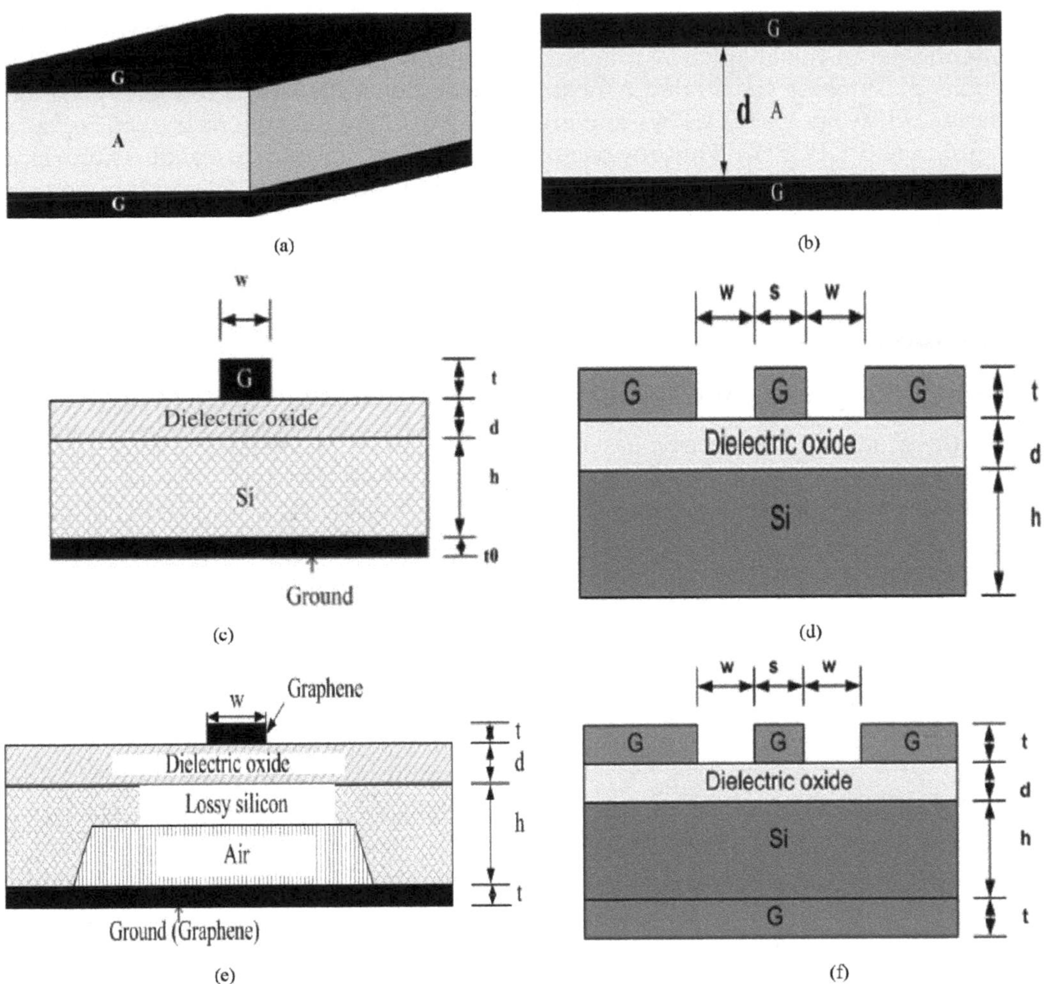

FIGURE 9.2
(a) Three-dimensional view of GPPW (b) Two-dimensional cross-sectional view of GPPW (c) 2-D cross-sectional view of graphene plasmonic nanostrip waveguide (d) 2-D cross-sectional view of graphene plasmonic coplanar waveguide (e) 2-D cross-sectional view of graphene plasmonic suspended nanostrip waveguide (f) 2-D cross-sectional view of graphene plasmonic graphene backed coplanar waveguide. Here, *G* stands for graphene material, and *A* stands for air or dielectric.

time approximation (RTA) formulation, which describes the effect of spatial dispersion in decreasing the confinement and losses of slow surface plasmons in the parallel plate waveguides [18]. To increase the confinement, a dielectric layer is inserted between the anisotropic graphene plates, and the value of static electric field is increased by enhancing the graphene conductivity [19]. Recent developments show that the characteristics of terahertz surface plasmons such as propagation length and localization length can be improved by the use of Kerr-type nonlinear media [20] and one-dimensional photonic crystal [21]. By changing the width of one plate of GPPW and composition of dielectric material between the plates, we can obtain several other waveguide structures, which have either planar or quasi-planar geometry as shown in Figure 9.2. These waveguides are mentioned in the chapter as graphene plasmonic nanostrip waveguide (GPNSW), graphene plasmonics suspended nanostrip waveguide (GPSNW), graphene plasmonics coplanar waveguide (GPCPW), and graphene plasmonics graphene backed coplanar waveguide (GB-GPCPW). These waveguide structures can be fabricated using rigorous processes such as exfoliation, spin coating, and CVD process, depending on their applications [23].

9.4 Modal Properties of Graphene Plasmonics Parallel Plate Waveguide

Firstly, the geometry of graphene-based plasmonic parallel plate waveguide is shown in Figure 9.2a, which consists of a dielectric material layer sandwiched between two large graphene layers. This wave-guiding structure supports TM, TE, and quasi-TEM modes [22,24]. Now, applying boundary conditions for tangential components of electric and magnetic fields, we can write:

$$\left.\begin{aligned}&(E_1 - E_2)\times n_{1-2} = 0\\ &(H_1 - H_2)\times n_{1-2} = \frac{4\pi}{c}\sigma(\omega,\mu,\gamma,T)E_{||}\end{aligned}\right\} \tag{9.5}$$

Here, n_{1-2} is the unit vector along the normal, oriented from region 1 to region 2, and $E_{||}$ is the electric field of the wave in the xz plane, which induces current in the graphene layers. Further, we obtain the equations for TM polarized waves as

$$\begin{bmatrix} 1 & -1 & -e^{q'd'} & 0 \\ i\frac{4\pi}{c}\frac{q}{k_0}\sigma+1 & -\varepsilon\frac{q}{q'} & \varepsilon\frac{q}{q'}e^{q'd} & 0 \\ 0 & \varepsilon\frac{q}{q'}e^{q'd} & -\varepsilon\frac{q}{q'} & i\frac{4\pi}{c}\frac{q}{k_0}\sigma+1 \\ 0 & -e^{q'd'} & -1 & 1 \end{bmatrix} \times \begin{bmatrix} E_1 \\ E_2^+ \\ E_2^- \\ E_3 \end{bmatrix} = 0 \tag{9.6}$$

and the equations for TE polarized waves as

$$\begin{pmatrix} 1 & -\varepsilon\frac{q}{q'} & \varepsilon\frac{q}{q'}e^{q'd} & 0 \\ i\frac{4\pi}{c}\frac{k_0}{q}\sigma-1 & 1 & e^{q'd} & 0 \\ 0 & e^{q'd} & 1 & i\frac{4\pi}{c}\frac{k_0}{q}\sigma-1 \\ 0 & \varepsilon\frac{q}{q'}e^{q'd} & -\varepsilon\frac{q}{q'} & 1 \end{pmatrix} \times \begin{pmatrix} H_1 \\ H_2^+ \\ H_2^- \\ H_3 \end{pmatrix} = 0 \tag{9.7}$$

Now, setting the determinants of the matrices to zero, we obtain the dispersion relations as

$$\begin{cases} 1+i\dfrac{4\pi}{c}\dfrac{q}{k_0}\sigma(\omega)=-\varepsilon\dfrac{q}{q'}\dfrac{e^{q'd}+1}{e^{q'd}-1} \\ 1+i\dfrac{4\pi}{c}\dfrac{q}{k_0}\sigma(\omega)=-\varepsilon\dfrac{q}{q'}\dfrac{e^{q'd}-1}{e^{q'd}+1} \end{cases} \quad \text{For TM}_1 \text{ and TM}_2 \tag{9.8}$$

$$\begin{cases} i\dfrac{4\pi}{c}\dfrac{k_0}{q}\sigma(\omega)-1=\dfrac{q'}{q}\dfrac{e^{q'd}-1}{e^{q'd}+1} \\ i\dfrac{4\pi}{c}\dfrac{k_0}{q}\sigma(\omega)-1=\dfrac{q'}{q}\dfrac{e^{q'd}+1}{e^{q'd}-1} \end{cases} \quad \text{For TE}_1 \text{ and TE}_2 \tag{9.9}$$

where, $k_0 = \omega/c$, c is the speed of light, $q = \sqrt{(\beta^2 - k_0^2)}$, $q' = \sqrt{(\beta^2 - \varepsilon_{k0}^2)}$, and $\beta = k_z$ is the magnitude of the wave vector component along the propagation direction. The propagating modes in graphene plasmonic coplanar waveguides are even and odd modes. The even mode provides low-loss THz propagation, so it is the preferred mode. Here, the TM mode has an evanescent decay in the case of metals with hyperbolic sine and cosine fields. The presence of a quasi-TEM mode can be evaluated by assuming that the graphene PPWG is having slight perturbations from the perfect PPWG [22]. Graphene SPPs, propagating on a metal-dielectric interface, possess many remarkably exceptional characteristics—for instance, higher field confinement, reduced losses, and greater propagation lengths as compared to inert metals—allowing its vast usage in the area of nanoplasmonics. It has extraordinary optical and electronic properties, used rigorously in the THz frequency regime. The plasmonic ring resonators have been studied for the microwave and IR frequencies for a long time. Now, we will study the application of GPPW for demultiplexing operation.

9.5 GPPWs for Demultiplexing Operation

Firstly, the metal-based split-ring resonators (SRRs) were utilized for enhancement of electric and magnetic field responses [25]. Several devices can be designed with the help of SRRs such as low-pass filters [26], metamaterials [27], etc. They can also be used to study the photon decay rate [28]. Additionally, SPP-based demultiplexers have been demonstrated for demultiplexing and filtering operations [29]. The wavelength-division-demultiplexer (WDM) using metal-insulator-metal (MIM) plasmonic nanodisk resonators has been designed with the help of FDTD-based electromagnetic simulations [30]. The graphene SRR WDM is discussed in the subsequent section.

SRRs have been progressively utilized for enhancement as well as localization of electric and magnetic field responses in nanoscale integrated circuit applications. The wavelength demultiplexing application is provided by input and output graphene-based nano-transmission lines connected with SRR. The demultiplexing operation in the proposed design varies with the material properties of the surrounding material and also on the chemical potential of graphene. The material properties of graphene-based complementary SRR (CSRR) WDM mainly decide the localization of transmission peaks.

The applied gate voltage in graphene changes the transmission peaks of SRR structures, which in turn characterizes the demultiplexing action of the device. The resultant transmittance simulations finally provide the demultiplexing characteristics. The curves between the transmission resonances in the frequency spectra have been plotted. Further, the dependence of sensitivity of the nano-device has been observed with variations in design dimensions.

9.5.1 Model Structure and Theoretical Analysis

Now, the geometry of the graphene-based plasmonic nanostrip waveguide along with its two-wire transmission line equivalent circuit is shown in Figure 9.3a. The desired geometry of nanostrip waveguide from GPPW can be obtained by varying the width of the upper plate while keeping lower plate unchanged [31,32]. The equivalent circuit representation requires knowledge of propagation constant γ ($\gamma = \alpha + j\beta$) and the characteristic impedance Z_0 of the nanostrip waveguide, which has been modeled with the help of full-wave EM solver CST Microwave Studio. In the full-wave simulations, we have used the following parameters to obtain the transmission line characteristics of this waveguide geometry: width of the nanostrip waveguide is w, thickness of silicon dioxide $d = 40$ nm, $h = 100$ nm, $t = t_0 = 0.5$ nm, and chemical potential μ of graphene strip: varied from 0.4 to 0.7. The aspect ratio is defined as $w/(d+h) = w/h'$. The variation of the normalized phase constant (β/k_0) and normalized attenuation constant (α/k_0) with frequency and aspect ratio is shown in Figure 9.3b–d.

The value of characteristic impedance of the graphene-based plasmonics nanostrip waveguide can be expressed with the relationships $Z_{VI} = V/I$ or $Z_{PV} = V^2/P$ or $Z_{PI} = P/I^2$, where P is the power and V and I are the voltage and currents. Among the three, the most suitable definition for nanostrip waveguide is Z_{PI}, which is computed on the basis of power and current. Here, the power flow can be obtained from the knowledge of electric and magnetic fields and current can be computed by integrating magnetic field lines along a known path. Next, the variation in characteristic impedance of the nanostrip waveguide with respect to aspect ratio is given in Figure 9.3e. It is clear from Figure 9.3e that with an increase in the aspect ratio, there is a decrease in characteristic impedance. The observed highest value of the characteristic impedance for graphene-based nanostrip waveguide is around ~3000 Ω, which is larger than its microwave counterpart, i.e., microstrip line. These large values of characteristic impedance play a major role in radical miniaturization of THz integrated circuits. Further, closed form expressions have been derived to obtain normalized phase constant and characteristic impedance of graphene-based plasmonics nanostrip waveguide as given in Equations (9.10 and 9.11).

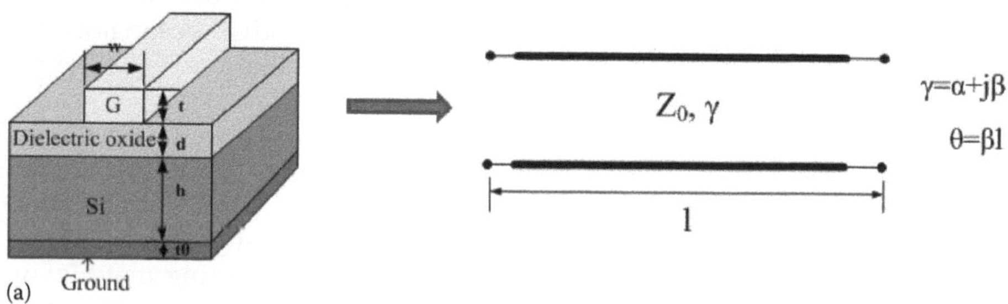

FIGURE 9.3
(a) Geometry of graphene plasmonics nanostrip waveguide along with its 2-wire transmission-line equivalent circuit.
(Continued)

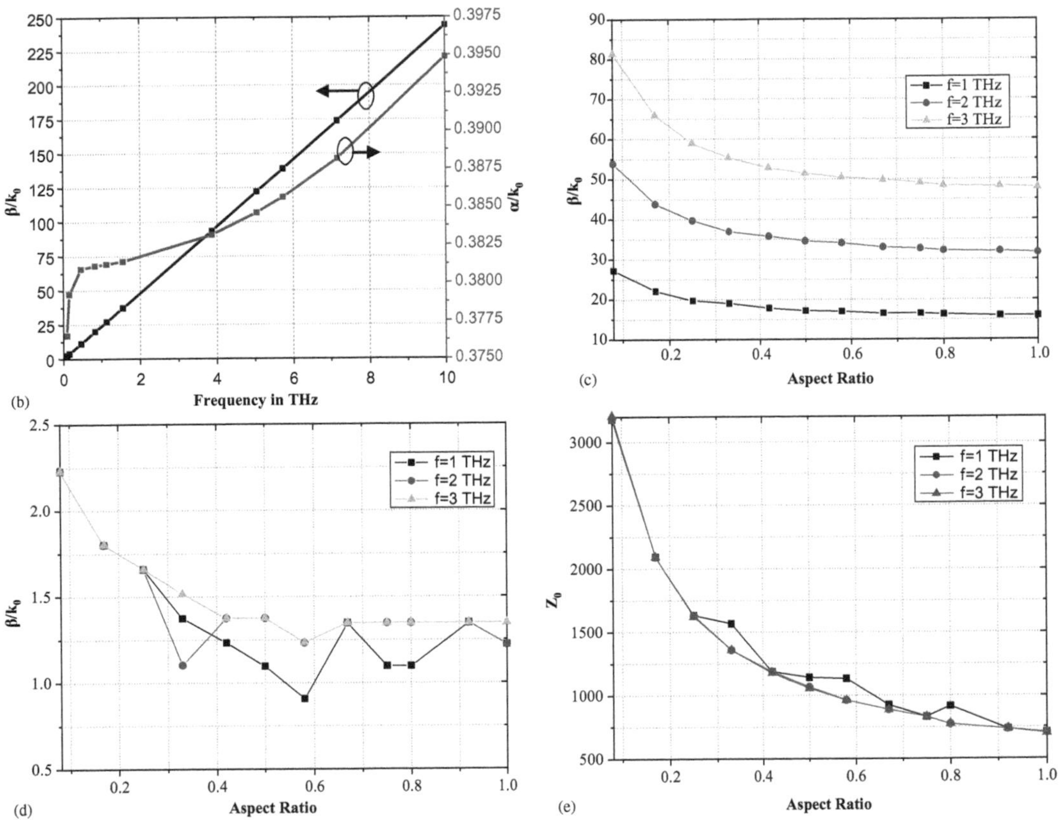

FIGURE 9.3 (Continued)
(b) Plot of β/k_0 and α/k_0 with frequency. (c) Plot of β/k_0 with aspect ratio. (d) Plot of α/k_0 with aspect ratio. (e) Plot of Z_0 with aspect ratio.

Then, the closed-form expression for evaluating normalized propagation constant $\left(\beta/k_o\right)$ is obtained as

$$\frac{\beta}{k_o} = A_2 f\left(A_1 (w/h)^2 + B_1 (w/h) + C_1\right)\left(\left(A_3 d^2 + B_3 d + C_3\right)\right)e^{-2.66\sigma} \tag{9.10}$$

Here, σ is conductivity of graphene, f is frequency of operation, h is thickness of silicon layer, and d is thickness of SiO_2 layer.

$$A_1 = \frac{\left(5.4533h^2 - 7.9759h - 605.088\right)}{\left(h^2 - 868.09h - 6063.3556\right)}$$

$$B_1 = \frac{\left(1856516h^2 + 32475202h + 499886728\right)}{\left(h^2 - 192087878h - 5114185973\right)}$$

$$C_1 = \frac{\left(0.8804h^2 - 4.6082h - 124.046\right)}{\left(h^2 - 1.2433h + 225.4\right)}$$

$$A_2 = \frac{\left(0.8703h^2 - 0.9625h - 35.7841\right)}{\left(h^2 + 3.01594h + 31.2524\right)}$$

$$A_3 = \frac{\left(0.6721h^2 - 8.9619h - 203.4982\right)}{\left(h^2 + 19.4096h - 173.1685\right)}$$

$$B_3 = \frac{\left(-6.4543h^2 - 304h - 119225.99\right)}{\left(h^2 - 161.2127h - 3131.4411\right)}$$

$$C_3 = \frac{\left(34719h^2 + 2862283h + 57340555\right)}{\left(h^2 - 257562.68h + 24617562\right)}$$

Also, the characteristic impedance, Z_0, is obtained with the help of fitting curve analytics as

$$Z_o = e^{\left(-1.5379(w/h) - 6.3263f + 2.6518\sigma + \frac{0.01475\beta}{k_o} + 8.0039\right)} \tag{9.11}$$

The computed and simulated results agree well as depicted in Figure 9.4a and b.

The device is composed of two graphene plasmonic nanostrip waveguides (GPNSW) and a ring resonator. Its schematic diagram is shown in Figure 9.5. Here, the ring resonator consists of graphene and the surrounding material is with variable refractive index. In case, the structure is placed underground, the graphene CSRR is covered by superficial sandy layer. The 2D simulations have been performed with the widths of GPNSWs at input and output side taken as d, and the side length of the ring as l. Also, the coupling width between GPNSWs and the modified ring is g. Next, the values of d, l and g are taken as 40 nm, 250 nm, and 10 nm, respectively. The resonant behavior of SRR is given by the following equation [33]:

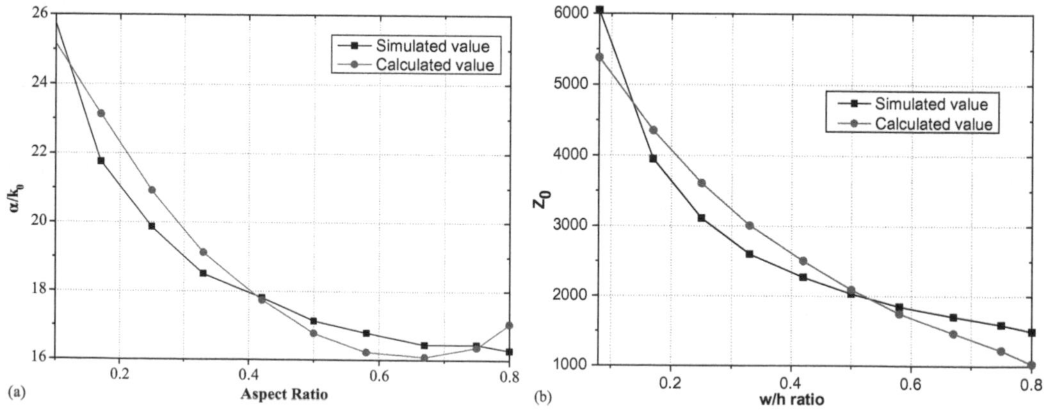

FIGURE 9.4
(a) Comparison of simulated and calculated values of normalized propagation constant with aspect ratio. (b) Comparison of simulated and calculated values of characteristic impedance with aspect ratio.

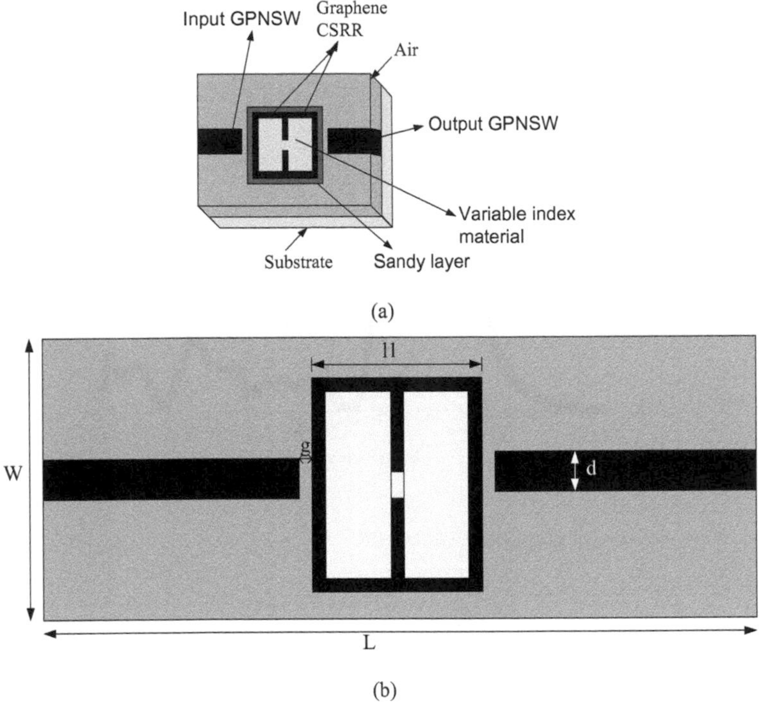

FIGURE 9.5
Schematic diagram of graphene-based CSRR. (a) 3-D view. (b) 2-D view.

$$\frac{J_n'(kR_m)}{J_n'(kR_{\text{inner}})} - \frac{N_n'(kR_m)}{N_n'(kR_{\text{inner}})} = 0 \tag{9.12}$$

Here, k is wave vector and J_n' and N_n' are the derivatives to the Bessel functions of the first kind and second kind of order n, respectively.

9.5.2 Transmission Properties of CSRR

The transmission characteristics of the graphene-based CSRR presented in the previous section are shown in Figure 9.6. The surface plasmonic resonances occur at some fixed frequencies in these characteristics. These transmission resonances have been shown with respect to the propagating frequencies. Here, the widths of input and output GPNSWs d have been considered to be 40 nm. Also, the split-ring radius is considered as 100 nm with the value of the coupling distance, g, between the waveguides and the ring resonator set to be 10 nm. There are three transmission resonances available in the frequency response curve shown in Figure 9.6. These transmission peaks are observed at 3.3 THz, 8 THz, and 9.5 THz.

Next, the transmission spectra of graphene-based CSRR has been observed further with respect to various design parameters such as chemical potential of graphene, refractive index of material, etc. [25]. Firstly, the transmission resonances have been obtained with

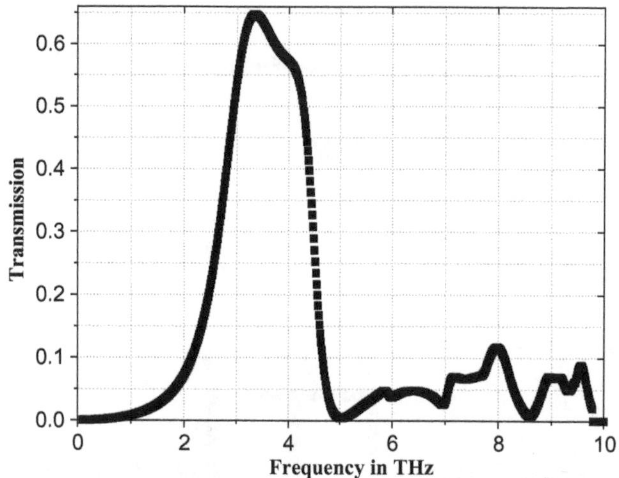

FIGURE 9.6
The transmission characteristics of graphene-based CSRR.

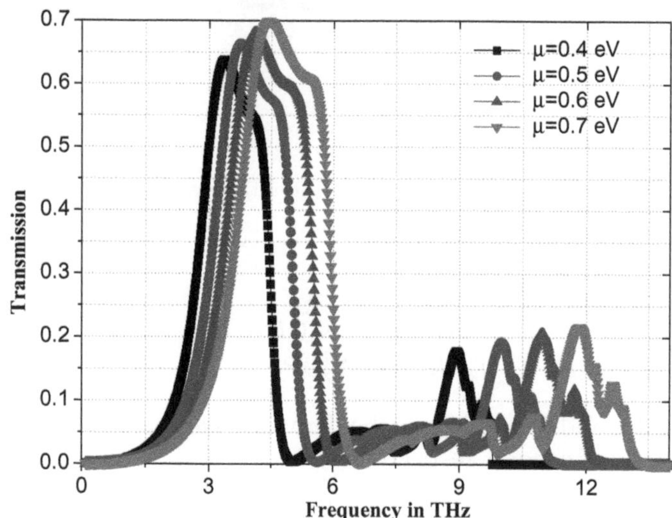

FIGURE 9.7
The transmission characteristics of graphene-based CSRR at specified values of chemical potential.

different values of chemical potential of graphene as shown in Figure 9.7. These resonances travel toward the right with the increase in the values of applied chemical potential resulting in sufficient increase in transmission. These values of chemical potential have been varied from 0.4 to 0.7 eV. The transmission characteristics show a linear variation with chemical potential.

Now, the impact of variation in values of structural parameters have been observed on the spectra of graphene-based CSRR. Firstly, the transmission peaks have been observed at different values of side lengths of the ring resonator, $l1$ = 200 nm, 225 nm, 250 nm, and 275 nm as shown in Figure 9.8. There is a relative change in the position of the peaks of the transmission spectrum with a change in values of the side lengths of the modified ring. With the increase in side lengths, the transmission resonances are obtained at lower values

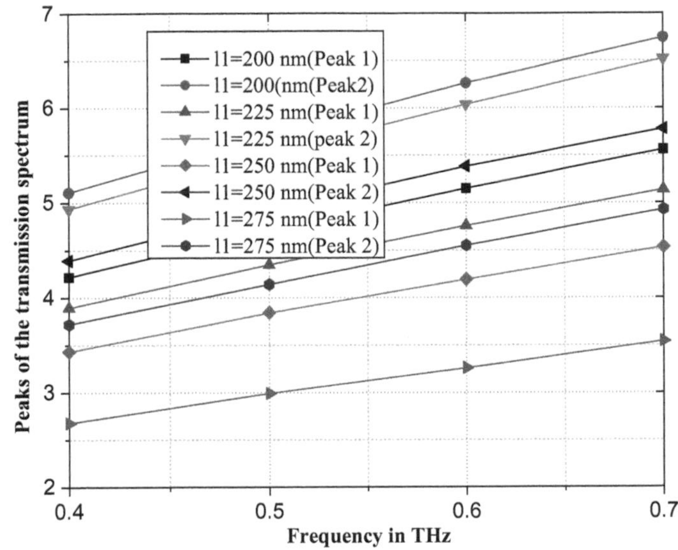

FIGURE 9.8
Transmission peaks with chemical potential at different side lengths of ring resonator.

of frequencies and consequently, higher values of wavelengths. As is clearly observed from the figure, as the values of transmission decrease, the transmission loss also increases.

Figure 9.9 gives the transmission characteristics of the structure at different widths of input and output GPNSWs, considering different values of the widths to be d = 40 nm, d = 50 nm, and d = 60 nm. With the increase in widths, the transmission resonances are obtained at higher values of frequencies and consequently, lower values of wavelengths. Thus, the transmission peaks can be tuned at desired frequencies or wavelengths,

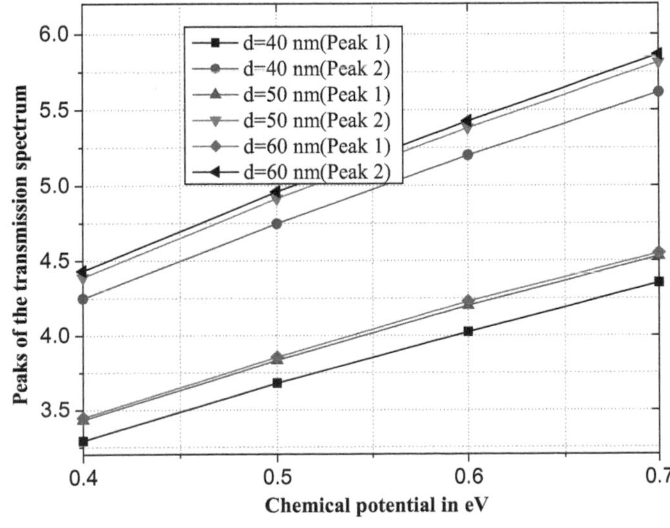

FIGURE 9.9
Transmission peaks with chemical potential at different widths of input and output waveguides of graphene CSRR.

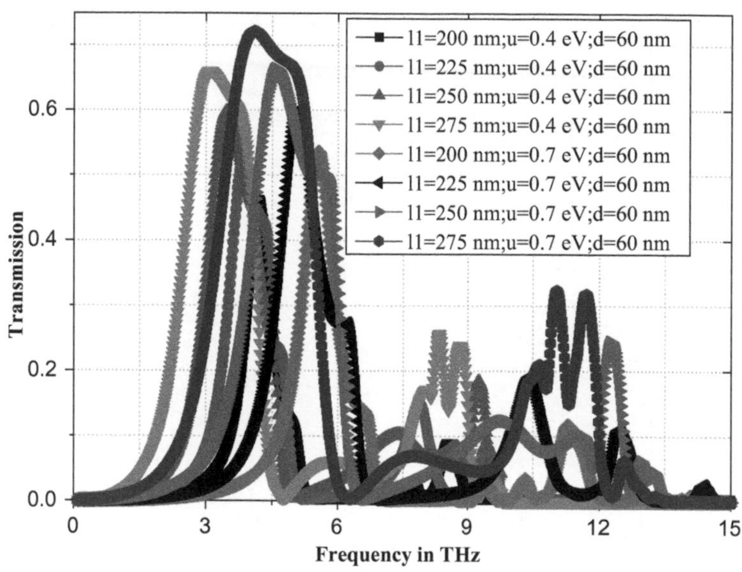

FIGURE 9.10
Transmission characteristics with wavelength at different radii of ring resonator for fixed values of chemical potential and widths.

increasing the radius of the ring resonator or decreasing the widths of the input and output waveguides, simultaneously suffering with the increase in transmission loss. Figure 9.10 shows the transmission characteristics of the structure for different lengths, *l*1 of ring resonator. The spectra is obtained at different values of chemical potential and widths of the input and output waveguides.

9.5.3 Transmission Properties of WDM

Graphene CSRR performs the action of channel demultiplexing, selecting a single channel from multiple ones [30]. Three such graphene resonators are shown in Figure 9.11a, which work at a different set of values of dielectric constant of surrounding material, *e* and chemical potential of graphene, μ. Here, resonator I works at $l1 = 600$ nm, $e1 = 0.25$, and $\mu 1 = 0.9$. Then, resonator II has $l2 = 620$ nm, $e2 = 0.5$, $\mu 2 = 0.95$, and resonator III has $l3 = 600$ nm, $e3 = 1.0$, $\mu 3 = 1.0$. This nano-device gives the transmission spectra, which is tunable with the variation in index of the material and electronic properties of graphene that can be varied with applied voltage or chemical doping. According to the transmission spectra in Figure 9.11b, resonator I works at 6.9 THz. Similarly, resonator II operates at frequency 6.6 THz, and resonator III works on 7.6 THz. Similarly, the demultiplexer can be used for selecting an appropriate channel among the transmitted ones.

Thus, the graphene CSRR coupled with GPNSWs has been utilized for wavelength demultiplexing of different channels. From the transmission spectra, one can infer that the peak positions vary linearly with the change in chemical potential and thus, applied voltage bias of graphene. The structural parameters have been stated further as increasing the radius of the ring resonator or decreasing the widths of the input and output waveguides, simultaneously suffering with the increase in transmission

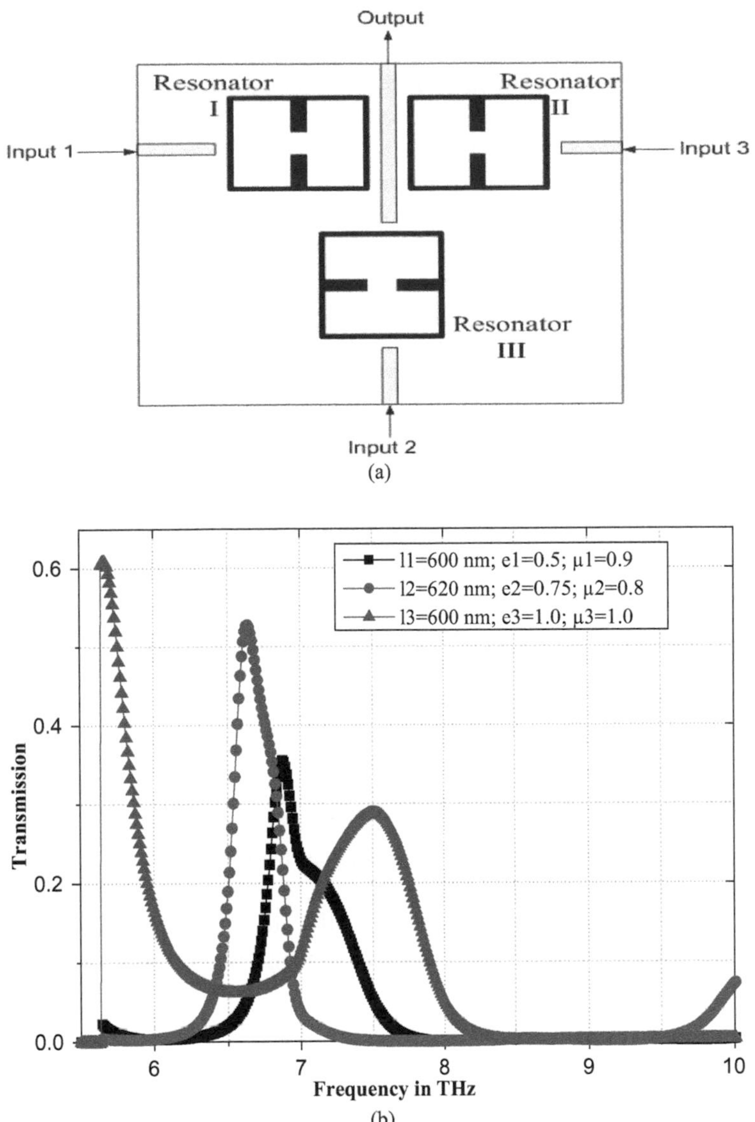

FIGURE 9.11
(a) Graphene-based CSRR WDM and (b) its transmission characteristics with frequency for different resonators.

loss. The nano-device is compact in size can be utilized effectively as a tunable band-pass filter (BPF) in a THz frequency regime.

9.6 GPPW for Sensing Operation

Pristine graphene material withstands excellent characteristics in the optical and THz frequency range [12–14]. The graphene-based plasmonic parallel plate waveguides have been a matter of interest for researchers due to their distinguishably extraordinary features [18–22].

Various graphene-based devices such as tunable filter [26], low-pass filters [34], band-pass filters [35], directional coupler [36], oscillators [37], nano-patch antenna [38], cloaks [39], phase-shifters [15], coplanar waveguides and their discontinuities [32], and millimeter wave microstrip mixers [40] have been studied to date. Wang et al. have proposed a surface plasmon polaritons (SPPs)-based refractive index sensor consisting of a ring resonator coupled with two metal-insulator-metal (MIM) waveguides [33]. The transmission characteristics have been analyzed to obtain the sensing potential of the device. The tuning range of SRR has been increased by optimizing the SRR modal characteristics [25]. Moreover, graphene-based SRRs [41–42] have been used for sensing [43] and tuning in hybrid metamaterials [44]. In this section, graphene-based suspended waveguide structures have been used in order to realize millimeter wave/THz frequency circuits such as diplexer, sensors, etc., providing supplementary benefits of tunability, high flexibility, high conductivity, and low-cost alternative to silver and gold nano-circuits.

First, we have analyzed a graphene-based suspended nanostrip waveguide using full-wave EM solver to evaluate its transmission line characteristics such as propagation constant and characteristic impedance. Subsequently, closed-form expressions to determine the propagation constant and characteristic impedance are also presented. The suspended structure has been used to design a filter-based diplexer operating at THz frequencies. Moreover, a sensor structure with two GPSNWs connected by a modified ring resonator has been designed. The chemical potential variation in graphene owes to the tuning capability of the graphene nano-device. The transmission peaks of SRR structures vary with the change in the dielectric constant of the material under sensing. The output transmission characteristics provide information about sensitivity of the graphene-based structure.

9.6.1 Model Structure and Theoretical Analysis

The proposed graphene-based suspended nanostrip line is shown in Figure 9.12a in which the substrate consists of a silicon-dioxide layer (ε_r = 3.9) over a silicon wafer (ε_r = 11.9, σ = 0.00025 S/m). The lossy model takes skin depth into account without spatial sampling, leading to lesser simulation times. Also, a layer of graphene is present on the back side of the silicon substrate, which acts as a ground. The graphene surface conductivity has been defined with the help of the Kubo formula for Drude model of its conductivity σ [24]. The values of chemical potential for graphene conduction in the THz frequency region lies in the range $0 < \mu_c < 1$ (units in eV). The design parameters are chemical potential, μ_c = 0.5 eV; relaxation time, τ = 0.5 ps; corresponding mobility, μ = 30,000 cm^{2}V^{-1}s^{-1}; and temperature, T = 300 K. The entire structure is embedded in a surrounding air box (ε_r = 1). Here, the thickness of different layers is taken as t = 0.5 nm, thickness of lossy silicon substrate: h = 100 nm, thickness of the silicon dioxide layer: d = 20 nm. Various plots have been drawn with respect to the w/h ratio for the waveguide structure, where w is the width of the graphene nanostrip, d is the height of dielectric material, and h is the height of the substrate as shown in Figure 9.12a. The thickness of the graphene layer and the ground plane is denoted by t. The chemical potential μ_c of the graphene strip and ground layer is 0.5 eV. The gate provided via silicon is used for tuning transmission through this structure. The schematic diagram of the suspended nanostrip line is shown in Figure 9.12a. Next, Figure 9.12b and c shows the curve between the phase constant β, attenuation constant α, and frequency (in THz) for given wavelengths, showing higher values of normalized phase constant (β/k_0) as compared to normalized attenuation constant (α/k_0) owing to the mode propagation through the waveguide structure. The characteristic impedance shows values in the range of KΩ (Figure 9.12d–f).

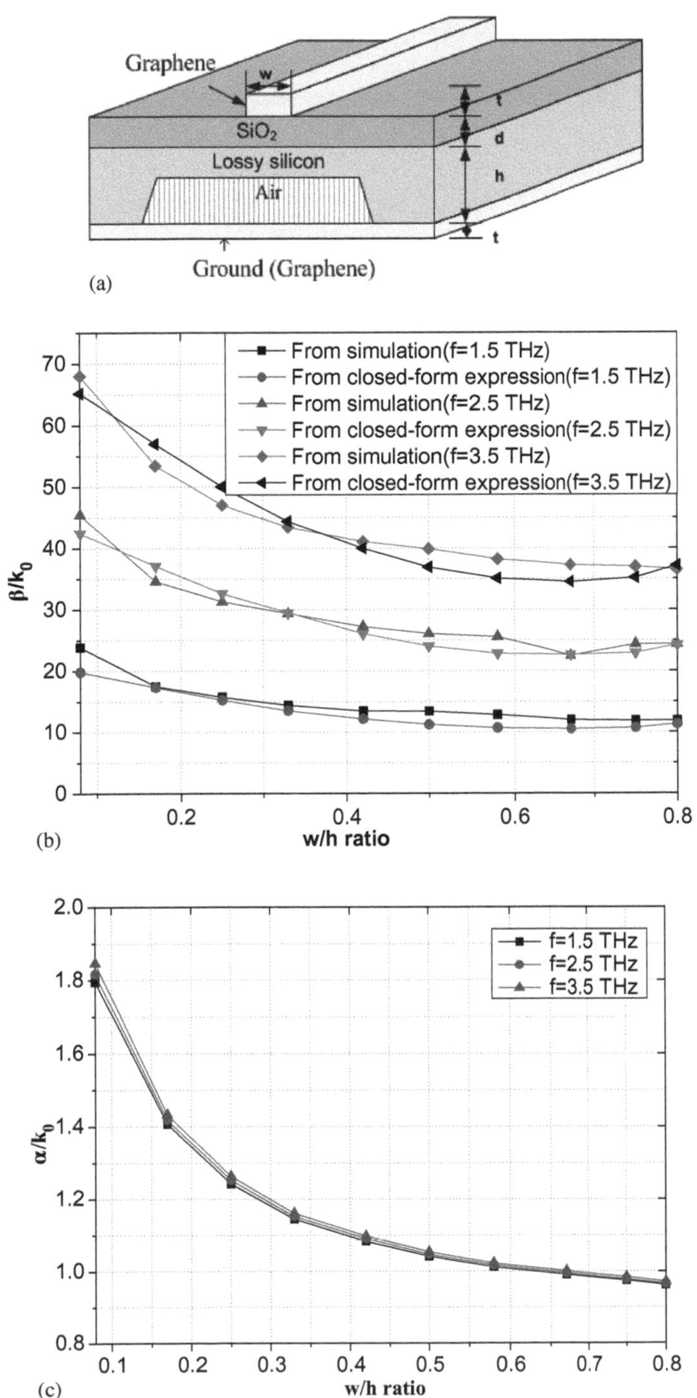

FIGURE 9.12
(a) Cross-sectional view of graphene suspended nanostrip transmission line. (b) Normalized phase constant, β/k_0 with w/h ratio. (c) Normalized attenuation constant, α/k_0 with w/h ratio. *(Continued)*

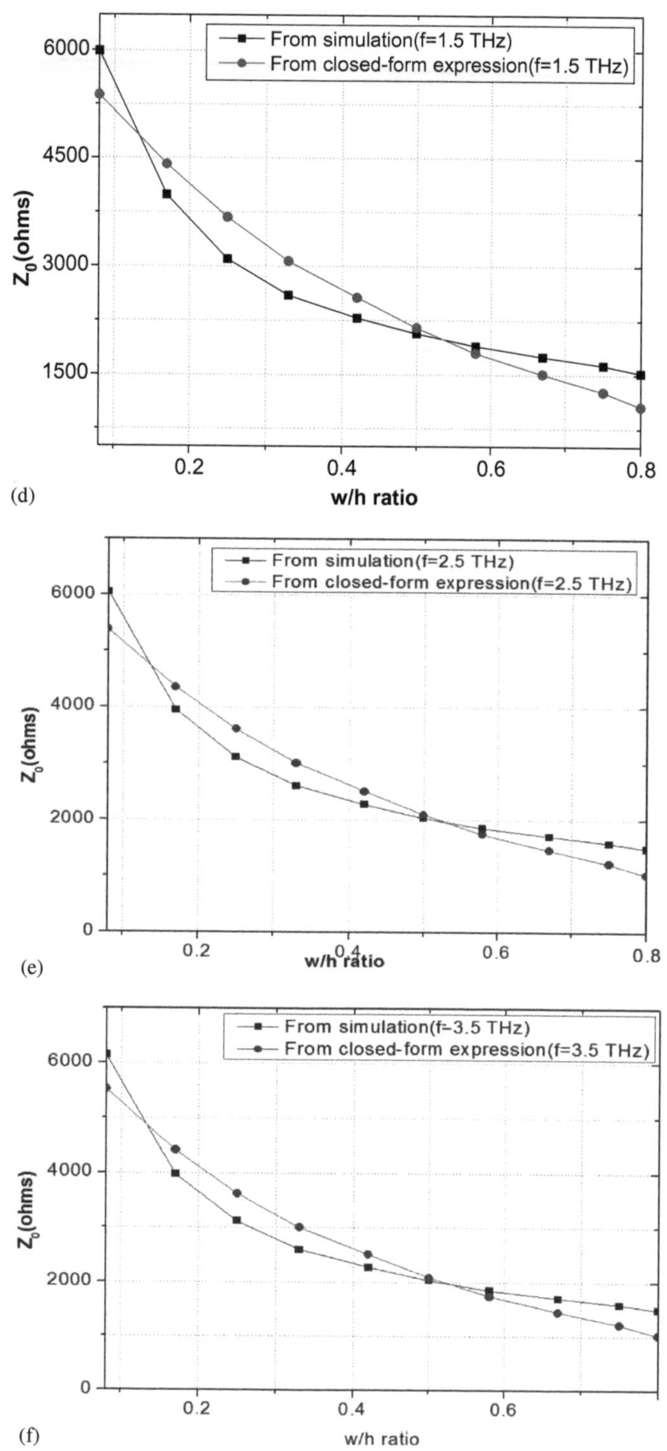

FIGURE 9.12 (Continued)
(d) Characteristic impedance, Z_0 with *w/h* ratio at frequency, $f = 1.5$ THz. (e) Characteristic impedance, Z_0 with *w/h* ratio at frequency, $f = 2.5$ THz. (f) Characteristic impedance, Z_0 with *w/h* ratio at frequency, $f = 3.5$ THz.

The formula obtained from curve-fitting results have been shown in terms of conductivity, σ; aspect-ratio w/h; thickness of the silicon layer, d_1; frequency, f; and thickness of SiO_2 layer, d_2.

$$\frac{\beta}{k_o} = A_2 f\left(A_1\left(w/h\right)^2 + B_1\left(w/h\right) + C_1\right)\left(\left(A_3 d_2^2 + B_3 d_2 + C_3\right)\right)e^{-12.855\sigma} \tag{9.13}$$

where

$$A_1 = \frac{\left(0.0000008186426 d_1^2 - 0.000005741778 d_1 + 0.00153540628\right)}{\left(d_1^2 + 0.000001428464 d_1 - 0.01028940621\right)}$$

$$B_1 = \frac{\left(0.00000729628 d_1^2 - 0.00316782490 d_1 + 0.25889104067\right)}{\left(d_1^2 + 0.00008387767 d_1 + 0.03249911566\right)}$$

$$C_1 = \frac{\left(0.00000226708 d_1^2 - 0.00072225229 d_1 + 0.02320356799\right)}{\left(d_1^2 + 0.00004741758 d_1 - 0.00077150191\right)}$$

$$A_2 = \frac{\left(0.00012925767 d_1^2 + 0.00672921587 d_1 + 0.70355612897\right)}{\left(d_1^2 - 0.000000774019 d_1 - 0.00358137783\right)}$$

$$A_3 = \frac{\left(0.000000212281 d_1^2 - 0.00008215508 d_1 - 0.01108422066\right)}{\left(d_1^2 + 0.00000619957 d_1 + 0.00025100378\right)}$$

$$B_3 = \frac{\left(0.00376644876 d_1^2 + 0.20308313729 d_1 + 3.1364\right)}{\left(d_1^2 + 0.00168112835 d_1 - 0.07306\right)}$$

$$C_3 = \frac{\left(128995 d_1^2 + 6700185 d_1 + 399353033\right)}{\left(d_1^2 + 6021.49 d_1 + 115070\right)}$$

The derived closed-form formulas for the normalized propagation constant provided in Equation 9.13 give the following expression for the characteristic impedance:

$$Z_o = e^{\left(0.1548(w/h) + 0.0026 f + 8.6959\sigma + \frac{0.0000000267\beta}{k_o} + 0.1275\right)} \tag{9.14}$$

The closed-form expressions can be compared, agreeing well with the simulated results (within 95% confidence intervals) as shown in Figure 9.12b, d–f.

9.6.2 Characteristics of Graphene Plasmonic Diplexer Using Coupled Line Resonators

The coupled line resonator design is shown in Figure 9.13a, in which two SIRs have been interconnected with input and output GPNSWs. The substrate material for the aforementioned design is SiO_2 with lossy silicon at the bottom. The substrate material dimensions

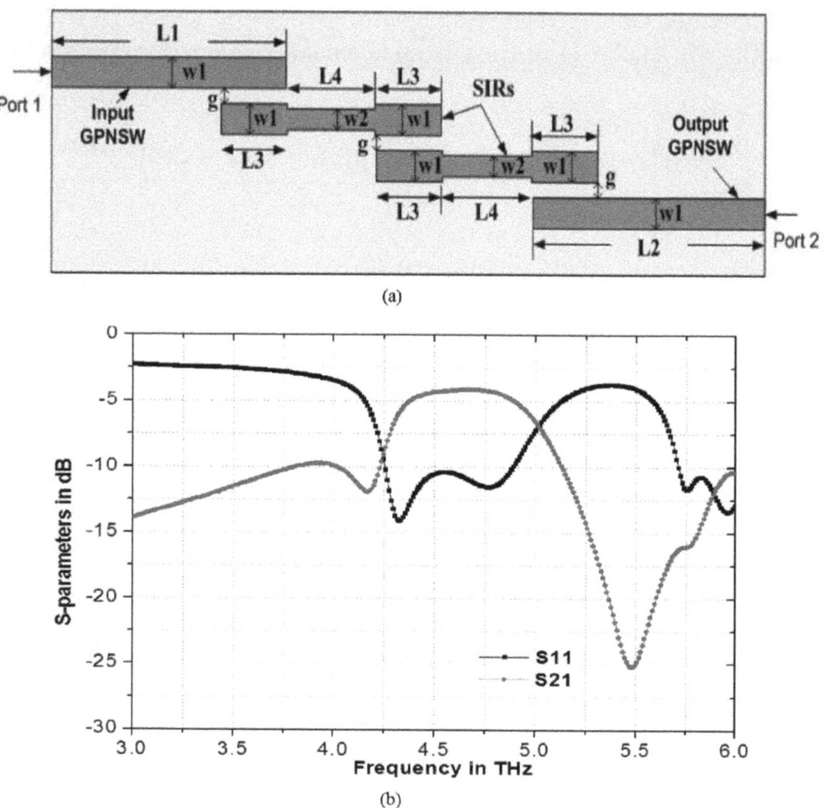

FIGURE 9.13
(a) Design of wideband graphene-based band-pass filter, where L_1 = 1300 nm, L_2 = 1500 nm, L_3 = 200 nm, L_4 = 200 nm, $w1$ = 50 nm, $w2$ = 40 nm, g = 40 nm. (b) Transmission and reflection coefficients of designed band-pass filter.

are taken as 480 nm × 2500 nm × 90 nm. The parametric features for the graphene layer are the same as selected in Section 9.6.2 with the thickness of the graphene layer as 0.5 nm. Here, graphene material also provides tunability, which is an additional benefit over its conventional metal counterparts. Its simulated results have been shown in Figure 9.13b, which depict the region of pass-band from 4.3 to 4.9 THz.

The geometry of a graphene plasmonic diplexer based on coupled line resonator is shown in Figure 9.14a. The signal at port 1 is divided into two parts with a 45° chamfered power divider circuit and propagated across a band-pass filter embedded in both of the output branches. The design parameters are L_1 = 1000 nm, L_2 = 200 nm, L_3 = 200 nm, L_4 = 1000 nm, W_1 = 400 nm, W_2 = 200 nm, W_3 = 50 nm, W_4 = 50 nm, and W_5 = 40 nm. Here, the chemical potential of graphene in the upper branch is 0.3 eV; in the lower branch, it is 0.25 eV, with the rest of the parameters chosen to be the same. Also, the simulated characteristics of the diplexer are shown in Figure 9.14. The values of isolation come out to be above 20 dB as shown in Figure 9.14b. The design of the circuit is as follows:

One can infer from the simulation results that the diplexer resonates at two frequencies. The first frequency band occurs at 4.85 THz, which is resonant frequency of the upper-branch resonator. The second one lies at 4.7 THz, resonant frequency of the second resonator, with an insertion loss of around −22 dB. Therefore, the two frequency bands can be transmitted concurrently through the graphene diplexer.

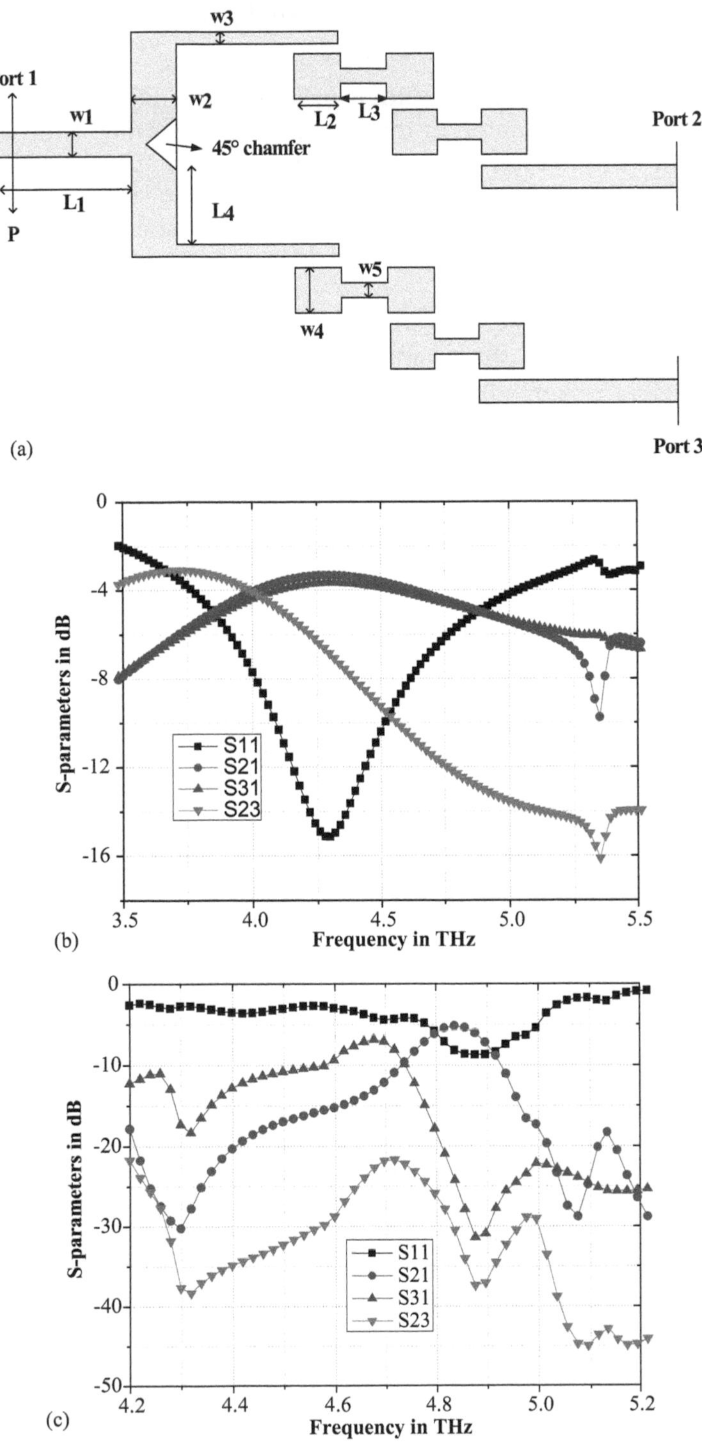

FIGURE 9.14
(a) Design of graphene-based diplexer. (b) Simulation results of graphene power splitter. (c) Simulation results of graphene band-pass filter-based diplexer.

9.6.3 Sensing Application of Graphene-Based Modified SRR in Suspended Nanostrip Waveguide

The geometry of graphene plasmonic SRR consists of two GPSNWs at the input and output and a modified ring resonator. In the geometry, the modified ring structure is composed of graphene with the rest of the material being the material under sensing. In the design, the widths of input and output GPSNWs are *d*, the side length of the modified ring is *l*1 and the coupling width between GPSNWs and the modified ring is *g* as shown in Figure 9.15a. For general cases, the values of *d*, *l*1, and *g* are set to be 40 nm, 250 nm, and 10 nm, respectively.

Next the material parameters, dielectric permittivity, ε (eps) and dielectric permeability, μ (mu) can be obtained from the s-parameters of the graphene-based SPPs waveguide sensor designed [45] as shown in Figure 9.15b. The fact can be validated by the following expressions [31]:

$$n = \frac{1}{kd}\cos^{-1}\left[\frac{1}{S21}\left(1 - S_{11}^2 + S_{21}^2\right)\right] \tag{9.15}$$

$$z = \sqrt{\frac{(1+S11)^2 - S_{21}^2}{(1-S11)^2 - S_{21}^2}} \tag{9.16}$$

$$\varepsilon = n/z \text{ and } \mu = nz \tag{9.17}$$

Here, *n* is refractive index, *z* is wave impedance, and *k* is wave vector.

Figure 9.16 shows the transmission spectrum for the aforementioned graphene-based CSRR. The surface plasmonic resonances occur at some fixed frequencies in these characteristics. These transmission resonances have been shown with respect to the propagating frequencies.

Next, the transmission spectra of graphene-based CSRR is observed further with respect to various design parameters such as chemical potential of graphene, refractive index of material, etc. Firstly, the transmission resonances have been obtained with different values of chemical potential of graphene as shown in Figure 9.16. These resonances travel toward the right with the increase in the values of applied chemical potential resulting in a sufficient increase in transmission. These values of chemical potential have been varied from 0.2 to 1.0 eV. The curves show linear variation. Then, the peaks of the transmission spectrum have been observed with frequency in Figure 9.17 at different values of refractive index of the material under sensing, which can be used for detection of materials of different refractive index depending on its sensitivity.

As the dielectric constant, eps is varied from 0.2 to 0.4, and peak I shifts by 6 nm. The sensitivity of the index sensor can be calculated from $d\lambda/dn$, resulting in the index sensitivity of 1341.64 nm/RIU for peak. Here, peak positions are expressed in nm scale, and RIU represents refractive index unit. The transmission resonances and dielectric constant vary linearly as shown in Figure 9.18. Thus, there is approximately linear shift in the transmission peaks with variation in the values of dielectric constant and refractive index. Now, the impact of variation in values of structural parameters have been observed on the transmission spectra of the graphene-based CSRR.

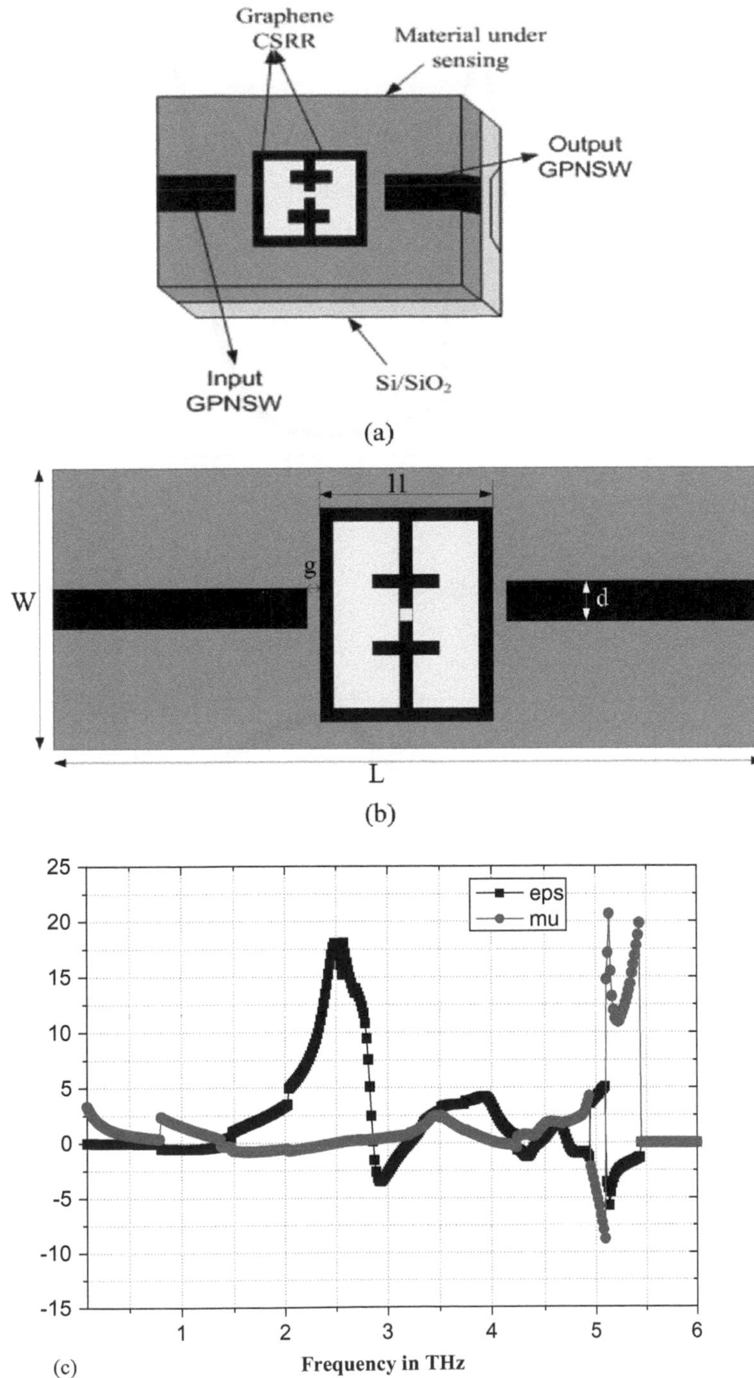

FIGURE 9.15
Schematic diagram of two waveguides with a modified ring resonator. (a) 3-D view (b) 2-D view showing graphene-based CSRR. (c) Dielectric permittivity, eps, and permeability, mu of graphene-based CSRR.

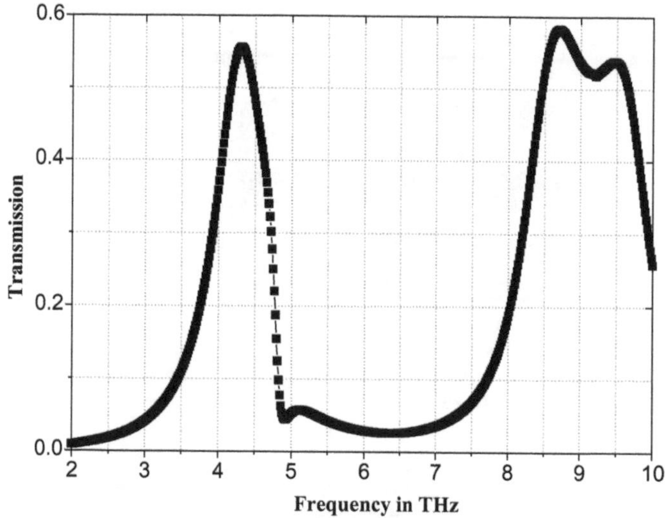

FIGURE 9.16
Transmission characteristics of graphene-based sensor.

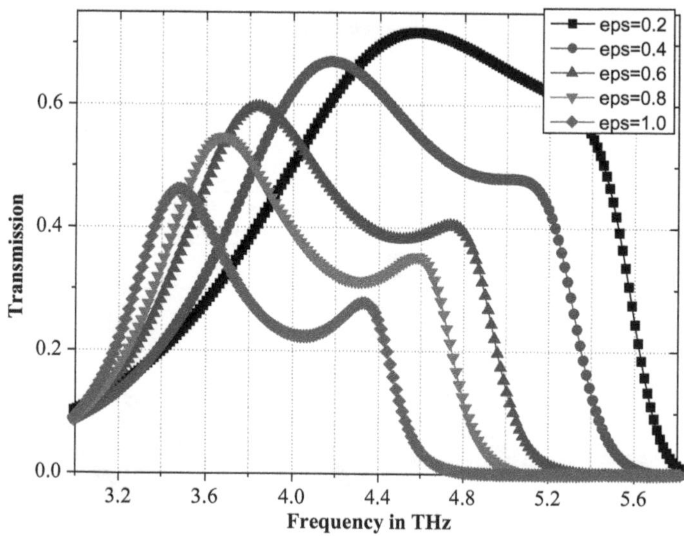

FIGURE 9.17
Transmittance characteristics of graphene-based sensor structure for different values of dielectric constant.

Firstly, the transmission peaks have been observed for different values of $l1 = 200$ nm, 225 nm, 250 nm, and 275 nm as shown in Figure 9.19. There is a relative change in the position of the transmission peaks with a change in the dimensions of the modified ring. As the side lengths increase, the transmission resonances are obtained at lower values of frequencies and consequently, higher values of wavelengths. As is clearly observed from the figure, the transmission loss will increase with a decrease in transmission values.

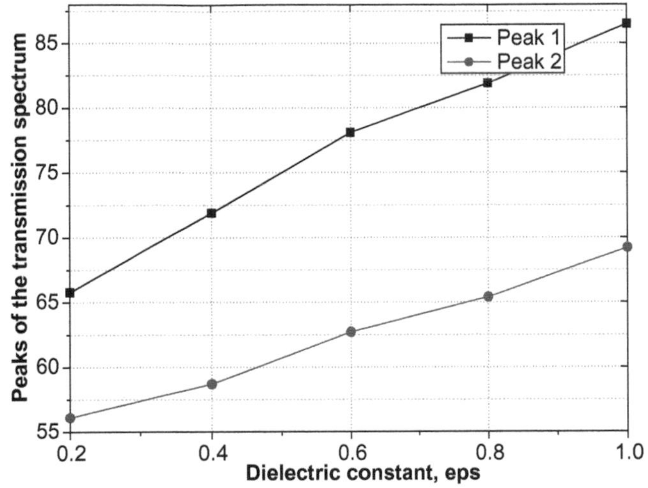

FIGURE 9.18
Transmission spectrum of graphene-based sensor with respect to dielectric constant showing the sensitivity to different refractive index material.

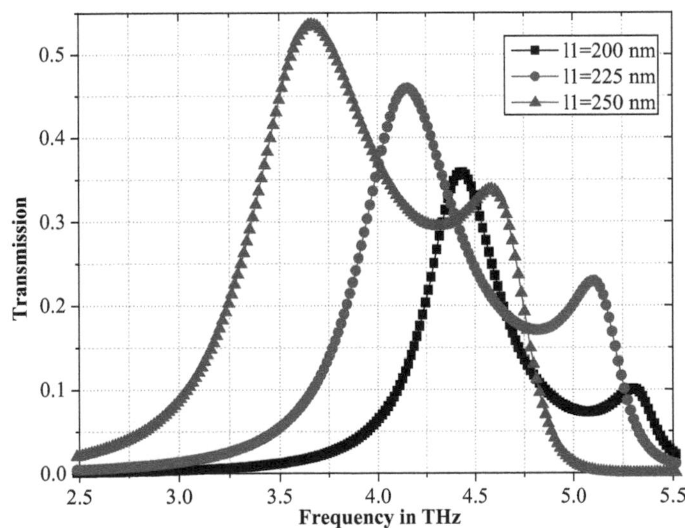

FIGURE 9.19
The transmission characteristics with frequency at different side lengths of ring resonator.

Figure 9.20 gives the frequency spectra of a proposed structure at different values of widths of input and output GPNSWs, with widths as $d = 10$ nm, 20 nm, 30 nm, and 40 nm. With the increase in widths, the transmission resonances are obtained at higher values of frequencies and consequently, lower wavelengths. The sensitivity of the sensor depends on dimensions of ring resonator and GPNSWs.

Figure 9.21 gives transmission spectra for different values of *l*1 with fixed values of chemical potential and widths of GPSNSWs. Next, the sensitivity of the structure with different dielectric constant values of the material under sensing at different values of the

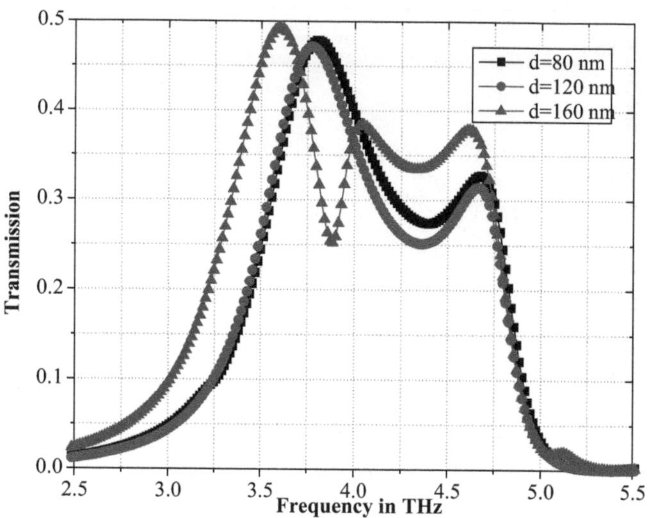

FIGURE 9.20
The transmission characteristics with frequency at different widths of GPNSWs.

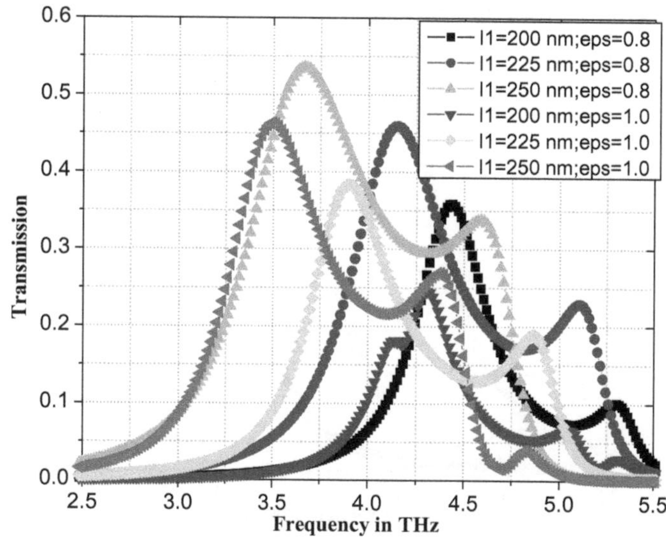

FIGURE 9.21
Transmission characteristics with frequency for different values of side lengths of modified ring resonator at specified values of chemical potential and widths.

side lengths of the ring resonator—$l1 = 200$ nm, 225 nm, and 250 nm—and different values of the widths $d = 80$ nm, $d = 120$ nm, and $d = 160$ nm are shown in Tables 9.1 and 9.2. It can be inferred that the sensitivity and tunability of the graphene-based SRR is better than other plasmonic waveguides. Also, Tables 9.1 and 9.2 depict the values of the sensitivities of the structure at different values of dielectric constant of the sensing material, eps with different values of $l1 = 200$ nm, 225 nm, and 250 nm for different widths of input and output nanostrip waveguides, $d = 80$ nm and $d = 120$ nm.

TABLE 9.1

Sensitivities of the Structure at Different Values of Dielectric Constant of the Sensing Material, eps with Different Values of Side Lengths of Ring Resonator, *l*1 = 200 nm, 225 nm, and 250 nm at *d* = 80 nm

*l*1 = 200 nm	λ(eps = 0.2)	λ(eps = 0.4)	Sensitivity(nm/RIU)
Peak 1	53880	60284	14319.7793
Peak 2	43248	51710	18921.6072
	λ(eps = 0.8)	λ(eps = 1.0)	
Peak 1	67015	71003	8917.439
Peak 2	63090	65694	5822.721
*l*1 = 225 nm	λ(eps = 0.2)	λ(eps = 0.4)	Sensitivity(nm/RIU)
Peak 1	58886	63983	11397.24
Peak 2	48236	52889	10404.42
	λ(eps = 0.8)	λ(eps = 1.0)	
Peak 1	78870	83594	10563.19
Peak 2	64175	66089	4279.834
*l*1 = 250 nm	λ(eps = 0.2)	λ(eps = 0.4)	Sensitivity(nm/RIU)
Peak 1	61648	69748	18112.15
Peak 2	43248	51649	18785.21
	λ(eps = 0.8)	λ(eps = 1.0)	
Peak 1	78870	83594	10563.19
Peak 2	64175	66089	4279.834

TABLE 9.2

Sensitivities of the Structure at Different Values of Dielectric Constant of the Sensing Material, eps with Different Values of Side Lengths of Ring Resonator, *l*1 = 200 nm, 225 nm, and 250 nm at *d* = 120 nm

*l*1 = 200 nm	λ(eps = 0.2)	λ(eps = 0.4)	Sensitivity(nm/RIU)
Peak 1	61648	69748	18112.1506
Peak 2	43248	51649	18785.2071
	λ(eps = 0.8)	λ(eps = 1.0)	
Peak 1	78870	83594	10563.1851
Peak 2	64175	66089	4279.8341
*l*1 = 225 nm	λ(eps = 0.2)	λ(eps = 0.4)	Sensitivity(nm/RIU)
Peak 1	62489	70183	17204.31
Peak 2	79250	83329	9120.921
	λ(eps = 0.8)	λ(eps = 1.0)	
Peak 1	43392	44236	1887.241
Peak 2	64891	66236	3007.511
*l*1 = 250 nm	λ(eps = 0.2)	λ(eps = 0.4)	Sensitivity(nm/RIU)
Peak 1	64647	71810	16016.95
Peak 2	55237	61044	12984.85
	λ(eps = 0.8)	λ(eps = 1.0)	
Peak 1	81413	85196	8459.045
Peak 2	64930	66048	2499.924

Therefore, in this section, the transmission line characteristics of graphene-based suspended nanostrip waveguides have been studied. Also, graphene plasmonic diplexer structure has been designed and analyzed for nanophotonic THz integrated circuit applications. Moreover, the graphene SRR interconnected with graphene input and output suspended nanostrip waveguides has been studied for sensing of materials of different refractive index. Here, the sensing characteristics have been analyzed with simulations. The locations of the transmission resonances vary linearly with the dielectric potential of the material under sensing. This nano-device can again be used as BPF at THz frequencies. It can be used for detection of materials of different refractive index or for detection of different human body parts such as skin, bone, mucosa, etc., for paleontology and study of fossils.

9.7 Conclusion

The GPNSW-based CSRR structure has been designed and analyzed for wavelength demultiplexing of different channels. The simulations of transmission characteristics provide the transmission peaks showing linear variation with chemical potential. Also, the effect of change in dimensions has been observed, resulting in tuning of the transmission peaks at desired frequencies or wavelengths. Moreover, increasing the side lengths of graphene CSRR or decreasing the widths of GPNSWs changes the transmission peaks with greater losses. The device is compact with ultra-high sensitivity. The transmission line characteristics of GPNSWs have been studied. Also, graphene plasmonic diplexer structure has been designed and analyzed for nanophotonic THz integrated circuit applications. Moreover, the SRR structure interconnected with graphene input and output suspended nanostrip waveguides is studied for sensing of materials of different refractive index. It can be used for detection of materials of different refractive index.

References

1. A. J. Seeds, H. Shams, M. J. Fice and C. C. Renaud, "Terahertz photonics for wireless communications," *Journal of Lightwave Technology*, vol. 33, no. 3, pp. 579–587, 2015.
2. T. Nagatsuma, G. Ducournau and C. C. Renaud, "Advances in terahertz communications accelerated by photonics," *Nature Photonics*, vol. 10, no. 6, pp. 371–379, 2016.
3. I. F. Akyildiz, J. M. Jornet and C. Han, "Terahertz band: Next frontier for wireless communications," *Physical Communication*, vol. 12, pp. 16–32, 2014.
4. T. Nagatsuma, S. Horiguchi, Y. Minamikata, Y. Yoshimizu, S. Hisatake, S. Kuwano, N. Yoshimoto, J. Terada and H. Takahashi, "Terahertz wireless communications based on photonics technologies," *Optics Express*, vol. 21, no. 20, pp. 23736–23747, 2013.
5. G. Ducournau, P. Szriftgiser, F. Pavanello, E. Peytavit, M. Zaknoune, D. Bacquet, A. Beck, T. Akalin and J. F. Lampin, "THz communications using photonics and electronic devices: The race to data-rate," *Journal of Infrared, Millimeter and Terahertz Waves*, vol. 36, no. 2, pp. 198–220, 2015.

6. T. P. McKenna, J. A. Nanzer and T. R. Clark, "Photonic millimeter-wave system for high-capacity wireless communications," *John Hopkins APL Technical Digest*, vol. 33, no. 1, pp. 1–6, 2015.
7. T. Kürner and S. Priebe, "Towards THz communications—Status in research, standardization and regulation," *Journal of Infrared, Millimeter, and Terahertz Waves*, vol. 35, no. 1, pp. 53–62, 2014.
8. S. Arnon and D. Kedar, "Non-line-of-sight underwater optical wireless communication network," *Journal of the Optical Society of America*, vol. 26, no. 3, pp. 530–539, 2009.
9. L. Rakotondrainibe, Y. Kokar, G. Zaharia and G. El Zein, "Toward a gigabit wireless communications system," *International Journal of Communication Networks and Information Security*, vol. 1, no. 2, pp. 36–42, 2009.
10. A. Mansour, R. Mesleh and M. Abaza, "New challenges in wireless and free space optical communications," *Optics and Lasers in Engineering*, vol. 89, pp. 95–108, 2017.
11. J. Singh, P. Gilawat and B. Shah, "Performance evaluation of 32 × 40 Gbps (1.28 Tbps) FSO link using RZ and NRZ line codes," *International Journal of Computer Applications*, vol. 85, no. 4, p. 32, 2014.
12. T. Low and P. Avouris, "Graphene plasmonics for terahertz to mid-infrared applications," *ACS Nano*, vol. 8, no. 2, pp. 1086–1101, 2014.
13. J. S. Gómez-Díaz and J. Perruisseau-Carrier, "Microwave to THz properties of graphene and potential antenna applications," *Proceedings of ISAP2012*, Nagoya, Japan, pp. 239–242, 2012.
14. A. N. Grigorenko, M. Polini and K. S. Novoselov, "Graphene plasmonics – Optics in flatland," *Nature Photonics*, vol. 6, pp. 749–758, 2012.
15. P. Y. Chen, C. Argyropoulos and A. Alu, "Terahertz antenna phase shifters using integrally-gated graphene transmission-lines," *IEEE Antennas & Propagation Society*, vol. 61, no. 4, pp. 1528–1537, 2013; H. Hajian, A. Soltani-Vala, M. Kalafi and P. T. Leung, "Surface plasmons of a graphene parallel plate waveguide bounded by Kerr-type nonlinear media," *Journal of Applied Physics*, vol. 115, p. 083104, 2014.
16. D. Correas Serrano, J. S. Gomez-Diaz, J. Perruissea-Carrier and A. Alvarez-Melcon, "A graphene-based plasmonic tunable low-pass filters in the THz band," *IEEE Transactions on Nanotechnology*, vol. 13, no. 6, pp. 1145–1153, 2014.
17. P. J. Burke, "An RF circuit model for carbon nanotubes," *IEEE Transactions on Nanotechnology*, vol. 2, no. 1, pp. 53–55, 2002.
18. D. Correas-Serrano, J. S. Gomez-Diaz, J. Perruisseau-Carrier and A. Álvarez-Melcón, "Spatially dispersive graphene single and parallel plate waveguides: Analysis and circuit model," *IEEE Transactions on Microwave Theory and Techniques*, vol. 61, no. 12, pp. 4333–4344, 2013.
19. A. Malekabadi, S. A. Charlebois and D. Deslandes, "Parallel plate waveguide with anisotropic graphene plates: Effect of electric and magnetic biases," *Journal of Applied Physics*, vol. 113, p. 113708, 2013.
20. J. S. Gomez-Diaz, J. R. Mosig and J. Perruisseau-Carrier, "Effect of spatial dispersion on surface waves propagating along graphene sheets," *IEEE Transactions on Antennas and Propagation*, vol. 61, no. 7, pp. 3589–3596, 2013.
21. H. Hajian, A. Soltani-Vala and M. Kalafi, "Optimizing terahertz surface plasmons of a monolayer graphene and a graphene parallel plate waveguide using one-dimensional photonic crystal," *Journal of Applied Physics*, vol. 114, no. 2013, pp. 0331021-8, 2013.
22. G. W. Hanson, "Quasi-transverse electromagnetic modes supported by a graphene parallel-plate waveguide," *Journal of Applied Physics*, vol. 104, pp. 1–5, 2008.
23. E. Kymakis, E. Stratakis, M. M. Stylianakis, E. Koudoumas and C. Fotakis, "Spin coated graphene films as the transparent electrode in organic photovoltaic devices," *Thin Solid Films*, vol. 520, pp. 1238–1241, 2011.
24. G. W. Hanson, "Dyadic green's functions and guided surface waves for a surface conductivity model of graphene," *Journal of Applied Physics*, vol. 103, pp. 1–8, 2008.
25. T. Wu, Y. Liu, Z. Yu, Y. Peng, C. Shu and H. He, "The sensing characteristics of plasmonic waveguide with a ring resonator," *Optics Communications*, vol. 323, no. 7, pp. 44–48, 2014.
26. Y. Gao, G. Ren, B. Zhu, L. Huang, H. Li, B. Yin and S. Jian, "Tunable plasmonic filter based on graphene split-ring," *Plasmonics*, vol. 1, no. 1, pp. 291–296, 2016.

27. P. Q. Liu, I. J. Luxmoore, S. A. Mikhailov, N. A. Savostianova, F. Valmorra, J. Faist and G. R. Nash, "Highly tunable hybrid metamaterials employing split-ring resonators strongly coupled to graphene surface plasmons," *Nature Communications*, vol. 6, pp. 8969, 1–7, 2015.
28. Y. P. Chen, W. E. I. Sha, L. Jiang and J. Hu, "Graphene plasmonics for tuning photon decay rate near metallic split-ring resonator in a multilayered substrate," *Optics Express*, vol. 23, no. 3, pp. 2798–2807, 2015.
29. C. Zhao and J. Zhang, "Plasmonic demultiplexer and guiding," *ACS*, vol. 4, no. 11, pp. 6433–6438, 2010.
30. G. Wang, H. Lu, X. Liu, D. Mao and L. Duan, "Tunable multi-channel wavelength demultiplexer based on MIM plasmonic nanodisk resonators at telecommunication regime," *Optics Express*, vol. 19, no. 4, pp. 3513–3518, 2011.
31. N. Joshi and N. P. Pathak, "Graphene-backed graphene plasmonic coplanar waveguide (GB-GPCPW) for terahertz integrated circuit applications," *IEEE Proceedings of AEMC*, vol. 103, pp. 1–2, 2015.
32. N. Joshi and N. P. Pathak, "Modeling of graphene coplanar waveguide and its discontinuities for THz integrated circuit applications," *Plasmonics*, vol. 12, no. 5, pp. 1545–1554, 2016.
33. B. Wang and G. P. Wang, "Plasmonic waveguide ring resonator at terahertz frequencies," *Applied Physics Letters*, vol. 89, no. 13, pp. 1–4, 2006; X. Gu, I. T. Lin and J. M. Liu, "Extremely confined terahertz surface plasmon-polaritons in graphene-metal structures," *Applied Physics Letters*, vol. 103, pp. 071103, 1–4.
34. D. Correas Serrano, J. S. Gomez-Diaz, J. Perruissea-Carrier and A. Alvarez-Melcon, "A graphene-based plasmonic tunable low-pass filters in the THz band," *IEEE Transactions on Nanotechnology*, vol. 13, no. 6, pp. 1145–1153, 2014.
35. H. Deng, Y. Yan and Y. Xu, "Tunable flat-top bandpass filter based on coupled resonators on a graphene sheet," *IEEE Photonics Technology Letters*, vol. 27, no. 11, pp. 1161–1164, 2015.
36. M. D. He, K. J. Wang, L. Wang, J. B. Li, J. Q. Liu, Z. R. Huang, L. Wang, L. Wang, W. D. Hu and X. Chen, "Graphene based terahertz tunable plasmonic directional coupler," *Applied Physics Letters*, vol. 105, pp. 081903, 1–5, 2014.
37. F. Rana, "Graphene terahertz plasmon oscillators," *IEEE Transactions on Nanotechnology*, vol. 7, no. 1, pp. 91–99, 2008.
38. I. Llaster, C. Kremers, A. C. Aparicio, J. M. Jornet, E. Alarcon and D. N. Chigrin, "Graphene based nano patch antenna for terahertz radiation," *Photonics and Nanostructures-Fundamentals and Applications*, vol. 10, no. 4, pp. 353–358, 2012.
39. P. Y. Chen and A. Alù, "Atomically thin surface cloak using graphene monolayers," *ACS Nano*, vol. 5, pp. 5855–5863, 2011.
40. G. Hotopan, S. Ver Hoeye, S. Vazquez, R. Camblor, M. Fernandez, F. Las Heras, P. Alvarez and R. Menendez, "Millimeter wave microstrip mixer based on graphene," *Progress in Electromagnetics Research*, vol. 118, pp. 57–69, 2011.
41. J. Wang, W. B. Lu, X. B. Li, X. F. Gu and Z. G. Dong, "Plasmonic metamaterial based on the complementary split ring resonators using graphene," *Journal of Physics D: Applied Physics*, vol. 47, no. 32, p. 325102, 2014.
42. S. Cakmakyapan, H. Caglayan and E. Ozbay, "Coupling enhancement of split ring resonators on graphene," *Carbon N. Y.*, vol. 80, no. 1, pp. 351–355, 2014.
43. H. J. Chen and K. Di Zhu, "Graphene-based nanoresonator with applications in optical transistor and mass sensing," *Sensors (Switzerland)*, vol. 14, no. 9, pp. 16740–16753, 2014.
44. P. Q. Liu, I. J. Luxmoore, S. A Mikhailov, N. A. Savostianova, F. Valmorra, J. Faist and G. R. Nash, "Highly tunable hybrid metamaterials employing split-ring resonators strongly coupled to graphene surface plasmons," *Nature Communications*, vol. 6, p. 8969, 2015.
45. D. R. Smith, D. C. Vier, T. Koschny and C. M. Soukoulis, "Electromagnetic parameter retrieval from inhomogeneous metamaterials," *Physical Review E*, vol. 71, pp. 036617, 1–11, 2005.

10

Analysis of CNTFET for SRAM Cell Design

Shashi Bala and Mamta Khosla

CONTENTS

10.1 Introduction

Over the past four decades, scaling of semiconductor devices, structure design novelty, and advancement in fabrication technology have led to the significant development of complementary metal oxide semiconductor (CMOS) technology. As a result, integrated circuits (ICs) give greater performance. On the other hand, if device scaling occurs continuously, it leads to short channel effects (SCEs) [1–3] including drain-induced barrier lowering (DIBL), gate-induced drain leakage (GIDL), high leakage current, low drive current, and subthreshold degradability. Such SCEs cause a serious problem by interfering with further scaling of devices, thus making it difficult for the devices to maintain their high performance. To overcome these problems, various semiconductor devices such as double gate (DG) [4,50], Fin FET, gate-all-around (GAA), Tunnel FET, carbon nanotube FET (CNTFET), Tunnel CNTFET, etc., are under research in the electronics industry and also have been analyzed to obtain the highest possible ON current and lowest possible OFF current as well as a maximum ratio of I_{ON}/I_{OFF} [5,51,52]. Also, the power reduction in the portable devices is a big challenge for researchers. In recent years, carbon nanotubes (CNTs) have become a good replacement for channel material in metal oxide semiconductor field effect transistor (MOSFET) due to their good electrical properties; therefore, carbon nanotube field effect transistors (CNTFETs) have become an alternative nanoscale transistor for future circuit design [6,53,54]. Some of the extensive advantages of CNTFETs are aggressive channel length scaling due to the absence of mobility degradation [8], variable bandgap with single material [7,55,56], ultra-thin body device that is possible due to smaller diameter (1–3 nm) [9,63], and compatibility of CNT with high-K materials [10] resulting in high ON current. CNTFETs give better electrostatic control, maximum packing density, and steep subthreshold slope over others [11,12,61], and they offer the highest packing density, better gate control, and steep subthreshold. Due to various challenges associated with CMOS technology for static random access memory (SRAM) design, a DG FinFET-based 10T SRAM cell was analyzed by [13,57,58], and performance of various CMOS logic structures were investigated by [14,59,60,62]. All CNTFET-based 6T-SRAM cells were designed using a model file (Stanford University), which is a benchmark model file of conventional CNTFET.

10.2 Carbon Nanotubes

A carbon nanotube (CNT) is a tube that is cylindrical in shape, made of rolled-up carbon, with a diameter in nanometer ranges and a length up to a few micrometers; the thickness is around one ten-thousandth of a human hair strand. When a graphene sheet is rolled up, it looks slightly similar to a chicken wire, which has a hexagonal network. The carbon particles exist at the tops of the hexagons. CNTs have a number of structures, depending on the length, thickness, and nature of helicity and the number of graphene sheet layers. Although CNTs are designed with a single graphene sheet, electrical characteristics vary by changing their chiral number, causing CNTs to act either as metals or semiconductors. CNTs normally have diameters ranging from <1 nm to 50 nm. They are normally a few microns in length; however, current innovations have made CNTs abundant longer, with length measured in centimeters [15] (Figure 10.1).

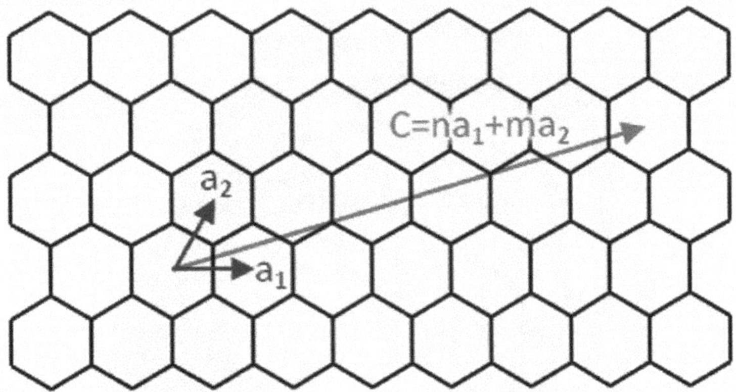

FIGURE 10.1
CNT is a single layer of graphene sheet rolled up into the shape of a cylinder. (From Zhu, S., and Xu, G., *Nanoscale*, 2, 2538–2549, 2010.)

10.2.1 Structure and Types of Carbon Nanotubes

CNTs have basically three types and can be described by their structures:

- Single-walled nanotubes (SWNTs)
- Double-walled nanotubes (DWNTs) and multi-walled nanotubes (MWNTs)

10.2.1.1 Single-Walled Nanotubes

A CNT is formed by rolling the sheet of graphene around an axis. A CNT can be simply defined as a cylindrical tube that has a diameter range from 1 nm up to a few nanometers. Figure 10.2 represents the SWNT. The carbon molecules are arranged in a hexagonal lattice with distance of 0.142 nm, and the space among the planes is 0.335 nm [16]. Graphene is cut into defined sheets, known as nanoribbon, according to some defined parameters called chiral vectors. In a chiral vector $C_h = na_1 + na_2 = (n, m)$, n, m are integers called chiral numbers, and a_1 and a_2 are primitive lattice vectors of graphene. The chiral number (n, m) of CNTs can be classified as the type of CNTs, zigzag CNTs, and armchair CNTs as presented in Figure 10.2a and b. The chiral number n and m are equal for armchair CNTs, while n or $m = 0$ for zigzag CNTs. For other values of chiral numbers, CNTs are known as chiral. CNTs exhibit metallic or semiconducting properties depending upon their various structures. By sustaining the condition $n - m = 3i$ (where i is an integer), armchair-type CNTs are always metallic, whereas zigzag CNTs are either metallic or semiconducting [17] (Figure 10.4). In CNTs, bandgap can be calculated using diameter (E_g inversely depends to the diameter), and the calculation of diameter depends on chiral vectors. Hence, all the calculations begin from a chiral vector, which determines its structure and properties. The relation of diameter (d) with chiral numbers and bandgap with diameter is given as [9] (Figure 10.3)

$$d = \frac{\sqrt{3}}{\Pi} a\left(\sqrt{n^2 + m^2 + nm}\right) \tag{10.1}$$

$$E_g = \frac{2aV_{pp\Pi}}{\sqrt{3}d} \tag{10.2}$$

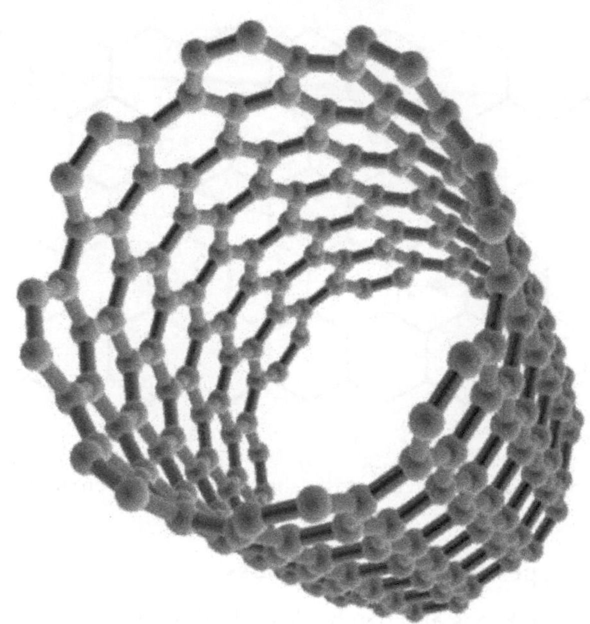

FIGURE 10.2
Schematic view of SWNT.

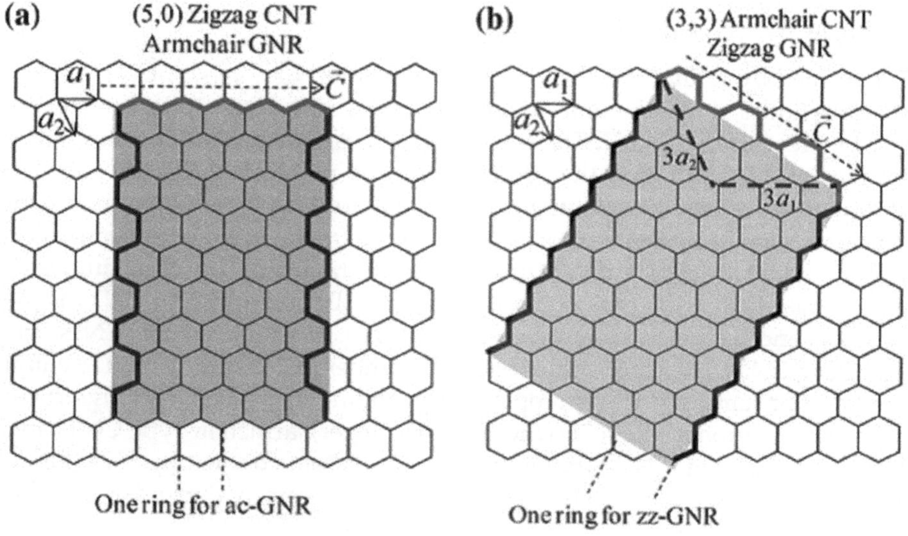

FIGURE 10.3
Types of SWNTs: (a) Zigzag (b) Armchair CNT.

10.2.1.2 Multi-Walled Nanotubes

DWNTs contain two and MWNTs contain more than two walls of graphene sheets (Figure 10.4). A DWNT is a superior type of MWNT, and interlayer space in multi-walled nanotubes is around 3.3 Å (330 pm). When DWNTs and SWNTs are compared,

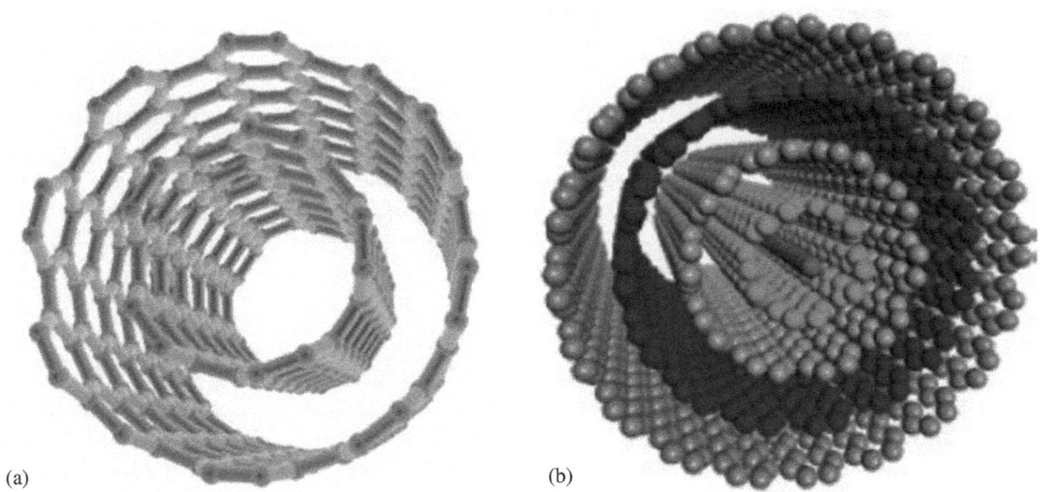

FIGURE 10.4
Basic structures of (a) double-walled, (b) multi-walled CNTs.

they have the same properties but the resistance to chemicals in MWNTs is expressively improved. DWNT is specifically essential when functionalization is essential (i.e., injecting of chemical functions at the surface of the CNTs) to enhance novel properties of the CNT. Covalent functionalization will break down some C=C paired bonds, generate some "holes" in the structure on the CNT, and thus adjust both its mechanical and electrical properties in the SWNT. In the DWNT, just the outer surface is improved. Via the CCVD technique in 2003, DWNT synthesis on the gram-scale was proposed for the first time through the selective reduction of oxide solutions in methane and hydrogen [22] (Table 10.1).

TABLE 10.1
Comparison of SWNT and MWNT

S. No.	SWNT	MWNT
1	Graphene sheet is rolled up in only single layer.	Graphene sheet is rolled up in multiple layers.
2	SWNT is more elastic.	MWNT is less elastic.
3	A catalyst is required for synthesis of SWNT.	There is no need for a catalyst for production of MWNT.
4	SWNT requires appropriate control for growth, so bulk synthesis is difficult.	Bulk synthesis is easy.
5	Purity is less.	Purity is good.
6	During functionalization, there is more of a chance to defect.	There is less of a chance to defect but if it occurred, then it would be difficult to improve.
7	Accumulation is less in the body.	Accumulation is more in the body.
8	Characterization is easy.	MWNT has a difficult structure.
9	SWNT can be simply twisted.	It is difficult to twist MWNT.

Source: Hirlekar, R. et al., *Asian J. Pharm. Clin. Res.*, 2, 17–27, 2009.

10.3 Properties of CNT

CNTs exhibit superb electrical and mechanical properties, which make them appropriate for use as channels in MOSFET. The CNTFET is similar to traditional MOSFET but with the channel made up of CNT. The channel is responsible for all conduction in MOSFE. The CNT exhibits unique properties such as large carrier mobility, thermal stability and high mechanical strength, compatibility with high dielectric constants, and excellent electrostatics due to a small diameter.

10.3.1 Electrical Conductivity

CNTs that behave like metallic substances work as strong conductive material. A chiral vector defines how to twist the sheet of graphene, and the CNT conductivity is well defied as interconnects. In concept, metallic CNTs have 1,000 times more electrical current density as compared to copper, which is 4×10^9 A/cm^2 [18]. Different CNTs, such as macroscopic structures, can be described by traditional electrical properties such as resistance, inductance, and capacitance, which ascend from the intrinsic structure of the CNT and its interface with other objects. Inside the CNTs, electrical transport is affected by scattering, lattice vibrations, and defects that lead to resistance, such as that in bulk materials. Most significantly, CNTs have a one-dimensional nature that leads to a new kind of quantized resistance related to their contacts with three-dimensional (3-D) macroscopic objects such as metal electrodes. The measured resistivity of the SWNT is 10^{-4} Ω cm at 27°C; representative SWNT wires are the best conductive carbon threads [19].

10.3.2 Thermal Properties

CNTs are estimated to be good thermal conductors due to presenting a "ballistic conduction" property, but they have good insulators depending on the axis of the carbon nanotube. It is expected that CNTs are capable of transfering up to 6000 $W \cdot m^{-1} \cdot K^{-1}$ at normal room temperature, compared to copper, a material that transfers 385 $W \cdot m^{-1} \cdot K^{-1}$ and is famous for better thermal conductivity. In the vacuum, probable temperature stability of CNTs is 2800°C; in air, the temperature stability is around 750°C. Thermal growth of graphite threads will be powerfully anisotropic, but CNTs are isotropic. This may be helpful for C–C compounds. It is estimated that low-defect tubes will have very low coefficients of thermal expansion [20,21].

10.3.3 Mechanical Properties

The tensile strength of CNTs makes them the strongest materials, and elastic modulus makes its stiffness materials yet discovered. This is due to sp^2 covalent bonds between the separate carbon particles. Because carbon-carbon ties bond, CNTs are predictable large modulus that make them strong beside their axes. The value of Young's modulus of single-walled tubes is predictable as much as 1 Tpa to 1.8 Tpa. In the case of CNTs, the value of an elastic modulus is high, which makes them suitable for tips of scanning microscopy (as an application of CNTs). The Young's modulus of SWNTs can be determined by the diameter and chirality of the tube. For MWNTs, stress is maintained in an outer graphite shell; for SWNT bundles, it has been proved that the reduced moduli are due to the weak inter-tube structure. A SWNT is almost 10–100 times stronger than steel [24]. The finest nanotubes of Young's modulus can be high as 1000 Gpa, which is about 5x higher compared to steel. Tensile strength is around 50x higher compared to steel, which is 63 Gpa. These properties of CNTs make them appropriate for applications such as aerospace. Arthur C. Clarke proposed that CNTs could be used in an Earth-to-space cable and space elevator. CNTs have extraordinary electronic properties. An especially outstanding fact is that CNTs have both metallic or semiconducting properties, depending on chiral numbers. Thus, a few CNTs have higher conductivities as compared to copper, while other CNTs are better compared to silicon. Great attention has been paid to the probability of nanoscale electronic devices being built from CNTs, and some improvement is being made in this area. There are a number of areas of technology where carbon nanotubes are already being used. These include flat-panel displays, scanning probe microscopes, and sensing devices. The unique properties of carbon nanotubes will unquestionably lead to many more applications [25–27] (Table 10.2).

10.3.4 Aspect Ratio

The virtue of high aspect ratio of CNTs leads to low CNT load and provides the same electrical conductivity compared with other conductive materials. Exclusive electrical conductivity in evaluation with conventional additive materials such as chopped carbon fiber, carbon black, or stainless steel fiber is governed by high aspect ratio of CNTs [29].

10.3.5 Absorbent

CNTs and CNT compounds possess huge mechanical strength, low weight, versatility, and perfect electrical properties, and they have emerged as perspective absorbing materials. Therefore, these become suitable candidates for use in water, air, and gas filtration. The SWNT-polyurethane compounds absorption frequency range extended from 6.4–8.2 (1.8 GHz) to 7.5–10.1 (2.6 GHz) and to 12.0–15.1 GHz (3.1 GHz) according to Wang et al. (2013). A great amount of work has been approved in order to replace the activated charcoal with CNTs in particular frequencies and applications [29].

TABLE 10.2

Mechanical Properties of Other Materials Compared with CNTs

Material	Young's Modulus (GPa)	Tensile Strength (GPa)	Density (g/cm^3)
SWNT	1054	150	N/A
MWNT	1200	150	2.6
Steel	208	0.4	7.8
Epoxy	3.5	0.005	1.25
Wood	16	0.008	0.6

Source: Varshney, K., *Int. J. Eng. Res.*, 2, 660–677, 2014.

10.4 Carbon Nanotube Field Effect Transistors

In simple words, CNTFET, when compared with traditional MOSFET, is FET with a channel made up of a single CNT or an array of CNTs instead of bulk silicon. The structures of CNTFETs are shown in Figure 10.5. Source and drain are connected through an array of CNTs as a channel instead of a bulk silicon substrate. The structure of the planer CNTFET shown in Figure 10.2a has only one difference from traditional MOSFET—the use of CNTs as channel. Because CNT is a cylindrical nanotube, coaxial geometry is also probable coaxial geometry of CNTFET, as shown in Figure 10.2b. The chosen geometry depends on its application [30]. The maximum of CNTFETs are fabricated with planer structure due to their compatibility and adequate ease with current manufacturing technologies. The top-gate geometry in planer technology provides the better performance [31]. However, coaxial geometry is most preferred due to capacitive coupling between the gate and the CNT surface, which is maximized by this geometry. When compared to other geometries, CNT encourages more charge in the channel at a given gate potential.

10.4.1 Types of CNTFET

There are four types of CNTFETs: Schottky barrier (SB) CNTFET, Conventional (C-CNTFET), Tunnel (T-CNTFET), and Partially gated (PG-CNTFET). These types are differentiated by channel—intrinsic CNT or doped CNT.

10.4.1.1 Schottky Barrier CNTFET

The SB-CNTFET channel consists of intrinsic CNT, and both source and drain sides have direct metal contact to the gate-controlled nanotube channel. The electrons and holes direct tunneling from metal into the channel (CNT), which depends on the height and width of the SB. The subthreshold slope of SB-CNTFET is always greater than

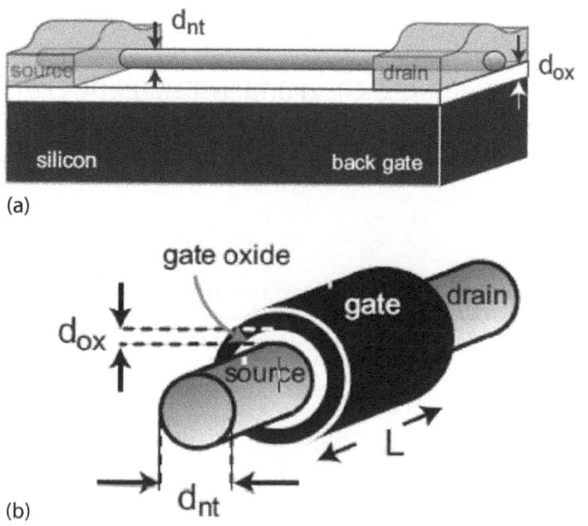

FIGURE 10.5
CNTFET structure for (a) planer geometry (b) coaxial geometry. (From Knoch, J., and Appenzeller, J., *Phys. Status Solidi*, 205, 679–694, 2008.)

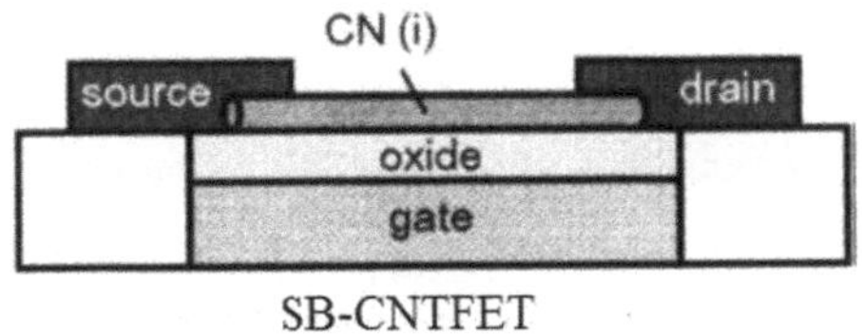

FIGURE 10.6
Structure of SB-CNTFET. (From Knoch, J. et al., *Solid State Electron.*, 49, 73–76, 2005.)

60 mV/dec, and that is the main drawback of this device [32]. Still, SB-CNTFETs are the most common CNT device layout because whole CNT is intrinsic, and controlled doping is not an easy task. The reason behind the challenge of controlled doping is that removing any atoms of carbon that really form the CNT and replacing with a dopant atom would destroy the actual properties of the nanotube. The structure of SB-CNTFET is shown in Figure 10.6.

10.4.1.2 Partially Gated CNTFET

The partially gated CNTFET shown in Figure 10.7 illustrates that the whole CNT (as channel) is intrinsic or uniformly doped. PG-CNTFET has uniform doping in the channel and works in depletion mode. The behavior of this device is *p*-type or *n*-type depending on the type of doping of the channel. This device gives better characteristics with ohmic contacts at the source and drain sides. The source exhaustion phenomenon, $I_{DS} = q{*}Q_L{*}V_T$, in this device is limited to the I_{ON}. Carrier density is represented by Q_L and V_T is thermal velocity [37]. When the whole channel is intrinsic, the SBs are formed at the sides of the source and drain; in partial gate, SB effects are not dominant. Also, this device works in enhancement mode.

10.4.1.3 Conventional

In conventional (C-CNTFET), the structure is the same as conventional MOSFET except for one difference—the channel region is intrinsic CNT while the source-drain are heavily doped. This device works as *p*-type or *n*-type CNTFET depending on the doping of the source and drain sides [35]. The structure of C-CNTFET is shown in Figure 10.8. The benefits of this structure are the minority carrier injected from the drain side (e.g., holes in an *n*-type) is suppressed because of the same doping profile in the source and drain regions, and the low I_{OFF} is obtained due to the unipolar device characteristics. Due to the absence of SBs, I_{OFF} is restricted by thermal emission, and direct tunneling does not take place in C-CNTFET, which results in a subthreshold slope that is 60 mV/dec.

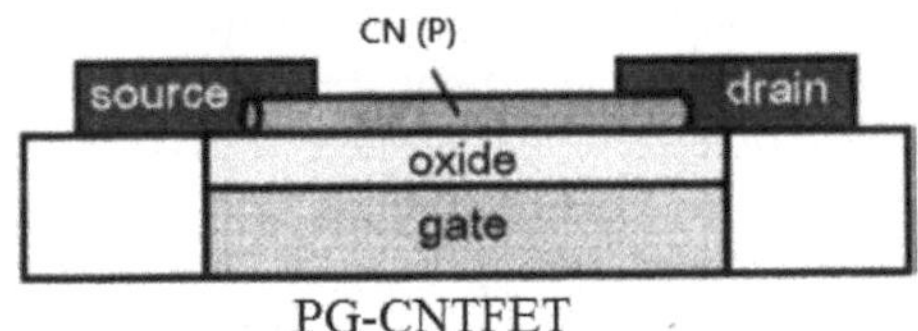

FIGURE 10.7
Structure of partially gated (PG-CNTFET). (From Knoch, J. et al., *Solid State Electron.*, 49, 73–76, 2005.)

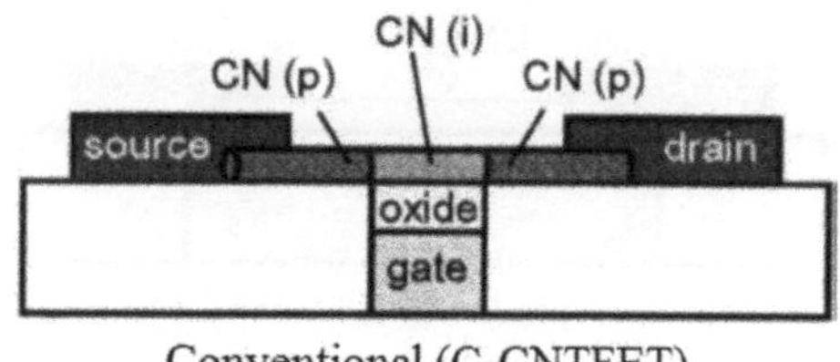

FIGURE 10.8
Structure of conventional (C-CNTFET). (From Knoch, J. et al., *Solid State Electron.*, 49, 73–76, 2005.)

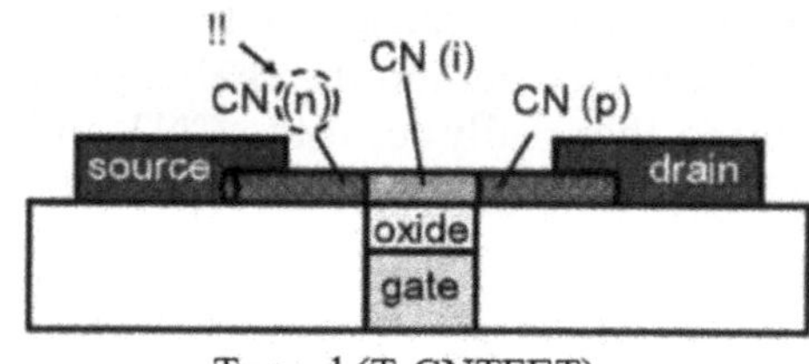

FIGURE 10.9
Structure of tunnel (T-CNTFET). (From Knoch, J. et al., *Solid State Electron.*, 49, 73–76, 2005.)

10.4.1.4 Tunnel CNTFET

The doping profile of T-CNTFET shown in Figure 10.9 is *n-i-p* or *p-i-n*, which is different from the *p-i-p* or *n-i-n* doping profile of C-CNTFET. To overcome the charge pile-up in the channel of C-CNTFET, gate-controlled T-CNTFET was proposed in [26]. The band-to-band tunneling is controlled by the gate, hence, the name gate-controlled T-CNTFET. As compared to conventional CNTFET, T-CNTFET gives superior results in terms of I_{OFF} and inverse subthreshold swing (SS). In silicon MOSFET, the use of the principle of band-to-band tunneling results in an SS value that is less than 60 mV/dec [29]. CNTs have many properties that make gate-controlled tunneling a practical method for nanoscale devices. Some of these include ballistic transport in the channel region of the device. In the case of a T-CNTFET, SS is independent of temperature and therefore has no negative temperature impact on characteristics of the device [30,33].

10.5 Why CNTFET Is Preferred over MOSFET

As conventional MOSFET has failed to deliver the adequate performance in the nanoscale domain, novel semiconductor devices have been proposed for low-power, high-performance applications. The design of CNTFET is a great success in the semiconductor industry given Moore's Law. As the cost of electronic systems for the consumer reduces, the cost of manufacturing for producers to sustain Moore's Law follows an opposite trend—i.e., research development. Manufacturing and test costs are increasing with each new chip generation. In this research work, the study of CNTFET for low-power SRAM will be carried out.

In CNTFET, the various performance parameters to be affected are drive current, threshold voltage, and transconductance. The design parameters that affect the aforementioned performance parameters are the following:

- Carbon nanotube diameter and chirality
- Bandgap of the graphene
- Schottky barrier contacts
- Oxide thickness
- Doping of CNT
- Density of states of CNT

10.6 Applications of CNTFET

The CNTFET does not suffer from short-channel effects and can be aggressively scaled down to achieve better performance. In CNTFET, the various performance parameters to be affected are drive current, threshold voltage, and transconductance. The main advantage of the CNTFET is the dependence of threshold voltage on diameter, which makes it more suitable for SRAM cell design.

10.6.1 Static Random Access Memory

SRAM is semiconductor memory and is a major element of embedded systems, reconfigurable hardware, field-programmable gate arrays (FPGAs), and microprocessors—just a few names of digital system. A design criterion that is most important is fast access time of memory and high density for design for a few years. SRAM is volatile in nature, and date is lost when power is switched off. SRAM is typically made up of six transistors for storage of each memory bit. For high density, the 6T SRAM should be sized appropriately. Figure 10.10 shows the 6T SRAM has two cross-coupled inverters;

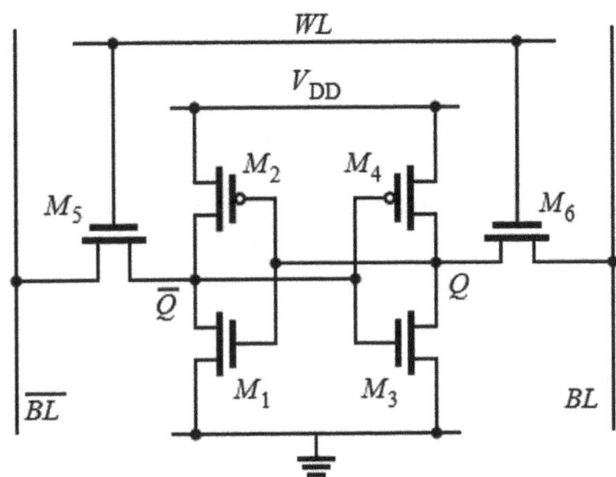

FIGURE 10.10
Schematic view of CMOS SRAM.

these inverters have stable "0" and stable "1" states. For controlling of read and write operations, there are two transistors—M_1 and M_6—which are known as access transistor and derive transistor. Word line (WL) enables the SRAM and controlling transistors M_1 and M_6. Noise margin in SRAM is improved by bit line [47].

10.6.2 Working of SRAM

SRAM works in three operation modes: write mode, hold mode, and read mode [48]. WL of SRAM is connected with V_{SS} when the read mode operation and data are retained without flipping until whenever power is on. The bit line is connected to the V_{DD} and when SRAM is in the read operation for reading "0" or "1," the bit line discharged by the access transistor.

10.6.2.1 Cell Ratio

Cell ratio (CR) is also called the β ratio and is presented in Equation 10.3.

$$CR = \frac{W_1/L_1}{W_5/L_5} \tag{10.3}$$

The ratio of drive transistor and load transistor is known as the CR, and Static Noise Margin (SNM) directly depends on it. If the CR increases, SNM also increases; this in turn increases the current in a memory cell [49].

10.6.2.2 Pull-up Ratio

Pull-up Ratio (PR) is also called α ratio and is presented in Equation 10.4.

$$PR = \frac{W_4/L_4}{W_6/L_6} \tag{10.4}$$

It is the ratio of the load transistor and the access transistor. When SNM increases, the PR value of the memory increases. Write operation is a success when it passes the current through M4 and M6.

10.6.2.3 Static Noise Margin

SNM is basically a measure of the side length of the largest square fitted inside two lobes, voltage transfer characteristic and reverse voltage transfer characteristics. PR, CR, supplied voltage, read margin, and write margin play crucial roles in determining the performance of SRAM. They determine stability of SRAM, and a high SNM indicates more stable SRAM.

10.6.2.4 Read Noise Margin

The read noise margin (RNM) determines cell stability during the read operation. Actually, the bit line and WL are pre-charged during the read operation. Voltage division across terminals causes low SNM; therefore, the state of the cell may change and wrong data may be read. Thus, RNM is defined as the cell's ability to maintain its state during the read operation. The read stability of the SRAM cell improves for higher RNM.

10.6.2.5 Write Noise Margin

The write operation occurs when the applied bit-line voltage changes the state of the cell and causes the write margin's minimum bit-line voltage to change its state. Its value depends upon cell designing parameters. Upon successful completion of the write operation, the write margin voltage is considered a maximum noise voltage. A cell shows poor write ability if it has lower WNM.

10.6.2.6 Data Retention Voltage

The minimum V_{DD} required to retain the state of the cell is data retention voltage (DRV). The particular value of decreasing V_{DD} at which the state of the cell has flipped is known as DRV, and its value is slightly greater than the threshold voltage.

10.6.3 CNTFET-Based SRAM Cell

Apart from analog applications, SRAM is one of most important components of digital systems. SRAM is preferred over dynamic random access memory [39] in fast applications due to the absence of refresh time. According to the International Technology Roadmap for Semiconductors (ITRS 2013), more than 90% of the chip area is occupied by memory. Also, memory directly affects the system's performance. With device dimensions scaling down, the threshold voltage also gets scaled down, which degrades the standby power of the SRAM cell. With excellent behavior of fabricated CNTFETs, these were supposed to be excellent options for future digital circuits [36–38]. Hence, logic circuits were designed and simulated based on CNTFET, which show considerable improvement over conventional MOSFET-based logic circuits.

The first CNTFET-based SRAM cell was proposed by Y. B. Kim et al. [43] and consisted of eight transistors; it showed excellent characteristics, proving CNTFET to be a promising future device for memory design. A lot of work has been carried out on 6T-SRAM cell design based on CNTFETs [39–42]. The main advantage of the CNTFET is dependence of threshold voltage on diameter, making it more suitable for SRAM cell design. Figure 10.11 shows the 6T-SRAM cell design using conventional CNTFET. The cell configuration is similar to the CMOS- based SRAM cell, replacing MOSFET with CNTFET. It consists of two cross-coupled

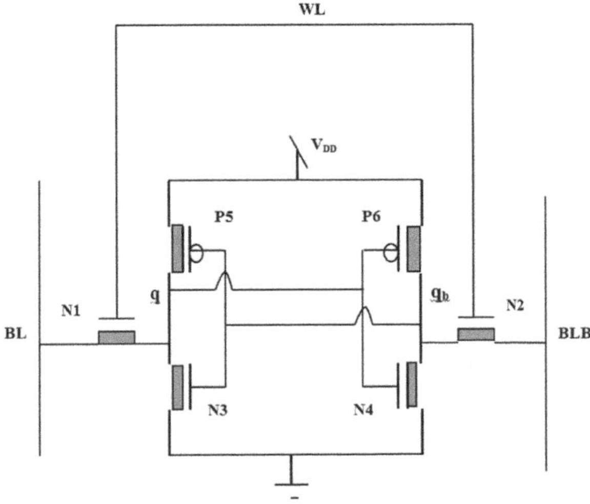

FIGURE 10.11
6T SRAM cell design using conventional CNTFET.

TABLE 10.3

Comparison of 6T SRAM Cell Using 16 nm CMOS and 10 nm CNTFET Technology Models

6T Cell for 16 nm CMOS of SRAM		
Parameters	25°C	100°C
SNM (volts)	0.117	0.105
Access time (ns)	2.483	4.83
Write margin (V)	0.2715	0.273
Power standby (pw)	10.65	22.912
6T Cell for 10 nm CNTFET of SRAM		
Parameters	25°C	100°C
SNM (volts)	0.178	0.171
Access time (ns)	2.389	1.627
Write margin (V)	0.2715	0.18
Power standby (pw)	76.53	48.5

Source: Pushkarna, A. et al., Comparison of performance parameters of SRAM designs in 16 nm CMOS and CNTFET technologies, in *Proceedings of 23rd IEEE International SOC Conference*, pp. 339–342, 2010.

inverters connected to bit lines (BL and BLB) through access transistors N1 and N2. For inverters, n-type CNTFETs are labeled as N3 and N4 and p-type PCNTFETs are labeled as P5 and P6. WL is used to activate the access transistors in order to read and write values at nodes q and q_b.

10.6.4 CNTFET-Based SRAM over CMOS-Based SRAM Cell

The SRAM circuit can be used to compare the two technologies, hence, researchers designed and simulated the SRAM and compared it with that of conventional MOSFET-based SRAM. The first CNTFET-based SRAM reported in literature [43] consists of eight transistors and shows improvement of 56% in SNM, a reduction of 48% in dynamic power consumption. While comparing with 8T FinFET-based SRAM, CNTFET-based SRAM shows very small write leakage current, thereby resulting in low dynamic power dissipation. In comparison to the 6T SRAM design with CMOS and FinFET-based SRAM, it has been found that CNTFET-based SRAM gives low static power dissipation [44]. Thus, the literature reveals that CNTFET-based SRAM is suitable for low-power applications.

The CNTFET does not suffer from short-channel effects and hence can be aggressively scaled down to achieve better performance. This leads to design of CNTFET-based SRAM with a 10 nm channel length, and it is compared with the 16 nm channel length CMOS-based SRAM. Simulation results demonstrate a clear superiority of CNTFET-based SRAM in terms of SNM and access time as given in Table 10.3. Also CNTFET is less sensitive to temperature variation, which is demonstrated through simulations [40]. With an increase in temperature, access time reduces in the case of CNTFET-based SRAM, which is in contrast to the case of the CMOS-based SRAM cell.

10.7 Simulation Results of Conventional CNTFET

Figure 10.8 shows the cross-sectional view of CNTFET, and parameters for simulation are adopted from [34,45]. Diameter (d) = 1 nm for (13,0) zigzag CNT; length of gate (L_G), source length (L_S), and drain length (L_D) are the same and equal to 20 nm; thickness of gate oxide (t_{ox}) = 1 nm and work function of the gate-metal = 4.1V; doping concentration for *N*-type = 5E-03 molar fraction; and *p*-type = −5E-03 molar fraction. The device has been simulated by using NanoTCAD ViDES [46].

Figure 10.12 shows the $I_{DS} - V_{GS}$ characteristic of CNTFET, which has been calibrated with existing device characteristics [45]. The results show quite good agreement with the existing device, which validates the performance of the device.

Figure 10.13 show $I_{DS} - V_{GS}$ characteristics of CNTFET at different drain biases (V_{DS}). There is the great impact of V_{DS} on the device characteristic. I_{OFF} diminishes with reduction of V_{DS}. There is an increase in device performance for low-power circuit application.

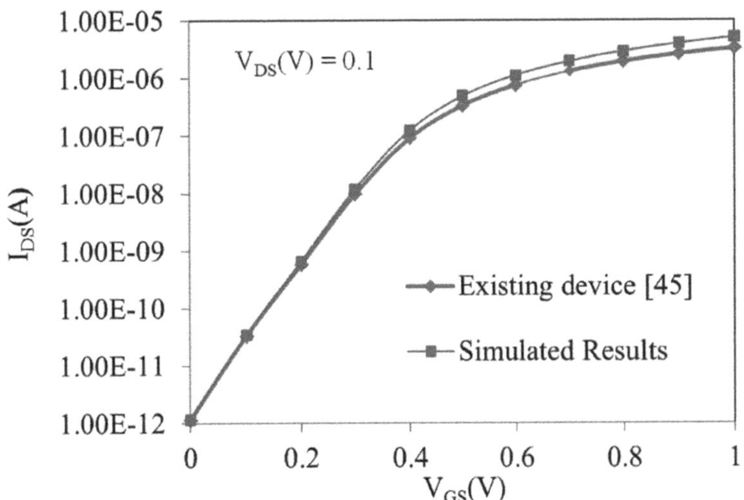

FIGURE 10.12
$I_{DS} - V_{GS}$ characteristic of CNTFET and the existing device. (From Yang, X., and Mohanram, K., *IEEE Electron Device Lett.*, 32, 231–233, 2011.)

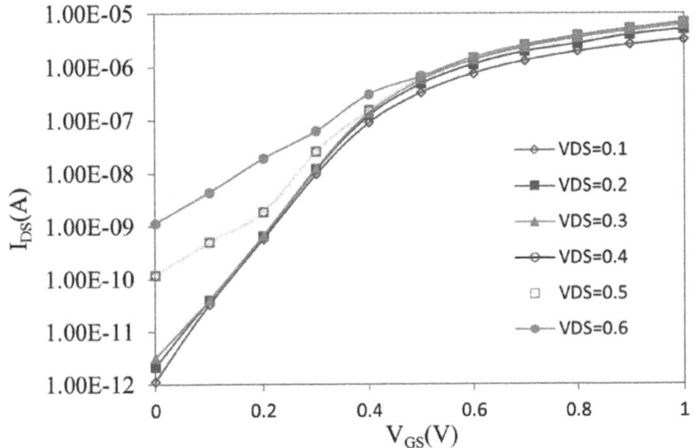

FIGURE 10.13
$I_{DS} - V_{GS}$ characteristics for different V_{DS}.

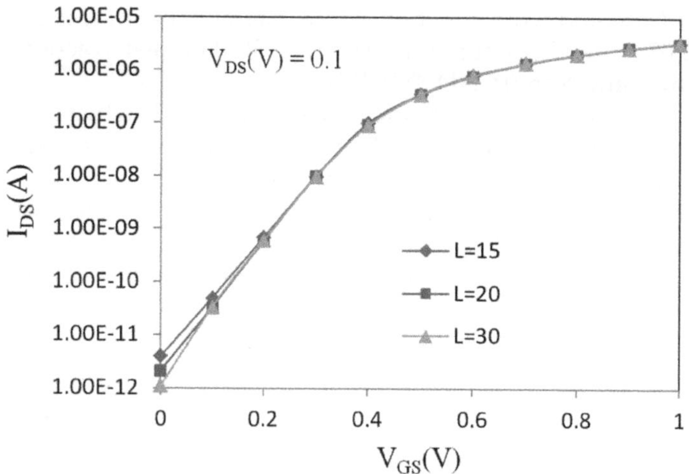

FIGURE 10.14
Effect of channel length on $I_{DS} - V_{GS}$ characteristics.

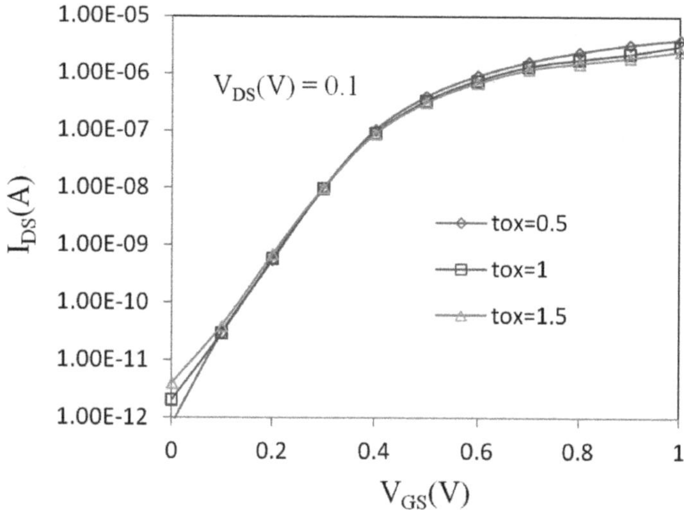

FIGURE 10.15
Effect of oxide thickness on $I_{DS} - V_{GS}$ characteristics.

Figure 10.14 shows the impact of channel length on $I_{DS} - V_{GS}$ characteristics. The results shown in Figure 10.14 illustrate the scaling of the channel length, and it can be observed that channel length affects the I_{OFF} directly and I_{OFF} reducing when the channel length increases.

Figure 10.15 shows the effect of oxide thickness on $I_{DS} - V_{GS}$ characteristics. This parameter is very sensitive for the device. Oxide thickness also plays a very important role in the scaling of the device as it leads to oxide capacitance, which is used to calculate quantum capacitance limit. It can be observed from the result that I_{ON} increases as the oxide thickness decreases.

Another important parameter is diameter of the CNT. The bandgap of CNT devices is inversely proportional to diameter. This can be observed from the result in Figure 10.16. The diameter has the direct effect on I_{OFF} because as diameter decreases I_{OFF}, the I_{ON}/I_{OFF} ratio increases.

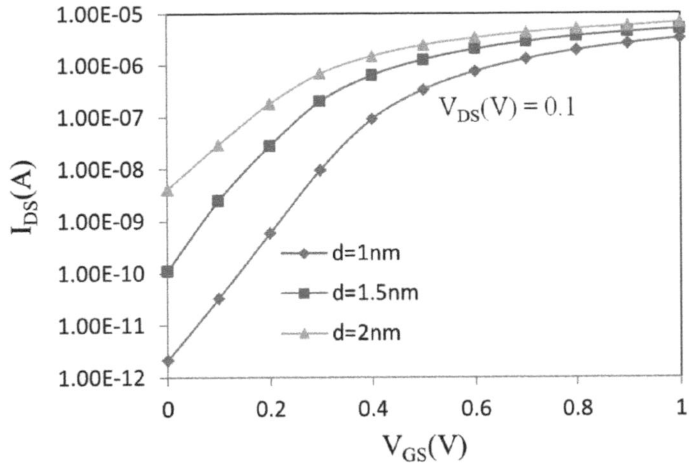

FIGURE 10.16
Effect of diameter on $I_{DS} - V_{GS}$ characteristics.

10.8 Conclusion

As conventional MOSFET has failed to deliver adequate performance in the nanoscale domain, novel semiconductor devices have been proposed for low-power applications. CNTFETs, with their excellent electrical properties, have shown promise as future devices. Among different types of CNTFETs, conventional CNTFET exhibits better performance. However, due to difficult doping techniques, SBCNTFET is preferred. The main disadvantages of SBCNTFET are ambipolar currents, which limit values of subthreshold slope, and I_{ON}/I_{OFF} ratio. Tunnel CNTFETs have low I_{ON}. It has been demonstrated that CNTFET accounts for low leakage current, making it a suitable option for low-power circuit design. A comparison shows that CNTFET-based SRAM provides stable design with the potential benefit of power reduction as compared to conventional CMOS SRAM design. Therefore, it can be inferred that power reduction is a potential advantage with good stability.

References

1. J. Tanaka, T. Toyabe, S. Ihara, S. Kimura, H. Noda, and K. Itoh, "Simulation of sub-0.1-μm MOSFET's with completely suppressed short-channel effect," *IEEE Electron Device Letters*, vol. 14, no. 8, pp. 396–399, 1993.
2. B. Raj, A. K. Saxena, and S. Dasgupta, "Nanoscale FinFET based SRAM cell design: Analysis of performance metric, process variation, underlapped FinFET and temperature effect," *IEEE Circuits and System Magazine*, vol. 11, no. 2, pp. 38–50, 2011.
3. B. Raj, A. K. Saxena, and S. Dasgupta, "Analytical modeling for the estimation of leakage current and subthreshold swing factor of nanoscale double gate finFET device," *Microelectronics International*, vol. 26, pp. 53–63, 2009.

4. S. K. Vishvakarma, V. Agrawal, B. Raj, S. Dasgupta, and A. K. Saxena, "Two dimensional analytical potential modeling of nanoscale symmetric double gate (SDG) MOSFET with ultra thin body (UTB)," *Journal of Computational and Theoretical Nanoscience*, vol. 4, no. 6, pp. 1144–1148, 2007.
5. Q. Zhang, W. Zhao, and S. Alan, "Low subthreshold swing tunnel transistors," *IEEE Electron Device Letter*, vol. 27, pp. 297–300, 2006.
6. A. Javey, M. Shim, and H. Dai, "Electrical properties and devices of large-diameter single-walled carbon nanotubes," *Applied Physics Letters*, vol. 80, no. 6, pp. 1064–1066, 2002.
7. P. L. McEuen, M. S. Fuhrer, and H. Park, "Single-walled carbon nanotube electronics," *IEEE Transactions on Nanotechnology*, vol. 1, no. 1, pp. 78–85, 2002.
8. T. Durkop, S. A. Getty, E. Cobas, and M. S. Fuhrer, "Extraordinary mobility in semiconducting carbon nanotubes," *Nano Letters*, vol. 4, no. 1, pp. 35–39, 2004.
9. M. S. Dresselhaus, G. Dresselhaus, and R. Saito, "Physics of carbon nanotubes," *Carbon*, vol. 33, no. 7, pp. 883–891, 1995.
10. A. Singh, M. Khosla, and B. Raj, "Circuit compatible model for electrostatic doped Schottky barrier CNTFET," *Journal of Electronic Materials*, vol. 45, no. 12, pp. 4825–4835, 2016.
11. B. Raj, A. K. Saxena, and S. Dasgupta, "A compact drain current and threshold voltage quantum mechanical analytical modeling for FinFETs," *Journal of Nanoelectronics and Optoelectronics (JNO)*, vol. 3, no. 2, pp. 163–170, 2008.
12. B. Raj, "Quantum mechanical potential modeling of FinFET," in *Towards Quantum FinFET*, Springer, Cham, Vol. 17, pp. 81–97, 2014.
13. S. Bhushan, S. Khandelwal, and B. Raj, "Analyzing different mode FinFET based memory cell at different power supply for leakage reduction," *Proceedings of Seventh International Conference on Bio-Inspired Computing*, 2013.
14. M. Pattanaik, B. Raj, S. Sharma, and A. Kumar, "Diode based trimode multi-threshold CMOS technique for ground bounce noise reduction in static CMOS adders," *Advanced Materials Research*, vol. 548, pp. 885–889, 2012.
15. S. Zhu and G. Xu, "Single-walled carbon nanohorns and their applications," *Nanoscale*, vol. 2, no. 12, pp. 2538–2549, 2010.
16. M. Paradise and T. Goswami, "Carbon nanotubes-production and industrial application," *Material Design*, vol. 28, pp. 1477–1489, 2007.
17. K. B. Kumar and M. Kumar Majumder, "Carbon nanotube: Properties and applications," in *Carbon Nanotube Based VLSI Interconnects*, Springer, New Delhi, pp. 17–37, 2015.
18. H. Dai, A. Javey, E. Pop, D. Mann, and Y. Lu, "Electrical transport properties and field-effect transistors of carbon nanotubes," *NANO: Brief Reports and Reviews*, vol. 1, no. 1, pp. 1–4, 2006.
19. R. P. Bandaru, "Electrical properties and applications of carbon nanotube structures," *Journal of Nanoscience and Nanotechnology*, vol. 7, pp. 1–29, 2007.
20. E. Pop, D. Mann, Q. Wang, K. Goodson, and H. Dai, "Thermal conductance of an individual single-wall carbon nanotube above room temperature," *Nano Letters*, vol. 6, no. 1, pp. 96–100, 2006.
21. H. Stahl, J. Appenzeller, R. Martel, P. Avouris, and B. Lengeler, "Intertube coupling in ropes of single-wall carbon nanotubes," *Physical Review Letters*, vol. 85, no. 24, pp. 5186–5189, 2000.
22. E. N. Ganesh, "Single walled and multi walled carbon nanotube structure, synthesis and applications," *International Journal of Innovative Technology and Exploring Engineering (IJITEE)*, vol. 2, no. 4, pp. 2278–3075, 2013.
23. R. Hirlekar, M. Yamagar, H. Garse, M. Vij, and V. Kadam, "Carbon nanotubes and its applications: A review," *Asian Journal of Pharmaceutical and Clinical Research*, vol. 2, no. 4, pp. 17–27, 2009.
24. P. J. Harris, "Carbon nanotubes and related structures: New materials for the twenty-first century," *American Journal of Physics*, vol. 72, p. 415, 2004.
25. M. Meo and M. Rossi, "Prediction of young's modulus of single wall carbon nanotubes by molecular-mechanics based finite element modelling," *Composites Science and Technology*, vol. 66, no. 11–12, pp. 1597–1605, 2006.

26. M.-F. Yu, B. S. Files, S. Arepalli, and R. S. Ruoff, "Tensile loading of ropes of single wall carbon nanotubes and their mechanical properties," *Physical Review Letters*, vol. 84, no. 24, pp. 5552–5555, 2000.
27. M. Nardelli, J.-L. Fattebert, D. Orlikowski, C. Roland, Q. Zhao, and J. Bernholc, "Mechanical properties, defects, and electronic behavior of carbon nanotubes," *Carbon*, vol. 38, no. 11, pp. 1703–1711, 2000.
28. K. Varshney, "Carbon nanotubes: A review on synthesis, properties and applications," *International Journal of Engineering Research*, vol. 2, no. 4, pp. 660–677, 2014.
29. K. Singh and B. Raj, "Temperature dependent modeling and performance evaluation of multi-walled CNT and single-walled CNT as global interconnects," *Journal of Electronic Materials*, vol. 44, no. 12, pp. 4825–4835, 2015.
30. J. Knoch and J. Appenzeller, "Tunneling phenomena in carbon nanotube field-effect transistors," *Physica Status Solidi*, vol. 205, no. 4, pp. 679–694, 2008.
31. S. J. Wind, J. Appenzeller, R. Martel, V. Derycke, and P. Avouris, "Vertical scaling of carbon nanotube field-effect transistors using top gate electrodes," *Applied Physics Letters*, vol. 80, no. 20, pp. 3817–3819, 2002.
32. J. Appenzeller, J. Knoch, V. Derycke, R. Martel, S. Wind, and P. Avouris, "Field-modulated carrier transport in carbon nanotube transistors," *Physical Review Letters*, vol. 89, no. 12, pp. 126801–126804, 2002.
33. J. Appenzeller, "Comparing carbon nanotube transistors—The ideal choice: A novel tunneling device design," *IEEE Transactions on Electron Devices*, vol. 52, no. 12, pp. 2568–2576, 2005.
34. J. Knoch, S. Mantl, and J. Appenzeller, "Comparison of transport in carbon nanotube field-effect transistors with Schottky contacts and doped source/drain contacts," *Solid State Electronics*, vol. 49, no.1, pp. 73–76, 2005.
35. A. Singh, M. Khosla, and B. Raj, "Comparative analysis of carbon nanotube field effect transistors," *2015 IEEE 4th Global Conference on Consumer Electronics (GCCE)*, IEEE, pp. 552–555, 2015.
36. B. Raj, J. Mitra, D. K. Bihani, V. Rangharajan, A. K. Saxena, and S. Dasgupta, "Process variation tolerant FinFET based robust low power SRAM cell design at 32nm technology," *Journal of Low Power Electronics (JOLPE)*, vol. 7, no. 2, pp. 163–171, 2011.
37. R. Martel, H. S. K. Chan, and P. Avouris, "Carbon nanotube field effect transistors for logic applications," *Proceedings of the Electron Devices Meeting*, pp. 751–754, 2001.
38. B. Raj, A. K. Saxena, and S. Dasgupta, "High performance double gate FinFET SRAM cell design for low power application," *International Journal of VLSI and Signal Processing Applications*, vol. 1, no. 1, pp. 12–20, 2011.
39. S. Lin, Y. Kim, and F. Lombardi, "A new SRAM cell design using CNTFETs," *Proceedings of the International SoC Design Conference*, vol. 1, pp. 168–171, 2008.
40. A. Pushkarna, S. Raghavan, and H. Mahmoodi, "Comparison of performance parameters of SRAM designs in 16nm CMOS and CNTFET technologies," *Proceedings of 23rd IEEE International SOC Conference*, pp. 339–342, 2010.
41. J. Guo, A. Javey, H. Dai, and M. Lundstrom, "Performance analysis and design optimization of near ballistic CN field-effect transistors," *Proceedings of Electron Device Meeting*, pp. 703–706, 2004.
42. K. Singh and B. Raj, "Influence of temperature on MWCNT bundle, SWCNT bundle and copper interconnects for nanoscaled technology nodes," *Journal of Materials Science: Materials in Electronics*, vol. 26, no. 8, pp. 6134–6142, 2015.
43. Y. B. Kim, Y. B. Kim, F. Lombardi, and Y. J. Lee, "A low power 8T SRAM cell design technique for CNTFET," *Proceedings of the SoC Design Conference*, vol. 1, pp. I-176, 2008.
44. Z. Zhang, M. A. Turi, and J. G. Delgado-Frias, "SRAM leakage in CMOS, FinFET and CNTFET technologies: Leakage in 8T and 6T SRAM cells," *Proceedings of the Great Lakes Symposium on VLSI*, pp. 267–270, 2012.
45. X. Yang and K. Mohanram, "Modeling and performance investigation of the double-gate carbon nanotube transistor," *IEEE Electron Device Letters*, vol. 32, no. 3, pp. 231–233, 2011.
46. G. Fiori and G. Iannaccone, "NanoTCAD ViDES," 2008.

47. V. K. Sharma, M. Pattanaik, and B. Raj, "INDEP approach for leakage reduction in nanoscale CMOS circuits," *International Journal of Electronics*, vol. 102, no. 2, pp. 200–215, 2015.
48. S. Birla, R. K. Singh, IACSIT Member, and M. Pattnaik, "SNM analysis of various SRAM topologies," *IACSIT International Journal of Engineering and Technology*, vol. 3, no. 3, pp. 304–309, 2011.
49. R. Keerthi and H. Chen, "Stability and SNM analysis of low power SRAM," *IEEE International Instrumentation and Measurement Technology Conference*, Victoria, Canada, pp. 1681–1684, May 12–15, 2008.
50. A. Singh, M. Khosla, and B. Raj, "Analysis of electrostatic doped Schottky barrier carbon nanotube FET for low power applications," *Journal of Materials Science: Materials in Electronics*, vol. 28, pp. 1762–1768, 2017.
51. A. Singh, M. Khosla, and B. Raj, "*Compact model for ballistic single wall CNTFET under quantum capacitance limit*," *Journal of Semiconductors* (*JoS*), vol. 37, pp. 104001–104008, 2016.
52. A. Javey, H. Kim, M. Brink, Q. Wang, A. Ural, J. Guo, P. McIntyre, P. McEuen, M. Lundstrom, and H. Dai, "High K dielectrics for advanced carbon nanotube transistors and logic," *Nature Mater*, vol. 1, no. 4, pp. 241–246, 2002.
53. K. Singh and B. Raj, "Comparison of temperature dependent performance and analysis of SWCNT bundle and copper as VLSI interconnects," *Research Cell: An International Journal of Engineering Sciences*, vol. 17, no. 1, pp. 583–595, 2016.
54. M. K. Majumder, B. K. Kaushik, & S. K. Manhas, "Analysis of delay and dynamic crosstalk in bundled carbon nanotube interconnects". *IEEE Transactions on Electromagnetic Compatibility*, vol. 56, no. 6, pp. 1666–1673, 2014.
55. A. Singh, M. Khosla, and B. Raj, "Comparative analysis of carbon nanotube field effect transistor and nanowire transistor for low power circuit design," *Journal of Nanoelectronics and Optoelectronics*, vol. 11, pp. 388–393, 2016.
56. L. Hong, "Circuit modeling and performance analysis of multi-walled carbon nanotube interconnects," *Proceedings of the IEEE Transactions on Electron Devices*, vol. 55, no. 6, pp. 1328–1337, 2008.
57. K. Singh and B. Raj, "Performance and analysis of temperature dependent multi-walled carbon nanotubes as global interconnects at different technology nodes," *Journal of Computational Electronics*, vol. 14, no. 2, pp. 469–476, 2015.
58. A. Singh, D. K. Saini, D. Agarwal, S. Aggarwal, M. Khosla, and B. Raj, "Modeling and simulation of carbon nanotube field effect transistor and its circuit application," *Journal of Semiconductors* (*JoS*), vol. 37, pp. 074001074006, 2016.
59. S. Singh, S. Yadav, J. Rahul, A. Srivastava, and B. Raj, "Impact of HfO_2 in graded channel dual insulator double gate MOSFET," *Journal of Computational and Theoretical Nanoscience*, vol. 12, no. 6, pp. 950–953, 2015.
60. A. K. Dogre, M. Pattanaik, B. Raj, and A. Naik, "A novel switched capacitor technique for NBTI tolerant low power 6T-SRAM cell design," *Journal of VLSI and Signal Processing (IOSR-JVSP)*, vol. 4, no. 2, pp. 68–75, 2014.
61. S. Kumar and B. Raj, "Estimation of stability and performance metric for inward access transistor based 6T SRAM cell design using n-type/p-type DMDG-GDOV TFET," *IEEE VLSI Circuits and Systems Letter*, vol. 3, no. 2, 2017.
62. N. Jain and B. Raj, "Parasitic Capacitance and Resistance Model Development and Optimization of Raised Source/Drain SOI FinFET Structure for Analog Circuit Applications". *Journal of Nanoelectronics and Optoelectronics*, vol. 13, no. 4, pp. 531–539, 2018.
63. A. Singh, M. Khosla, and B. Raj, "Design and analysis of electrostatic doped Schottky barrier CNTFET based low power SRAM," *International Journal of Electronics and Communications*, vol. 80, pp. 67–72, 2017.

11

Design of Ternary Logic Circuits Using CNFETs

Chetan Vudadha and M. B. Srinivas

CONTENTS

11.1 Introduction

Integrated circuit (IC) technology has enabled rapid advances in design and implementation of innovative devices and systems that have changed the way we live and communicate. Integration of more transistors increases the computing power and helps in building efficient systems [1]. The number of transistors that can be integrated on a chip has been doubling every 1–2 years as predicted by Gordon Moore, an industry pioneer, in the 1960s [2]. This prediction, famously known as Moore's Law, has been proven correct, time and again. This has been made possible mainly due to the continuous scaling or miniaturization of components that are integrated onto a chip [3]. For example, in CMOS technology, the gate length of a Metal Oxide Semiconductor Field Effect Transistor (MOSFET) has been scaling by a factor of 0.7 every 2 years. Over the last few years, FinFET-based devices [4] (a variation of MOSFET) have been fabricated at 22 nm, and the 14 nm technology is expected to be reached in the near future [5]. CMOS technology scaling beyond the deep sub-micron/nano range, while enabling higher integration of VLSI designs, has caused various reliability issues. Some of the issues of CMOS scaling beyond the nanometer range are increased leakage current, processes variations, etc. [6]. These non-idealities have caused the I-V characteristics of MOSFETs to be different from what is expected. It has become more difficult to improve performance by technology scaling.

This has led to the emergence of alternate computing paradigms (reversible computing [7], multi-valued logic (MVL) computing [8]) coupled with emerging devices such as carbon nanotube field effect transistor (CNFET) [9], quantum dot gate field effect transistor (qFET) [10], etc. Researchers have also been investigating new materials and devices in sub-10 nm, which could possibly replace MO S-based transistors. Based on the ITRS road map [5], some of the emerging devices that have the characteristics to replace traditional MOS-based devices are CNFET [11,12], nanowire field effect transistor (NWFET) [13], graphene transistor [14,15], and III-V compound semiconductor [16,17].

One of the computing paradigms that has received considerable attention over the last few decades is multi-valued logic (MVL) [18]. Three-valued logic or ternary logic, which is a special case of MVL, has attracted considerable interest over the last couple of decades. A recent survey presents various contemporary aspects related to MVL [8]. Some of the advantages of MVL include reduced interconnect complexity, less device count, etc. This is due to the fact that more information is embedded per digit. For example, it is possible to represent a 14-digit (N-digit) binary number using only 9 ($\log_3(2^N - 1)$) ternary digits. Ternary logic is a special case of MVL with three significant states. There have been many CMOS-based implementations for ternary logic [19,20]. It has been shown that the performance of CMOS-based designs is enhanced by adding MVL blocks to binary designs [21,22]. A design for ternary memory units and sequential circuits has been presented in [23]. A CMOS-based ternary Wallace tree multiplier has been implemented in [24]. Apart from the works that focus on novel designs [23–26], there have been many works that focus on synthesis of MVL logic circuits [27–29].

The CMOS implementations of MVL are mainly classified as current-mode circuits [30], which require transistor biasing and voltage-mode circuits [22], which require additional voltage sources to create multi-threshold transistors. Due to the problems in MOS-based devices and nonavailability of appropriate devices, design of efficient MVL circuits has long remained a concern [18]. However, the emergence of several new device technologies [9,10,31] has led to renewed interest in ternary and quaternary logic in particular.

CNFETs have been used widely in the implementation of ternary logic circuits. CNFET is one of the promising alternatives to MOSFET due to its unique one-dimensional band structure that suppresses backscattering and makes near-ballistic operation a realistic possibility [32–35]. CNFETs use a single-walled CNT as a conducting channel, which is conducting or semiconducting depending on the angle of atom arrangement along the tube, also called a chirality vector. Unlike in MOS technology, where body biasing is used to control threshold voltages, in CNFET technology the threshold voltage is controlled by changing the diameter of a CNT, which in turn depends on the chirality vector [36]. This dependence makes CNTFET suitable for implementation of MVL circuits.

The combination of ternary logic and CNFETs has the capability to achieve efficient realizations of digital systems. There have been many CNFET-based design [36–40] and synthesis techniques [41] that are used to realize ternary logic circuits. The existing work on CNFET-based ternary logic circuits is relatively recent, and there is scope to explore new design techniques to realize efficient ternary circuits.

The chapter presents different design approaches to implement ternary logic circuits using CNFETs. In this chapter, Sections 11.2 and 11.3 present an overview of ternary logic and CNFETs. A brief review of CNFET-based ternary logic circuits is presented in Section 11.4. Different design approaches to implement ternary circuits are presented in Section 11.5. In Section 11.6, we present the implementation of basic ternary circuits using different approaches. This section also presents the HSPICE simulation results. Finally, conclusions are presented in Section 11.7.

11.2 Ternary Logic

Binary logic, when given a significant third value, is called ternary logic or three-valued logic; functions realized with three values are called ternary logic functions. The values 0, 1, and 2 form the nomenclature to denote the ternary values in this paper. A function $f(X)$ is defined as a ternary logic function mapping $\{0,1,2\}^n$ to $\{0,1,2\}$ where X is given by $X_1,\ldots,X_n$. When $X_i, X_j \varepsilon \{0,1,2\}$, the basic operations of ternary logic can be defined as

$$X_i + X_j = \max\{X_i, X_j\} \tag{11.1}$$

$$X_i \cdot X_j = \min\{X_i, X_j\} \tag{11.2}$$

where Equations (11.1) and (11.2) indicate OR and AND operations for ternary logic [36]. Another important logic function in ternary logic is a ternary inverter. Table 11.1 shows the outputs of different ternary inverters that are used in ternary logic. Corresponding to each of the outputs, three inverters are defined—namely, negative ternary inverter (NTI), standard ternary inverter (STI), and positive ternary inverter (PTI), respectively. The logic values assumed for different voltage levels are shown in Table 11.2; voltages 0, $V\,dd/2$, and $V\,dd$ correspond to logic values 0, 1, and 2, respectively.

TABLE 11.1

Ternary Inverters

Input (x)	NTI (x)	STI (x)	PTI (x)
0	2	2	2
1	0	1	2
2	0	0	0

Source: Lin, S. et al., *IEEE T. Nanotechnol.*, 10, 217–225, 2011.

TABLE 11.2

Logic Symbols

Voltage Level	Logic Value
0	0
$V_{dd}/2$	1
V_{dd}	2

Implementation of ternary logic circuits requires transistors with different threshold voltages. Hence, CNFET technology, where the threshold voltage of the transistor can be modified by changing its physical dimensions, is suitable to implement ternary logic circuits [36]. The following section presents a brief overview of CNFET.

11.3 Carbon-Nanotube FET

A single-walled carbon nanotube (SWCNT) is obtained by rolling up a sheet of graphite along a rolled-up vector $C = na + mb$, as shown in Figure 11.1, where m and n are positive integers that specify the chirality of the tube and ${}^0a^0$ and ${}^0b^0$ are lattice unit vectors [42]. The angle of atom arrangement along the tube—also called a chiral angle, roll-up vector, or chirality vector in a single-walled CNT (SWCNT)—is represented by an integer pair (n, m). The value of (n, m) determines if the CNT is metallic or semiconducting.

SWCNT is further classified into three groups, depending on the angle of atom arrangement, i.e., chirality vector, along which the CNT is rolled. The three groups of CNT are named as armchair CNT if CNT has $n = m$, zigzag CNT if $n = 0$ or $m = 0$, and chiral CNT if m and n are different and nonzero. All armchair CNTs behave as conductors. On the other hand, zigzag and chiral CNTs show metallic (conducting) behavior when $n = m$ or $n - m = 3i$, where i is an integer; otherwise, they show semiconducting behavior. Hence, zigzag and chiral CNTs are used in realizing a CNTFET [43]. The chirality vector (n, m) also sets the diameter of the CNT.

CNFET is a transistor that makes use of semiconducting carbon nanotubes as channel material between two metal electrodes that act as source and drain contacts. The operating principle of CNFET is similar to that of MOS transistors. As shown in Figure 11.2, this three- (or four-) terminal device consists of a semiconducting nanotube, acting as conducting channel, bridging the source and drain contacts. The device is turned on or off electrostatically via the gate. The drain current is directly proportional to the number of CNTs connected between the source and the drain and their respective diameters [44,45].

Three types of CNTFET devices have been reported in the literature. They are known as Schottky barrier CNTFET (SB-CNTFET), MOSFET-like CNTFET (MCNTFET), and band-to-band tunneling CNTFET (T-CNTFET). Due to the similarities of M-CNTFET with MOSFET in terms of operation and intrinsic attributes, CNTFET has been used in

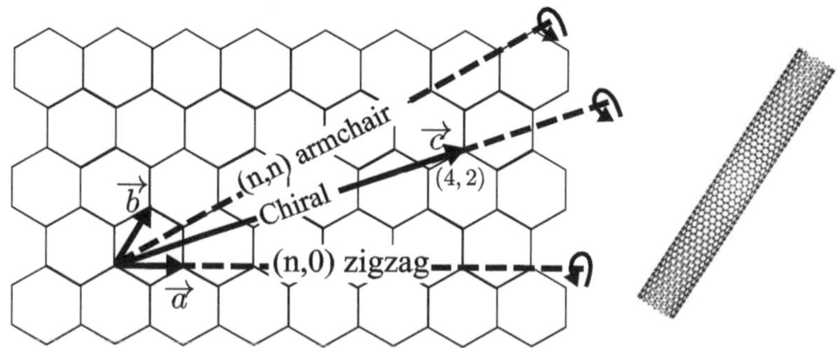

FIGURE 11.1
Unrolled sheet of graphite and the rolled lattice structure of CNT. (From Saito, R. et al., *Physical Properties of Carbon Nanotubes*, Imperial College Press, London, UK, 1998.)

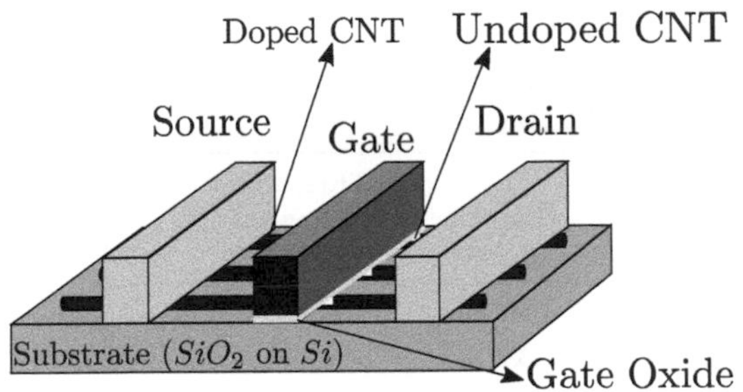

FIGURE 11.2
3D view of carbon nanotube field effect transistor (CNFET).

implementation of logic circuits [45]. The gate width of CNTFET can be approximated using the following equation [44]:

$$W \approx \min(W_{\min}, N \times S) \tag{11.3}$$

In Equation (11.3), $W_{\min}$ is the minimum gate width, N is the number of tubes, and S is the distance between the centers of two adjoining CNTs under the same gate, also called a Pitch. The diameter of CNT, D_{CNT}, depends on the chirality vector (n, m) and can be calculated by the following equation:

$$D_{CNT} = \frac{\sqrt{3}a_0}{\pi}\left(\sqrt{n^2 + m^2 + mn}\right) \tag{11.4}$$

where $a_0 = 0.142$ nm is the interatomic distance between each carbon atom and its neighbor. The threshold voltage, which is the voltage needed to turn ON the device electrostatically via the gate, can be approximated to the first order as the half bandgap and can be calculated by Equation (11.5) [44].

$$V_{th} \approx \frac{E_g}{2e}\frac{1}{\sqrt{3}}\frac{aV\pi}{eD_{CNT}} = \frac{0.43}{D_{CNT}(\text{nm})} \tag{11.5}$$

In the previous equation, V_π(= 3.033 eV) is the carbon π–π bond energy in the tight bonding model, a(= 0.249 nm) is the carbon-carbon atom distance, and e is the unit electron charge. If the chirality vector of CNT changes, then the threshold voltage of the CNTFET will also change. Assuming the m value in the chirality vector is always zero, the ratio of the threshold voltages of two CNTFETs with different chirality vectors can be represented by the following equation:

$$\frac{V_{th1}}{V_{th2}} = \frac{D_{CNT2}}{D_{CNT1}} = \frac{n_2}{n_1} \tag{11.6}$$

Equation (11.6) shows that threshold voltage of the CNFET is inversely proportional to the diameter of CNT, which as previously mentioned, depends on its chirality vector. It is the threshold voltage controllability of CNFET that makes it well suited for the implementation of multi-valued logic circuits. The relationship between chirality, CNT diameter, and threshold voltage can be derived from relations presented in [44] and is shown in Table 11.3.

TABLE 11.3

Relation Between Chirality, CNT Diameter, and Threshold Voltage

Chirality	Diameter of CNT	Threshold Voltage of N-CNTFET	Threshold Voltage of P-CNTFET
(19,0)	1.487 nm	0.289V	−0.289V
(17,0)	1.330 nm	0.328V	−0.328V
(16,0)	1.253 nm	0.348V	−0.348V
(14,0)	1.100 nm	0.398V	−0.398V
(13,0)	1.018 nm	0.428V	−0.428V
(11,0)	0.861 nm	0.506V	−0.506V
(10,0)	0.783 nm	0.559V	−0.559V

Source: Deng, J., and Wong, H. S. P., *IEEE Trans. Electron Devices*, 54, 3186–3194, 2007.

There are many CNTFET device models in the literature [44–48]. Stanford CNFET device models available at [49], which are based on work presented in [44,45], have been widely used for the implementation of CNFET-based circuits. The technology parameters of CNTFET along with their brief description and numeric value are given in Table 11.4.

The effect of chirality variations on an N-CNFET is studied with the help of the *I*–*V* characteristics of the transistor, which are simulated in HSPICE using the CNTFET model in [49]. The CNFET is configured to have three CNTs (all with the same chirality) and a default pitch value equal to 20 nm. Figure 11.3 shows the $I - V$ characteristics for a V_{GS} of

TABLE 11.4

Technology Parameters for CNFET Model

Parameter	Description	Value
L_{ch}	Physical channel length	32 nm
Lgeff	Mean free path in the intrinsic CNT channel region	100 nm
L_{ss}	Length of doped CNT source-side extension region	32 nm
L_{dd}	Length of doped CNT drain-side extension region	32 nm
E_{fi}	Fermi level of the doped S/D tube	0.6 eV
Kgate	Dielectric constant of high-k top gate dielectric material	16
Tox	Thickness of high-k top gate dielectric material	4.0 nm
Csub	Coupling capacitance between the channel region and the substrate	40 pF/m
Vfbn & *Vfbp*	Flat-band voltage for n-CNTFET and p-CNTFET, respectively	0 eV, 0 eV
L_channel	Physical gate length	32 nm
Pitch	Distance between the centers of two adjacent CNTs	20 nm
Leff	Mean free path in p+/n+ doped CNT	15 nm
phi_M	Work function of source/drain metal contact	4.6 eV
phi_S	CNT work function	4.5 eV

Source: Deng, J., and Wong, H. S. P., *IEEE Trans. Electron Devices*, 54, 3186–3194, 2007; *IEEE Trans. Electron Devices*, 54, 3195–3205, 2007; Stanford University, *Stanford University CNFET Model Website*, Stanford University, Stanford, CA, 2008.

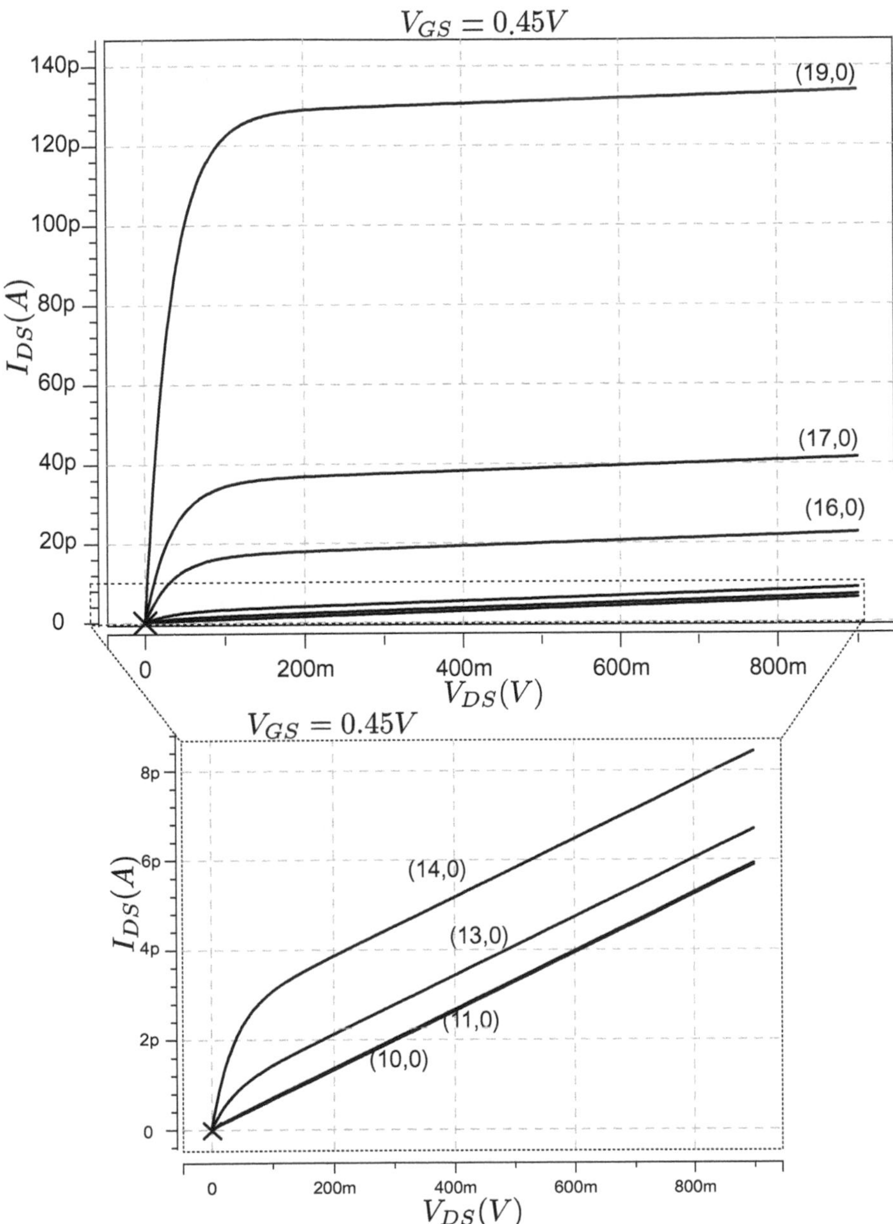

FIGURE 11.3
I-V characteristics of N-CNFET.

0.45V, where the x-axis indicates the drain-to-source voltage (V_{DS}) and the y-axis indicates the drain current (I_{DS}). As seen from this figure, for a fixed V_{GS}, the drain current (I_{DS}) is proportional to the diameter of CNTs, which in turn is proportional to the value of n in the chirality vector (see Table 11.3). There have been advances in the manufacturing processes of well-controlled CNTs [50,51]. While techniques exist to synthesize CNFETs of desired chirality [52,53], those with three chiralities—i.e., (19,0), (13,0), and (10,0)—are normally used in the implementation of CNFET-based ternary logic circuits [36].

11.4 Ternary Logic Circuits Using CNFETs

CNFETs have been used widely in the implementation of ternary logic circuits. The threshold voltage of CNFETs depends on the diameter of the CNT, which in turn depends on the chirality vector. This dependence makes CNTFET suitable for implementation of MVL circuits. While interest in design of CNFET-based logic circuits waned over recent years due to complex fabrication technology and reliability issues, recent demonstration of a CNFET-based processor/computer by Stanford University researchers [54] has reignited this interest.

CNFET-based ternary logic circuits using resistive loads have been presented in [55]. The disadvantage of this approach, however, is that it needs large off-chip resistances. A more efficient design methodology, which eliminates the need for large resistances by employing an active load with p-type CNFETs, has been presented in [36,37]. This work presented designs for ternary NTI, PTI, STI, NAND, and NOR gates, which were simulated using HSPICE with the Stanford CNTFET model of [49]. Ternary inverters are an integral part of many design approaches. Figure 11.4 shows the implementation of basic ternary inverters.

Recently, there have been many implementations of CNFET-based ternary arithmetic circuits (Adders [38,39,56–59] and ALU [40]) that focus on optimizing the design parameters. Apart from novel designs, there have been efforts to develop synthesis algorithms for CNFET-based ternary logic circuits. Recently, a synthesis technique for ternary logic circuits, which exploits the advantages of CNFET, has been presented in [41]. This technique combines the cube representation [60] and the unary operators [61] to arrive at a 3:1 multiplexer-based synthesis procedure. This work also presents a procedure for obtaining expressions for two and three variable functions using the unary operators that have been presented. Ternary functions with three (and more) variables are handled by a decomposition procedure based on work presented in [62].

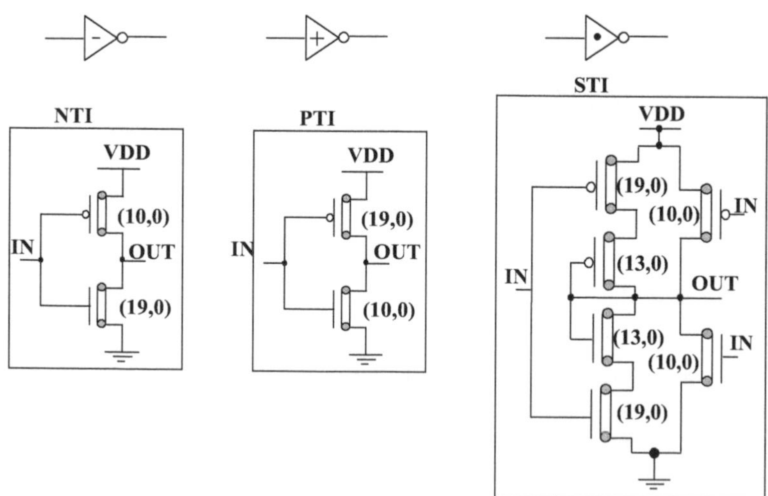

FIGURE 11.4
CNFET-implementation of basic ternary inverters. (From Lin, S. et al., *IEEE Trans. Nanotechnol.*, 10, 217–225, 2011.)

11.5 Design Approaches to Implement Ternary Logic Circuits

11.5.1 Decoder-Encoder Based Approach

CNFET-based ternary logic circuits using resistive loads have been presented in [55]. The disadvantage of this approach, however, is that it needs large off-chip resistances. A more efficient design methodology, which eliminates the need for large resistances by employing an active load with p-type CNFETs, has been presented in [36,37]. This approach is also called a decoder-encoder-based approach [36] to design ternary logic circuits and can be divided into three main stages. In the first stage, a ternary decoder is used to convert a ternary signal into mutually exclusive unary functions that will have two logic levels, logic 0 and logic 2. The relation between ternary input X and decoder outputs (indicated by X^0,X^1,X^2) is given by

$$X^k = \begin{cases} 2, & \text{if } X = k \\ 0, & \text{if } X \neq k \end{cases} \tag{11.7}$$

These decoder outputs can take only two logic values, i.e., logic 2 and logic 0, corresponding to logic 1 and logic 0 in binary logic. The outputs of ternary decoders are combined using binary logic gates in the second stage. In the third and final stage, the outputs of the second stage are combined using an encoder to generate the ternary outputs. The ternary encoder consists of a level shifter and a ternary or gate.

Consider the example of a ternary function represented in Table 11.5, where A and B are ternary inputs and F is ternary output. The output function F is equal to logic 2 for the input signals $A = 1$ and $B = 0$ or $A = 2$ and $B = 0$. It is equal to logic 1 for the signal values $A = 0$ and $B = 1$ or $A = 0$ and $B = 2$. F is equal to logic 0 in all the remaining input cases. Using Table 11.5, output function F may be written as

$$F = 2 \cdot \sum (3,6) + 1 \cdot \sum (1,2) + 0 \cdot \sum (0,4,5,7,8) \tag{11.8}$$

$$F = 2 \cdot (A^1B^0 + A^2B^0) + 1 \cdot (A^0B^1 + A^0B^2) \tag{11.9}$$

TABLE 11.5

Truth Table (Example 1)

Decimal Equivalent	*A*	*B*	*F*
0	0	0	0
1	0	1	1
2	0	2	1
3	1	0	2
4	1	1	0
5	1	2	0
6	2	0	2
7	2	1	0
8	2	2	0

where A^k and B^k ($k = 0,1,2$) represent the unary outputs of the ternary signals A and B. F^2 and F^1 represent the unary signals of output F, which are combined at the final stage using an encoder to generate ternary output F. The final encoder consists of a level shifter and a ternary or gate. The ternary logic gates are usually represented with a (dot) on the gate. The implementation of function F following the methodology in [36] is shown in Figure 11.5. The implementation of the decoder and encoder used in [36] are shown in Figure 11.6.

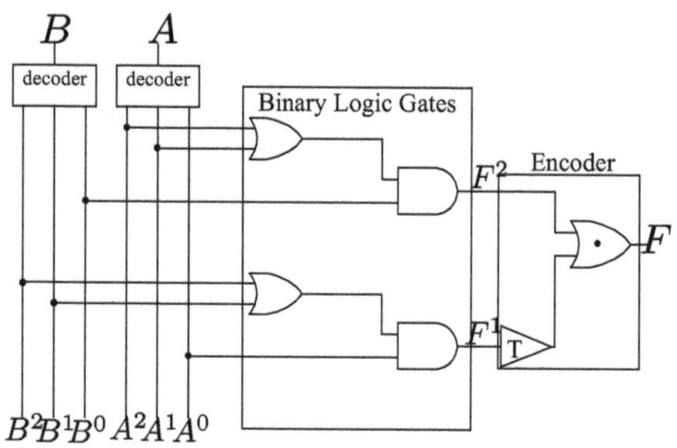

FIGURE 11.5
Realization of a ternary function using existing decoder-encoder-based approach.

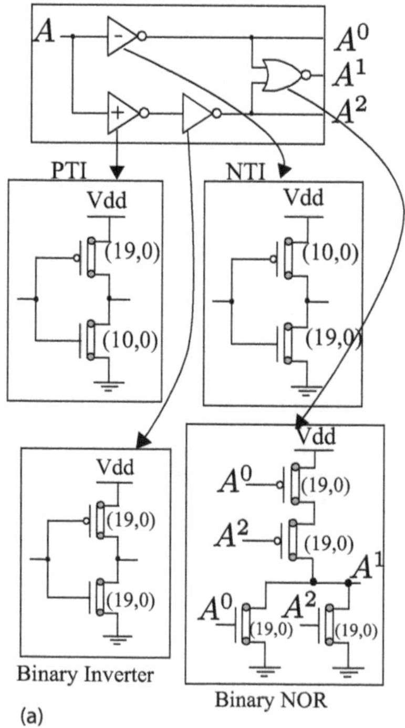

FIGURE 11.6
(a) CNFET-based implementation of decoder. *(Continued)*

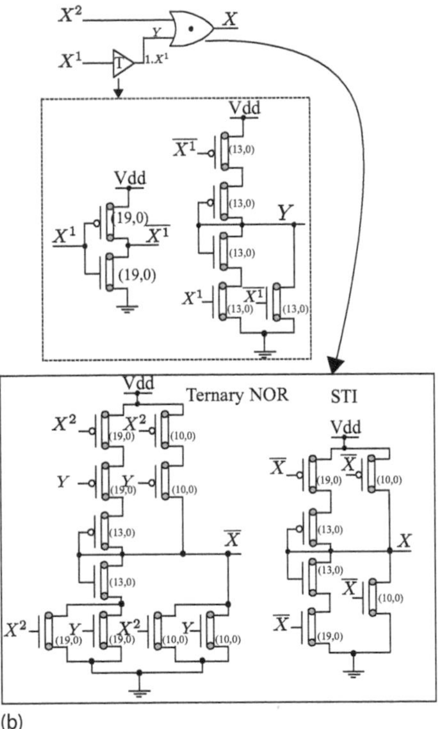

FIGURE 11.6 (Continued)
(b) CNFET-based implementation of encoder.

11.5.2 3:1 Multiplexer-Based Approach

Another approach uses unary operators and 3:1 multiplexers for implementing ternary logic circuits and has been presented in [59,63]. To illustrate this approach, consider a ternary function represented in Table 11.5, where A and B are ternary inputs and F is ternary output. Since it has two inputs, one of the inputs is chosen as a select line for a 3:1 multiplexer and the other input is used in generation of unary operators. If input A is chosen as the select signal for the 3:1 multiplexer and $A = 0$, then (0,1,2) is transformed into (0,1,1) with respect to input B. Similarly, when $A = 1$ or $A = 2$, (0,1,2) is transformed into (2,0,0) with respect to input B. The transformations, also called unary operators, are implemented in [59], such that they have low power consumption. Figure 11.7 shows the implementation of example ternary function using the CNFET-based 3:1 multiplexers and unary operators as presented in [59]. The 3:1 multiplexer requires 18 CNFETs for implementation as shown in Figure 11.8.

11.5.3 Decoderless Approach

A novel approach, which avoids the use of decoders for every input, has been presented in [64]. The details of this approach are presented in the following subsections.

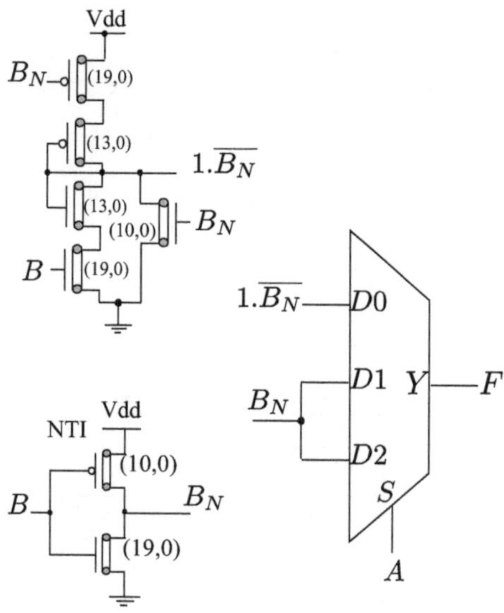

FIGURE 11.7
Implementation of function represented by Table 11.5 using approach. (From Srinivasu, B., and Sridharan, K., *IET Circuits Devices Syst.*, 1–13, 2016.)

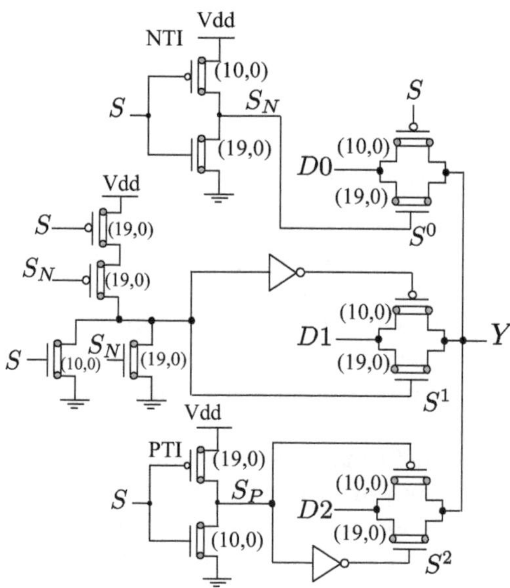

FIGURE 11.8
Transistor-level implementation of 3:1 multiplex. (From Srinivasu, B., and Sridharan, K., *IET Circuits Devices Syst.*, 1–13, 2016.)

11.5.3.1 Overview

As seen from the decoder-encoder-based approach, each of the inputs is converted into a mutually exclusive binary signal using ternary decoders. As the number of inputs increases, there will be an increase in the number of decoders needed, thus resulting in more area and power consumption. In the example discussed earlier, two decoders were needed for two inputs, *A* and *B*, in the implementation of the function *F*. The decoderless approach obviates the need for a decoder by optimally grouping the terms in Equation (11.9). Considering the same example, the function *F* can be represented as

$$F = 2 \cdot ((A^1 + A^2)B^0) + 1 \cdot (A^0(B^1 + B^2)) = 2.F^2 + 1.F^1 \tag{11.10}$$

F^2 and F^1 are the unary signals of output function *F*. F^2 and F^1 are signals that can take the values of logic 2 (logic high) or logic 0 (logic low). In the previous equations, F^2 is equal to logic 2 when B^0 is logic 2 and $(A^1 + A^2)$ is logic 2. The term B^0 will be equal to logic 2 only when input ternary signal *B* is equal to logic 0. The other term of F^2, i.e., $(A^1 + A^2)$ will be logic 2 only when the input ternary signal *A* is equal to logic 1 or logic 2. The complete term $(A^1 + A^2)B^0$ of the function F^2 can be realized using NTI gates and CNFET transistors, as shown in Figure 11.9a.

The realization of F^2 consists of one NTI gate and four additional transistors—M1, M2, M3, and M4. M1 and M2 are p-type CNFETs, whereas M3 and M4 are *n*-type CNFETs. The chiralities of the respective transistors are also shown in the figure. In accordance with the chiralities, M1, M2, M3, and M4 transistors have threshold voltages equivalent to −0.428, −0.559, 0.428, and 0.289*V*, respectively. Input *A* is through an NTI gate and connected to the base of M1 and M3. Since M1 has a threshold voltage of −0.428*V*, M1 is *ON* only when the output of NTI is logic 0 (i.e., when *A* is logic 1 or logic 2). M3 has a threshold voltage of 0.428*V* and thus M3 is *ON* only when output of NTI gate is logic 2 (i.e., when *A* is logic 0). The base of transistors M2 and M4 are connected to input *B*. Since M2 has a threshold voltage of −0.559*V*, M2 is *ON* when *B* is logic 0. The threshold voltage of M4 is 0.289*V*, so M4 is

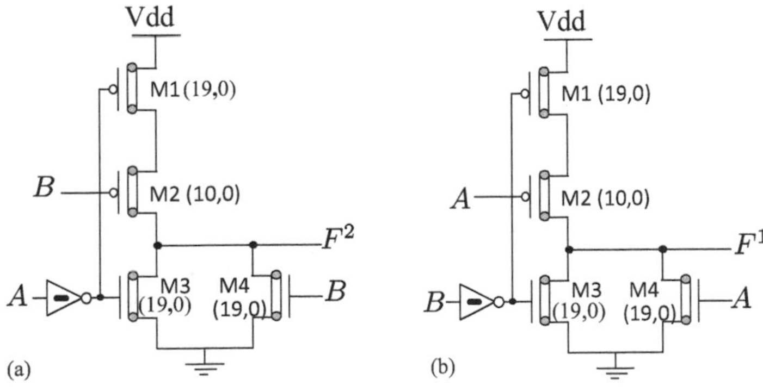

FIGURE 11.9
Implementation of (a) F^2 and (b) F^1 in decoderless approach.

ON when *B* is logic 1 or logic 2. The overall output F^2 is logic 2 when both M1 and M2 are *ON* and F^2 is logic 0 when either M3 or M4 is *ON*. Similar analysis can be done to the circuit implementation of F^1, which is shown in Figure 11.9b.

As seen from the previous example, the use of a decoder circuit for each ternary input can be avoided using the decoderless approach. This results in fewer transistors as well as less delay and power consumption when compared to the existing decoder/encoder approach. The previous analysis can be described more formally as follows:

Proposition 1.1: *Without the loss of generality, let F (A, B) be any ternary function with A and B as inputs. If it is possible to represent the unary terms of the inputs A and B using any combination of transistors, PTI and NTI gates, then there is no need to use a decoder for each input at the initial stage.*

11.5.3.2 Steps Involved

Based on the idea presented in the previous section, an approach for implementation of CNFET-based ternary circuits without a decoder is presented. The steps involved in this approach are described in the following section and later on illustrated with an example.

1. At first, any function *F* is represented using a *K*-map. The terms are grouped at this stage, and appropriate equations are determined.
2. The equations can be used to determine the components (such as NTI and PTI, etc.) required for CNFET-based implementation. Also, the number of transistors required for implementation of unary signals of *F*, that is, F^2, F^1, and F^0 is determined. Since unary signals are mutually exclusive, implementation of any two out of three functions is enough. The selection of functions is made in such a way that the least number of transistors is required for implementation.
3. Based on the unary functions chosen, the optimized last stage encoder circuit is used to generate the final ternary output.

11.5.3.3 Function Simplification

Any given function *F* is first represented using *K*-map and equations in terms of unary functions. The terms in the *K*-map can be grouped, or the equations can be simplified such that the terms in the final simplified equation can be realized using the ternary inputs directly. This is achieved by using a combination of NTI gates, binary inverters, and transistors. As an example, consider a function *F*(*A*, *B*) represented by the truth table shown in Table 11.6. The output is logic 0 when $A = B$, is logic 1 when $A < B$, and logic 2 when $A > B$. The function can be represented using the *K*-map as shown in Figure 11.10. To get the unary functions corresponding to three levels, the 1s, 2s, and 0s are grouped separately. This results in minimized functions for F^2, F^1, and F^0. The grouping of terms can be done using the *K*-map shown in Figure 11.10.

The functions can also be derived using the simplification of equation as illustrated in the following:

$$F = 2 \cdot \sum(3,6,7) + 1 \cdot \sum(1,2,5) + 0 \cdot \sum(0,4,8) \tag{11.11}$$

$$F = 2 \cdot (A^1B^0 + A^2B^0 + A^2B^1) + 1 \cdot (A^0B^1 + A^0B^2 + A^1B^2) + 0 \cdot (A^0B^0 + A^1B^1 + A^2B^2) \tag{11.12}$$

TABLE 11.6
Truth Table (Example 2)

Decimal Equivalent	*A*	*B*	*F*
0	0	0	0
1	0	1	1
2	0	2	1
3	1	0	2
4	1	1	0
5	1	2	1
6	2	0	2
7	2	1	2
8	2	2	0

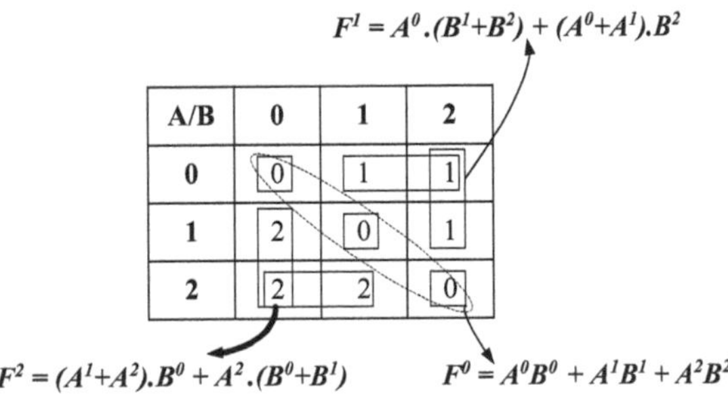

FIGURE 11.10
K-map simplification for function *F*.

$$F = 2\cdot(A^1 + A^2)B^0 + A^2(B^0 + B^1)) + 1\cdot(A^0(B^1 + B^2) + B^2(A^0 + A^1)) + 0\cdot(A^0B^0 + A^1B^1 + A^2B^2) \quad (11.13)$$

$$F = 2 \cdot F^2 + 1 \cdot F^1 + 0 \cdot F^0 \quad (11.14)$$

11.5.3.4 Implementation of Unary Functions

Functions F^2, F^1, and F^0 in Equation (11.13) represent the unary functions of output *F* and are mutually exclusive. Hence, any two of the three unary functions are enough to generate the final output *F*. As mentioned earlier, the two functions that are to be implemented are chosen in such a way that realization of the circuit results in the least number of transistors. After the simplification is done, the resulting expressions contain either individual unary terms of the inputs, that is, A^0, B^0, A^1, etc., or group terms such as $(A^1 + A^2)$, $(B^0 + B^1)$, etc. These terms can be realized using PTI gates, NTI gates, and transistors.

For example, A^0 indicates that the output function is logic 2 when input *A* is logic 0, i.e., a connection to *V DD* is made possible when *A* is logic 0. This pull-up functionality is realized by designing a *p*-type CNFET with a threshold voltage of –0.559*V*, i.e., chirality equivalent to (10,0). Such a transistor will be *OFF* when gate voltage is greater than 0.341*V* and hence for logic 1 (0.45*V*) and logic 2 (0.9*V*), it will be in the *OFF* state. The values of input for which the output remains logic 0 are considered for designing the *n*-type CNFET structure. In the case of A^0, the output remains logic 0 when the input is at logic 1 or logic 2. This pull-down functionality is realized by *n*-type CNFET with threshold voltage of 0.289*V* equivalent to a chirality of (19,0). Such a transistor will be *ON* when gate voltage is greater than 0.289*V* and hence for logic 1(0.45*V*) and logic 2(0.9*V*), it remains in the *ON* state. A similar analysis can be done to design *p*-type and *n*-type structures for different terms of a simplified equation. Figures 11.11 and 11.12 show the realization of different unary terms (individual and group) that might appear in any simplified equation. To realize the output unary functions, the *p*-type CNFET structures corresponding to the input unary terms are placed in a series to realize the AND (·) function and in parallel to realize the OR (+) function. On the contrary, *n*-type CNFET structures are placed in a series to realize the OR (+) function and in parallel to realize the AND (·) function.

Figures 11.11 and 11.12 also show the number of transistors required to implement the individual and group unary terms. This is used to calculate the number of transistors required for the implementation of unary functions F^2, F^1, and F^0.

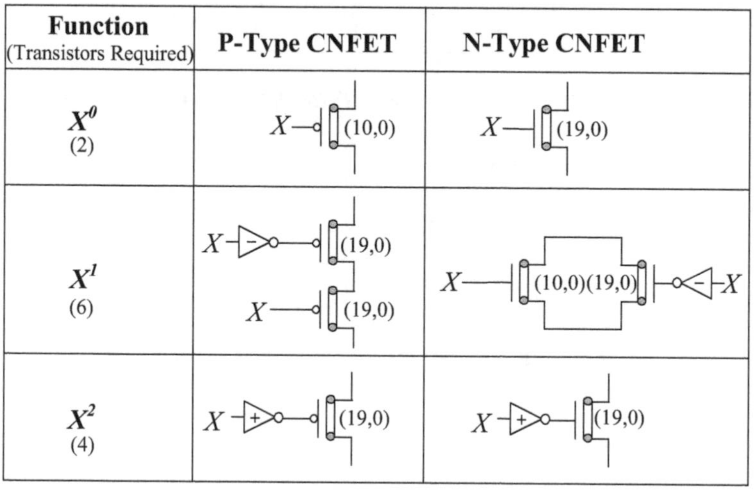

Function (Transistors Required)	P-Type CNFET	N-Type CNFET
X^0 (2)	X (10,0)	X (19,0)
X^1 (6)	X − (19,0) X (19,0)	X (10,0)(19,0) − X
X^2 (4)	X + (19,0)	X + (19,0)

FIGURE 11.11
Transistor-level realization of X^0, X^1, and X^2 unary terms.

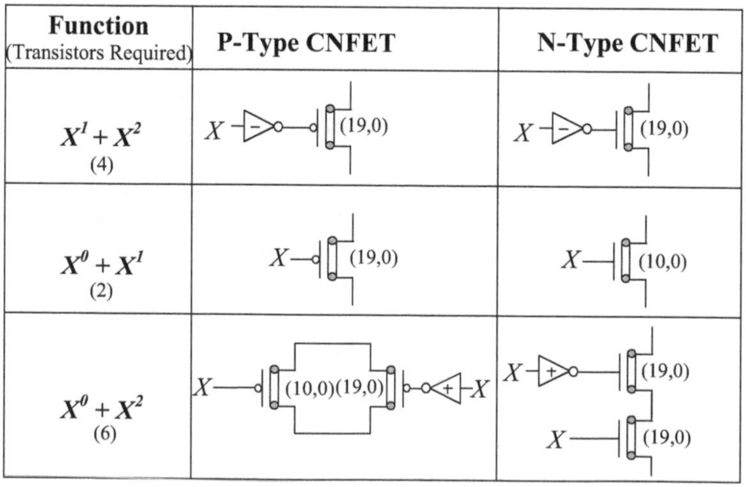

Function (Transistors Required)	P-Type CNFET	N-Type CNFET
$X^1 + X^2$ (4)	X − (19,0)	X − (19,0)
$X^0 + X^1$ (2)	X (19,0)	X (10,0)
$X^0 + X^2$ (6)	X (10,0)(19,0) + X	X + (19,0) X (19,0)

FIGURE 11.12
Transistor-level realization of $X^0 + X^1$, $X^1 + X^2$ and $X^0 + X^2$ unary terms.

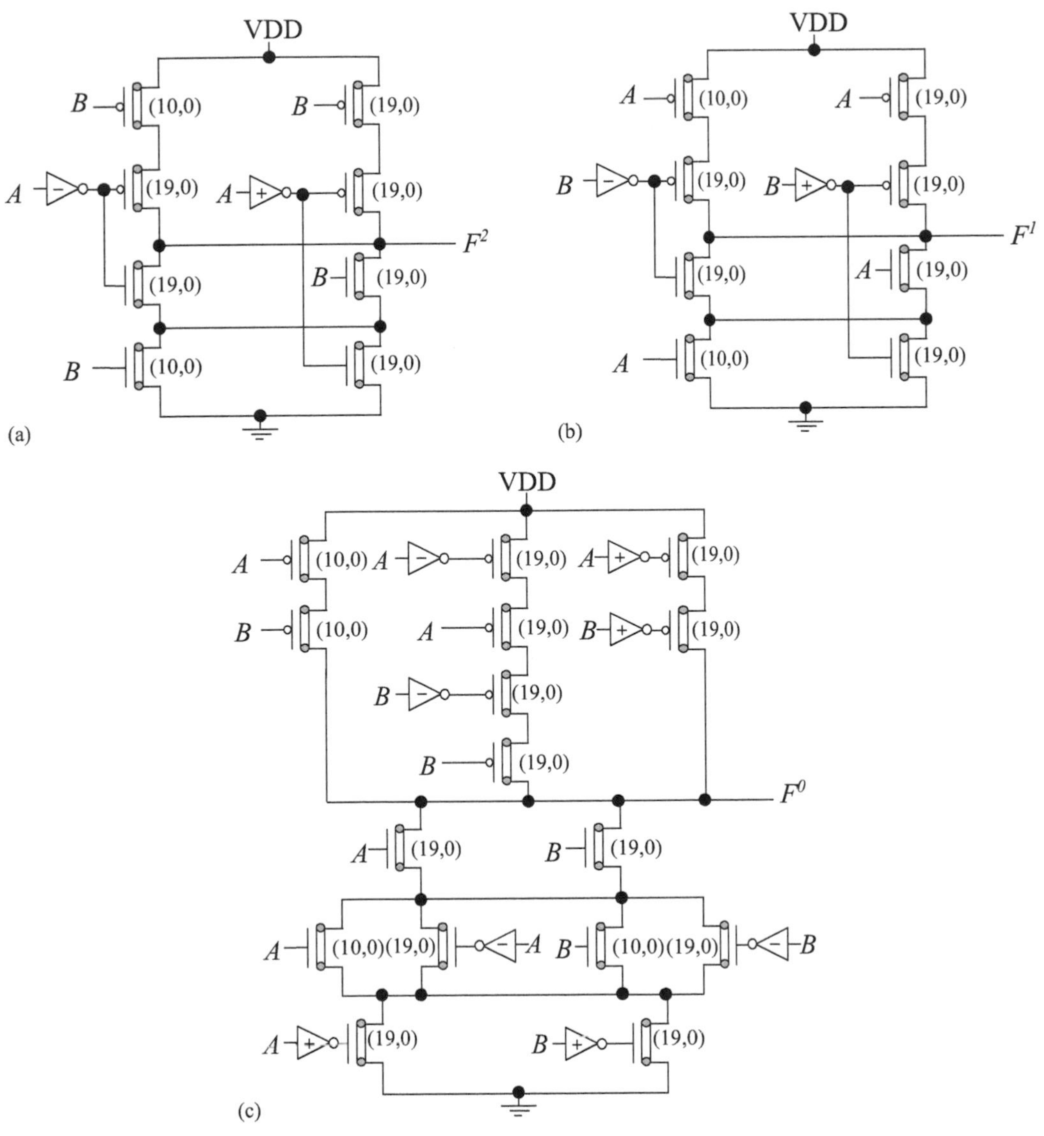

FIGURE 11.13
Implementation of (a) F^2, (b) F^1, and (c) F^0 using decoderless approach.

The implementation of functions F^2, F^1, and F^0 is shown in Figure 11.13. The simplified function of F^2 contains $(A^1 + A^2)B^0$ as the first term; therefore, the *p*-type structure corresponding to B^0 is placed in a series with the p-type structure corresponding to $(A^1 + A^2)$. The resulting structure is then placed in a series with the structure corresponding to $A^2(B^0 + B^1)$ to get the final *p*-type CNFET structure. A process similar to complementary logic style is followed to design the n-type structure resulting in the circuit shown in Figure 11.13. The realization of all unary functions F^2, F^1, and F^0 results in 12, 12, and 24 transistors, respectively.

11.5.3.5 Low-Power Encoder

In the final stage, an encoder is used to combine the unary signals and a final ternary output is generated. An encoder generates logic 2 and logic 0 by a direct connection to *V DD* and *GND*. However, for generation of *logic* 1 at output, a direct path from *V DD* to *GND* is created. One of the major disadvantages of the existing ternary adder designs [39,57] is that they use encoders that have a low resistance path between *V DD* and *GND* to generate *logic* 1. This results in a large static current and hence large static power consumption. The encoder design in the existing approach [36] (shown in Figure 11.5) uses a level shifter and a ternary OR gate. Hence, it requires a large number of transistors and also has large power consumption. The power consumption is mainly because of multiple direct paths that exist from *V DD* to *GND*, while generating logic 1. An improved encoder, which requires fewer transistors when compared to encoder in [36] and has lower power consumption when compared to encoders in [36,39,57], is presented in this section. Figure 11.14 shows the encoder [64], which has unary functions F^2 and F^1 as inputs. Since the functions F^2 and F^1 are unary function and can be either logic 0 or logic 2, simple binary inverters can be used to get their complements. Implementation of a simple binary inverter is also shown in Figure 11.14.

Existing encoder and the encoder shown in Figure 11.14 uses F^1 and F^2 as inputs to generate ternary output *F*. Hence, if any other pair of functions, i.e., F^2 and F^0 or F^1 and F^0, are implemented to reduce the number of transistors as explained in Section 11.5.3.4, F^1 or F^2 have to be generated using binary NOR gate. To avoid this, two more encoders are presented in [64] that take combinations of (F^2, F^0) or (F^1, F^0) as inputs. Figure 11.15 shows encoder designs for the function pairs (F^1, F^0) and (F^2, F^0). As in the case of (F_2, F_1)-based encoder, simple binary inverters can be used to get the complement of unary functions for the encoders presented. For the example under consideration (Equation 11.13), implementation of F^2 and F^1 results in the least number of transistors and hence the encoder design presented in Figure 11.14 is used in the final stage. As seen from the decoderless approach, any ternary function can be realized without decoders thus resulting in reduced area. Although the decoderless approach is explained using an example of a 2-input function, the same methodology can be extended to any number of inputs. However, care should be taken to avoid stacking of multiple transistors.

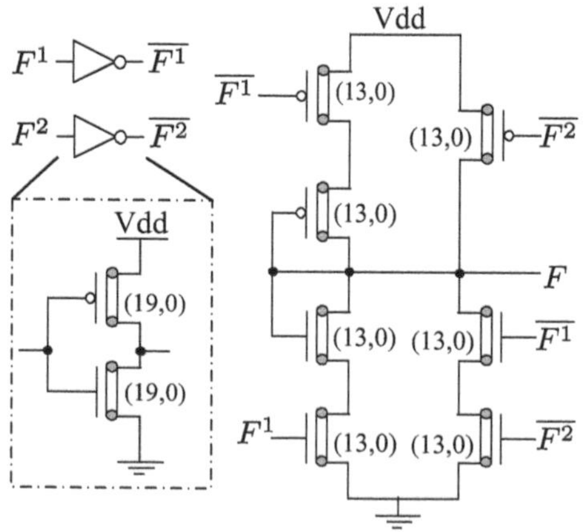

FIGURE 11.14
Encoder with F^2 and F^1 as inputs ((F^2, F^1)-Encoder).

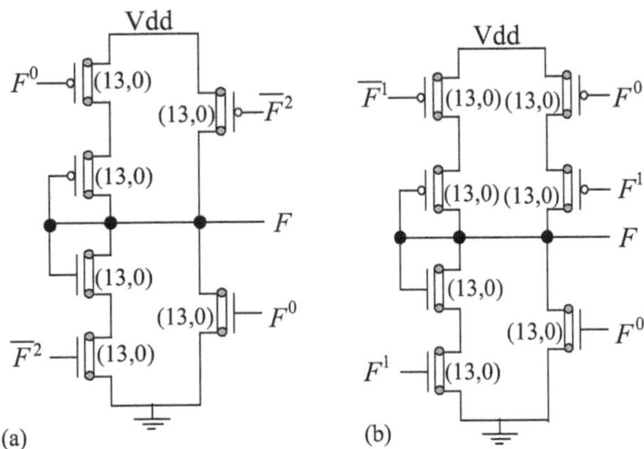

FIGURE 11.15
Encoder design for (a) (F^2, F^0) and (b) (F^1, F^0) combination.

11.5.4 2:1 Multiplexer-Based Approach

In this section, an approach that uses 2:1 multiplexers for implementation of ternary logic circuits is presented. This is based on the work presented in [65].

11.5.4.1 Basic Idea

The basic idea involved in this approach is presented using the following proposition:

Proposition 1.2: *A 3:1 multiplexer can be implemented using two 2:1 multiplexers.*

Proof. Consider the expression for a 3:1 multiplexer (shown in Figure 11.7), which is represented as $Y = S^0 \cdot D_0 + S^1 \cdot D_1 + S^2 \cdot D_2$, where D_0, D_1 and D_2 represent inputs while S represents select signal. S^0, S^1, and S^2 are related to S according to Equation (11.7) and can be generated as shown in Figure 11.8:

$$Y = S^0 \cdot D_0 + S^1 \cdot D^1 1 + S^2 \cdot D_2 \tag{11.15}$$

$$Y = S^0 \cdot D_0 + (S^1 + S^0) \cdot (S^1 + S^2) \cdot D_1 + S^2 \cdot (S^1 + S^2) \cdot D_2 \tag{11.16}$$

$$\because S^1 = (S^1 + S^0) \cdot (S^1 + S^2),\; S^2 \cdot S^2 = S^2, \text{ and } S^2 \cdot S^1 = 0$$

$$Y = S^0 \cdot D^0 + (S^1 + S^2) \cdot ((S^1 + S^0) \cdot D^1 + S^2 \cdot D^2) \tag{11.17}$$

$$Y = S^0 \cdot D_0 + \overline{S^0} \cdot (\overline{S^2} \cdot D^1 + S^2 \cdot D^2) \tag{11.18}$$

$\because$ $(S^1 + S^2) = S^0$, $(S^1 + S^0) = S^2$, where S^0, S^1, and S^2 represent the binary NOT of signals S^0, S^1, and S^2, respectively, as given by Equation (11.19):

$$\overline{S^k} = \begin{cases} 2 & \text{if } S^k = 0 \\ 0 & \text{if } S^k = 2 \end{cases} \tag{11.19}$$

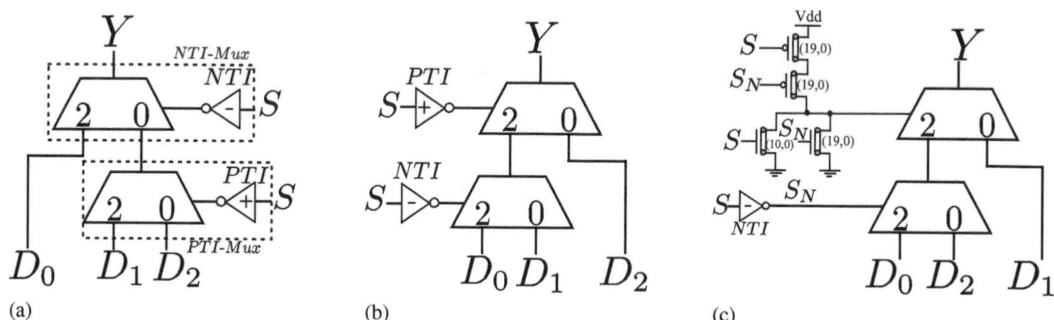

FIGURE 11.16
(a) A 3:1 multiplexer operation using 2:1 multiplexers for Equation 11.18. (b) A 3:1 multiplexer operation using 2:1 multiplexers for Equation 11.20. (c) A 3:1 multiplexer operation using 2:1 multiplexers for Equation 11.21.

Alternatively, Equation (11.15) can also be represented as Equations (11.20) and (11.21).

$$Y = \overline{S^2} \cdot (S^0 \cdot D_0 + \overline{S^0} \cdot D^1) + S^2 \cdot D_2 \tag{11.20}$$

$$Y = \overline{S^1} \cdot (S^0 \cdot D_0 + \overline{S^0} \cdot D_2) + S^1 \cdot D_1 \tag{11.21}$$

The relation in Equations (11.18), (11.20), and (11.21) are similar to relation for a 2:1 multiplexer, $Y = \overline{S} \cdot D_0 + S \cdot D_1$, where D_0 and D_1 are inputs and S is the select line. Hence, a 3:1 multiplexer can be implemented using two 2:1 multiplexers as shown in Figure 11.16.

11.5.4.2 Ternary Circuits Using CNFET-Based 2:1 Multiplexers

As evident from the Figure, Equations (11.18) and (11.20) are less complex to implement, when compared to 11.21 because generation of S^0, $\overline{S^0}$, S^2, and $\overline{S^2}$ requires PTI and NTI whereas implementation of S^1 and $\overline{S^1}$ requires complex *NOR*-like structure in addition to an NTI. Hence, circuits shown in Figure 11.18a and b are used for implementing ternary functions. These circuits use two types of 2:1 multiplexers, namely PTI-Mux and NTIMux, which are implemented using CNFETs as shown in Figure 11.17a and b.

A 3:1 multiplexer presented in [41] requires 18 transistors. However, 2:1 multiplexers with inverters (NTI-Mux and PTI-Mux), which are equivalent to one 3:1 multiplexer, require only 12 transistors. Further, PTI-Mux and NTI-Mux are used in the implementation of ternary logic circuits. To illustrate a 2:1 multiplexer-based approach, consider the same example ternary function represented in Table 11.5, which was used to show a 3:1 multiplexer-based approach. The K-map for this ternary function is shown in Table 11.7.

For this ternary function, *A* and *B* are ternary inputs and *F* is ternary output. In a 2:1 multiplexer-based approach, similar to a 3:1 multiplexer-based approach, one of the inputs is chosen as the select line and the second input is used to realize the unary operators. But unlike the 3:1 multiplexer-based approach, the 2:1 multiplexer approach implements the unary operators using 2:1 multiplexers, PTI and NTI. Figure 11.8 shows the implementation of ternary function shown in Table 11.7, where A is chosen as the select line and unary operator is realized using a 2:1 multiplexer. This implementation requires 12 CNFETs when compared to the 3:1 multiplexer-based implementation, which requires 25

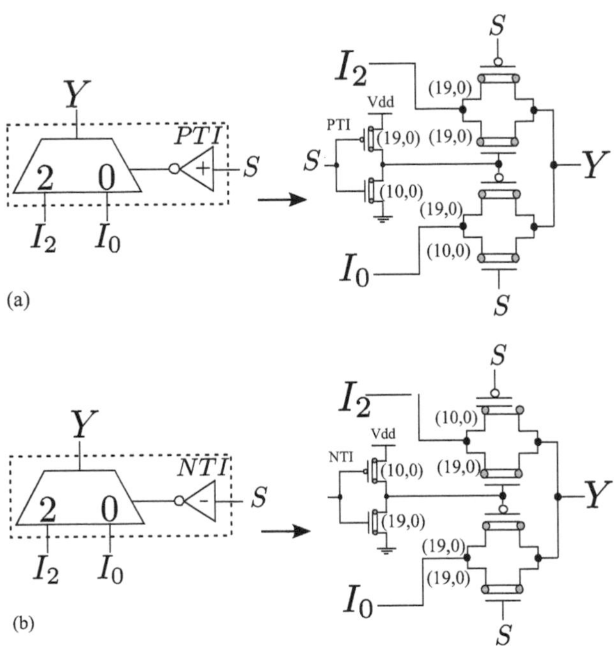

FIGURE 11.17
CNFET-based implementation of (a) PTI-Mux and (b) NTI-Mux.

TABLE 11.7
Ternary Function (Example 1)

A/*B*	0	1	2
0	0	1	1
1	2	0	0
2	2	0	0

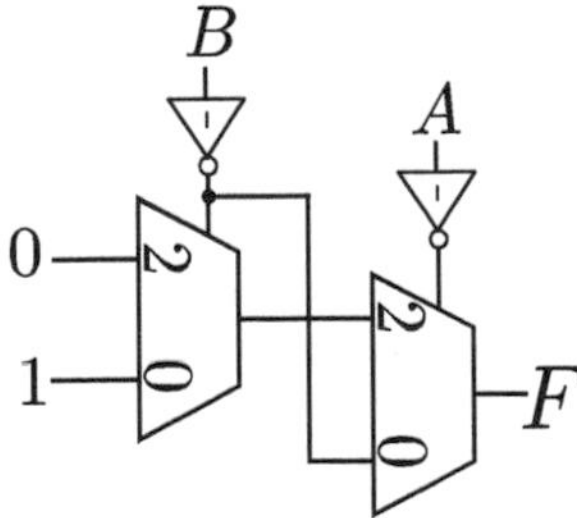

FIGURE 11.18
2:1 multiplexer-based implementation for ternary function in Table 11.7.

transistors. Since both decoders and encoders are avoided in this approach, it results in low area and power consumption when compared to design approaches presented earlier. However, since this approach uses transmission gates for realizing multiplexers, it results in ternary circuits with large propagation delay.

11.6 Implementation and Simulation

For a comparison of different approaches, basic ternary logic functions—namely half adder and 1-digit multiplier—have been implemented using different approaches, and circuit parameters such as delay, power, and number of transistors have been compared to understand relative performance. The functionality of the basic circuits is represented by Table 11.8.

As seen from this table, the half adder has two outputs (*Sum* and *Carry*), and the 1-digit multiplier has two outputs (*Product* and *Carry*). As an example, the circuit implementations for half adder using different approaches are shown in Figures 11.19 through 11.22.

TABLE 11.8

Truth-Table for Basic Ternary Circuits

Inputs		Half-Adder		1-Digit Multiplier	
A	**B**	**Sum**	**Carry**	**Product**	**Carry**
0	0	0	0	0	0
0	1	1	0	0	0
0	2	2	0	0	0
1	0	1	0	0	0
1	1	2	0	1	0
1	2	0	1	2	0
2	0	2	0	0	0
2	1	0	1	2	0
2	2	1	1	1	1

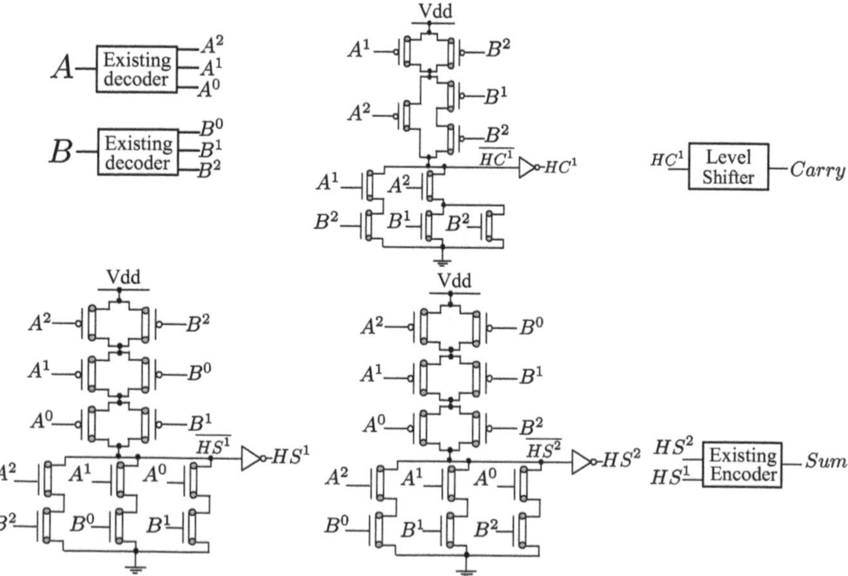

FIGURE 11.19
Half adder using existing decoder-encoder-based approach (all CNETs have chirality as (19,0)).

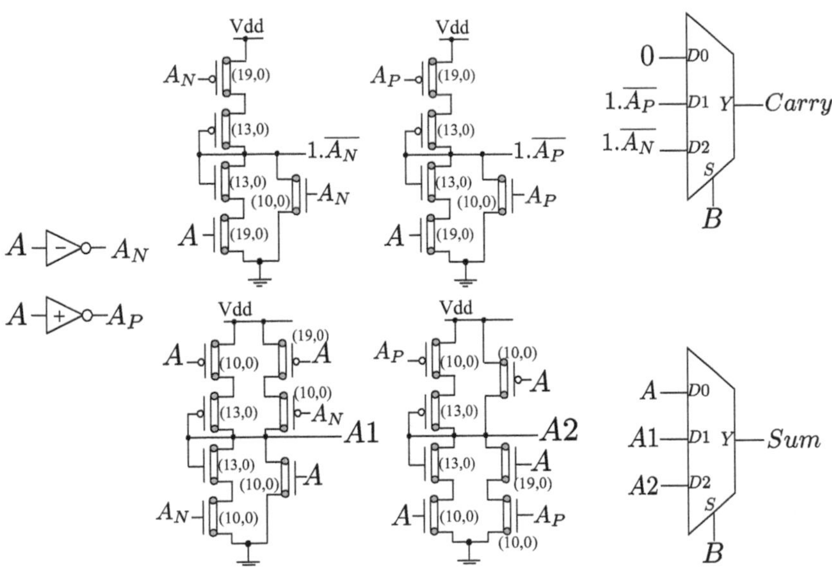

FIGURE 11.20
Half adder using existing 3:1 multiplexer-based approach.

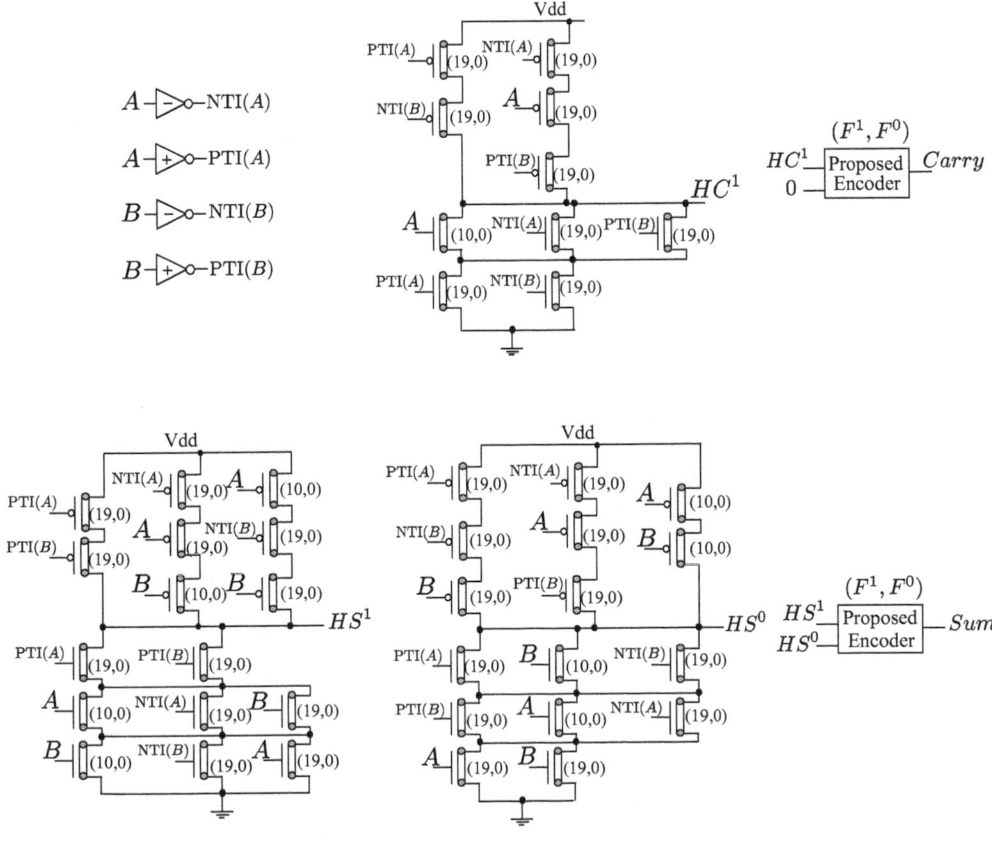

FIGURE 11.21
Half adder using decoderless approach.

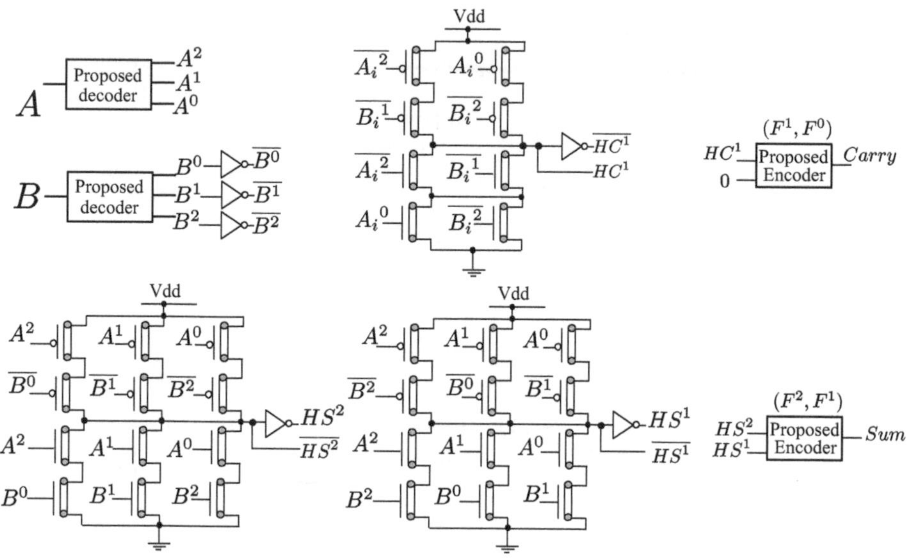

FIGURE 11.22
Half adder using 2:1 multiplexer-based approach.

These circuits along with the circuits for the multiplier have been implemented using the CNTFET model available at [49] and simulated using HSPICE. The following subsections explain the simulation environment and the results obtained.

11.6.1 Simulation Environment

All the circuits are simulated in HSPICE using the CNTFET model of [44,45,49] at 0.9V power supply and room temperature. The CNFETs used in the implementation are configured to have three tubes and a default pitch value equal to 20 nm. All the other parameters are set to their default values as presented in Table 11.4. In this work, ternary logic values 0, 1, and 2 correspond to voltages 0, $V\,dd/2$, and $V\,dd$, respectively.

For binary logic gates, the logic values 0 and 1 correspond to voltages 0 and $V\,dd$. Binary gates are implemented using transistors, which are connected in complementary logic style and have chirality of (19,0). Ternary circuits are implemented using different approaches, and the design parameters are compared. For fair comparison, all the circuits have been simulated with the same test pattern. Power consumption results are obtained by simulating the circuits with random input patterns at a switching frequency of 500 MHz. Propagation delay results for different circuits are obtained by finding worst-case Fan-Out of 4 (FO4) delay of the critical path. FO4 delay is calculated by loading the output node with four STI gates (implementation of STI gate is presented in [36]).

11.6.2 Results and Discussion

Table 11.9 summarizes the simulation results of different parameters for various ternary functions, namely ternary half adder and multiplier, which are implemented using different approaches. The existing decoder-encoder-based approach uses a decoder for each input and a complex encoder for each ternary output resulting in large propagation

TABLE 11.9

Simulation Results for Basic Circuits

Design Approach	Half-Adder	1-Digit Multiplier
Power Consumption (in μW)		
Decoder-Encoder Based [36]	1.24 (100%)	0.780 (100%)
3:1 Multiplexer based [41,59]	1.62 (131%)	0.779 (100%)
Decoderless	0.745 (60%)	0.372 (48%)
2:1 Multiplexer based	0.121 (10%)	0.069 (9%)
Propagation (FO4) Delay (in ps)		
Decoder-Encoder Based [36]	42.4 (100%)	35.7 (100%)
3:1 Multiplexer based [41,59]	47.0 (111%)	26.4 (74%)
Decoderless	43.0 (101%)	25.3 (71%)
2:1 Multiplexer-based	44.2 (104%)	17.5 (49%)
Power-Delay Product (PDP) (in $\times 10^{-17}$*J)*		
Decoder-Encoder-Based [36]	5.25 (100%)	2.78 (100%)
3:1 Multiplexer-based [41,59]	7.62 (145%)	2.06 (74%)
Decoderless	3.20 (61%)	0.94 (34%)
2:1 Multiplexer-based	0.53 (10%)	0.12 (4%)
Number of CNFETs		
Decoder-Encoder-Based [36]	88 (100%)	66 (100%)
3:1 Multiplexer-based [41,59]	52 (59%)	40 (61%)
Decoderless	72 (82%)	50 (76%)
2:1 Multiplexer-based	36 (41%)	30 (45%)

delay, power consumption, and transistor count when compared to other approaches. The problem of increased transistor count is addressed in the existing 3:1 multiplexer-based approach [41,59]. But the disadvantage of this approach is that it uses a complex implementation for realizing unary operators, resulting in large power consumption when compared to the decoder-encoder-based approach in [36].

Implementation using a decoderless approach shows a reduction of up to 18% in transistor count and up to 40% in power consumption when compared to the existing approach in [36]. This approach also shows a reduction in propagation delay (up to 7%) and power consumption (up to 54%), but it requires more transistors (up to 38% more) for implementation when compared to the 3:1 multiplexer-based approach [41,59]. However, for some circuits—e.g., ternary half adder—the decoderless approach results in circuits with large transistor stacking leading to higher propagation delay.

Ternary circuits implemented using a 2:1 multiplexer-based approach show a reduction of up to 30% transistor, 34% in propagation delay, and 91% power consumption when compared to an existing 3:1 multiplexer-based approach [41,59]. This is because unlike in the existing 3:1 multiplexer-based approach, where unary operators are implemented using complex circuits that have multiple direct paths between *V DD* and *GND*, in the 2:1 multiplexer-based approach, unary operators are implemented using 2:1 multiplexers resulting in low power consumption. The 2:1 multiplexer-based approach has the least power consumption and transistor count when compared to other approaches. However, since the 2:1 multiplexer-based approach uses transmission gates, ternary circuits designed using this approach cannot drive large load capacitance.

11.7 Conclusions

The combination of ternary logic and CNFETs has the capability to achieve efficient realizations of digital systems. The existing work on CNFET-based ternary logic circuits is relatively recent, and there is scope to explore new design approaches to realize efficient ternary circuits. This chapter presented different design approaches to implement basic ternary logic circuits. These approaches lead to ternary circuits that are optimized for one or more design parameters. Simulation results indicate that basic ternary circuits, namely half adder and 1-digit multiplier, which are implemented using a 2:1 multiplexer-based approach, result in up to 91% reduction in power consumption, up to 45% reduction in delay, and 30% reduction in transistor count when compared to other design approaches.

References

1. N. Weste and D. Harris, *CMOS VLSI Design: A Circuits and Systems Perspective*, 4th ed. Boston, MA: Addison-Wesley Publishing Company, 2010.
2. E. Mollick, "Establishing Moore's law," *IEEE Annals of the History of Computing*, vol. 28, no. 3, pp. 62–75, 2006. doi:10.1109/MAHC.2006.45.
3. M. Horowitz, "Scaling, power and the future of CMOS," in *Proceedings of the 20th International Conference on VLSI Design Held Jointly with 6th International Conference: Embedded Systems*, ser. VLSID'07. Washington, DC: IEEE Computer Society, p. 23, 2007. doi:10.1109/VLSID.2007.140.
4. X. Huang, W.-C. Lee, C. Kuo, D. Hisamoto, L. Chang, J. Kedzierski, E. Anderson et al., "Sub-50 nm p-channel finfet," *IEEE Transactions on Electron Devices*, vol. 48, no. 5, pp. 880–886, 2001.
5. "International technology roadmap for semiconductors 2.0, 2015 edition, executive report," Available: http://www.itrs.net, Technical Report, 2015.
6. K. Roy, S. Mukhopadhyay, and H. Mahmoodi-Meimand, "Leakage current mechanisms and leakage reduction techniques in deep-submicrometer CMOS circuits," *Proceedings of the IEEE*, vol. 91, no. 2, pp. 305–327, 2003.
7. E. P. DeBenedictis, J. K. Mee, and M. P. Frank, "The opportunities and controversies of reversible computing," *Computer*, vol. 50, no. 6, pp. 76–80, 2017.
8. V. Gaudet, "A survey and tutorial on contemporary aspects of multiple-valued logic and its application to microelectronic circuits," *IEEE Journal on Emerging and Selected Topics in Circuits and Systems*, vol. 6, no. 1, pp. 5–12, 2016.
9. S. J. Tans, A. R. M. Verschueren, and C. Dekker, "Room-temperature transistor based on a single carbon nanotube," *Nature*, vol. 393, no. 6680, pp. 49–52, 1998.
10. S. Karmakar, J. A. Chandy, and F. C. Jain, "Design of ternary logic combinational circuits based on quantum dot gate fets," *IEEE Very Large Scale Integration Systems*, vol. 21, no. 5, pp. 793–806, 2013. doi:10.1109/TVLSI.2012.2198248.
11. A. Javey, J. Guo, Q. Wang, M. Lundstrom, and H. Dai, "Ballistic carbon nanotube field-effect transistors," *Nature*, vol. 424, pp. 654–657, 2003.
12. J. Deng, N. Patil, K. Ryu, A. Badmaev, C. Zhou, S. Mitra, and H. S. P. Wong, "Carbon nanotube transistor circuits: Circuit-level performance benchmarking and design options for living with imperfections," in *2007 IEEE International Solid-State Circuits Conference. Digest of Technical Papers*, pp. 70–588, February 2007.
13. Y. Cui, Z. Zhong, D. Wang, W. U. Wang, and C. M. Lieber, "High performance silicon nanowire field effect transistors," *Nano Letters*, vol. 3, no. 2, pp. 149–152, 2003. doi:10.1021/nl0258751.

14. X. Wang, Y. Ouyang, X. Li, H. Wang, J. Guo, and H. Dai, "Room temperature all-semiconducting sub-10-nm graphene nanoribbon field-effect transistors," *Physical Review Letters*, vol. 100, p. 206803, 2008. doi:10.1103/PhysRevLett.100.206803.
15. M. C. Lemme, T. J. Echtermeyer, M. Baus, and H. Kurz, "A graphene field-effect device," *IEEE Electron Device Letters*, vol. 28, no. 4, pp. 282–284, 2007.
16. T. Ashley, A. R. Barnes, L. Buckle, S. Datta, A. B. Dean, M. T. Emery, M. Fearn et al., "Novel insb-based quantum well transistors for ultra-high speed, low power logic applications," in *Proceedings of the 7th International Conference on Solid-State and Integrated Circuits Technology*, vol. 3, pp. 2253–2256, 2004.
17. C. I. Kuo, H. T. Hsu, C. Y. Wu, E. Y. Chang, Y. Miyamoto, Y. L. Chen, and D. Biswas, "A 40-nm-gate InAs/In0.7Ga0.3As composite-channel HEMT with 2200 ms/mm and 500-GHz ft," in *2009 IEEE International Conference on Indium Phosphide Related Materials*, pp. 128–131, 2009.
18. S. L. Hurst, "Multiple-valued logic: Its status and its future," *IEEE Transactions on Computers*, vol. C-33, no. 12, pp. 1160–1179, 1984.
19. P. C. Balla and A. Antoniou, "Low power dissipation MOS ternary logic family," *IEEE Journal of Solid-State Circuits*, vol. 19, no. 5, pp. 739–749, 1984.
20. A. Heung and H. T. Mouftah, "Depletion/enhancement CMOS for a lower power family of three-valued logic circuits," *IEEE Journal of Solid-State Circuits*, vol. 20, no. 2, pp. 609–616, 1985.
21. D. A. Rich, "A survey of multivalued memories," *IEEE Transactions on Computers*, vol. 35, no. 2, pp. 99–106, 1986. doi:10.1109/TC.1986.1676727.
22. Y. Yasuda, Y. Tokuda, S. Zaima, K. Pak, T. Nakamura, and A. Yoshida, "Realization of quaternary logic circuits by n-channel MOS Devices," *IEEE Journal of Solid-State Circuits*, vol. 21, no. 1, pp. 162–168, 1986.
23. I. Jordan and H. Mouftah, "Design of ternary cos/mos memory and sequential circuits," *IEEE Transactions on Computers*, vol. 26, pp. 281–288, 1977.
24. D. Mateo and A. Rubio, "Design and implementation of a 5 times; 5 trits multiplier in a quasi-adiabatic ternary CMOS logic," *IEEE Journal of Solid-State Circuits*, vol. 33, no. 7, pp. 1111–1116, 1998.
25. I. Halpern and M. Yoeli, "Ternary arithmetic unit," *Electrical Engineers, Proceedings of the Institution of*, vol. 115, no. 10, pp. 1385–1388, 1968.
26. S. Kawahito, M. Kameyama, T. Higuchi, and H. Yamada, "A 32 * 32-bit multiplier using multiple-valued MOS current-mode circuits," *IEEE Journal of Solid-State Circuits*, vol. 23, no. 1, pp. 124–132, 1988.
27. S. Y. H. Su and P. T. Cheung, "Computer minimization of multivalued switching functions," *IEEE Transactions on Computers*, vol. C-21, no. 9, pp. 995–1003, 1972.
28. H. M. Wang, C. L. Lee, and J. E. Chen, "Algebraic division for multilevel logic synthesis of multi-valued logic circuits," in *Multiple-Valued Logic, 1994. Proceedings, Twenty-Fourth International Symposium on*, pp. 44–51, 1994.
29. M. Hawash, M. Lukac, M. Kameyama, and M. Perkowski, "Multiple-valued reversible benchmarks and extensible quantum specification (XQS) format," in *2013 IEEE 43rd International Symposium on Multiple-Valued Logic*, 2013, p. 41.
30. K. W. Current, "Current-mode CMOS multiple-valued logic circuits," *IEEE Journal of Solid-State Circuits*, vol. 29, no. 2, pp. 95–107, 1994.
31. M. Klein, J. A. Mol, J. Verduijn, G. P. Lansbergen, S. Rogge, R. D. Levine, and F. Remacle, "Ternary logic implemented on a single dopant atom field effect silicon transistor," *Applied Physics Letters*, vol. 96, no. 4, p. 043107, 2010. doi:10.1063/1.3297906.
32. A. Rahman, J. Guo, S. Datta, and M. S. Lundstrom, "Theory of ballistic nanotransistors," *IEEE Transactions on Electron Devices*, vol. 50, no. 9, pp. 1853–1864, 2003.
33. Y.-M. Lin, J. Appenzeller, J. Knoch, and P. Avouris, "High-performance carbon nanotube field-effect transistor with tunable polarities," *IEEE Transactions on Nanotechnology*, vol. 4, no. 5, pp. 481–489, 2005. doi:10.1109/TNANO.2005.851427.
34. A. Akturk, G. Pennington, N. Goldsman, and A. Wickenden, "Electron transport and velocity oscillations in a carbon nanotube," *IEEE Transactions on Nanotechnology*, vol. 6, no. 4, pp. 469– 474, 2007.

35. H. Hashempour and F. Lombardi, "Device model for ballistic CNFETs using the first conducting band," *IEEE Design & Test*, vol. 25, no. 2, pp. 178–186, 2008. doi:10.1109/MDT.2008.34.
36. S. Lin, Y. B. Kim, and F. Lombardi, "CNTFET-Based design of ternary logic gates and arithmetic circuits," *IEEE Transactions on Nanotechnology*, vol. 10, no. 2, pp. 217–225, 2011.
37. S. Lin, Y. B. Kim, and F. Lombardi, "A novel CNTFET-based ternary logic gate design," in *2009 52nd IEEE International Midwest Symposium on Circuits and Systems*, pp. 435–438, 2009.
38. P. Keshavarzian and R. Sarikhani, "A novel CNTFET-based ternary full adder," *Circuits, Systems, and Signal Processing*, vol. 33, no. 3, pp. 665–679, 2014.
39. R. F. Mirzaee, K. Navi, and N. Bagherzadeh, "High-Efficient circuits for ternary addition," *VLSI Design*, vol. 2014, 534587, 15p, 2014. doi:10.1155/2014/534587.
40. S. L. Murotiya and A. Gupta, "Hardware-Efficient low-power 2-bit ternary ALU design in CNTFET technology," *International Journal of Electronics*, vol. 103, no. 5, pp. 913–927, 2016. doi:10.1080/00207217.2015.1082199.
41. B. Srinivasu and K. Sridharan, "A synthesis methodology for ternary logic circuits in emerging device technologies," *IEEE Transactions on Circuits and Systems I: Regular Papers*, vol. 64, no. 8, pp. 2146–2159, 2017.
42. R. Saito, G. Dresselhaus, and M. S. Dresselhaus, *Physical Properties of Carbon Nanotubes*. London, UK: Imperial College Press, 1998.
43. J. Appenzeller, "Carbon nanotubes for high-performance electronics - Progress and prospect," *Proceedings of the IEEE*, vol. 96, no. 2, pp. 201–211, 2008.
44. J. Deng and H. S. P. Wong, "A compact SPICE model for carbon-nanotube field-effect transistors including nonidealities and its application-part I: Model of the intrinsic channel region," *IEEE Transactions on Electron Devices*, vol. 54, no. 12, pp. 3186–3194, 2007.
45. J. Deng and H. S. P. Wong, "A compact SPICE model for carbon-nanotube field-effect transistors including nonidealities and its application-Part II: Full device model and circuit performance benchmarking," *IEEE Transactions on Electron Devices*, vol. 54, no. 12, pp. 3195–3205, 2007.
46. G. Gelao, R. Marani, R. Diana, and A. G. Perri, "A semiempirical spice model for n-type conventional cntfets," *IEEE Transactions on Nanotechnology*, vol. 10, no. 3, pp. 506–512, 2011.
47. S. Fregonese, H. C. d'Honincthun, J. Goguet, C. Maneux, T. Zimmer, J. P. Bourgoin, P. Dollfus, and S. Galdin-Retailleau, "Computationally efficient physics-based compact cntfet model for circuit design," *IEEE Transactions on Electron Devices*, vol. 55, no. 6, pp. 1317–1327, 2008.
48. R. Marani, G. Gelao, and A. G. Perri, "Modelling of carbon nanotube field effect transistors oriented to spice software for a/d circuit design," *Microelectronics Journal*, vol. 44, no. 1, pp. 33–38, 2013. http://www.sciencedirect.com/science/article/pii/S0026269211001613.
49. Stanford University, *Stanford University CNFET Model Website*. Stanford University, Stanford, CA, 2008. http://nano.stanford.edu/model.php?id=23.
50. Y. Li, W. Kim, Y. Zhang, M. Rolandi, D. Wang, and H. Dai, "Growth of single-walled carbon nanotubes from discrete catalytic nanoparticles of various sizes," *Journal of Physical Chemistry B Materials*, vol. 105, no. 46, pp. 11424–11431, 2001.
51. A. Lin, N. Patil, K. Ryu, A. Badmaev, L. G. D. Arco, C. Zhou, S. Mitra, and H. S. P. Wong, "Threshold voltage and on-off ratio tuning for multiple-tube carbon nanotube fets," *IEEE Transactions on Nanotechnology*, vol. 8, no. 1, pp. 4–9, 2009.
52. Y. Ohno, S. Kishimoto, T. Mizutani, T. Okazaki, and H. Shinohara, "Chirality assignment of individual single-walled carbon nanotubes in carbon nanotube field-effect transistors by micro-photocurrent spectroscopy," *Applied Physics Letters*, vol. 84, no. 8, pp. 1368–1370, 2004. http://scitation.aip.org/content/aip/journal/apl/84/8/10.1063/1.1650554.
53. B. Wang, C. H. P. Poa, L. Wei, L.-J. Li, Y. Yang, and Y. Chen, "(n, m) Selectivity of single-walled carbon nanotubes by different carbon precursors on Co-Mo catalysts," *Journal of the American Chemical Society*, vol. 129, no. 29, pp. 9014–9019, 2007.
54. M. M. Shulaker, G. Hills, N. Patil, H. Wei, H.-Y. Chen, H. S. P. Wong, and S. Mitra, "Carbon nanotube computer," *Nature*, vol. 501, no. 7468, pp. 526–530, 2013. doi:10.1038/nature12502.
55. A. Raychowdhury and K. Roy, "Carbon-nanotube-based voltage-mode multiple valued logic design," *IEEE Transactions on Nanotechnology*, vol. 4, no. 2, pp. 168–179, 2005.

56. S. A. Ebrahimi, P. Keshavarzian, S. Sorouri, and M. Shahsavari, "Low power CNTFET-based ternary full adder cell for nanoelectronics," *International Journal of Soft Computing and Engineering (IJSCE)*, vol. 2, no. 2, pp. 291–295, 2012.
57. K. Sridharan, S. Gurindagunta, and V. Pudi, "Efficient multiternary digit adder design in CNTFET technology," *IEEE Transactions on Nanotechnology*, vol. 12, no. 3, pp. 283–287, 2013.
58. S. L. Murotiya and A. Gupta, "Design of high speed ternary full adder and three-Input XOR circuits using CNTFETs," in *2015 28th International Conference on VLSI Design*, pp. 292–297, 2015. http://ieeexplore.ieee.org/lpdocs/epic03/wrapper.htm?arnumber=7031749.
59. B. Srinivasu and K. Sridharan, "Carbon nanotube FET-based low-delay and low-power multi-digit adder designs," *IET Circuits, Devices & Systems*, vol. 11, pp. 1–13, 2016. http://digital-library.theiet.org/content/journals/10.1049/iet-cds.2016.0013.
60. S. L. Hurst, "An extension of binary minimization techniques to ternary equations," *Computers Journal*, vol. 11, no. 3, pp. 277–286, 1968.
61. M. H. Moaiyeri, A. Doostaregan, and K. Navi, "Design of energy-efficient and robust ternary circuits for nanotechnology," *IET Circuits, Devices & Systems*, vol. 5, no. 4, pp. 285–296, 2011.
62. T. Sasao, "Multiple-valued decomposition of generalized boolean functions and the complexity of programmable logic arrays," *IEEE Transactions on Computers*, vol. C-30, no. 9, pp. 635–643, 1981.
63. B. Srinivasu and K. Sridharan, "Low-complexity multiternary digit multiplier design in CNTFET technology," *IEEE Transactions on Circuits and Systems II: Express Briefs*, vol. 63, no. 8, pp. 753–757, 2016.
64. C. Vudadha, P. S. Phaneendra, and M. B. Srinivas, "An efficient design methodology for CNTFET-based ternary logic circuits," in *IEEE International Symposium on Nanoelectronic and Information Systems (iNIS)*, pp. 278–283, December 2016.
65. C. Vudadha, S. Katragadda, and P. S. Phaneendra, "2:1 multiplexer based design for ternary logic circuits," in *2013 IEEE Asia Pacific Conference on Postgraduate Research in Microelectronics and Electronics (PrimeAsia)*, pp. 46–51, December 2013.

[illegible]

Section V

Modeling of Emerging Non-Silicon Transistors

12

Different Analytical Models for Organic Thin-Film Transistors: Overview and Outlook

W. Boukhili and R. Bourguiga

CONTENTS

12.1 Introduction

The evolution of modern device microelectronics has led to the realization of devices with several high-quality materials in order to fulfill different requirements. We must go back to the end of the 1970s to explain the birth of plastic electronics. Indeed, it was in 1977 that Alan J. Heeger, Alan G. MacDiarmid, and Hideki Shirakawa demonstrated the existence of polymers of good conductivity. Their research was awarded the Nobel Prize in Chemistry in 2000. For a polymer to conduct electricity, it must have at least alternating single and double bonds between its carbon atoms. To improve this conduction, it can be "doped," which consists of removing electrons (by oxidation) or adding (by reduction). In fact, organic semiconductors (OSCs) based on π-conjugated small molecules and conductive polymers have widely been studied in recent years. It would be fair to state that the field of organic semiconductors has witnessed the sought-after technological revolution of plastic electronic devices. In addition, OSCs can be manufactured and processed at room temperature, making their production easier and cheaper than for conventional silicon and inorganic semiconductors. They can be transparent, flexible, and developed over large areas or non-planar geometries; completely plastic devices can be realized. In contrast, inorganic materials are relatively expensive to process, difficult to functionalize, and normally allow low throughput methodologies. During these last two decades, organic thin-film transistors (OTFTs) have attracted a great deal of attention due to their potential application ranging from flexible radio-frequency identification tags (RFID), electronic paper, intelligent textiles, bioelectronics to organic circuits due to their inherent advantages such as lightweight feature, flexibility, large area capability, structural flexibility, low temperature process, and ease of manufacture compared to general Si technology. The improvement and development of electronic devices is the result of several technological achievements. This design approach is mainly time-consuming and expensive. With the evolution of calculation units and the development of physical simulators, a new technical development of electronic components has grown in importance the last two decades in industries and research laboratories. This technique is called "behavioral numerical simulation" or "physical modeling." This type of simulation uses the different laws of components physics and can reproduce the electrical and optoelectronic characteristics associated with physical structures and specific bias conditions. The development of an electronic component today includes several stages: physical modeling, realization, electrical measurements, and electrical optoelectronics and possibly thermal modeling. These four steps are complementary, allow characterization, and fully understand the operation of the component. Effectively, analytical modeling plays a fundamental role in the conception of the new device as it suggests a mathematical tool for analyzing experimental observations and reproducing the measurement data of the components, while also providing the capability of predicting performance modeling that can guide further development and improvement efforts. Significant progress in the development of efficient and stable OTFTs is opening the possibility for their large-scale deployment. In fact, the electrical characteristics of OTFTs have been intensively studied and largely clarified. Thus, many models have been introduced and developed for modeling of the OTFTs and extending the understanding of OTFTs' operation. Furthermore, the modeling of OTFTs has left its infancy and has reached a level whereby a good agreement has been reached with the experimental characteristics.

This chapter presents an up-to-date review of the several models commonly used to reproduce the current-voltage characteristics of the OTFTs. A review of the principal techniques used for the modeling of OTFTs' current measurements is proposed. The consistency, accuracy, and sensitivity of the effects of the modeling procedure are examined through analytical modeling and experimental measurements. The validity and the application of these different analytical models for different types of transistors have been made and evaluated in [1–6].

As a matter of fact, OTFTs are one of the most widely used tools for the analysis of the mobility of organic semiconducting materials. While the working principle is well understood and characterized, there is still no general analytical solution for the current-voltage characteristics.

12.2 Device Description and Standard Model

OTFTs are a variety of metal-insulator-semiconductor (MIS) structures in which the semiconductor is organic. An OTFT is composed of an organic semiconductor provided with two electrodes: the source (S) and the drain (D), whose role is to inject or collect charges. A third electrode, the gate (G), is separated from the semiconductor by an insulating layer for modulating the current between the source and the drain. So, OTFTs are three metal

TABLE 12.1

The Set of Parameters that Gave a Good Agreement Between the Experimental Data and Those Obtained by the Model 1

Parameters	m	λ (V^{-1})	α_s	σ_0(AV^{-1})	V_{th}(V)	R_s (Ω)	$g_{ch,\,sat}$(AV^{-1})
	2.66	2.88×10^{-2}	0.066	0.78×10^{-12}	10.88	4.75×10^{7}	5.6×10^{-7}

electrodes devices in which the current flow going between the source and drain is modulated by a gate bias. A major difference with commonly used inorganic transistors is that no inversion layer is formed, but the conduction occurs by means of the majority carriers, which accumulate at the organic semiconductor/insulator interface. It is possible to examine several architectures of thin film transistors. Table 12.1 shows the four basic architectures using a "classical" stack of the different layers of the transistor. They are classified as bottom gate or top gate depending on whether the gate is respectively below or above the semiconductor, and top-contact or bottom-contact depending on the position of the source and drain metal contacts relative to the semiconductor. From a performance point of view of the transistors, these different architectures will not be equivalent. As an example, it appears that depending on the device, we will be in the presence of either a semiconductor-insulator interface or a semiconductor insulating interface which, though equivalent materials, will not necessarily have the same properties. The reasoning also applies to the conductive electrodes semiconductor interface. Each geometry has its advantages and disadvantages depending on the materials, the deposition technologies used, and the final properties of the transistor. Possible transistor structures are shown in the following table.

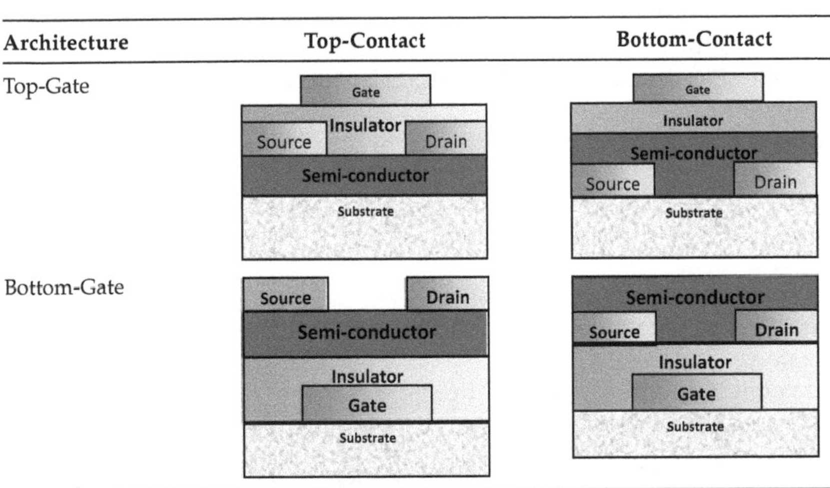

Different Architectures of Thin-Film Transistor

The operating of the TFTs is based on the fact that a current flows between the two source and drain electrodes when a bias voltage is applied between them. The current intensity is modulated by the voltage V_G applied to the gate electrode. Note also that the source is taken as the reference potential. In such structure, in the absence of bias, the existence of two pn junctions prevents the passage of current between the source and the drain; the transistor is off. When the gate is positively biased at a voltage greater than the threshold voltage V_{th} of the transistor, the MOS capacitor operates in inversion: an n-type conducting channel will take place between the source and the drain. A positive bias applied with respect to the drain makes it possible to circulate the current in the conductive channel; the transistor is on.

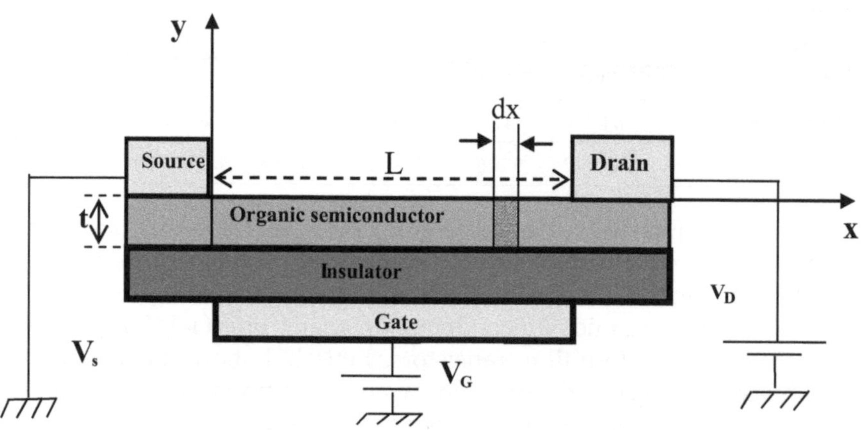

FIGURE 12.1
Geometry structure of OTFTs.

This chapter features the different model formulations used for organic thin-film transistors. Firstly, we present the derivation of the analytical expression of the drain current in the linear and saturation regimes by the systematic model that is currently used, which is called the Sze model [7]. The Sze model makes it possible to predict the evolution of the electrical characteristics according to the geometry and properties of the materials. It is established for the basilar structure for the TFTs as shown in Figure 12.1. It is a freestanding top-contact structure. The OTFTs consist essentially of a substrate, having at its ends two ohmic contacts forming two electrodes, between which flows the main current, called source and drain (Figure 12.1); a third electrode isolated from the substrate serves as control and is called the gate Figure 12.1.

The surface-induced charge at the abscissa x of the channel is given by the expression:

$$q\Delta n(x) = \frac{C_i}{t}\left(V_G - V(x)\right) \tag{12.1}$$

where:

C_i is the insulation capacity per unit area
t is the thickness of the semiconductor film
V_G is the gate voltage
$V(x)$ is the voltage at the abscissa x of the channel counted from the source
$\Delta n(x)$ is the variation of the induced charge at the abscissa x of the channel

The drain current I_D can be expressed as a function of the conductivity (σ_0) at zero gate voltage, the conductivity variation $\Delta\sigma(x)$, and the electric field $E(x)$ at the abscissa x:

$$I_D = tW\left(\sigma_0 + \Delta\sigma(x)\right)E(x) \tag{12.2}$$

with W as the channel width.

By introducing the mobility of the carriers μ, n_0 the charge density at zero gate voltage, we obtain:

$$I_D = W\mu qt\left(n_0 + \Delta n(x)\right)E(x) = W\mu qt\left(n_0 + \Delta n(x)\right)E(x)$$

$$I_D = W\mu C_i\left(\frac{qtn_0}{C_i} + V_G - V(x)\right)\frac{dV(x)}{dx}$$

Consequently, integrating over the entire channel length (L):

$$I_D \int_0^L dx = W\mu C_i \int_0^{V_d} \left(\frac{qn_0 t}{C_i} + V_G - V(x) \right) dV(x)$$

$$I_D = \frac{W}{L} \frac{C_i}{} \left((V_G - V_{th}) V_D - \frac{V_D^2}{2} \right) \tag{12.3}$$

where $V_{th} = -\frac{qn_0 d}{C_i}$ is the threshold voltage.

The threshold voltage depends on the charge density n_0, the thickness of the semiconductor t, and the capacitance of the insulator C_i.

Finally, in accumulation mode, we can distinguish two ranges of regime:

- Linear regime (low V_D):

$$I_D = \frac{WC_i}{L} \mu_{FET} \left(V_G - V_{th} - \frac{V_D}{2} \right) V_D \tag{12.4}$$

- Saturation regime (high V_D):

$$I_D = \frac{WC_i}{L} \mu_{FET} \left(V_G - V_{th} \right)^2 \tag{12.5}$$

12.2.1 Technologies Manufacturing Steps of the OTFTs

The technological part in the complete preparation of the organic thin-film transistors has been a long period of testing and optimization for each manufacturing step.

- **Si/SiO_2 substrate for the top-contact bottom-gate structure**

  Silicon dioxide can be obtained by thermal oxidation; it is a chemical reaction between silicon and the oxidizing substance, usually oxygen (O_2, called dry oxidation) or water vapor (H_2O, called wet oxidation). The reaction is carried out in an oven at high temperatures in the range of 800°C–1100°C, depending on the thickness and speed we want to achieve.

- **Si/SiO_2/ITO/Au substrate for the bottom-contact bottom-gate structure**

  Si/SiO_2/ITO/Au substrates are 15 × 15 mm² commercial substrates (Fraunhofer Institute for Photonic Microsystems IPMS) on which the electrodes are already deposited. N^+ doped silicon, silicon oxide with a thickness of 230 ± 10 nm constitutes the insulator. The source and drain electrodes consist of ITO/Au (10 nm/30 nm) and are inter-digitized. Treatment is done in an ultrasonic bath with acetone to remove the photo resists (15 min). This first treatment is followed by other successive cleaning procedures in an ultrasonic bath in isopropanol (15 min), acetone (15 min), and drying under nitrogen. Subsequent exposure to UV-ozone treatment for 20 min or O_2 plasma for 3 min (average power) is performed to remove organic residues from the surface. The gold constituting the source and drain metal for this structure adheres poorly on the insulators; therefore, a thin layer of attachment is necessary. A very thin ITO layer of the order of 3 nm is deposited before the gold layer to promote the adhesion of the Au layer on the SiO_2 surface.

- **Evaporation of source and drain metal contacts**

  The drain and source contacts are obtained by vacuum evaporation of gold through a mask. Evaporation of the metal electrodes on a substrate (which is deposited on the semiconductor layer) is not a procedure that requires special care because the parameters, such as the thickness of the layer or electrode deposition rate, generally do not influence device performance.
- **Deposition of the organic active layer by thermal evaporation under vacuum**

  An important difference between polymers and small molecules is the deposition technique in the form of thin films. Indeed, the small molecules are deposited from their gaseous phases by evaporation under vacuum because they have good thermal stability and a low molecular weight, while the polymers are deposited in solution by the technique of the spin coating.
- **Gate deposition**

  The gate electrode of the transistor is deposited on the rear face of the substrate. Below the substrate, a silver layer of 300 nm is deposited by evaporation under vacuum, on the back side of the silicon followed by annealing at 45°C in a furnace under nitrogen for 30 min.

  After manufacture, the final transistors with top-contact bottom-gate structure and bottom- contact bottom-gate structure are shown by the following photos that are made by an optical microscope.

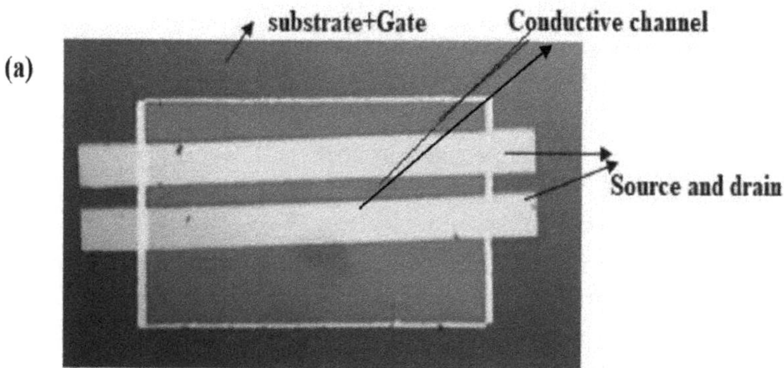

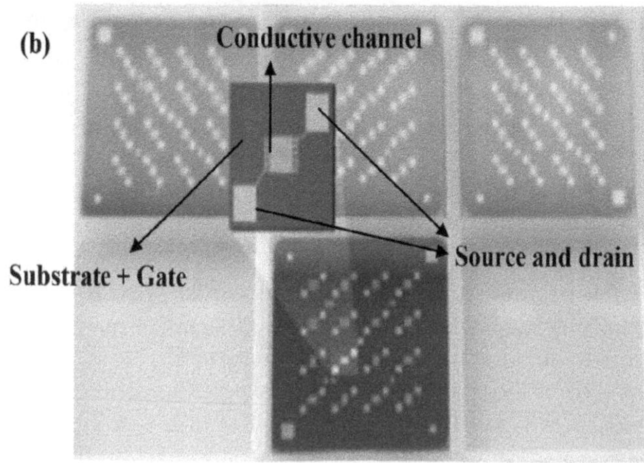

Photos by an optical microscope of the OTFTs: (a) top-contact bottom-gate structure and (b) bottom-contact bottom-gate structure

- **Electrical characterization of organic transistors**

  To study the operation of the fabricated transistors, we carried out a series of electrical measurements. These measurements were made using a semiconductor-based electronic component analyzer (Agilent 4156C HP 5156 parameter analyzer). The measuring equipment consists of a vacuum chamber, moving spikes, and connection cables. Although the analyzer is connected to a computer, the electrical measurements were taken with the configuration shown in the following photo. Two sets of measurements were performed: transfer curves and output curves. The output characteristic is the variation of the drain current as a function of the drain voltage (I_D vs. V_D) for different gate voltages (V_G).

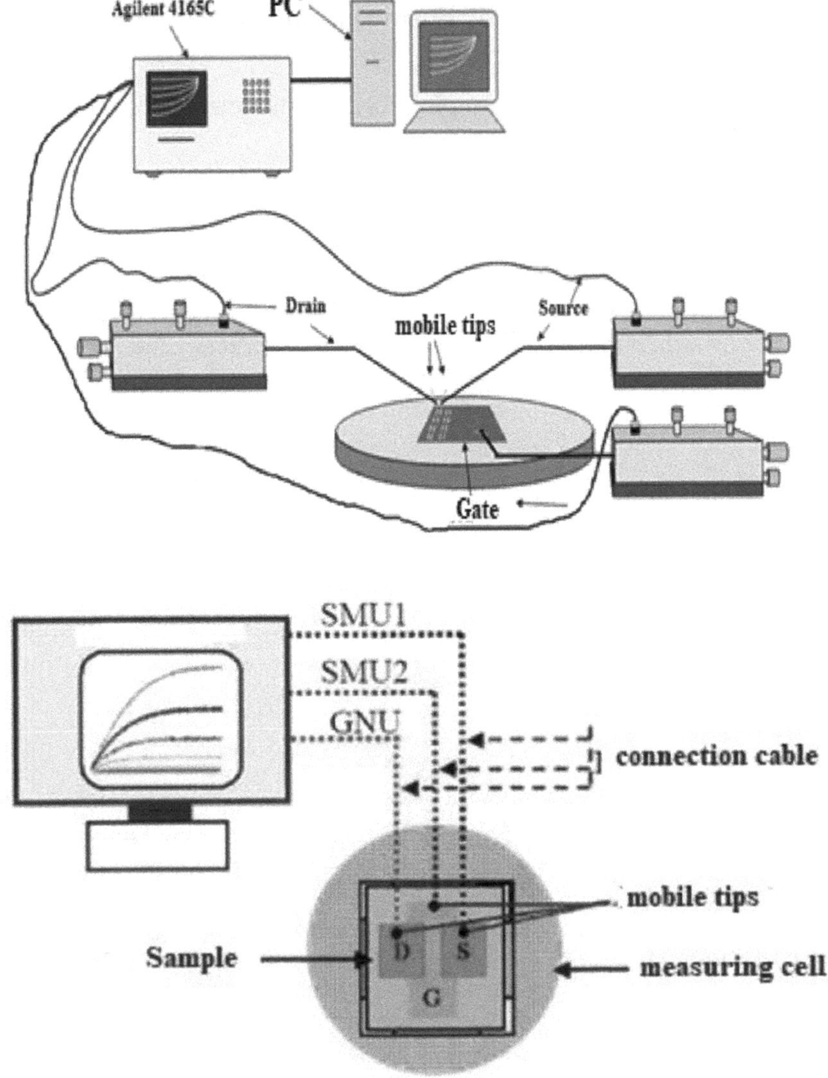

Photo of a simplified diagram of the electrical measurement bank of organic transistors

Improving the performance of the transistors requires, in addition to the care of synthesis and elaboration of thin-film layers, a better understanding of carrier transport mechanisms in this type of component. To do so, we must develop simple models in order to predict and reproduce the evolution of the electrical characteristics of thin-film transistors. The theoretical modeling step is essential in the design of electronic components as a tool for validating designer choices. It therefore makes it possible to know the future electrical characteristics of the components before the manufacturing step. But this theoretical modeling requires a mathematical model describing the physical phenomena of the transistor as well as a set of parameters to be determined.

12.3 Different Analytical Models Formulation for Current-Voltage Characteristics

In this section, we describe different simple analytical models used for OTFTs.

12.3.1 Model 1

12.3.1.1 Derivation of Model 1

In this first model, the formulation of the expression of the total drain current taken into account of the intrinsic conductivity of the organic semiconductor, and therefore an intrinsic current, must be added in the calculation of the total drain current. Taking into account the intrinsic conductivity of organic semiconductor, the total drain current in the linear and saturation regimes can be expressed as follows:

$$I_D = I_{\text{accum}} + I_{\text{intrinsic}} \tag{12.6}$$

where I_{accum} is the accumulation drain current (at $V_G \neq 0V$ and $V_D \neq 0V$) due to the accumulation of majority charge carriers at the conductive channel of the OTFT that can be calculated by the following equation:

$$I_{\text{accum}} = \frac{g_{\text{ch, eff}} V_D \left(1+\lambda V_D\right)}{\left[1+\left(V_D \big/ V_{D,\text{sat}}\right)^m\right]^m} \tag{12.7}$$

λ is the channel length modulation parameter and m is a control transition parameter from the linear to the saturation regime in the output characteristics of the organic TFTs.

$I_{\text{intrinsic}}$ is the intrinsic current (at $V_G = V_D = 0V$) due to the intrinsic conductivity of the organic semiconductor. The intrinsic current due to the contribution of the intrinsic conductivity of the organic active layer is generally given by the following expression [1,5,6]:

$$I_{\text{intrinsic}} = \sigma_0 V_D \tag{12.8}$$

where σ_0 is the intrinsic conductivity of the organic active layer and V_D is the drain voltage.

Therefore, the final expression of the total drain current is given by

$$I_D = \frac{g_{ch,\,eff} V_D \left(1+\lambda V_D\right)}{\left[1+\left(V_D \big/ V_{D,\,sat}\right)^m\right]^m} + \sigma_0 V_D \tag{12.9}$$

where α_s is the saturation modulation parameter and δ is the transition width parameter. λ is the channel length modulation parameter and m is a control transition parameter from the linear to the saturation regime.

Figure 12.2 shows a graphical method to extract the different parameters indicated in the final expression of the drain current: R_s, m, ΔI, $I_{D,\,sat}$, $g_{ch,\,sat}$, and λ [1,5,6]. The parameter m is determined using the following relation:

$$m = \frac{\ln(2)}{\ln\left(\dfrac{I_{D,sat}}{I_{D,sat} - \Delta I}\right)} \tag{12.10}$$

where ΔI is the difference between the saturation current $I_{D,sat}$ and the current value obtained by the projection on the current-voltage curve of the intersection of the two slopes of the current-voltage characteristics in the linear and saturation regimes (Figure 12.2).

The parameter λ was determined from the slope of the current-voltage characteristic in the saturated state as shown in Figure 12.1 using the following equation:

$$g_{ch,sat} = \lambda\, I_{D,sat}$$

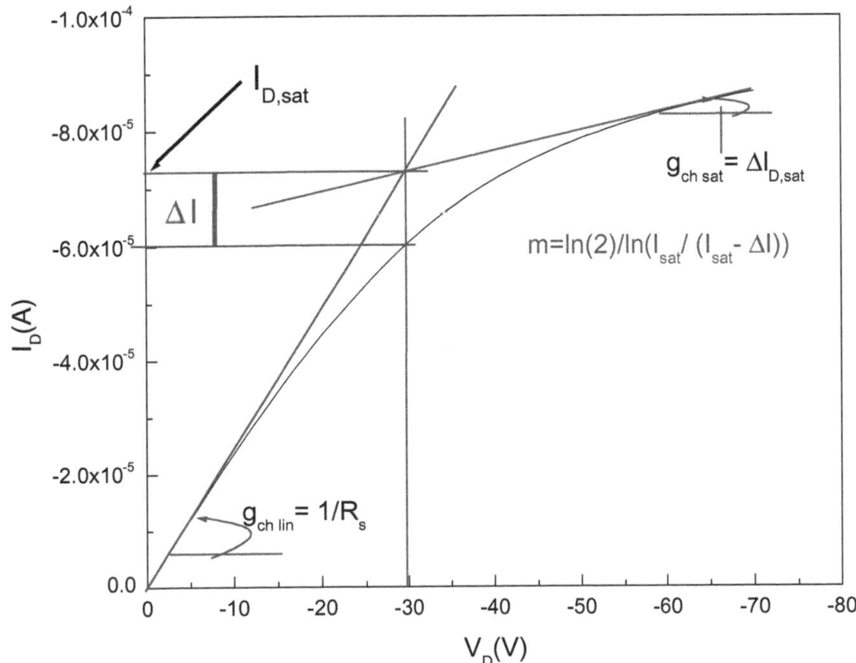

FIGURE 12.2
Graphical method to extract the different parameters.

12.3.1.2 Validation of Model 1

Using model 1, we have been able to reproduce the output electrical characteristics of our OTFTs. The obtained results are in good agreement with the experiment, as shown in Figure 12.3.

The set of parameters that gave a good agreement between the experimental data and those obtained by the model 1 are summarized in Table 12.1.

12.3.2 Model 2

12.3.2.1 Formulation of Model 2

In the disordered amorphous structure of organic materials, the charge transport occurs by variable range hopping (VRH) of charge carriers between strongly localized states [8–29]. The VRH model was proposed by Vissenberg and Matters in order to model the electrical characteristics of OTFTs [8]. It considers the hopping percolation of charge carriers between the localized states. In order to obtain an analytical expression of the drain current based on the VRH model, it is assumed that the current transport is parallel to the insulator–semiconductor interface [18].

Indeed, in combination with percolation [14–18] and VRH theory, the conductivity expression is defined as follows [8]:

$$\sigma(\delta,T)=\sigma_0\left[\frac{\pi N_t\delta\left(\frac{T_c}{T}\right)^3}{(2\alpha)^3 B_c\Gamma\left(1-\frac{T}{T_c}\right)\Gamma\left(1+\frac{T}{T_c}\right)}\right]^{\frac{T_c}{T}} \tag{12.11}$$

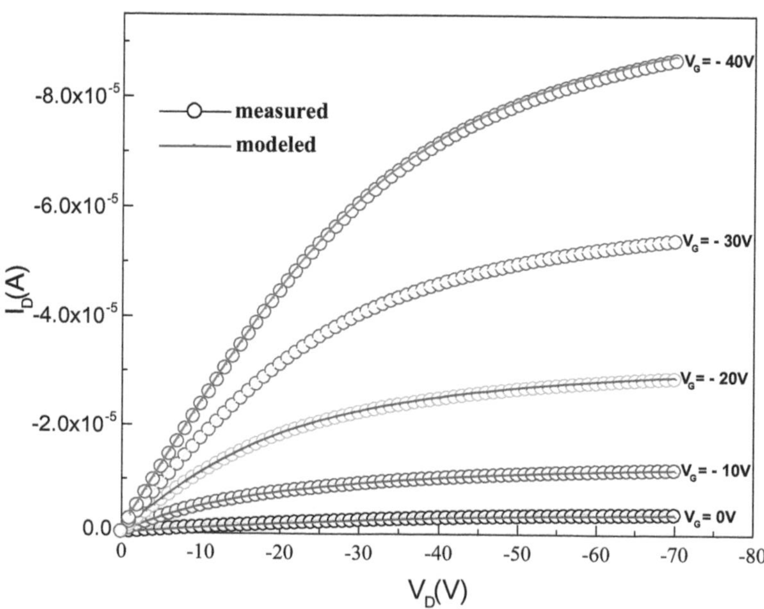

FIGURE 12.3
The good agreement between experimental (circle line) and that obtained by model 1 (full line) output characteristics of OTFTs.

where σ_0 is the percolation prefactor of the conductivity, which it has a meaning of a material-specific conductivity, α is an effective overlap parameter that governs the tunneling process between two localized states, and B_c is the critical number of bonds per site in the percolating network for a three-dimensional amorphous system, $B_c \cong 2$ [8].

$\Gamma(x) = \int_0^{+\infty} dy \exp^{(-y)} y^{(x-1)}$ is the gamma function; T_c is the characteristic temperature. N_t is the number of states per unit volume, and δ is the fraction of the localized states occupied by a carrier.

According to the developed theory in [8], the sheet conductance can be expressed as follows:

$$g_{sh} = \sigma_0 \sqrt{\frac{2\varepsilon_0\varepsilon_s K_B T_c}{\delta_0 N_t}} \frac{K_B T}{q(T-2T_c)} \frac{W}{L} \left[\left(\frac{V_G - V_{th}}{\dfrac{(2\alpha)^3 B_c 2\varepsilon_0\varepsilon_s K_B T_c}{C_i^2 \left(\dfrac{T_0}{T}\right)^3 \sin\left(\pi \dfrac{T_c}{T}\right)}} \right)^{\frac{2T_c}{T}-1} - 1 \right] \tag{12.12}$$

For the purpose of refining this last expression, we assume that:

$$\beta = \sigma_0 \sqrt{\frac{2\varepsilon_0\varepsilon_s K_B T_c}{\delta_0 N_t}} \frac{K_B T}{q(T-2T_c)} \text{ and } \rho = \frac{(2\alpha)^3 B_c 2\varepsilon_0\varepsilon_s K_B T_c}{C_i^2 \left(\dfrac{T_c}{T}\right)^3 \sin\left(\pi \dfrac{T_c}{T}\right)}$$

Therefore, the final expression of the sheet conductance is given by:

$$g_{sh} = \beta \frac{W}{L} \left[\left(\frac{V_G - V_{th}}{\rho} \right)^{\frac{2T_c}{T}-1} - 1 \right] \tag{12.13}$$

Using this expression for the simplified sheet conductance, the drain current I_D can be calculated by [15–18]:

$$I_D = \frac{W}{L} \int_{V_G - V_{th} - V_D}^{V_G - V_{th}} g_{sh}(V)\, dV \tag{12.14}$$

Finally, we obtain the following expressions of the drain current [8,14–18]:

$$I_D = \beta \frac{W}{L} \left[\left(\frac{V_G - V_{th}}{\rho} \right)^{\frac{2T_c}{T}} - \left(\frac{V_G - V_{th} - V_D}{\rho} \right)^{\frac{2T_c}{T}} \right] \tag{12.15}$$

in the linear regime if $|V_G - V_{th}| > |V_D|$

$$I_D = \beta \frac{W}{L}\left(\frac{V_G - V_{th}}{\rho}\right)^{\frac{2T_c}{T}} \tag{12.16}$$

in the saturation regime if $|V_G - V_{th}| < |V_D|$

12.3.2.2 Validation of Model 2

Here, we validated model 2 for OTFTs. Therefore, the modeled data by model 2 show a good agreement with the experimental data in terms of the output characteristics, as shown in Figure 12.4.

The set of parameters that gave a good agreement between the experimental data and those obtained by model 2 are summarized in Table 12.2.

12.3.3 Model 3

12.3.3.1 Formulation of Model 3

In the frame of the conventional crystalline semiconductor MOSFETs theory, the standard TFTs equations of drain current as function of the drain voltage (V_D) and the gate voltage (V_G) in the linear and saturation regimes are given by the following equations [20,21]:

TABLE 12.2

The Set of Parameters that Gave a Good Agreement Between the Experimental Data and Those Obtained by the Model 2

Parameters	$\sigma_0(Sm^{-1})$	$\alpha^{-1}(m)$	$2T_c/T$
	6×10^6	1.2	1.98

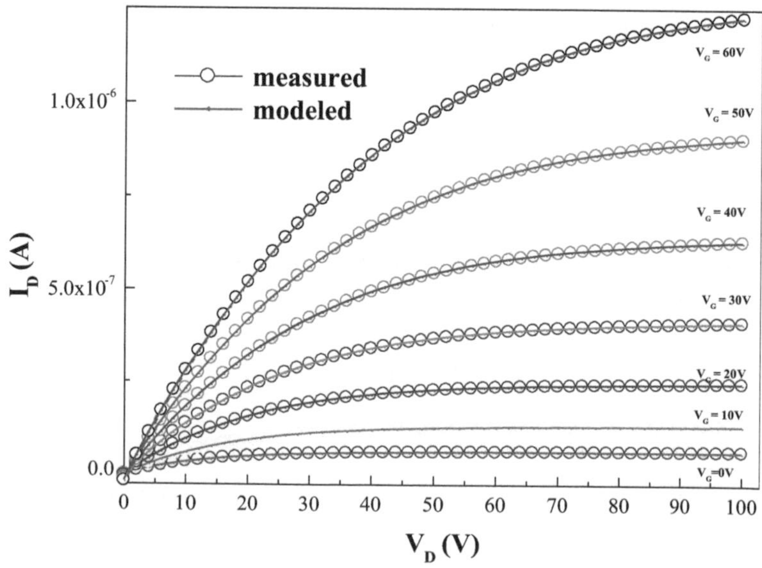

FIGURE 12.4
The good agreement between experimental (circle line) and that obtained by model 2 (full line) output characteristics of OTFTs.

$$I_D = \begin{cases} \frac{W}{L} C_{i\ \mathrm{FET}} \left(V_G - V_{th} - \frac{V_D}{2} \right) V_D & \text{linear regime if} \quad |V_G - V_{th}| > |V_D| \\ \frac{W}{2L} C_{i\ \mathrm{FET}} \left(V_G - V_{th} \right)^2 & \text{saturated regime if} \quad |V_G - V_{th}| < |V_D| \end{cases} \quad (12.17)$$

The resistances of both drain and source contacts represent a particularly relevant parameter for OTFTs, influencing its overall performance in terms of conductance [22,23]. In addition, the adhesion to substrate, particularly for bottom contacts, needs to be taken into account. Indeed, the gate voltage V_G is not equal to the gate-to-source voltage V'_G because the source terminal is not grounded, but its potential is raised by the amount $R_S I_D$ by the current I_D flowing through R_S. Moreover, the drain voltage V_D is not equal to the drain-to-source voltage V'_D because the source terminal is not grounded and the drain terminal is connected to V_D through R_D, so that the gate and drain voltage expressions are written as follows [4,20–27]:

$$\begin{cases} V'_G = V_G - R_S I_D = V_G - \frac{R_C}{2} I_D \\ V'_D = V_D - \left(R_S + R_D \right) I_D = V_D - R_C I_D \end{cases} \quad (12.18)$$

Figure 12.5 depicts an equivalent circuit of the studied OTFTs in which the contact resistance effects that modified the drain current have been taken into account in series with channel resistance R_{ch} [4].

According to this equivalent circuit and the expressions of contacts resistance, the drain current is treated as a function V'_G and V'_D and can be written as follows:

$$I_D = \begin{cases} \frac{W}{L} C_i \mu_{FET} \left(V'_G - V_{th} - \frac{V'_D}{2} \right) V'_D & \text{linear regime} \quad |V'_G - V_{th}| > |V'_D| \\ \frac{W}{2L} C_i \mu_{FET} \left(V'_G - V_{th} \right)^2 & \text{saturated regime} \quad |V'_G - V_{th}| < |V'_D| \end{cases} \quad (12.19)$$

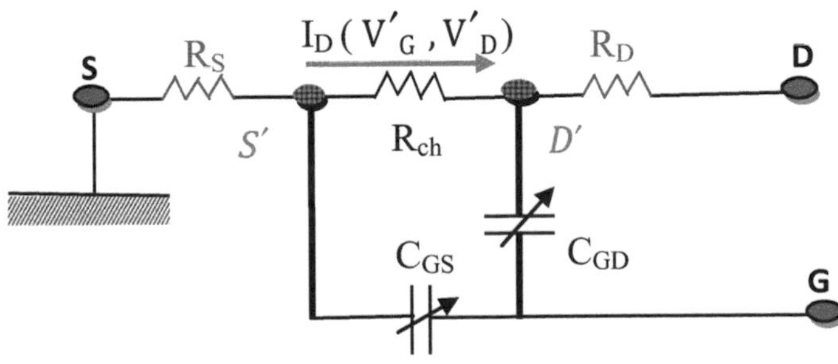

FIGURE 12.5
An equivalent circuit of OTFTs.

Unlike crystalline field effect devices, carrier mobility in OTFTs is gate bias dependent. In an attempt to take into account the mobility dependence with gate voltage in the parameter extraction, several groups have used an empirical relation of the field-effect mobility as [20–27]:

$$\mu_{FET} = \mu_0 \left(\left| \frac{V_G - V_{th}}{V_{aa}} \right| \right)^{\gamma} \tag{12.20}$$

where μ_0 is the voltage-independent mobility and is often considered as the band mobility for the material of the TFT under analysis; V_{th} is the threshold voltage.

V_{aa} and γ are empirical parameters defining the variation of mobility with gate voltage. Parameter γ is associated with the conduction mechanism of the device, and it depends on doping density and dielectric permittivity of the organic semiconductor material.

Finally, we obtain the following expressions for the drain current I_D in both regimes [4,22–27]:

$$I_{D,\text{lin}} = \frac{\left(\left(W\left(\mu_0 / V_{aa}^{\gamma}\right)C_i\left(V_G - V_{th}\right)^{\gamma}\right) / L\left\{V_D\left(V_G - V_{th}\right) - (1/2)V_D^2\right\}\right)}{1 + \left(\left(W\left(\mu_0 / V_{aa}^{\gamma}\right)C_i\left(V_G - V_{th}\right)^{\gamma}\right) / L\right)\left\{-V_D R_c / 2 + V_D\left(V_G - V_{th}\right)\right\}} \tag{12. 21}$$

$$I_{D,\text{sat}} = \frac{4L}{WC_i\left(\mu_0 / V_{aa}^{\gamma}\right)\left(V_G - V_{th}\right)^{\gamma} R_C^2} \left\{ 1 + \frac{W\left(\mu_0 / V_{aa}^{\gamma}\right)C_i R_c}{2L}\left(V_G - V_{th}\right)^{\gamma+1} - \sqrt{1 + \frac{W\left(\mu_0 / V_{aa}^{\gamma}\right)C_i R_c}{L}\left(V_G - V_{th}\right)^{\gamma+1}} \right\}$$

12.3.3.2 Validation of Model 3

Using model 3, we have reproduced the output characteristics of OTFTs. A very good agreement is obtained between the calculations by model 3 and experimental output characteristics, as shown in Figure 12.6.

The set of parameters that gave a good agreement between the experimental data and those obtained by model 3 are summarized in Table 12.3.

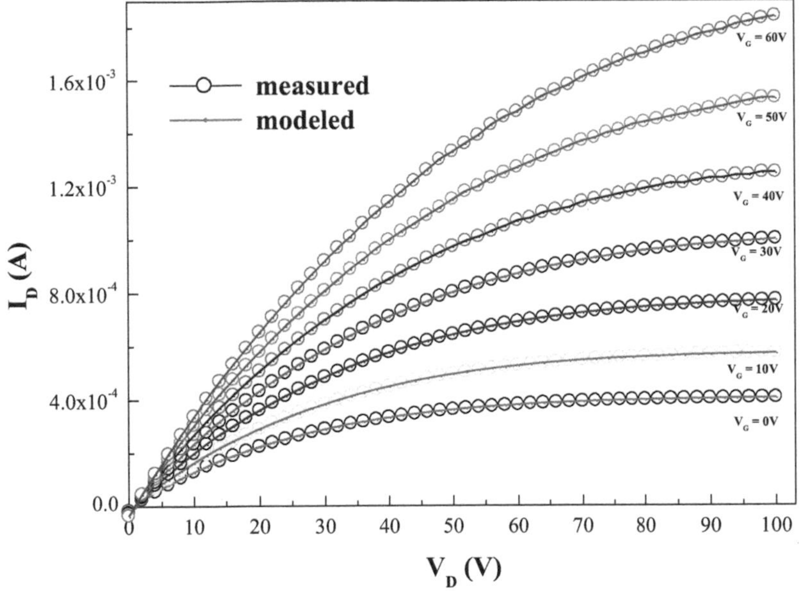

FIGURE 12.6
The good agreement between experimental (circle line) and that obtained by model 3 (full line) output characteristics of OTFTs.

TABLE 12.3
The Set of Parameters that Gave a Good Agreement Between the Experimental Data and Those Obtained by the Model 3

Parameters	$V_{aa}(V)$	γ	μ_0 ($cm^2 V^{-1} s^{-1}$)	$V_{th}(V)$	R_c (KΩ)
	9	0.22	2×10^{-4}	−10	9×10^7

12.3.4 Model 4

12.3.4.1 Formulation of Model 4

In the frame of the assumption used in [3,13,29], the developed expression of the field effect mobility is given by

$$\mu_{FET}(T,T_0) = \mu_0 A(T,T_0) \frac{C_i^{2\left(\frac{T_0}{T}-1\right)}}{\varepsilon_{DBP}^{\left(\frac{T_0}{T}-1\right)}} (V_{GS} - V_{th})^{2\left(\frac{T_0}{T}-1\right)} \tag{12.22}$$

where ε is the dielectric constant of the organic semiconductor, T_0 is the characteristic temperature, k_B is the Boltzmann constant, and q is the electronic charge.

The function $A(T,T_0)$ is given by

$$A(T,T_0)=\frac{qk_BTN_V\exp\left[-\frac{E_{F0}-E_V}{kT}\right]}{\left[\pi qk_BTg_{d0}\exp\left(-\frac{E_{F0}-E_V}{kT_0}\right)\right]^{\frac{T_0}{T}}}\left[\frac{\sin\left(\frac{\pi T}{T_0}\right)}{2k_BT_0}\right]^{\frac{T_0}{T}}$$

where g_{d0} is the density of localized states at the valence band, which is described by an exponential type distribution, N_V is the valence band state density, and μ_0 is taken to be one ($\mu_0 = 1\ \text{cm}^2\,\text{V}^{-1}\,\text{s}^{-1}$) and used only for dimensional purposes.

In the frame of the VRH, the total device resistance is obtained from the inverse of the sheet conductance and can be expressed as [3,13,29]:

$$R_{\text{tot}}=\frac{1}{g_{\text{sh}}}=\sqrt{\frac{\delta_0 N_t}{2\varepsilon_0\varepsilon_s K_B T_0}}\,\frac{q(T-2T_0)}{\sigma_0 K_B T}\,\frac{L}{W}\left[\left(\frac{\dfrac{V_G-V_{\text{th}}}{2(2\alpha)^3 B_c\varepsilon_0\varepsilon_s K_B T_0}}{C_i^2\left(\dfrac{T_0}{T}\right)^3\sin\left(\pi\dfrac{T_0}{T}\right)}\right)^{\frac{2T_0}{T}-1}-1\right]^{-1} \tag{12.23}$$

where σ_0 is the percolation prefactor of the conductivity, α is an effective overlap parameter that governs the tunneling process between two localized states, B_c is the critical number of bonds per site in the percolating network for a three-dimensional amorphous system, and $B_c \cong 2$ [3,13,29]. N_t is the number of states per unit volume, and δ is the fraction of the localized states occupied by a carrier.

The modeling of the current-voltage characteristics (output and transfer characteristics) of TFTs was performed using the following expression [3,13,29]:

$$I_D = K\mu_{\text{FET}}\frac{(V_{\text{GS}}-V_{\text{th}})(1+\lambda V_D)}{\left[1+R_cK\mu_{\text{FET}}(V_{\text{GS}}-V_{\text{th}})\right]\left[1+\left[\dfrac{V_D}{\alpha_s(V_{\text{GS}}-V_{\text{th}})}\right]^m\right]^{\frac{1}{m}}} \tag{12.24}$$

where λ is the channel length modulation parameter, α_s is the saturation modulation parameter, and m is a control transition parameter from the linear to the saturation regime in the output characteristics of the OTFTs.

12.3.4.2 Validation of Model 4

A very good agreement has been obtained between the calculations by model 4 and experimental output characteristics, as shown in Figure 12.7.

The set of parameters that gave a good agreement between the experimental data and those obtained by model 4 are summarized in Table 12.4.

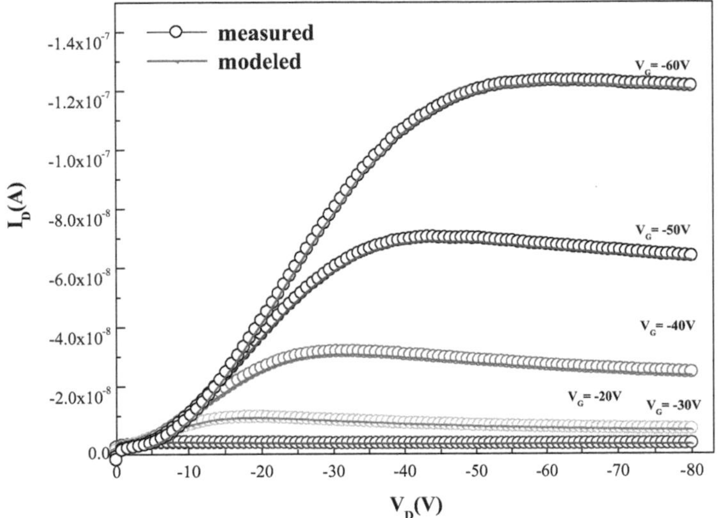

FIGURE 12.7
The good agreement between experimental (circle line) and that obtained by model 4 (full line) output characteristics of OTFTs.

TABLE 12.4

The Set of Parameters that Gave a Good Agreement Between the Experimental Data and Those Obtained by the Model 4

Parameters	M	λ (V^{-1})	$V_{th}(V)$	$R_c(\Omega)$
	2.58	1.8×10^{-2}	12	7×10^9

12.4 Conclusions

In this chapter, organic thin-film transistors have been theoretically studied and modeled, focusing particularly on the different analytical models of the electrical characteristics of the devices. We first introduced the general standard model that describes current-voltage equations in thin-film transistors, focusing on the geometry and properties of the active layer materials. We also presented a detailed description of the various manufacturing steps that we used to fabricate OTFTs as well as the electrical characterization step. Finally, we presented the formulation of the different models used in OTFTs. The future ultimate goal is to provide a common analytical model that can be used in the modeling of any organic thin-film transistor.

References

1. W. Boukhili, M. Mahdouani, M. Erouel, J. Puigdollers, R. Bourguiga, "Reversibility of humidity effects in pentacene based organic thin-film transistor: Experimental data and electrical modeling." *Synth. Met.* 199 (2015) 303–309.
2. W. Boukhili, M. Mahdouani, R. Bourguiga, J. Puigdollers, "Experimental study and analytical modeling of the channel length influence on the electrical characteristics of small-molecule thin-film transistors." *Superlattices Microstruct.* 83 (2015) 224–236.

3. W. Boukhili, M. Mahdouani, R. Bourguiga, J. Puigdollers, "Temperature dependence of the electrical properties of organic thin-film transistors based on tetraphenyldibenzoperiflanthene deposited at different substrate temperatures: Experiment and modeling." *Microelectron. Eng.* 150 (2016) 47–56.
4. W. Boukhili, M. Mahdouani, R. Bourguiga, J. Puigdollers, "Temperature dependence of the electrical properties of organic thin-film transistors based on tetraphenyldibenzoperiflanthene deposited at different substrate temperatures: Experiment and modeling." *Microelectron. Eng.* 160 (2016) 39–48.
5. M. Mahdouani, W. Boukhili, R. Bourguiga, "Negative output differential resistance effect in organic thin film transistors based on pentacene: Characterization and modeling." *Mater. Today Commun.* 13 (2017) 367–377.
6. W. Boukhili, C. Tozlu, M. Mahdouani, S. Erten-Ela, R. Bourguiga, "Illumination and dipole layer effects on the density of state distribution in n-type organic thin film phototransistors based on naphthalene bis-benzimidazole: Experiment and modeling." *Microelectron. Eng.* 179 (2017) 37–47.
7. S. M. Sze, *Physics of Semiconductors Devices*, New York: Wiley, p. 442, 1998.
8. M. C. J. M. Vissenberg, M. Matters, "Theory of the field-effect mobility in amorphous organic transistors." *Phys. Rev. B* 57 (1998) 12964.
9. H. Bässler, "Charge transport in disordered organic photoconductors – a Monte-Carlo simulation study." *Phys. Status Solidi B* 175 (1993) 15–56.
10. M. Schwoerer, H. C. Wolf, *Organic Molecular Solids*, Weinheim, Germany: Wiley-VCH, 2007.
11. P. M. Borsenberger, R. Richert, H. Bässler, "Dispersive and nondispersive charge transport in a molecularly doped polymer with superimposed energetic and positional disorder." *Phys. Rev. B* 47 (1993) 4289.
12. M. Estrada, I. Mejia, A. Cerdeira, J. Pallares, L. Marsal, B. Iniguez, "Mobility model for compact device modeling of OTFTs made with different materials." *Solid State Electron.* 52 (2008) 787.
13. L. Li, H. Kosina, "An analytical model for organic thin film transistors." *Conference on Electron Devices and Solid-State Circuits (EDSSC)*, Hong Kong, IEEE (2005) pp. 571–574.
14. V. Ambegaokar, B. I. Halperin, J. S. Langer, "Hopping conductivity in disordered systems." *Phys. Rev. B* 4 (1971) 2612.
15. M. Sahimi, *Applications of Percolation Theory*, London, UK: Taylor & Francis Groups, 1994.
16. G. Pike, C. Seager, "Percolation and conductivity: A computer study. I." *Phys. Rev. B* 10 (1974) 1421–1434.
17. G. Horowitz, P. Delannoy, "An analytical model for organic based thin film transistors." *J. Appl. Phys.* 70 (1991) 469.
18. E. Calvetti, L. Colalongo, Zs. M. Kovács-Vajna, "Organic thin film transistors: a DC/dynamic analytical model." *Solid-State Electron.* 49 (2005) 567.
19. G. Horowitz, R. Hajlaoui, H. Bouchriha, R. Bouirguiga, M. Hajlaoui, "Organic field effect transistors." *Adv. Mater.* 10 (1998) 923–927.
20. R. F. Pierret, *Semiconductor Device Fundamentals*, Boston, MA: Addison-Wesley Publishing Company, 1996.
21. S. M. Sze, *Physics of Semiconductor Devices*, 2nd ed., New York: John Wiley & Sons, 1981.
22. P. V. Necliudov, M. S. Shur, D. J. Gundlach, T. N. Jackson, "Modeling of organic thin film transistors of different designs." *J. Appl. Phys.* 88 (2000) 6594–6597.
23. A. Cerdeira, M. Estrada, B. Iniguez, J. Pallarès, L. F. Marsal, "Modeling and parameter extraction procedure for nanocrystalline TFTs." *Solid State Electron.* 48 (2004) 103–109.
24. P. Mittal, B. Kumar, Y. S. Negi, B. K. Kaushik, R. K. Singh, "Channel length variation effect on performance parameters of organic field effect transistors." *Microelectron. J.* 43 (2012) 985–994.
25. D. Hong, G. Yerubandi, H. Q. Chiang, M. C. Spiegelberg, J. F. Wager, "Electrical modeling of thin-film transistors." *Crit. Rev. Solid State Mater. Sci.* 33 (2008) 101–132.
26. M. Estrada, A. Cerdeira, J. Puigdollers, L. Resendiz, J. Pallares, L.F. Marsal, C. Voz, B. Iniguez, "Accurate modeling and parameter extraction method for organic TFTs." *Solid State Electron* 49 (2005) 1009–1016.
27. B. Kumar, B. K. Kaushik, Y. S. Negi, S. Saxena, G. D. Varma, "Analytical modeling and parameter extraction of top and bottom contact structures of organic thin film transistors." *Microelectron. J.* 44 (2013) 736–743.
28. M. Sahimi, *Applications of Percolation Theory*, London, UK: Taylor & Francis Groups, 1994.
29. G. Pike, C. Seager, "Percolation and conductivity: A computer study. I." *Phys. Rev. B* 10 (1974) 1421–1434.

13

A Fundamental Overview of High Electron Mobility Transistor and Its Applications

D. Nirmal and J. Ajayan

CONTENTS

13.1 Introduction

The demand for high-speed semiconductor devices is increasing in recent years with the rapid development of the high-frequency wireless and fiber optical communication systems. High electron mobility transistors (HEMTs) are emerging as excellent candidates for millimeter, sub-millimeter, and microwave applications. These devices are also considered the most promising candidates for future Terahertz (THz) applications because of their low noise, high frequency, high breakdown voltage, low cost, higher level of integration, and high-power handling capabilities. For manufacturing HEMTs, usually III–V compound semiconductors are used: binary (GaAs, GaN, InSb, InP), ternary (AlGaAs, AlGaN, InGaAs, AlInSb), and even quaternary (InGaAsP) semiconductors. The advantage of these materials is that the bandgap and lattice constants can be modified by varying the chemical composition, and these materials have higher mobility compared to conventional silicon material. The HEMT is also known as a heterojunction field effect transistor (HFET), modulation doped field-effect transistor (MODFET), two-dimensional electron gas field effect transistor (TEGFET), or selectively doped heterojunction transistor (SDHT).

The HEMT was first demonstrated by Takashi Mimura and colleagues at Fujitsu Labs in 1980 [1] in the AlGaAs/GaAs material system. The HEMT was invented based on the concept of modulation doping, and the modulation doped structure creates an ultra-high mobility two-dimensional electron gas (2DEG) at the heterojunction interface between two different materials having different bandgaps, which was first demonstrated by Ray Dingle and collaborators at Bell Labs in 1978 [2]. The HEMT uses the concept of bandgap engineering, and it shows outstanding electron transport properties of 2DEG systems such as higher electron mobilities, higher saturation velocities, and higher sheet electron densities in III–V compound semiconductors. Over the last three decades, HEMTs have been demonstrated in several material systems such as AlGaAs/GaAs, AlGaN/GaN, and InAlAs/InGaAs. The revolution in atomic layer precision growth capabilities of deposition techniques such as molecular beam epitaxy (MBE) and metal organic chemical vapor deposition (MOCVD) also played a major role in the development of HEMT. To allow conduction in conventional metal oxide semiconductor transistors (MOSFETs), semiconductors (channel region) are doped with impurities, which donate mobile electrons (or holes). However, these carriers are slowed down through collisions with the impurities (dopants) used to generate them in the channel region. HEMTs avoid these collisions through the use of high-mobility electrons generated using the heterojunction of a highly doped wide-bandgap *n*-type donor-supply layer (e.g., AlGaAs) and a non-doped narrow-bandgap channel layer (e.g., GaAs) with no dopant impurities.

The electrons generated in the ultra-thin highly doped n-type wide bandgap donor-supply layer (e.g., AlGaAs) drop completely into a non-doped narrow-bandgap channel layer (e.g., GaAs) to form a depleted AlGaAs layer. In other words, in HEMTs, the narrow bandgap channel layer usually has higher electron affinity than the wide bandgap donor supply layer. Therefore, free electrons in the wide bandgap donor supply layer are transferred to the undoped narrow bandgap channel layer where these mobile electrons form a two-dimensional high mobility electron gas within 10 nm of the heterojunction interface. The heterojunction created by different bandgap materials forms a quantum well in the conduction band of the channel layer where the electrons can move with very high velocity without colliding with any impurities because the channel layer is undoped and from which these trapped electrons cannot escape. The effect of this is to create a very thin layer of highly mobile conducting electrons with very high concentration, giving the channel very low resistivity—in other words, "high electron mobility." The energy band diagram of a GaAs/AlGaAs HEMT is shown in Figure 13.1.

Two major innovations in the 1980s would greatly expand the high-frequency capabilities of HEMTs and would open up many new application areas in wireless and fiber optical communications, radar systems, sensing, space, and military. The first innovation was the demonstration of pseudomorphic HEMT (PHEMT) by Ketterson and colleagues at the University of Illinois in 1985 [3]. The PHEMT demonstrated by Ketterson has an AlGaAs/InGaAs/GaAs quantum-well structure where the enhanced electron transport properties in InGaAs together with the tight quantum-well confinement and large conduction band discontinuity (ΔEC) between the channel layer and barrier layer resulted in improved device scalability and performance. The second innovation was planar doping, otherwise called delta doping (δ-doping) of the barrier layer, which was introduced in the AlGaAs layer [4]. This δ-doping technique helps to reduce the thickness of the barrier layer, yielding improvements in transconductance, drain current, channel aspect ratio, and device scalability. HEMTs show higher breakdown voltage characteristics due to the

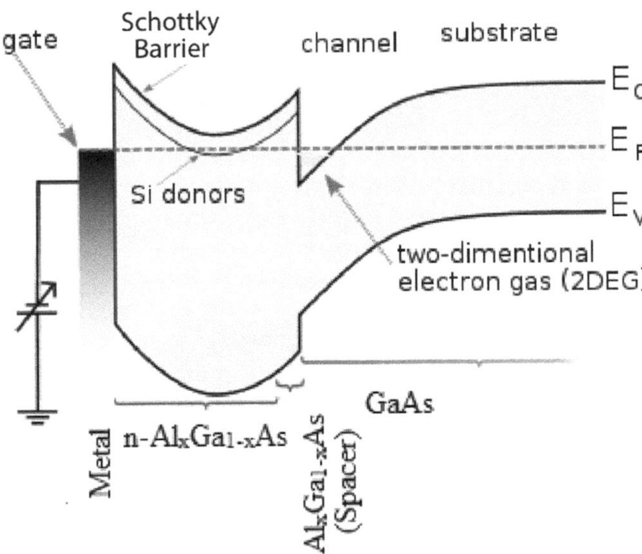

FIGURE 13.1
Energy band diagram of GaAs/A1GaAs HEMT, at equilibrium [Wikipedia].

removal of dopants from directly underneath the gate. The combination of δ-doping and the high mobility pseudomorphic InGaAs channel propelled PHEMTs to the 20 nm gate length regime and the achievement of record noise and power performance up to very high frequencies [5].

13.2 Types of HEMTS

According to the lattice constant utilized in the layers of a heterojunction HEMT structure, there are three types of HEMTs, namely lattice-matched HEMT (LHEMT), pseudomorphic HEMT (PHEMT), and metamorphic HEMT (MHEMT).

Lattice-Matched HEMT (LHEMT): If the lattice constants of the two semiconductor materials employed on both sides of the heterostructure interface are the same, then the type of HEMT is called LHEMT [6–9]. Examples for LHEMTs are HEMTs constructed on AlGaAs/GaAs and InAlAs/InGaAs/InP material systems. The schematic diagram for LHEMT constructed using the AlGaAs/GaAs material system is shown in Figure 13.2.

Pseudomorphic HEMT (PHEMT): Pseudomorphic HEMT is also known as a non-lattice-matched HEMT because lattice constants of the two semiconductor materials employed in both sides of the heterostructure interface are slightly different. PHEMT consists of an extremely thin doped wideband electron supply layer compared to an undoped narrow-band channel layer. The outstanding properties of the PHEMT [10–14] in terms of power, noise, and low-loss switching at very high frequencies have made this device technology a success in the

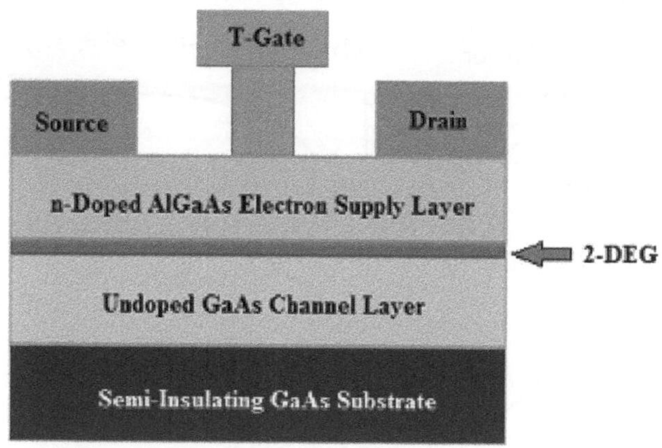

FIGURE 13.2
Example for LHEMT constructed in A1GaAs/GaAs material system.

commercial arena [5,13]. The most important applications of PHEMTs are cell phones, wideband wireless communications, fiber optical communications, satellite communications, radar systems, space, and defense. ExamPHEMTs are HEMTs constructed on AlGaAs/InGaAs/GaAs and InAlAs/InGaAs/InP material systems. The schematic diagram for PHEMT constructed using InAlAs/InGaAs/InP material system is shown in Figure 13.3.

Metamorphic HEMT (MHEMT): If the lattice constants of the two semiconductor materials employed on both sides of the heterostructure interface are

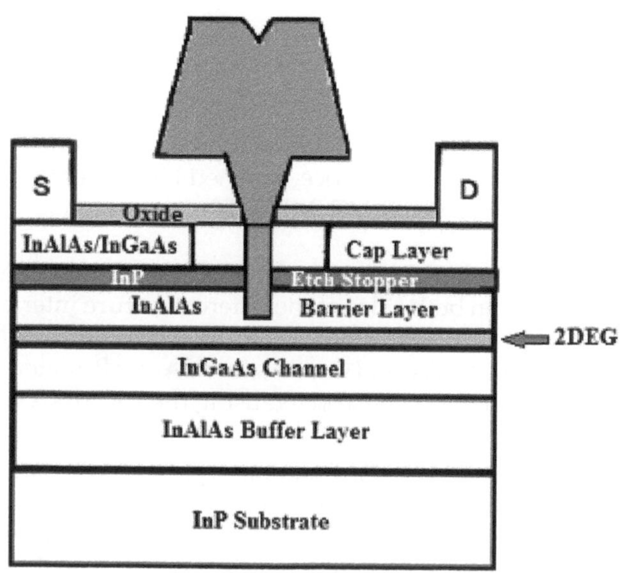

FIGURE 13.3
Example for PHEMT constructed in InA1As/InGaAs/InP material system. (From Kim, D. H., *IEEE Electron Device Lett.*, 29, 830–833, 2008.)

significantly different, a buffer layer can be inserted between the two semiconductor materials for compensating the large difference in lattice constants—this type of HEMT is called MHEMT [15–27]. Recent works on Enhancement mode (E-Mode) InAlAs/InGaAs HEMTs on InP substrate have demonstrated superior gain, low noise, and high-frequency performances over HEMTs on any other material systems such as GaAs. This is mainly due to the better electron transport properties such as higher sheet charge densities (2DEG), higher saturation velocities, and higher electron mobilities of the InGaAs channel. However, InP HEMT technology is not commercial yet due to the high cost of InP substrate, fragility, and limited wafer size (4″ diameter). Transistors and MMICs on GaAs substrates have become essential components in commercial low-noise wide bandwidth wireless communication systems mainly due to low cost, large wafer size (6″ diameter), better mechanical strength, and well matured processing technology. In recent years, MHEMTs have received much attention due to their capability of combining advantages such as high frequency, high gain, low noise, and high efficiency of the high-performance InP-based HEMT structure and the lower cost due to large wafer size, improved reproducibility, greater ease of wafer handling, better established packaging technology, and high mechanical strength GaAs substrate. MHEMTs have been considered a cost-effective alternative to the high-performance InP HEMTs. The schematic diagram for MHEMT constructed using InAlAs/InGaAs/GaAs material system is shown in Figure 13.4.

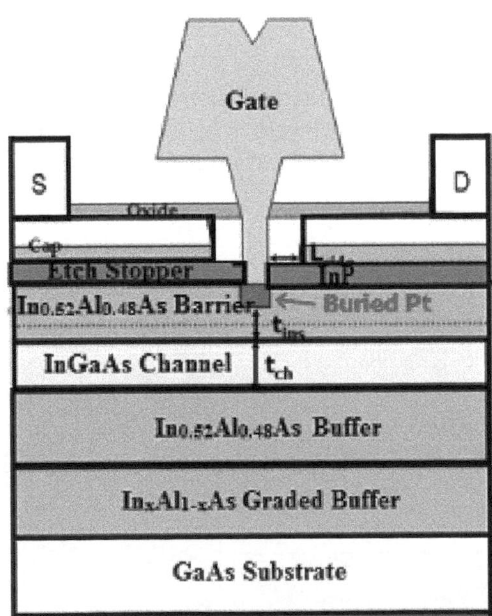

FIGURE 13.4
Example for MHEMT constructed in InA1As/InGaAs/GaAs material system. (From Kim, D. H., f_T = 688GHz and f_{max} = 800GHz in Lg = 40nm $In_{0.7}Ga_{0.3}As$ MHEMTs with g_m_max > 2.7mS/µm, in *Proceedings of IEDM Technical Digest*, p. 319, 2011.)

13.3 HEMT Structure and Principle of Operation

A simple HEMT structure consists of four layers, namely a substrate layer (InP or GaAs or GaN or SiC or Si or Sapphire); a buffer layer, which depends on the type of the substrate used; a high-mobility channel layer (GaAs, GaN, InGaAs, InAs, InSb); and a Schottky barrier layer. The buffer layer isolates the defects from the substrate and creates a smooth surface. In some advanced HEMTs based on InP or GaAs substrates, a barrier layer is used between the channel and buffer layers, which have multiple roles such as reducing lattice mismatch, and also it helps to introduce δ-doping to improve the performance of the device. A spacer layer, if used, separates the 2DEG from dopants of the electron supply layer thereby reducing electron scattering effects, which enhances electron mobility. Cap layer is used to facilitate the formation of non-alloyed ohmic contacts with very low resistance. The generic HEMT structure is shown in Figure 13.5.

An example for practical HEMT structure is shown in Figure 13.6. The epitaxial layer structure consists of a semi-insulating InP substrate; InAlAs buffer layer, which isolates the defects from the InP substrate and creates a smooth surface; a $In_{0.53}Ga_{0.47}As$ channel layer having high mobility; a InAlAs spacer layer, which separates the 2DEG from dopants of the electron supply layer thereby reducing electron scattering effects, which results in increased mobility; an InAlAs Schottky barrier layer; an InP etching stopper layer, which is used to reduce the access resistance; and a Si-doped InGaAs cap layer, which helps to facilitate the formation of non-alloyed ohmic contacts with very low resistance. A Si-doped plane was inserted between the Schottky barrier layer and the spacer layer to supply the electrons for current conduction. The δ-doping technique was used to improve the transconductance, increase the drain current, reduce access resistance, and also enhance device linearity.

13.3.1 Principle of Operation

In HEMTs, the conduction channel is a two-dimensional electron gas formed at the interface between the barrier and channel layers with different bandgaps instead of a three-dimensional conduction channel in conventional FETs. The fundamental characteristic

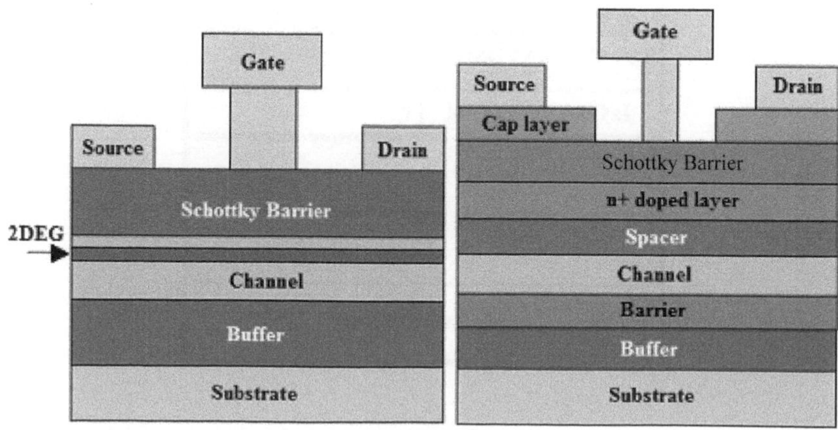

FIGURE 13.5
Generic HEMT structures.

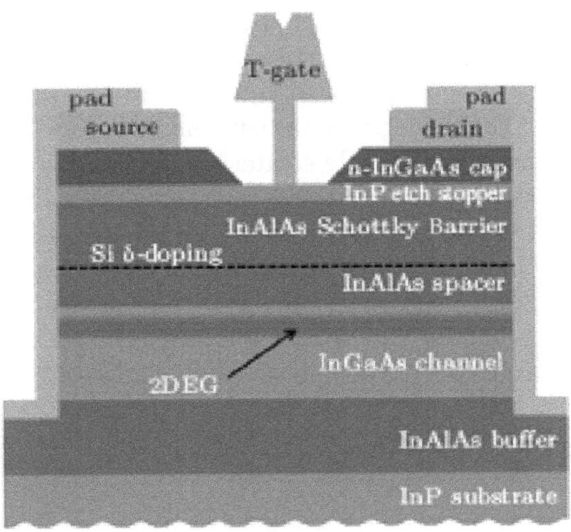

FIGURE 13.6
Structure of InPHEMT. (From Li-Dan, W., *Chin. Phys. B*, 23, 038501, 2014.)

of the HEMT is the conduction band offsets (ΔEC) between the semiconductor materials, which forms the barrier and channel layers. The conduction band of barrier layer is higher than the conduction band of channel layer. Therefore, a potential well is formed at the interface between barrier layer and channel layer due to this conduction band offset, and this potential well can contain a large number of electrons to form a 2DEG channel at the heterointerface. The energy-band diagram of a generic HEMT structure is shown in Figure 13.7.

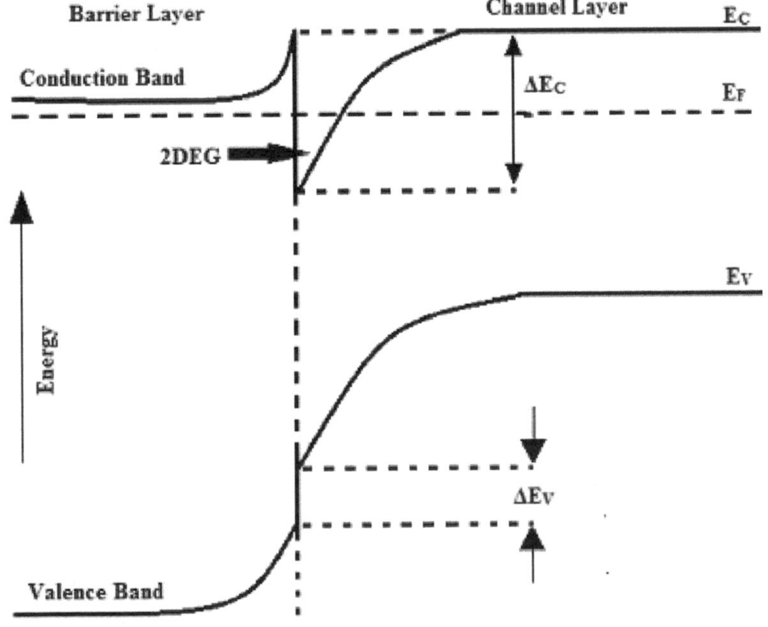

FIGURE 13.7
Energy band diagram of general HEMT showing 2DEG.

The electrons in the 2DEG channel are provided by the modulation doped barrier layer. The electrons are separated from the ionized donors by the potential barrier at the heterointerface and confined in a two-dimensional conduction channel, which can reduce different types of scattering mechanisms such as Coulomb scattering, phonon scattering, dislocation scattering, dipole scattering, charged impurities scattering, and interface roughness scattering between the carriers and the ionized donors. The scattering effects can also be reduced by inserting an undoped spacer layer between the barrier and channel layers. This technique ensures a high electron density and mobility, which makes HEMTs promising for high-frequency and high-power applications. In HEMTs, by employing a Schottky barrier (metal/semiconductor) gate above the doped barrier layer, the 2DEG sheet charge concentration can be controlled by applying an appropriate bias voltage.

13.4 Challenges in HEMT Technology

The challenges in HEMT technology are depicted in Figures 13.8 and 13.9. For the realization of high-speed, high-power HEMTs, substrates with high thermal stability, high breakdown field, high frequency of operation, and lower cost are required. The various substrates used for realizing HEMTs are sapphire, SiC, Si, GaN, InP, and GaAs. Among these, GaN is suitable for power electronic applications due to its material properties such as high thermal stability, high voltage operation, and higher breakdown voltage. However, GaN-based HEMTs suffer from low current density due to relatively low sheet charge density compared to InP substrate HEMTs. InP substrate-based HEMTs exhibit excellent gain and linearity, very high current gain cut-off frequency, very low noise, and the highest maximum frequency of oscillation for any transistors. InP substrates are suitable for future high-speed logic applications due to low voltage operation and high frequency of operation above 600 GHz. HEMTs on InP substrates are considered the most promising candidates for future THz applications. However, InP substrates suffer from relatively large substrate cost, low wafer size, and they are fragile compared to GaAs and GaN substrates. The properties of various substrates are given in Table 13.1.

Reduction of parasitic resistances (RS, RG, and RD) and capacitances (CGS and CGD) is essential in HEMT technology because these parasitics limit the high-frequency operation of the HEMTs. Various gate shapes such as T-gate, Y-Gate, rectangular gate, and double finger gate have been developed to minimize gate resistance RG. For higher stability and better reliability of monolithic microwave integrated circuits (MMICs) using HEMTs, devices must be coated with a passivating layer with a high dielectric constant. Silicon nitride (Si_3N_4) is one of the promising materials for HEMT passivation. However, due to the high dielectric constant of Si_3N_4 ($\epsilon_r = 7$), the C_{gd} increases, which reduces the cut-off frequency of the devices. In order to achieve high frequency of operation in a HEMT, it is essential to scale down the gate length and minimization of gate resistance. InP HEMTs are suitable for high-speed applications, and conventional InP HEMTs use a T-shaped gate structure because this structure provides a small gate contact area on the Schottky barrier layer. Therefore, the scalability can be improved, and this T-gate structure also provides a large cross-sectional area at the electrode, which helps to minimize the gate resistance. The main limitation of the T-gate structure is that of poor physical strength at the junction between a large top and narrow stem. In order to solve this problem, Fujitsu has developed

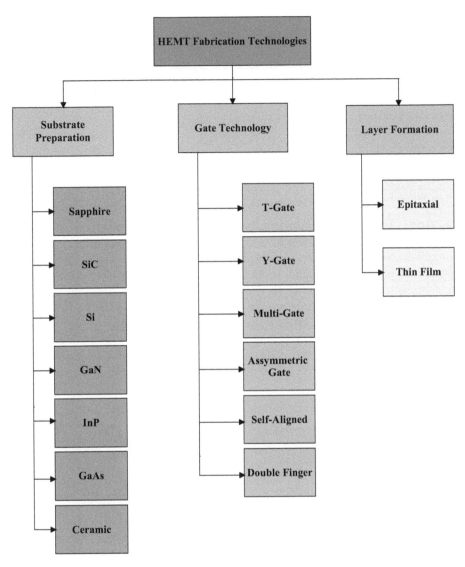

FIGURE 13.8
HEMT fabrication challenges.

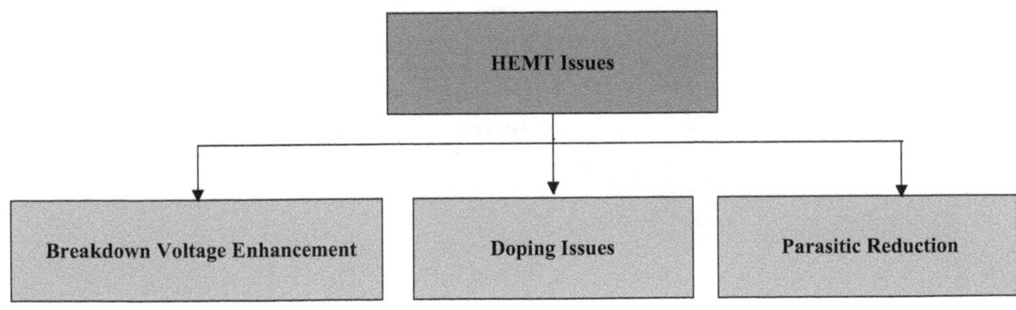

FIGURE 13.9
HEMT issues.

TABLE 13.1
Suitability of GaAs, Si, SiC, GaN, and InP Substrates for High-Power and High-Frequency Applications

Properties of Substrates	GaAs	Si	SiC	GaN	InP
Suitability for high power	High	Low	High	Excellent	Low
Suitability for high frequency	High	Low	Medium	High	Excellent
Cost of the substrates	Low	Low	Low	Low	High
HEMT structure feasibility	Excellent	Low	Low	Excellent	Excellent
Gain at high frequency	High	Low	Low	High	Excellent
Noise at high frequency	High	High	High	High	Excellent

the Y-gate structure, which is shown in Figure 13.10. This structure effectively prevents the gate top from peeling off, resulting in a high yield in gate electrode fabrication. In order to increase the cut-off frequency of InP HEMTs, a reduction of parasitic gate to drain feedback capacitance is essential; Fujitsu used benzocyclobutene (BCB, $\epsilon_r = 2.8$) for inter-layer dielectric film. Introducing a cavity in the periphery of the gate further reduces the C_{gd}, which results in high cut-off frequency. This cavity structure is also developed by Fujitsu, which is shown in Figure 13.11. Self-aligned gate technology enables the downscaling of HEMTs into the ultra-short sub-50 nm gate length regime. Asymmetric gate technology is used to improve the breakdown voltage of HEMTs. T-gate with cavity structure is the widely used gate structure for InP HEMTs.

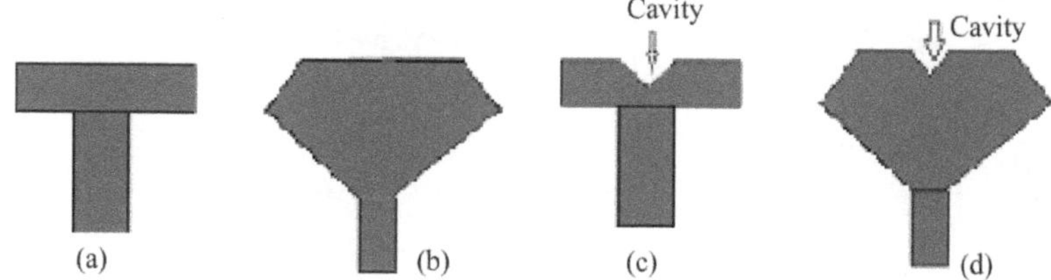

FIGURE 13.10
Gate structure (a) T-gate (b) Y-gate (c) T-gate with cavity (d) Y-gate with cavity.

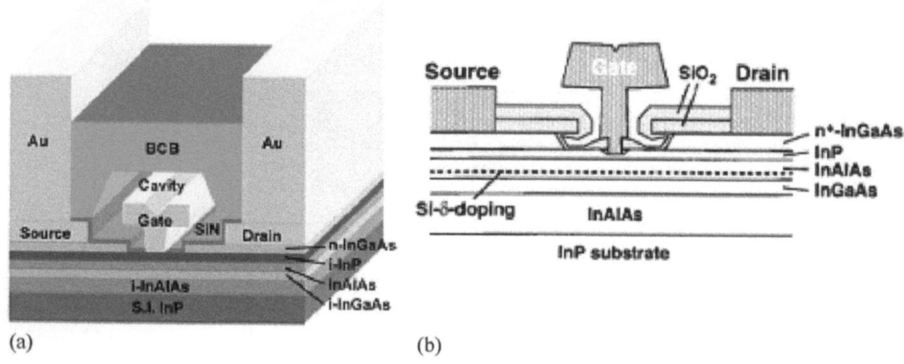

FIGURE 13.11
Gate structure (a) T-gate with cavity (b) double finger gate technology. (From Yamashita, Y., *IEEE Electron Device Lett.*, 22, 367–369, 2001.)

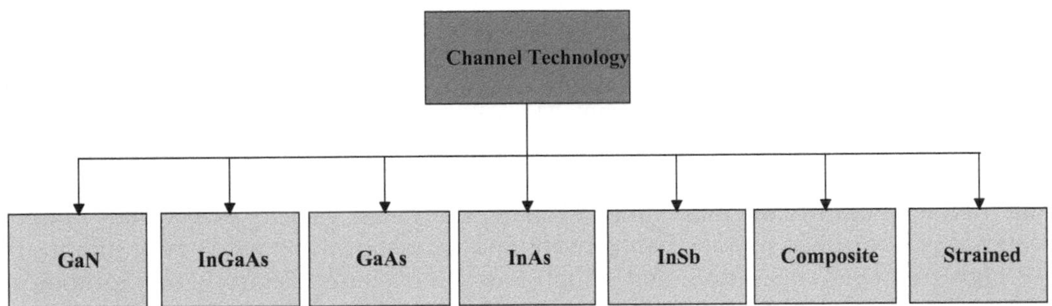

FIGURE 13.12
HEMT channel technologies.

TABLE 13.2
Properties of III–V Compound Semiconductors

Sl. No	Properties	GaAs	InP	$In_{0.53}Ga_{0.47}As$	GaN	InAs	InSb
1	Lattice constant (A^0)	5.653	5.869	5.869	–	6.058	–
2	Electron effective mass (m^*/m)	0.067	0.077	0.041	–	0.023	0.013
3	Electron mobility (cm^2/Vs)	8500	4600	12000	–	33000	78000
4	Hole mobility (cm^2/Vs)	400	150	300	–	460	–
5	Saturation velocity at low field ($\times 10^7$ cm/S)	2.1	2.5	3.1	–	7.7	>8
6	Bandgap (eV)	1.42	1.35	0.74	–	0.35	0.175
7	N_C ($\times 10^{17}$ cm^{-3})	4.7	5.7	2.1	–	8.7	–
8	N_V ($\times 10^{18}$ cm^{-3})	9.0	11	7.7	–	6.6	–

Molecular beam epitaxy (MBE) and metal organic chemical vapor deposition (MOCVD) techniques are widely used for the epitaxial growth of layers in a HEMT. The various channel technologies used for HEMTs are shown in Figure 13.12. The properties of these channel materials are given in Table 13.2. Among these channel materials, InGaAs, InAs, and InSb have higher electron mobility and velocity of saturation. In order to further enhance the mobility of electrons, various device architectures using composite channel or strained channel techniques have been demonstrated by various research groups.

An example for a composite channel is $In_xGa_{1-x}As/InAs/In_xGa_{1-x}As$. Introducing horizontal or vertical strain in the channel also improves the mobility, and this strain can be introduced in the channel layer by adding layers with slightly different lattice constants.

13.5 AlGaN/GaN HEMT

AlGaN/GaN HEMT is one of the most exciting device structures at the present time, and these HEMTs have excellent potential for high-speed, high-power microwave applications [29–34] such as microwave amplifiers, mixers, oscillators, and other nonlinear circuits due to their excellent characteristics such as high breakdown voltage, high electron charge density, and high electron mobility [35–38]. Modulation doping in the AlGaN/GaN material system was first demonstrated in 1992 [39], and the first AlGaN/GaN HEMT was reported in 1993 [40]. The enormous interest in recent years in the AlGaN/GaN material system stems from several factors such as the wide bandgap of GaN semiconductor yields a high breakdown voltage, which enables the high voltage and high temperature operation of HEMTs; the saturation velocity of electrons in GaN material is two times higher than conventional Si material, which is very important for high frequency operation; and a high two-dimensional electron concentration is induced at the AlGaN/GaN interface and yields high current. In the last several years, AlGaN/GaN HEMTs have been produced in large volume and have started to appear in radar systems, cellular phones, cable TV amplifiers, WiMax base stations, satellite communications, and a variety of military systems. Standard AlGaN/GaN HEMTs are depletion-mode (D-mode) devices due to the large 2-DEG induced by the polarization charge at the AlGaN/GaN interface [41]. However, enhancement-mode (E-mode) AlGaN/GaN HEMTs are suitable for high-speed, high-power applications as they can greatly simplify circuit designs and improve system reliability [42]. The main advantage of E-mode HEMTs over D-mode HEMTs is that they eliminate the need for a negative supply voltage on a chip.

13.5.1 Surface Potential-Based Drain Current Model

In order to calculate the drain current as a function of surface potential, an accurate estimation of surface potential is required. The main challenge in calculating surface potential in AlGaN/GaN HEMT is due to the complicated variation of the Fermi level E_f with applied bias voltages. Sourabh Khandelwal et al. [43] have developed an analytical model for estimating surface potential in AlGaN/GaN HEMTs, which is given below. For deriving the surface potential, assume a triangular profile for potential well. The self-consistent solution of the Schrodinger's and Poisson's equations is given by [44]

$$n_s = DV_{\text{th}}\left\{\ln\left[\exp\left(\frac{E_f - E_0}{V_{\text{th}}}\right)+1\right]+\ln\left[\exp\left(\frac{E_f - E_1}{V_{\text{th}}}\right)+1\right]\right\} \tag{13.1}$$

$$E_0 = \gamma_0 n_s^{2/3} \tag{13.2}$$

$$E_1 = \gamma_1 n_s^{2/3} \tag{13.3}$$

$$n_s = \frac{\varepsilon}{qd}\left(V_{g0} - E_f - V_x\right) \tag{13.4}$$

where n_s is the density of 2DEG; E_f is the position of Fermi level; E_0 and E_1 are the positions of first and second energy levels, respectively; D represents the density of states; q is the charge of an electron; V_{th} is the thermal voltage; d and ϵ are the thickness and permittivity of AlGaN layer, respectively; $C_g = \epsilon/d$ is the gate capacitance per unit area; $V_{g0} = V_g - V_{\text{off}}$; and V_x is the channel voltage at any point x in the channel. γ_0 and γ_1 are experimentally determined parameters whose values [45] are 2.12e-12 and 3.73e-12 $\text{Vm}^{4/3}$, respectively.

To calculate E_f and n_s, the variation of E_f with respect to gate voltage V_g is divided into three different regions: (1) $V_g < V_{\text{off}}$, the sub-V_{off} region, where $|E_f|$ is comparable to $|V_{g0}|$; (2) $V_g > V_{\text{off}}$, and $E_f < E_0$, the moderate 2-DEG region; and (3) $V_g > V_{\text{off}}$ and $E_f > E_0$, the strong 2-DEG region. The regional expressions for $V_g > V_{\text{off}}$ valid in regions (2) and (3) can be combined to obtain a single expression for E_f as [46]

$$E_{f,\text{above}} = V_{g0}\left(1 - H(V_{g0})\right) \tag{13.5}$$

where

$$H\left(V_{g0}\right) = \frac{V_{g0} + V_{\text{th}}\left[1 - \ln\left(\beta V_{g0n}\right)\right] - \frac{\gamma_0}{3}\left(\frac{C_g V_{g0}}{q}\right)^{2/3}}{V_{g0}\left(1 + \frac{V_{\text{th}}}{V_{g0d}}\right) + \frac{2\gamma_0}{3}\left(\frac{C_g V_{g0}}{q}\right)^{2/3}} \tag{13.6}$$

Here

$V_{g0n(d)} = V_{g0n(d)}\alpha_{n(d)} / \sqrt{V_{g0n(d)}^2 + \alpha_{n(d)}^2}$ is obtained using interpolation function, $\beta = C_g/qDV_{\text{th}}$, $\alpha_n = e/\beta$ and $\alpha_d = 1/\beta$. In the sub-V_{off} region, $|E_f|\rangle\rangle E_0$ and E_1. Therefore, $E_f - E_0$ and $E_f - E_1 \approx E_f$, and

$$n_{s,\text{sub-}V_{\text{off}}} = 2DV_{\text{th}} \exp\left(V_{g0} / V_{\text{th}}\right) \tag{13.7}$$

From Equation (13.4), by assuming $V_x = 0$, the expression for E_f in the sub-V_{off} region can be expressed as

$$E_{f,\text{sub-}V_{\text{off}}} = V_{g0} - \frac{2qDV_{\text{th}}}{C_g} \exp\left(\frac{V_{g0}}{V_{\text{th}}}\right) \tag{13.8}$$

For obtaining a compact model, a single unified expression for E_f is desired, which can be obtained by combining (13.5) and (13.8); we get

$$E_{f,\text{unified}} = V_{g0} - \frac{2V_{\text{th}} \ln\left(1 + \exp\left(V_{g0} / 2V_{\text{th}}\right)\right)}{1 / H\left(V_{g0,p}\right) + \left(C_g / qD\right)\exp\left(-V_{g0} / 2V_{\text{th}}\right)} \tag{13.9}$$

Here, $V_{g0,p}$ is equal to V_{g0} above V_{off} and is on the order of thermal voltage when $V_g < V_{\text{off}}$.

$$E_{f1} = E_{f,\text{unified}} - \frac{p}{q}\left(1 + \left(\frac{pr}{2q^2}\right)\right) \tag{13.10}$$

where

$$p = \frac{C_g}{q} V_{\text{gef}} - \sum_{i=0}^{1} DV_{\text{th}} \ln\left(\xi_i + 1\right) \tag{13.11}$$

$$q = \frac{-C_g}{q} - \sum_{i=0}^{1} \frac{D}{1+\xi_i^{-1}} \left(1+(2/3)\kappa_i V_{\text{gef}}^{-1/3}\right) \tag{13.12}$$

$$\xi_i = \exp\left(\frac{E_{f,\text{unified}} - \kappa_i V_{\text{gef}}^{2/3}}{V_{\text{th}}}\right) \tag{13.13}$$

$$V_{\text{gef}} = V_g - V_{\text{off}} - E_f \tag{13.14}$$

$$r = \sum_{i=0}^{1} \left(\frac{\frac{2}{9} V_{\text{gef}}^{-4/3} D\kappa_i \left(1+\xi_i^{-1}\right) + \frac{D}{V_{th}} \left(1+\frac{2}{3}\kappa_i V_{\text{gef}}^{-1/3}\right)^2}{\left(1+\xi_i^{-1}\right)^2} \right) \tag{13.15}$$

$$\kappa_i = \gamma_i \left(\frac{C_g}{q}\right)^{2/3} \tag{13.16}$$

Surface potential can be computed as

$$\psi(x) = E_{f1} + V_x \tag{13.17}$$

The drift diffusion model-based drain current can be computed as [47]

$$I_{\text{ds}} = \frac{\mu_{\text{eff}} C_g}{\sqrt{1+\theta_{\text{sat}}^2 \psi_{\text{ds}}^2}} \frac{W}{L} \left(V_{g0} - \psi_m + V_{\text{th}}\right)\left(\psi_{\text{ds}}\right)\left(1+\lambda V_{\text{ds}}\right) \tag{13.18}$$

Here, $\Psi_m = (\Psi_d + \Psi_s)/2$ and $\Psi_{ds} = (\Psi_d - \Psi_s)$ with Ψ_d and Ψ_s are drain and source surface potentials, respectively; $\theta_{sat} = \mu_{eff}/V_{sat}L$, V_{sat} is the saturation velocity; and λ is the channel length modulation parameter.

$$\mu_{\text{eff}} = \frac{\mu_0}{1+\text{UA}\left(V_{g0} - \psi_m\right) + \text{UB}\left(V_{g0} - \psi_m\right)^2} \tag{13.19}$$

where μ_0 is the low-field carrier mobility and UA and UB model the vertical field dependence of carrier mobility and are extracted from experimental data.

13.5.2 Current Collapse in AlGaN/GaN HEMTs

Several research groups have demonstrated that AlGaN/GaN HEMTs exhibit a breakdown voltage of several hundred volts with ultra-low dynamic ON resistance, which is very much essential for power electronic applications. Although a lot of progress has been achieved in AlGaN/GaN HEMTs with the innovation in the device structures with the development of the epitaxial growth techniques of GaN-based materials, there are still several challenging issues—such as low threshold voltage (V_{th}) in normally off devices, off-state high stress-induced current collapse, and self-heating effects—that need to be addressed [48,49]. Out of all of these, current collapse is the most important issue caused by injected carriers that originate from carriers supplied by the gate leakage current at off-state captured by the traps at the AlGaN barrier surface, AlGaN/GaN heterointerface, and GaN substrate. In recent years, the trap density in the 2DEG at the heterointerface has been greatly reduced with the improvement of quality on the epitaxial growth in wafers. Also, the GaN substrate traps are far from the channel to exert any significant influence, although their density is relatively large. Therefore, AlGaN surface traps act as the main source of influence on the 2DEG conductivity and have an important impact on the current collapse phenomenon in AlGaN/GaN power HEMT devices.

Current collapse is mainly due to the charge trapping at defects in the AlGaN/GaN layers and the passivation interface. When the applied voltage is very high, the channel electrons are accelerated by the electric field, and some of the accelerated electrons are trapped in the device. These trapped electrons deplete the 2DEG channel even after turn-ON state, and the dynamic ON-resistance is increased with the applied voltage. Therefore, the electric field affects the current collapse phenomena, and the field plate (FP) structure becomes a solution for suppressing the collapse phenomena due to the relaxation of electric-field concentration. The dynamic ON-resistance increase due to the current collapse phenomena can be greatly reduced by the single-gate FP and dual-FP structures compared with the source FP structure, because the gate edge electric field was reduced by the gate FP electrode. The dual-FP structure was found to be slightly more effective to suppress the current collapse phenomena than the single-gate FP structure, because the two-step FP structure relaxes the electric-field concentration at the FP edge. AlGaN/GaN HEMTs using FP techniques are shown in Figure 13.13.

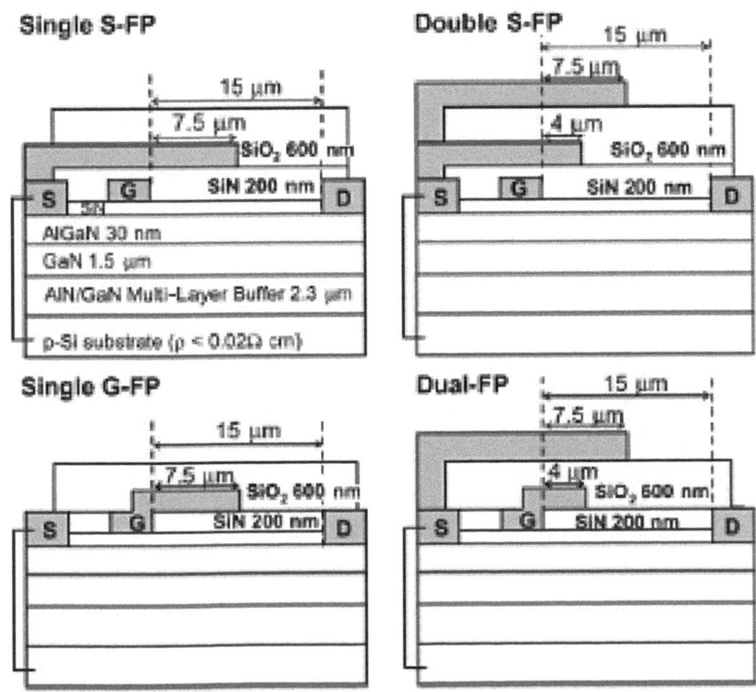

FIGURE 13.13
A1GaN/GaNHEMT structures with field plate techniques. (From Saito, W., *IEEE Electron Device Lett.*, 31, 659, 2010.)

13.6 Double-Gate HEMT

Rapid advancement in monolithic microwave integrated circuit (MMIC) technology and solid-state semiconductor technology has given rise to the need for extremely high cut-off frequencies (f_T) and high maximum oscillation frequency (fmax) for various applications. An fmax of above 1 THz with maximum unilateral gain available at 1.2 THz and maximum stable gain (MSG) at 1.1 THz has already been reported for fabricated sub-50 nm InAlAs/InAlAs/InP HEMT [48]. In order to further improve the frequency and noise performance for future high-speed low noise applications, transistors need to be scaled down to below 22 nm gate length regime. However, the performance of these extremely small gate length devices will be seriously affected by several short channel effects (SCEs) such as reduced transconductance, increased subthreshold swing, and the shift in threshold voltage toward being more negative. These SCEs can be minimized by making the lateral field much greater than the field in the longitudinal or the channel direction, which is accomplished by increasing the donor layer doping concentration and reducing the donor layer thickness. However, mobility degradation, velocity overshoot effect, and gate leakage current impose a fundamental limit on the increase in the doping concentration and reduction in the donor layer thickness in a SG-HEMT. Further miniaturization for ultra-high frequency and low-noise applications requires modification in the conventional HEMT structure. Thus, the need for the evolution of the standard HEMT design has led to the development of a DG-HEMT structure based on a transferred substrate technique in which two gates are placed on each side of the conducting InGaAs channel. The schematic of InAlAs/InGaAs DG-HEMT is shown in Figure 13.14.

InGaAs channel-based HEMTs on InP substrate have demonstrated state-of-the art performance for ultra-high frequency applications with a high fmax of 1 THz reported for a sub-50-nm InAlAs/InGaAs/InP HEMT [12,48]. However, the conventional single-gate (SG) HEMTs have already reached their limit of scaling. Therefore, in order to achieve further performance enhancement required for future low-noise millimeter-wave and

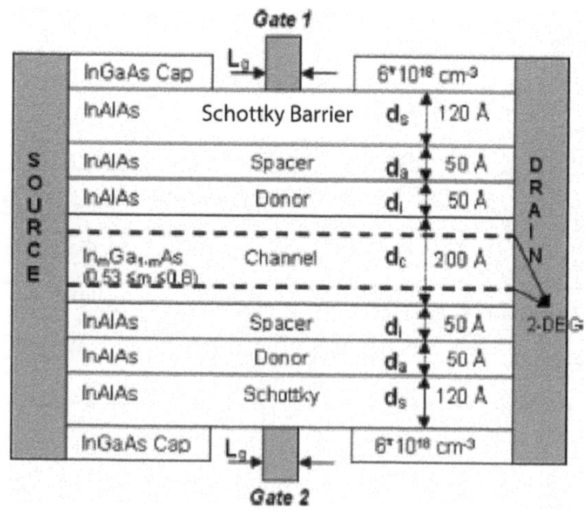

FIGURE 13.14
Structure of InAlAs/InGaAs DG-HEMT. (From Wichmann, N., 100 nm InAlAs/InGaAs double-gate HEMT using transferred substrate, in *IEDM Technical Digest*, pp. 1023–1026, December 2004.)

sub-millimeter wave applications, alternative device architectures are being explored. InAlAs/InGaAs Double-gate (DG) HEMTs with sub-100 nm gate length have emerged as the most promising candidate for the future low-noise microwave applications with improved scalability [49–51]. Sub-100-nm-gate-length InGaAs channel DG-HEMT fabricated through the transferred substrate technique has demonstrated superior high frequency and low-noise performance as compared to its SG counterpart [52–54].

13.6.1 Impact of Temperature and Indium Concentration on the Sheet Charge Density of Symmetric InAlAs/InGaAs DG-HEMT

The variation of sheet charge density (n_s) of a 2DEG formed at the heterointerface with the applied gate-to-source voltage V_{gs} is expressed as [56]

$$n_s = \frac{\varepsilon_m}{qd}\left(V_{gs} - V_{th} - E_f\right) \tag{13.20}$$

where q is the electron charge and $d = d_s + d_a + d_i$ is the total InAlAs layer thickness, ϵ_m is the indium mole fraction (m) dependent permittivity of $In_mGa_{1-m}As$, which is given by [57]

$$\varepsilon_m = \left(12.9 - 1.64m\right)\varepsilon_0 \tag{13.21}$$

Here, ϵ_0 is the permittivity of free space.

The variation of Fermi potential (E_f) with n_s is given by [58]

$$E_f = k_1 + k_2\sqrt{n_s} + k_3 n_s \tag{13.22}$$

where k_1, k_2, and k_3 are temperature-dependent constants and their values are given in [55].

The threshold voltage of the device (V_{th}) is given by [55]

$$V_{th} = \left(\phi_b - \Delta E_C - \frac{qN_d d_d^2}{2\varepsilon_d}\right) \tag{13.23}$$

where $d_d = d_s + d_a$, ϕ_b is the Schottky barrier height, ΔE_C is the temperature (T)-dependent conduction band discontinuity at the $In_{0.52}Al_{0.48}As/In_mGa_{1-m}As$ heterointerface, ϵ_d is the permittivity of the $In_{0.52}Al_{0.48}As$ layer, and N_d is the donor layer doping concentration. The dependence of temperature (T) and channel indium mole fraction (m) on ΔE_C at the $In_{0.52}Al_{0.48}As/In_mGa_{1-m}As$ heterojunction can be expressed as [59–63]

$$\Delta E_C(m,T) = 0.73\left(E_g^{In_{0.52}Al_{0.48}As}(T) - E_g^{In_mGa_{1-m}As}(m,T)\right) \tag{13.24}$$

$$E_g^{In_{0.52}Al_{0.48}As}(T) = 0.52E_g^{InAs}(T) + 0.48E_g^{AlAs}(T) - 0.17472 \tag{13.25}$$

$$E_g^{In_mGa_{1-m}As}(T) = mE_g^{InAs}(T) + (1-m)E_g^{GaAs}(T) - 0.477m(1-m) \tag{13.26}$$

where the temperature-dependent bandgap of semiconductors can be computed as [61]

$$E_g^{\text{InAs}}(T) = 0.417 - 0.276\times10^{-3}\left(\frac{T^2}{T+93}\right) \tag{13.27}$$

$$E_g^{\text{AlAs}}(T) = 3.099 - 0.885\times10^{-3}\left(\frac{T^2}{T+530}\right) \tag{13.28}$$

$$E_g^{\text{GaAs}}(T) = 1.519 - 0.5405\times10^{-3}\left(\frac{T^2}{T+204}\right) \tag{13.29}$$

The variation of ϕ_b with T and bandgap of semiconductors can be computed as [57,61]

$$\phi_b(T) = 0.56\text{eV} + 0.5\left(E_g^{\text{In}_{0.52}\text{Ga}_{0.48}\text{As}}(at\ 300K) - E_g^{\text{In}_{0.52}\text{Ga}_{0.48}\text{As}}(T)\right) \tag{13.30}$$

The dependence of temperature (T) and channel indium composition (m) and position in the channel (x) on the sheet carrier concentration (n_s) in a 2-DEG is given by [55]

$$n_s(x,m,T) = \frac{\left(\sqrt{k_2(T)^2 + 4k_4(T)\left|V_{\text{off}}(m,T)\right|w(T)} - k_2(T)\right)^2}{4k_4(T)^2} \tag{13.31}$$

where

$$V_{\text{off}}(m,T) = V_{\text{th}}(m,T) + k_1(T) \tag{13.32}$$

$$k_4(T) = k_3(m,T) + \frac{qd}{\varepsilon(m,T)} \tag{13.33}$$

$$w(x,m,T) = \frac{V_{\text{gs}} - V_{\text{off}}(m,T) - V(x)}{\left|V_{\text{off}}(m,T)\right|} \tag{13.34}$$

$V(x)$ is the potential at any point x in the channel. Figure 13.15 reveals the effect of temperature (T) and indium mole fraction (m) on the performance of InAlAs/InGaAs SG-HEMT/DG-HEMTs. The figure clearly shows that the drain current of both SG-HEMT and DG-HEMT decreases with the increase in temperature, and this is mainly due to the decrease in the carrier mobility and saturation velocity with the increase in temperature. Also, both the HEMTs with higher indium mole fraction exhibit higher drain current. For achieving high cut-off frequency, the HEMT structure must be designed with reduced parasitics and low gate lengths. The parasitics include gate-to-source capacitance (C_{GS}), drain-to-source capacitance (C_{DS}), source resistance, and drain resistance.

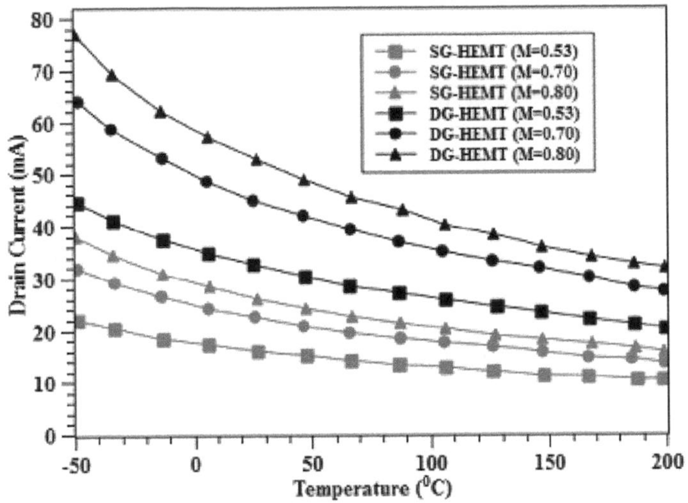

FIGURE 13.15
Effect of temperature (T) and indium mole fraction (m) on the performance of InA1As/InGaAs SG-HEMT/DG-HEMTs. (From Bhattacharya, M., *IEEE Trans. Nanotechnol.*, 12, 965–970, 2013.)

13.7 Small-Signal Equivalent Circuit of AlGaN/GaN HEMT

The complete small-signal equivalent circuit for an AlGaN/GaN HEMT including intrinsic and extrinsic parasitics is shown in Figure 13.16. The extrinsic parasitics include gate resistance R_g, drain resistance R_d, source resistance R_s, gate inductance L_g, source inductance L_s, drain inductance L_d, gate pad capacitance C_{pg}, and drain pad capacitance C_{pd}.

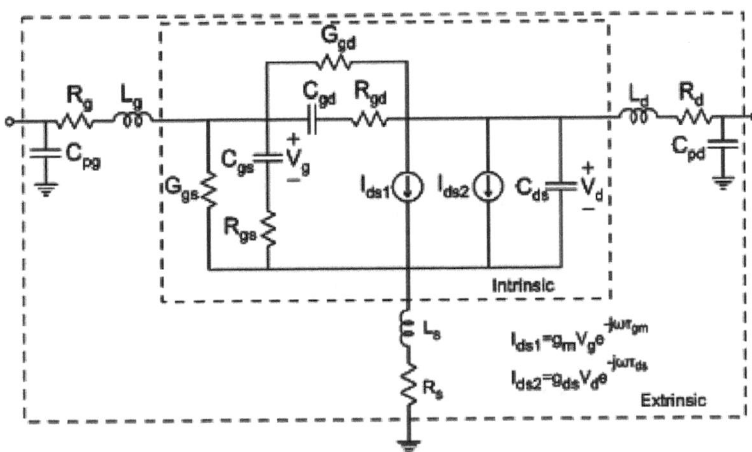

FIGURE 13.16
Small-signal equivalent circuit of A1GaN/GaN HEMT. (From Brady, R. G., *IEEE Trans. Microw. Theory Tech.*, 56, 1535, 2008.)

13.7.1 HEMT Parameters

The important geometrical parameter of the HEMT is the gate length L_g, gate width W_g, thickness of active layer, gate-to-channel distance, gate-to-source length (L_{gs}), and gate-to-drain (L_{gd}) terminal spacing. The geometrical dimension L_g is critical in determining the maximal frequency limits for the device. The drain current flowing through the device is directly proportional to the gate width W_g. Therefore, for low-noise, low-power applications, relatively small gate width devices are utilized, in contrast to large gate width devices used in power electronic applications. The electrical properties of the HEMTs are characterized by the following parameters: drain-to-source saturation current, I_{DSAT}, transconductance g_m, output conductance g_d, current gain cut-off frequency f_T, and maximum frequency of oscillation f_{max}. The DC behavior of HEMTs is characterized by output current I_{DSAT}, which is a function of L_g, W_g, drain-to-source current (V_{DS}), and gate-to-source current (V_{GS}). Transconductance of the HEMTs is one of the most important indicators of device quality for millimeter and microwave applications and is defined as

$$g_m = \frac{\partial I_D}{\partial V_{GS}}\bigg|_{V_{DS}=\text{Constant}}$$

The output conductance (g_d) is also a very important parameter in analog applications, and it plays a major role in determining optimum matching properties. The output conductance (g_d) is defined as

$$g_d = \frac{\partial I_D}{\partial V_{DS}}\bigg|_{V_{GS}=\text{Constant}}$$

The current gain (h_{21}) can be expressed as

$$h_{21} = \frac{\partial I_{DS}}{\partial I_{GS}}$$

The current gain cut-off frequency (f_T) of the HEMTs is defined as

$$f_T = \frac{g_m}{2\pi\tau}$$

where τ is the total delay encountered by the electron while traveling from source to drain, which is determined by internal and external parasitics of the device.

13.8 Applications of HEMT

The high-speed, low-DC-power consumption and low-noise characteristics of InP HEMTs make them an excellent choice for space-based millimeter-wave receivers. InP HEMTs are also used as low-noise amplifier (LNA) monolithic microwave integrated circuits (MMICs) and efficient driver amplifiers.

The high-performance enhancement mode MHEMTs are promising candidates for sub-millimeter wave applications such as high-resolution radars for space research, remote atmospheric sensing, imaging systems, and low-noise wide-bandwidth amplifiers for future communication systems.

AlGaN/GaN HEMTs have the excellent potential for high-speed, high-power microwave applications such as microwave amplifiers, mixers, oscillators, and other nonlinear circuits due to their excellent characteristics such as high breakdown voltage, high electron charge density, and high electron mobility.

References

1. T. Mimura, "A new field effect transistor with selectively doped GaAs/n-$Al_xGa_{1-x}As$ hetero junctions," *Jpn. J. Appl. Phys.*, vol. 19, no. 5, p. L225, 1980.
2. R. Dimgle, "Electron mobilities in modulation doped semiconductor hetero junction super lattice," *Appl. Phys. Lett.*, vol. 33, no. 7, p. 665, 1978.
3. A. Ketterson et al., "High transconductance InGaAs/GaAs pseu-domorphic MODFET," *IEEE Electron Dev. Lett.*, vol. 6, p. 628, 1985.
4. P. C. Chao et al., "High performance 0.1 µm gate length planar doped HEMTs," *IEDM*, 1987, p. 410.
5. J. A. del Alamo, "The high electron mobility transistor at 30: Impressive accomplishments and exciting prospects," *2011 International Conference on Compound Semiconductor Manufacturing Technology*, May 16–19, 2011, Indian Wells, CA.
6. A. Endoh, "High f_t 50nm-gate lattice-matched InAlAs/InGaAs HEMTs," in *Proc. 12th Int. Conf. InP Rel. Mat.*, 2000, pp. 87–90.
7. Y. Yamashita, "High f_t 50-nm-gate InAlAs/InGaAs high electron mobility transistors latticed-matched to InP substrates," *Jpn. J. Appl. Phys.*, vol. 39, pp. L838–L840, 2000.
8. T. Suemitsu, "30-nm-gate InP-based lattice-matched high electron mobility transistors with 350 GHz cutoff frequency," *Jpn. J. Appl. Phys.*, vol. 38, pp. L154–L156, 1999.
9. Y. Yamashita, "Ultra-short 25-nm-gatelattice-matched InAlAs/InGaAs HEMTs within the range of 400 GHz cutoff frequency," *IEEE Electron Device Lett.*, vol. 22, pp. 367–369, 2001.
10. L. D. Nguyen, "50-nm self-aligned-gate psedomorphic AlInAs/GaInAs high electron mobility transistors," *IEEE Trans. Electron Devices*, vol. 39, pp. 2007–2014, 1992.
11. K. Shinohara, "Ultrahigh-speed pseudomorphic InGaAs/InAlAs HEMTs with 400-GHz cut-off frequency," *IEEE Electron Device Lett.*, vol. 22, pp. 507–509, 2001.
12. Y. Yamashita, "Pseudomorphic $In_{0:52}Al_{0:48}As/In_{0:7}Ga_{0:3}As$ HEMTs with an ultrahigh f_T of 562GHz," *IEEE Trans. Electron Devices*, vol. 23, no. 10, pp. 573–575, 2002.
13. D. H. Kim, "30-nm InAs pseudomorphic HEMTs on an InP substrate with a current-gain cut-off frequency of 628 GHz," *IEEE Electron Device Lett.*, vol. 29, no. 8, pp. 830–833, 2008.
14. D. H. Kim, "30-nm InAs PHEMTs with f_T = 644 GHz and fmax = 681 GHz," *IEEE Electron Device Lett.*, vol. 31, no. 8, p. 806, 2010.
15. C.-J. Hwang, "An ultra low power MMIC amplifier using 50nm δ-doped $In_{0.52}Al_{0.48}As/In_{0.53}Ga_{0.48}As$ metamorphic HEMT," *IEEE Electron Device Lett.*, vol. 31, no. 11, p. 1230, 2010.
16. W. Ha, "Enhancement mode metamorphic HEMT on GaAs substrate with 2 S/mm g_m and 490 GHz f_T," *IEEE Electron Device Lett.*, vol. 29, no. 5, pp. 419–421, 2008.
17. K.-S. Lee, "35 nm zigzag T-gate $In_{0.52}Al_{0.48}As/In_{0.53}Ga_{0.47}As$ metamorphic GaAs HEMTs with an ultra high f_{max} of 520 GHz," *IEEE Electron Device Lett.*, vol. 28, no. 8, pp. 672–675, 2007.
18. K. Elgaid, "50-nm T-gate metamorphic GaAs HEMTs with f_T of 440 GHz and noise figure of 0.7 dB at26 GHz," *IEEE Electron Device Lett.*, vol. 26, no. 11, pp. 784–786, 2005.
19. C.-I. Kuo, "RF performance improvement of metamorphic high electron mobility transistor using $(In_xGa_{1-x}As)m/(InAs)n$ superlattice-channel structure for millimeter-wave applications," *IEEE Electron Device Lett.*, vol. 31, no. 7, p. 677, 2010.

20. K. C. Sahoo, "Novel metamorphic HEMT with highly doped InGaAs source/drain regions for high frequency applications," *IEEE Trans. Electron Devices*, vol. 57, no. 10, p. 2594, 2010.
21. H. Maher, "A 200GHz true E-mode low noise MHEMT," *IEEE Trans. Electron Devices*, vol. 54, no. 7, p. 1626, 2007.
22. H. Li, "Metamorphic AlInAs/GaInAs HEMTs on GaAs substrates by MOCVD," *IEEE Electron Device Lett.*, vol. 29, no. 6, p. 561, 2008.
23. A. Leuther, "20 nm metamorphic HEMT technology for terahertz monolithic integrated circuits," *Proceedings of the 9th European Microwave Integrated Circuits Conference*, p. 84, October 2014.
24. M. Boudrissa, "Enhancement mode $Al_{0.66}In_{0.34}As$/ $Ga_{0.67}In_{0.33}As$ metamorphic HEMT: Modelling and measurements," *IEEE Trans. Electron Devices*, vol. 48, no. 6, p. 1037, June 2001.
25. C. C. Huang, "Comprehensive temperature dependent studies of metamorphic high electron mobility transistor with double and single δ-doped structures," *IEEE Trans. Electron Devices*, vol. 58, no. 12, p. 4276, 2011.
26. W.-C. Hsu, "Performance improvement in tensile-strained $In_{0.5}Al_{0.5}As/In_xGa_{1-x}As/In_{0.5}Al_{0.5}As$ metamorphic HEMT," *IEEE Trans. Electron Devices*, vol. 53, no. 3, pp. 406–412, 2006.
27. D. H. Kim, "f_T = 688GHz and f_{max} = 800GHz in Lg = 40nm $In_{0.7}Ga_{0.3}As$ MHEMTs with g_m_max>2.7mS/μm," in *Proceedings of IEDM Technical Digest*, p. 319, 2011.
28. W. Li-Dan, "100-nm T-gate InAlAs/InGaAs InP-based HEMTs with f_T = 249 GHz and f_{max} = 415 GHz," *Chin. Phys. B*, vol. 23, no. 3, p. 038501, 2014.
29. M. Akita, "High-frequency measurements of AlGaN/GaN HEMTs at high temperatures," *IEEE Electron Device Lett.*, vol. 22, no. 8, pp. 376–377, 2001.
30. Y.-S. Lee, "A high-efficiency class-E GaN HEMT power amplifier for WCDMA applications," *IEEE Microw. Wireless Compon. Lett.*, vol. 17, no. 8, pp. 622–624, 2007.
31. U. K. Mishra, "GaN-based RF power devices and amplifiers," *Proc. IEEE*, vol. 96, no. 2, pp. 287–305, 2008.
32. Y. Okamoto, "100W C-band single chip GaN FET power amplifier," *Electron. Lett.*, vol. 42, no. 5, pp. 283–285, 2006.
33. K. Shinohara, "Scaling of GaN HEMTs and Schottky diodes for submillimeter-wave MMIC applications," *IEEE Trans. Electron Devices*, vol. 60, no. 10, p. 2982, 2013.
34. D. C. Dumka, "AlGaN/GaN HEMTs on diamond substrate with over 7 W/mm output power density at 10 GHz," *Electron Lett.*, vol. 49, no. 20, p. 1298, 2013.
35. R. T. Kemberley, "Impact of wide bandgap microwave devices on DoD systems," *Proc. IEEE*, vol. 90, no. 6, pp. 1059–1064, 2002.
36. L. M. Tolbert, "Wide band-gap semiconductors for utility applications," in *Proceedings of the Power Energy System*, February 2003, pp. 317–321.
37. R. J. Trew, "SiC and GaN transistors—Is there one winner for microwave power applications?" *Proc. IEEE*, vol. 90, no. 6, pp. 1032–1047, 2002.
38. M. Nakajima, "Low-frequency noise characteristics in ion-implanted GaN-based HEMTs," *IEEE Electron Device Lett.*, vol. 29, no. 8, p. 827, 2008.
39. M. Asif Khan et al., "Observation of a two-dimensional electron gas in low pressure metal-organic chemical vapor deposited GaN-$Al_xGa_{1-x}N$ heterojunctions," *Appl. Phys. Lett.*, vol. 60, p. 3027, 1992.
40. M. Asif Khan et al., "High electron mobility transistor based on a GaN-$Al_xGa_{1-x}N$ heterojunction," *Appl. Phys. Lett.*, vol. 63, p. 1214, 1993.
41. J. P. Ibbetson, "Polarization effects, surface states, and the source of electrons in AlGaN/GaN heterostructure field effect transistors," *Appl. Phys. Lett.*, vol. 77, no. 2, pp. 250–252, 2000.
42. W. B. Lanford, "Recessed-gate enhancement-mode GaN HEMT with high threshold voltage," *Electron. Lett.*, vol. 41, no. 7, pp. 449–450, 2005.
43. S. Khandewal, "Analytical modeling of surface-potential and intrinsic charges in AlGaN/GaN HEMT devices," *IEEE Trans. Electron Devices*, vol. 59, no. 10, pp. 2856–2860, 2012.
44. S. Kola, J. M. Golio, and G. N. Maracas, "An analytical expression for Fermi Level versus sheet carrier concentration for HEMT modeling," *IEEE Electron Device Lett.*, vol. 9, no. 3, pp. 136–138, 1988.

45. H. K. Kwon, "Radiative recombination of two dimensional electrons in a modulation doped AlGaN/GaN single heterostructure," *Appl. Phys. Lett.*, vol. 75, no. 18, pp. 2788–2790, 1999.
46. S. Khandelwal, "A physics based analytical model for 2DEG charge density in AlGaN/GaN HEMT devices," *IEEE Trans. Electron Devices*, vol. 58, no. 10, pp. 3622–3625, 2011.
47. S. Khandelwal, "A surface-potential-based drain current model for study of non-linearities in AlGaAs/GaAs HEMTs," in *Proceedings of the Compound Semiconductor IC Symposium*, October 2012, pp. 1–4.
48. W. Saito, "Field-plate structure dependence of current collapse phenomena in high-voltage GaN-HEMTs," *IEEE Electron Device Lett.*, vol. 31, no. 7, p. 659, 2010.
49. H. Huang et al., "Effects of gate field plates on the surface state related current collapse in AlGaN/GaN HEMTs," *IEEE Trans. Power Electron.*, vol. 29, no. 5, p. 2164, 2014.
50. R. Lai, "Sub 50 nm InP HEMT device with fmax greater than 1 THz," *IEDM*, 2007, pp. 609–611.
51. B. G. Vasallo, "Comparison between the noise performance of double-and single-gate InP based HEMTs," *IEEE Trans. Electron Devices*, vol. 55, no. 6, pp. 1535–1540, 2008.
52. B. G. Vasallo, "Comparison between the dynamic performance of double-and single-gate InP based HEMTs," *IEEE Trans. Electron Devices*, vol. 54, no. 11, pp. 2815–2822, 2007.
53. N. Wichmann, "InGaAs/InAlAs double-gate HEMTs on transferred substrate," *IEEE Electron Device Lett.*, vol. 25, no. 6, pp. 354–356, 2004.
54. B. G. Vasallo, "Monte Carlo comparison between InP-based double-gate and standard HEMTs," in *Proceedings of the 1st European Microwave Integrated Circuits Conference*, September 2006, pp. 304–307.
55. N. Wichmann, "100 nm InAlAs/InGaAs double-gate HEMT using transferred substrate," in *IEDM Technical Digest*, December 2004, pp. 1023–1026.
56. N. Wichmann, "Double gate HEMTs on transferred substrate," in *Proceedings of the International Conference on Indium Phosphide and Related Materials*, 2003, pp. 118–121.
57. M. Bhattacharya, "Impact of temperature and indium composition in the channel on the microwave performance of single-gate and double-gate InAlAs/InGaAs HEMT," *IEEE Trans. Nanotechnol.*, vol. 12, no. 6, pp. 965–970, 2013.
58. D. Delagebeaudeuf, "Metal-(n) AlGaAs-GaAs two dimensional electron gas FET," *IEEE Trans. Electron Devices*, vol. ED-29, no. 6, pp. 955–960, 1982.
59. M. Shur, *Physics of Semiconductor Devices*. Englewood Cliffs, NJ: Prentice-Hall, 1990.
60. N. Dasgupta, "An analytical expression for sheet-carrier concentration vs gate-voltage for HEMT modeling," *Solid-State Electron.*, vol. 36, no. 2, pp. 201–203, 1993.
61. I. Vurgaftman, "Band parameters for III-V compound semiconductors and their alloys," *Appl. Phys. Rev.*, vol. 89, no. 11, pp. 5815–5864, 2001.
62. J.-H. Huang, "Flat band current voltage temperature method for band-discontinuity determination and its application to strained $\mathrm{In}x\mathrm{Ga}_{1-x}\mathrm{As}/\mathrm{In}_{0.52}\mathrm{Al}_{0.48}\mathrm{As}$ hetero-structures," *J. Appl. Phys.*, vol. 76, no. 5, pp. 2893–2903, 1994.
63. Y. P. Varshni, "Temperature dependence of the energy gap in semiconductors," *Physica*, vol. 34, pp. 149–154, 1967.
64. R. G. Brady, "An improved small-signal parameter-extraction algorithm for GaN HEMT devices," *IEEE Trans. Microw. Theory Tech.*, vol. 56, no. 7, p. 1535, 2008.

Section VI

Emerging Nonvolatile Memory Devices and Applications

14

Spintronic-Based Memory and Logic Devices

Jyotirmoy Chatterjee, Pankaj Sethi, and Chandrasekhar Murapaka

CONTENTS

14.1 Introduction

Conventional microelectronic devices operate based on the displacement of charge in semiconductor materials. Besides charge, the electron possesses another degree of freedom known as "spin," which is associated with magnetic moment. Unlike the electron charge, spin can have two directions. It can take one of the two states relative to the magnetic field, generally known as "up" and "down." These two directions can be assigned to two binary data bits: "1" and "0." Spin-based devices, principally known as "spintronic" devices, have remarkable advantages over the existing technology. The spin switching is much faster than the charge transfer. The lower power consumption and high endurance make them attractive for future technological applications. Above all, the spin-based devices are fascinating as they exhibit nonvolatile data retention even after the power is removed. The individual discovery of giant magnetoresistance (GMR) in 1988 by Albert Fert and Peter Grünberg was a major breakthrough [1,2] for the research work in the spintronics field. In 2007, they were awarded Nobel Prize in Physics for the recognition of the impact of their discovery in science and technology. Soon after the discovery of GMR, low field GMR sensors called spin valves that were suitable for hard disk drive (HDD) read heads were invented in 1991 [3]. They were introduced in the read head in 1998 by IBM and were the first GMR-based product. This is the most successful spintronic device to date that has been used in the HDD industry. Tunneling magnetoresistance (TMR) was discovered more than a decade before the GMR discovery by Jullière in 1975, using a $Fe/GeO_x/Co$ tunnel junction [3]. However, the TMR value obtained was 14% at 4.2K. Two decades later, room temperature TMR of 11.8% was demonstrated by Moodera in 1995 by using $CoFe/Al_2O_3/Co$ magnetic tunnel junction (MTJ) [4]. In 2001, Butler demonstrated giant TMR in Fe/MgO/Fe MTJ [5]. He predicted a symmetry-based spin-filtering effect and 1,000% TMR for this trilayer system. Soon after, Yuasa and Parkin experimentally demonstrated giant TMR (200%) using crystalline MTJ of Fe/MgO/Fe and CoFe/MgO/CoFe [6,7]. These key achievements in the spintronics research field revolutionized the HDD industry with very high densities. The discovery of spin-transfer torque (STT) [4,5] opened the avenue for the electrical current manipulation of spin rather than magnetic field. The spintronics community reenergized from the demonstration of STT. One of the key memory devices proposed based on the STT phenomenon is spin-transfer torque magnetic random access memory (STT-MRAM). This is the only nonvolatile memory demonstrated to have semi-infinite endurance versus any other memory technology proposed so far [8]. Longer endurance and a smaller footprint (can be scaled down below 20 nm) make STT-MRAM a potential candidate as an embedded and stand-alone memory. However, for ultrafast memory application to replace SRAM, two-terminal STT-MRAM fails due to the risk of dielectric breakdown as relatively larger write-current are required for switching. More recently, spin-orbit torque-induced ultrafast magnetization manipulation inspired to develop an alternative memory device, named as spin-orbit torque MRAM (SOT-MRAM) [9–11]. These memory cells have similar properties to STT-MRAM with added advantages of reliability due to decoupled read and write paths. In addition, IBM proposed another alternative memory technology, named as racetrack memory, which is associated with spin-torque driven domain wall (DW) motion in ferromagnetic nanowires [12]. This chapter is dedicated to the fundamental understanding and operating principles of the aforementioned spintronic memory and logic devices.

14.1.1 Gaint Magneto Resistance

The GMR or TMR ratio is defined as R_{AP}-R_P/R_P, where R_{AP} and R_P are the resistances of "ferromagnetic/metallic or insulator/ferromagnetic" trilayers when the magnetizations are in antiparallel and parallel orientations, respectively. The underlying physics of GMR and TMR effects are different from each other. In the case of GMR, the spin-dependent scattering of electrons results in different resistances for parallel and antiparallel magnetization orientations of the two ferromagnetic layers, which is explained by Mott's two-current model [13]. According to this model, in the low-temperature limit, one can consider that conduction of spin-up (↑) and spin- down (↓) electrons takes place in two independent parallel channels. Figure 14.1 shows current-perpendicular-to-plane (CPP) geometry of ferromagnetic metal (Co)/normal metal (Cu)/ferromagnetic metal (Co) multilayer for GMR. For Co, majority electron, spin up (↑) has low density of states at Fermi energy compared to minority, spin- down (↓) electrons. Actually, the Fermi energy (E_F) lies above the $3d_{\uparrow}$ sub-band. It is well known from Fermi's Golden Rule that the scattering probability of electrons is proportional to the density of states. When the magnetizations of both the Co layers are in parallel orientation, the minority electrons undergo relatively high scattering compared to majority electrons while traveling through the multilayers under an applied bias. If we denote resistance of the ferromagnetic layer for majority electrons as r and minority electrons as R, then $r < R$ and the total resistance for parallel alignment is $R_P = 2Rr/(R + r)$. In the case of antiparallel alignment, spin up (↑) becomes minority and spin down (↓) becomes majority in the second Co electrode. Therefore, spin-up (↑) electrons are weakly scattered as majority electrons in the first Co layer but strongly scattered as minority electrons in the second Co layer. Similarly, spin- down (↓) electrons are strongly scattered as minority electrons in the first Co layer but weakly scattered as majority electrons in the second Co layer. Hence, the total resistance becomes, $R_{AP} = (R + r)/2$. Note that in the calculation of total resistance, the resistance of the Cu layer has been neglected for simplification. Then, GMR can be simply expressed as:

$$\text{GMR} = \frac{R_{AP} - R_P}{R_P} = \frac{(R-r)^2}{4Rr} \tag{14.1}$$

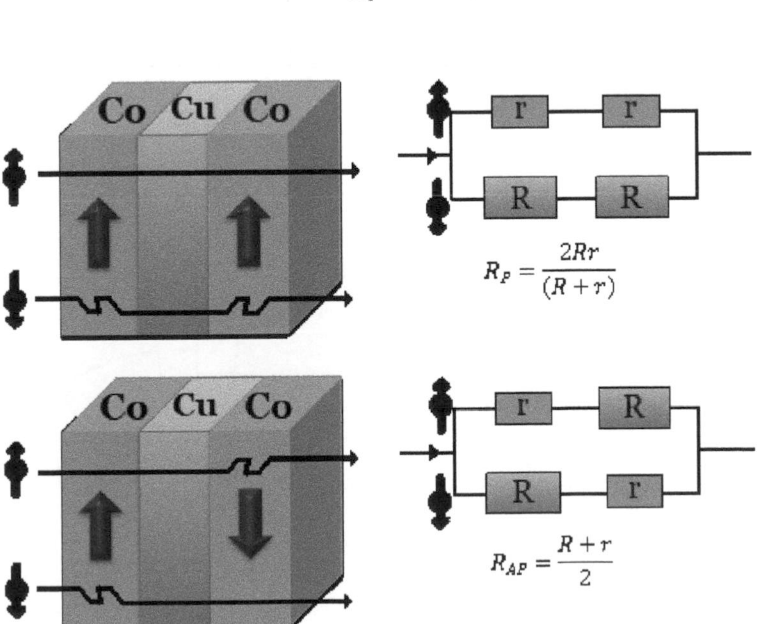

FIGURE 14.1
Illustration of GMR using two-current model. This explains the conduction of spin-up and spin-down electrons in Co/Cu/Co multilayer for CPP geometry.

14.1.2 TMR

Tunneling magnetoresistance ratio (TMR) is a parameter to measure transport property of magnetic tunnel junctions (MTJs) with ferromagnet/insulator/ferromagnet trilayers. This configuration is similar to CPP-GMR geometry except an insulating oxide barrier replaces the nonmagnetic spacer layer. Therefore, the transport mechanism across MTJ is a ballistic effect, governed by a quantum mechanical tunneling effect across the potential barrier created by the oxide barrier. Initially, a simple theoretical model was proposed by Jullière to explain the TMR effect [3].

14.1.2.1 Jullière Model

The schematics showing two ferromagnets and an oxide barrier are illustrated in Figure 14.2 to describe the Jullière model. Under an applied bias, there is a net electron tunneling current from ferromagnetic electrode 1 to 2. This tunneling current is

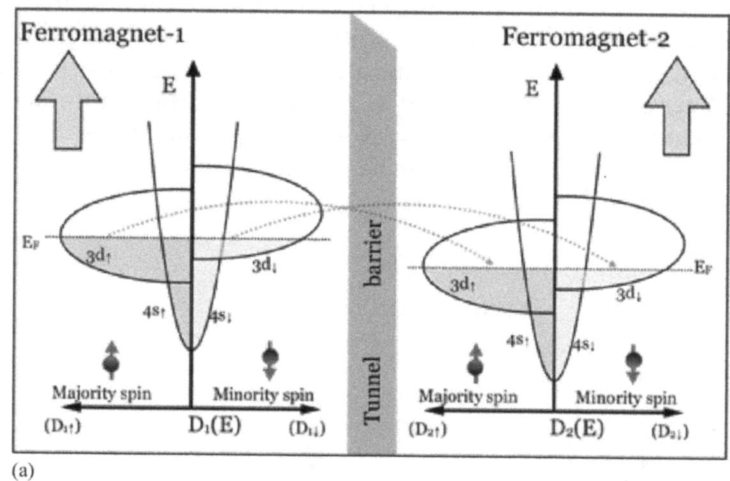

(a)

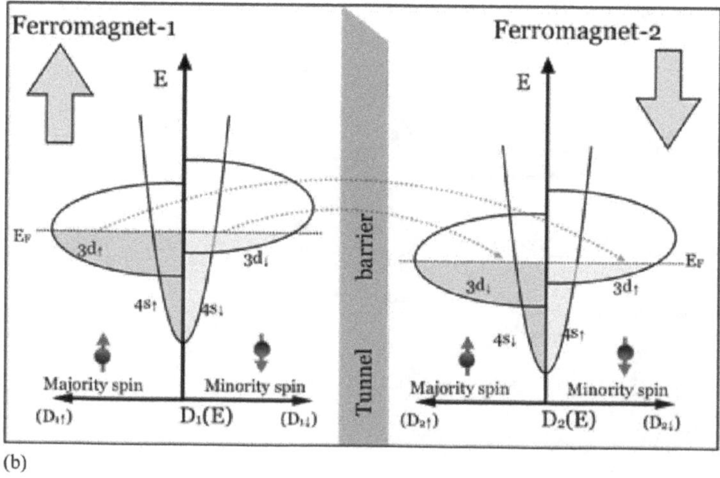

(b)

FIGURE 14.2
Schematic illustration of spin-dependent electron tunneling process across the ferromagnet/tunnel barrier/ferromagnet tunnel junction for (a) parallel and (b) antiparallel magnetization orientation of the ferromagnets.

proportional to the number of occupied states in ferromagnet-1 and number of unoccupied states in ferromagnet-2. Under low-temperature limit and low applied bias, this current is proportional to the product of density of states at the Fermi energy [$D_1(E_F)$*$D_2(E_F)$]. When the magnetization orientations of both ferromagnetic electrodes are parallel, the majority electrons (spin-up, ↑, parallel to magnetization direction) from ferromagnet-1 tunnel to the unoccupied majority spin states at Fermi level of ferromagnet-2.

Similarly, the minority electrons (spin-down, ↓, antiparallel to magnetization direction) from ferromagnet-1 also tunnel to the unoccupied minority states at Fermi energy of ferromagnet-2. Therefore, the conductance (G_P) in the case of parallel magnetization orientation can be written as

$$G_P \propto D_{1\uparrow}(E_F)D_{2\uparrow}(E_F)+D_{1\downarrow}(E_F)D_{2\downarrow}(E_F) \tag{14.2}$$

In contrast, for antiparallel magnetization orientation of the electrodes, the majority and minority electrons from ferromagnet-1 tunnel to the unoccupied minority and majority states at Fermi energy of ferromagnet-1, respectively. Therefore, the conductance (G_{AP}) in the case of antiparallel magnetization orientation can be written as

$$G_{AP} \propto D_{1\uparrow}(E_F)D_{2\downarrow}(E_F)+D_{1\downarrow}(E_F)D_{2\uparrow}(E_F) \tag{14.3}$$

Let's define a parameter a_i as a fraction of majority electrons in the i-th electrode, where i is 1 or 2 electrode. Then, a_i can be written as

$$a_i = \frac{n_{i\uparrow}(E_F)}{n_{i\uparrow}(E_F)+n_{i\downarrow}(E_F)} = \frac{D_{i\uparrow}(E_F)}{D_{i\uparrow}(E_F)+D_{i\downarrow}(E_F)} \tag{14.4}$$

Therefore, the spin polarization of ferromagnetic electrodes P_1 and P_2 are expressed by

$$P_1 = \frac{D_{1\uparrow}(E_F)-D_{1\downarrow}(E_F)}{D_{1\uparrow}(E_F)+D_{1\downarrow}(E_F)} = 2a_1 - 1 \tag{14.5}$$

$$P_2 = \frac{D_{2\uparrow}(E_F)-D_{2\downarrow}(E_F)}{D_{2\uparrow}(E_F)+D_{2\downarrow}(E_F)} = 2a_2 - 1 \tag{14.6}$$

Using the Equation 14.4, the conductance in parallel (G_P) and antiparallel (G_{AP}) alignment can be written as

$$G_P \propto a_1a_2 + (1-a_1)(1-a_2) \tag{14.7}$$

$$G_{AP} \propto a_1(1-a_2)+(1-a_1)a_2 \tag{14.8}$$

Using the Equations 14.5 through 14.8, TMR can be written as

$$\text{TMR} = \frac{G_P - G_{AP}}{G_{AP}} = \frac{R_{AP}-R_P}{R_P} = \frac{2P_1P_2}{1-P_1P_2} \tag{14.9}$$

The Jullière model accurately predicted TMR values of MTJs consisting of Co, Ni, and CoFe alloy with amorphous aluminum oxide barrier. However, this model could not predict the sign of spin polarization for Co and Ni. The giant TMR values obtained for MTJ with MgO crystalline tunnel barrier cannot be explained by this model. This model only considers the properties of the ferromagnetic electrodes by the expression of spin polarization. The role of the insulating barrier for the calculation of tunneling current was neglected, which plays an important role and must be accounted in the calculation of tunneling conductance.

14.1.2.2 Slonczewski Model

Slonczewski performed a detailed quantum mechanical calculation to calculate the tunneling conductance of a magnetic tunnel junction as a function of relative angle between magnetizations (θ) of two ferromagnetic metals [14]. He assumed the free electron of the two identical magnetic metals participate for tunneling across the insulator. This insulator is considered as a rectangular potential barrier with height of ϕ and width of a, as shown in Figure 14.3. Basically, the free electron with energy E propagates as a plane waves with wave vector k to the barrier. The electron wave function decays exponentially with a decay coefficient $\kappa = \sqrt{2m_e(\phi - E)/\hbar^2}$ in the tunnel barrier. Finally, the transmitted electron propagates again as a plane wave in the second ferromagnetic metals. Solving the Schrodinger equation, the transmission probability of electrons with particular spin states can be written as

$$T^{\sigma\sigma'} = \frac{16\kappa^2 k_\sigma k_{\sigma'} e^{-2\kappa a}}{\left(k_\sigma^2 + \kappa^2\right)\left(k_{\sigma'}^2 + \kappa^2\right)} \tag{14.10}$$

where σ, σ' represent majority ($\uparrow$) or minority ($\downarrow$) spins of the left and right ferromagnetic electrodes, respectively.

Therefore, the conductance in parallel and antiparallel alignment is written as

$$G_P \propto G^{\uparrow\uparrow} + G^{\downarrow\downarrow} \propto T^{\uparrow\uparrow} + T^{\downarrow\downarrow} \propto \frac{16\kappa^2 e^{-2\kappa a}\left[k_\uparrow^2\left(\left(k_\downarrow^2 + \kappa^2\right)^2\right) + k_\downarrow^2\left(\left(k_\uparrow^2 + \kappa^2\right)^2\right)\right]}{\left(k_\uparrow^2 + \kappa^2\right)^2\left(k_\downarrow^2 + \kappa^2\right)^2} \tag{14.11}$$

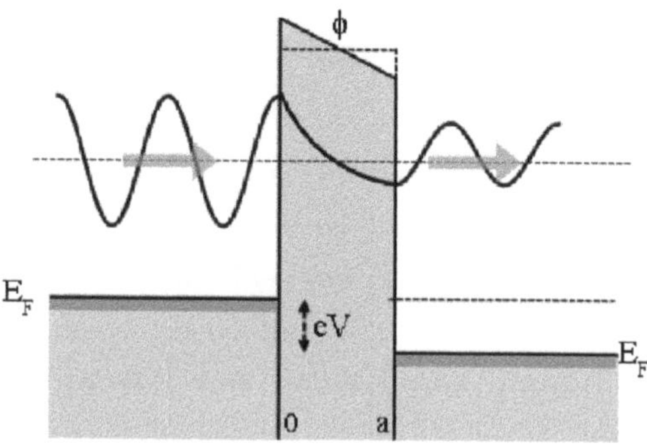

FIGURE 14.3
Schematic illustration of electron wave functions when electron tunneling occurs across a potential barrier.

$$G_{AP} \propto G^{\uparrow\downarrow} + G^{\downarrow\uparrow} \propto T^{\uparrow\downarrow} + T^{\downarrow\uparrow} \tag{14.12}$$

$$G_P - G_{AP} \propto 16\kappa^2 e^{-2\kappa a}\left[\frac{(k_\uparrow - k_\downarrow)(\kappa^2 - k_\uparrow k_\downarrow)}{(k_\uparrow{}^2 + \kappa^2)(k_\downarrow{}^2 + \kappa^2)}\right]^2 \tag{14.13}$$

Therefore,

$$\text{TMR} = \frac{G_P - G_{AP}}{G_P} = \frac{2P^2}{1+P^2} \tag{14.14}$$

Using Equations 14.11, 14.13, and 14.14, spin polarization can be written as

$$P = \frac{(k_{F\uparrow} - k_{F\downarrow})(\kappa^2 - k_{F\uparrow}k_{F\downarrow})}{(k_{F\uparrow} + k_{F\downarrow})(\kappa^2 + k_{F\uparrow}k_{F\downarrow})} = P_J\frac{(\kappa^2 - k_{F\uparrow}k_{F\downarrow})}{(\kappa^2 + k_{F\uparrow}k_{F\downarrow})} \tag{14.15}$$

In the previous expression of spin polarization, $k_{F\uparrow}$ and $k_{F\downarrow}$ are the wave vectors for majority and minority electrons at the Fermi energy, which are the main contributors of tunneling conductance. Only the electrons in a very narrow energy range below the Fermi energy of the emitting left electrode will contribute significantly to the tunneling current. Other electrons further below the Fermi energy feel a larger potential barrier height and therefore have a negligible contribution to the tunneling current. Slonczewski's spin polarization, expressed by Equation 14.15, signifies that not only the spin-polarization of ferromagnet is important but also the barrier plays an important role for calculating the effective spin polarization. In fact, for different bands, the parameters, $k_{F\uparrow}, k_{F\downarrow}$, and κ will be different and therefore yield different spin polarization. In the case of large barrier height, ($\kappa >> k_{F\uparrow}, k_{F\downarrow}$) the Equation 14.15 simply becomes a Jullière equation of spin polarization,

$$P \approx P_J = \frac{(k_{F\uparrow} - k_{F\downarrow})}{(k_{F\uparrow} + k_{F\downarrow})} \tag{14.16}$$

14.1.2.3 Spin-Filtering Effect by Coherent Tunneling

Butler and Mathon first theoretically predicted giant TMR for crystalline Fe(001)/MgO(001)/Fe(001) MTJ by first principle calculation [5,15]. In crystalline bcc Fe, there are different Bloch states with particular symmetries: Δ_1, Δ_2, and Δ_5. During tunneling, the electron wave function conserves its symmetry by coupling with the evanescent states with the same orbital symmetry in crystalline MgO. Figure 14.4a shows the band dispersion of bcc Fe along [001] direction. From the first principle calculations, it has been observed that both majority-spin and minority-spin Δ_2 and Δ_5 bands exist at the Fermi energy, resulting in low spin polarization corresponding to these bands. However, minority-spin Δ_1 band does not exist at the Fermi level, signifying 100% spin polarization. Among these states, Δ_1 evanescent state has the lowest decay rate compared with the other two states, shown in Figure 14.4b. Therefore, in the case of parallel magnetization orientation, Δ_1 majority electrons dominantly tunnel across the barrier. On the other hand, for antiparallel alignment,

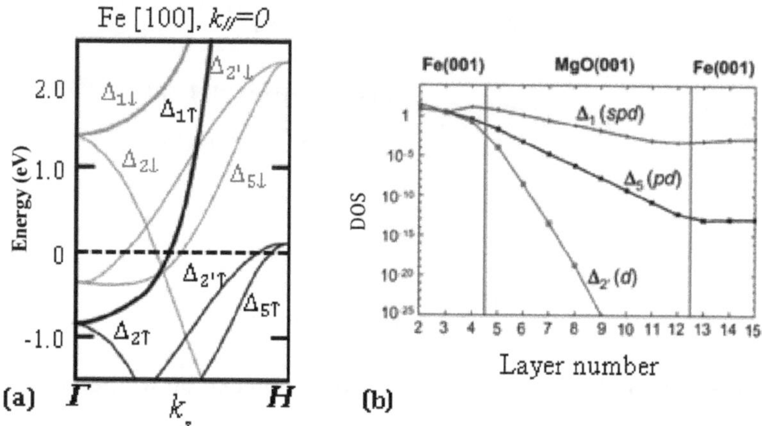

FIGURE 14.4
(a) Majority-spin and minority-spin band dispersion of bcc Fe along [001] direction. (b) Tunneling density of states (DOS) of majority-spin states for $k_{//}=0$ in Fe (001)/MgO(001)/Fe(001) with parallel magnetization orientation. (Redrawn from Butler, W., et al., *Phys. Rev. B*, 63, 1–12, 2001.)

the tunneling probability of different states will be negligibly small compared to that of Δ_1 channel in parallel alignment. This event is called spin-filtering effect as it only allows predominant tunneling of Δ_1 channel, which produces a giant TMR ratio about 1000% for Fe(001)/MgO(001)/Fe(001).

14.1.3 Spin-Transfer Torque

The pioneering works of Slonczewski [16] and Berger [17] predicted that the magnetization of a ferromagnetic layer can be controlled by a spin polarized electrical current. The mechanism by which the magnetization of a ferromagnetic electrode is manipulated by spin-polarized electrons is called spin-transfer torque (STT). This phenomenon can be viewed as the reciprocal effect of GMR or TMR, in which relative orientation of the magnetizations of two ferromagnetic electrodes in spin valves or magnetic tunnel junctions influence the electron transport. The STT effect has been illustrated in Figure 14.5 using a ferromagnet/non-magnet/ferromagnet spin-valve structure.

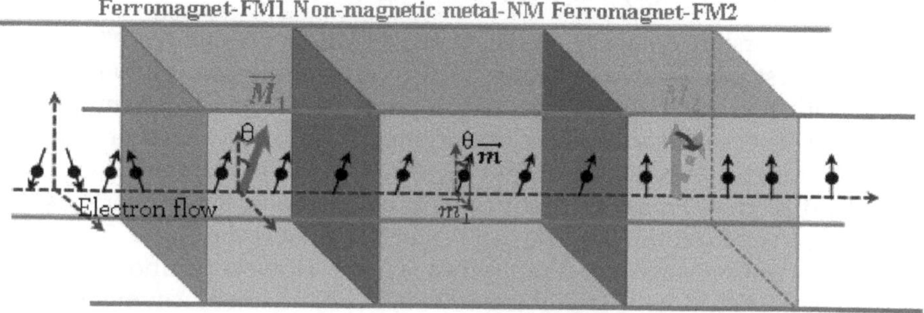

FIGURE 14.5
Schematic illustration of angular momentum transfer between the local spins and conduction electrons due to spin-transfer torque effect.

When initially unpolarized electrons flow through ferromagnetic materials (FM1), the majority of them are spin-polarized according to the direction of magnetization ($\vec{M}_1$), due to the spin-dependent scattering effect as explained in Section 14.1. These spin-polarized electrons traverse through the nonmagnetic layer without losing the polarization, as its thickness is less than the spin-diffusion length. After penetrating into other ferromagnetic layer (FM2), the spins of the electrons are quickly realigned along the magnetization of FM2 ($\vec{M}_2$) within a very short interval of thickness (≤1 nm) from the NM/FM2 interface. As a result, the electrons lose their transverse component of magnetic moment ($\vec{m}_\perp$), which is transferred to the local moments in FM2 to conserve the total angular momentum of the system. This spin transfer from the spin-polarized electrons to the local magnetization of FM2 exerts a torque on $\vec{M}_2$ and tends to align it along the direction of spin polarization of the incoming electrons and therefore along the magnetization of FM1. This phenomenon is known as spin-transfer torque (STT), and the analytical expression is derived below.

The transverse component of magnetic moment of electron is expressed as

$$\vec{m}_\perp = -\frac{g\mu_B}{\hbar}\vec{s}_\perp = \frac{g\mu_B}{2}\left[\hat{m}_2 \times \left(\hat{m}_2 \times \hat{m}_1\right)\right] \tag{14.17}$$

In this equation, $\hat{m}_1$ and $\hat{m}_2$ are the unit vectors representing the direction of magnetizations of FM1 and FM2; $\vec{s}_\perp$ is the transverse angular momentum of electron, which acts perpendicular to $\vec{M}_2$ in the plane intersecting both $\vec{M}_1$ and $\vec{M}_2$.

The number of spins reaching FM2 per unit time:

$$\frac{dN_s}{dt} = P\frac{dN_e}{dt} = P\frac{I}{e} = P\frac{\text{JA}}{e} \tag{14.18}$$

Therefore, STT, which is the rate of transferring spin angular momentum to FM2, is

$$\Gamma_{\text{STT}} = \frac{1}{\gamma}\frac{d\vec{m}_2}{dt} = P\frac{\text{JA}}{e}\frac{\hbar}{2}\left[\hat{m}_2 \times \left(\hat{m}_2 \times \hat{m}_1\right)\right] \tag{14.19}$$

The STT is also known as Slonczewski torque or in-plane torque or damping-like torque. In addition to STT, there can be a field-like torque or out-of-plane torque directed perpendicular to the plane of $\vec{M}_1$ and $\vec{M}_2$.

14.1.3.1 Spin-Transfer Torque and Magnetization Dynamics

Under the influence of STT, the magnetization dynamics of a ferromagnetic material are governed by Landau-Liftshitz-Gilbert (LLG) equation with the additional term of STT.

$$\frac{d\vec{M}}{dt} = -\gamma\left(\vec{M} \times \vec{H}_{\text{eff}}\right) + \frac{\alpha}{M_s}\left(\vec{M} \times \frac{d\vec{M}}{dt}\right) - P\frac{Jg\mu_B}{2M_S e}\left[\frac{\vec{M}}{M_S} \times \left(\vec{M} \times \hat{p}\right)\right] \tag{14.20}$$

where $\hat{p}$ is the unit vector of the moment of spin-polarized electron flux; $\vec{H}_{\text{eff}}$ is an effective field combining applied, anisotropy, and dipolar field; and α is the damping constant. The first, second, and third terms of the right-hand side of the LLG equation are the precessional torque due to effective field, damping torque, and Slonczewski torque, respectively. In the absence of STT and damping, the magnetization continuously precesses around $\vec{H}_{\text{eff}}$

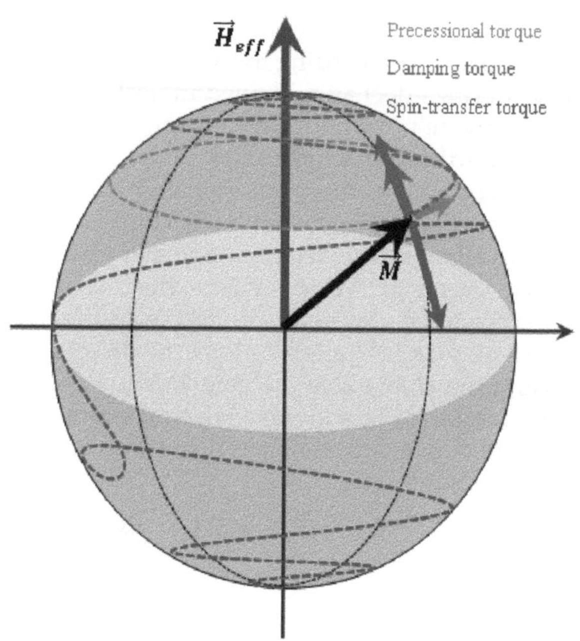

FIGURE 14.6
Schematic representations of magnetization dynamics under the influence of various torques acting individually or collectively.

with a frequency $\omega = \gamma H_{\text{eff}}$. When damping is present, the damping torque acts toward $\vec{H}_{\text{eff}}$ from $\vec{M}$, shown by the blue arrow. Therefore, precessional motion under the influence of field torque will be damped, as illustrated by the blue spiral line in Figure 14.6 and finally realigned along its initial position. When STT is present, two phenomena can occur. In the first case, when $\hat{p}$ and $\vec{H}_{\text{eff}}$ are parallel, STT and damping torques are in the same direction and damp the precession. In the other case, when $\hat{p}$ and $\vec{H}_{\text{eff}}$ are antiparallel, STT exerts in the opposite direction and counteracts the damping torque. At sufficiently large current density, the effect of damping is nullified, and a stable precession occurs at a certain cone angle. This sustained precession can generate magnetization oscillations in radiofrequency range and acts as a spin-transfer torque oscillator. When the current density is larger than some critical value, the magnetization spirals away from the precessional motion and eventually reverses in the direction opposite to $\vec{H}_{\text{eff}}$. After the pioneering theoretical work on STT, the phenomenon was first experimentally observed in 2000 in Co/Cu/Co spin valves [18,19]. It is noteworthy that thermal fluctuation at room temperature creates a small angle between $\hat{p}$ and $\vec{M}$, which is supposed to be either 0° or 180° depending on parallel or antiparallel magnetization orientations of storage and reference layer. This random thermal fluctuation initially stimulates the magnetization reversal by spin-transfer torque.

14.1.4 Introduction to STT-MRAM

Among all memory technologies, STT-MRAM is particularly promising because of its potential to function as a universal memory in the memory hierarchy. Together with non-volatility, other properties such as low power consumption, high write speed (few nanoseconds), semi-infinite endurance (>10^{16} write cycles), and high retention (>10 years) make STT-MRAM a potential candidate to replace embedded FLASH memory, L2/L3 cache, possibly in future

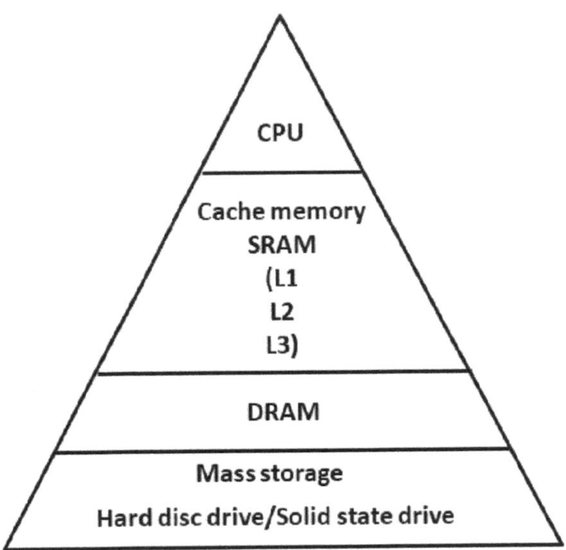

FIGURE 14.7
Present memory hierarchy used in computers. From top to bottom, the storage density of the memory increases and speed increases as we go from bottom level to top.

DRAM and to fill up the memory gap between DRAM and solid-state drive (storage class memory) of current memory hierarchy represented in Figure 14.7. The recent advancement of research proves that embedded-STT-MRAM would become a viable memory option for mobile, Internet of things (IoT) and computer applications [20–22]. Moreover, the radiation hardness properties of the materials used in the stack of STT-MRAM makes them useful for space and military applications. The research area of MRAM is gradually becoming successful and finding its position in the technology market.

14.1.4.1 Different Families of MRAM

The field of MRAM started even before the advent of TMR and STT, using the effect of GMR and AMR. After the discovery of TMR in Al_2O_3 as a tunnel barrier, MTJs were used as an active memory element written by magnetic field. A crossbar architecture of memory array was used, where the magnetic field was produced by two sets of orthogonal current lines (Stoner Wolfarth switching). This was the first generation of MRAM classified as field-written MRAM. The first MRAM, named Stoner-Wohlfarth MRAM, had a half-select problem, which is the accidental writing of other unselected memory cells along the word or bit lines (Figure 14.8).

Savtchenko solved this problem with a novel design named toggle MRAM. The memory cell in this design is aligned in 45° with respect to both bit and word lines. The free layer is made up of a synthetic antiferromagnetic layer, which is written by applying the current to word and bit lines in four steps. In 2006, the first commercial product of 4-Mb toggle MRAM was launched by Freescale, which found its position for transportation, industrial, and space applications [23]. In this technology, the main bottleneck was scaling down the cell because the write current increases as $1/F^2$, where F is the lithographic feature size. Another successful candidate of the field MRAM family is the thermally assisted MRAM (TAS-MRAM), which allows high retention of the storage layer with low writing current. The free layer of the memory cell consists of a ferromagnetic layer exchanged biased with

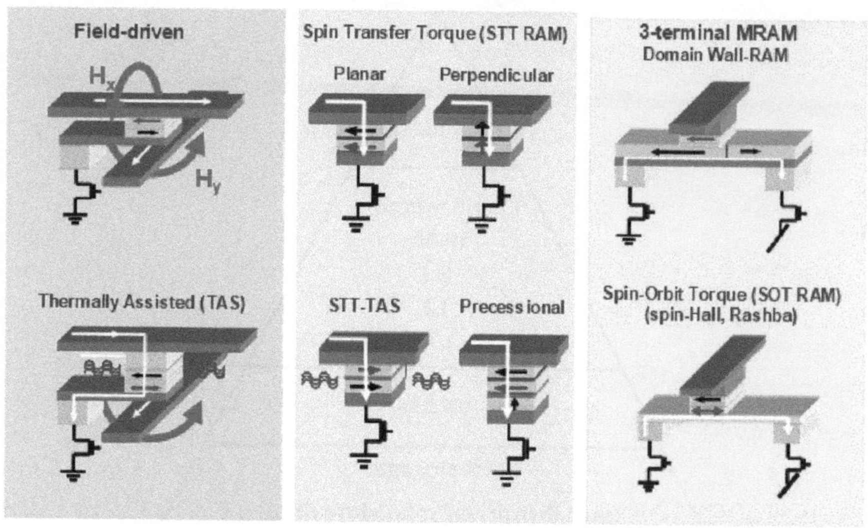

FIGURE 14.8
Schematic representations of magnetization dynamics under the influence of various torques acting individually or collectively.

an antiferromagnetic layer with a lower blocking temperature. Applying a heat pulse reduces the exchange bias of the free layer of the selected bits and then with a lower current in the field line switches the magnetization. However, the discovery of giant TMR and current-induced magnetization reversal by STT soon replaced this technology by STT-MRAM due to better downsize scalability, bit selectivity, and simpler design. These features will be discussed in detail later in the following paragraphs.

The most recent members of the family of MRAM are three-terminal STT-MRAM [24]. In this family, the two candidates are domain wall motion-based MRAM and spin-orbit torque MRAM (SOT-MRAM). These devices are still under research and needs to be more mature to reach industrial production. However, they are very interesting as the read and write path for this memory cell are separated, and the write current does not flow through the tunnel barrier. This makes the cell extremely endurant during writing compared to two-terminal STT-MRAM cells. The writing is performed by STT-induced domain wall motion or by spin-orbit torque. Though the device footprint is expected to be bigger than STT-MRAM, it is also interesting for ultrafast switching (order of picosecond), which makes them a potential candidate for replacement of L1 or L2 Cache SRAM.

14.1.4.2 In-Plane versus Perpendicular STT-MRAM

At the beginning of the introduction of STT-MRAM, the MTJs used were with in-plane magnetization. The shape of the bit was elliptical, providing shape anisotropy to retain the written bit information. However, after finding perpendicular magnetic anisotropy (PMA) material, the elliptic bits were replaced by circular bit with perpendicular MTJs (pMTJs). pSTT-MRAM is more useful than in-plane STT-MRAM for the following reasons explained below.

For in-plane STT-MRAM, the critical switching current (I_{co}) and thermal stability factor (Δ) are expressed by

$$I_{co} = \frac{4\alpha e}{\eta\hbar} k_B T \left[\frac{\pi M_s^2 V}{k_B T} + \Delta \right] \tag{14.21}$$

$$\Delta = \frac{K_{\text{eff}} V}{k_B T} = \frac{\pi^2 (M_s t)^2 w (\text{AR} - 1)}{k_B T} \tag{14.22}$$

where t is the thickness of storage layer, w is the width of elliptical cell, AR is the aspect ratio, α and η are Gilbert damping constant and spin-torque efficiency, respectively.

For out-of-plane STT-MRAM, the critical switching current (I_{co}) and thermal stability factor (Δ) are expressed by

$$I_{co} = \frac{4\alpha e}{\eta\hbar} k_B T \Delta \tag{14.23}$$

$$\Delta = \frac{K_{\text{eff}} V}{k_B T} = \left(\frac{K_i}{t} - 2\pi M_s^2 \right) \frac{\pi D^2 t}{4 k_B T} T \tag{14.24}$$

where D is the diameter of circular bit and K_{eff}, K_i are effective perpendicular magnetic anisotropy and interfacial anisotropy constant, respectively.

The critical switching current for the in-plane memory cell is much higher than out-of-plane because of the extra term of demagnetizing field. This is due to the fact that during STT-induced switching of in-plane magnetized MTJs, in its precessional motion, the storage layer magnetization has to go out-of-plane against the demagnetizing field, which costs lots of energy. On the other hand, the energy barrier, which determines the memory retention, is given by the in-plane shape anisotropy. In other words, for in-plane MTJs, the energy barrier for switching is much higher than the barrier for retention, which is quite unfavorable in terms of power consumption. In contrast, these barriers are identical for out-of-plane magnetized MTJs. Therefore, for the same value of Δ, the in-plane memory cell consumes more energy for magnetization reversal. Moreover, easier manufacturability of the circular shape memory element compared to elliptical makes pSTT-MRAM more attractive to realize a sub-20 nm memory cell.

14.1.4.3 Properties of Material for Designing STT-MRAM

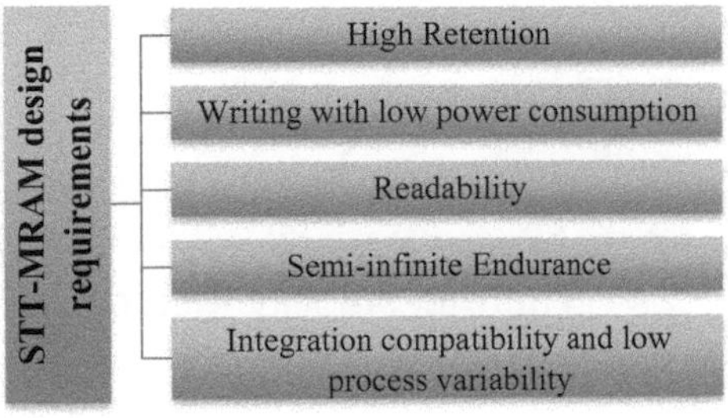

The design of the material stack for high-performance STT-MRAM devices is mainly based on five important requirements as shown in the aforementioned flowchart. In this section, we will elaborate on the importance of these five properties. The first requirement is that the data retention time should be large (at least 10 years) of the stored information. The larger the retention time, the larger is the required thermal stability factor. The required values of thermal stability factor (Δ) can be estimated from the following formula of probability of at least one bit failure in 10 years' time.

$$P(t)=1-\exp\left[\frac{-Nt}{\tau_0}\exp\left(-\Delta\frac{300}{T}\right)\right] \tag{14.25}$$

In this equation, τ_0, N, and t represent attempt time of the order of 1 ns, number of bits on the chip, and retention time, respectively. T is the temperature in degree Kelvin.

Figure 14.9 demonstrates the required values of Δ at room temperature for a memory array to reach a particular value of bit failure rate. For example, the required value of Δ at room temperature is 88 and 94 to obtain 10 years' retention time with probability of bit failure lower than 10^{-4}, respectively, for a 1 Gb and 64 Gb memory array operating at 80°C. Note that in this calculation, it was assumed that the energy barrier remains unchanged as a function of temperature. It is possible to obtain this value of Δ by using perpendicular magnetic anisotropic material. At present, a composite-type storage layer with two MgO interfaces is used to increase the PMA and therefore Δ of memory cell [25].

There is a dilemma between writing current and thermal stability factor. The critical switching current (I_{co}) is proportional to Δ as shown in Equations 14.21 and 14.23. Therefore, improving Δ also increases I_{co}: this is a classical dilemma between retention and writability. By choosing proper material, which increases spin polarization and reduces damping, as well as optimizing the structure of the stack, it is possible to reduce the switching current for high values of Δ. Heusler alloys, because of their half-metallic nature, are expected to have 100% spin polarization and also low damping. Researchers are widely searching

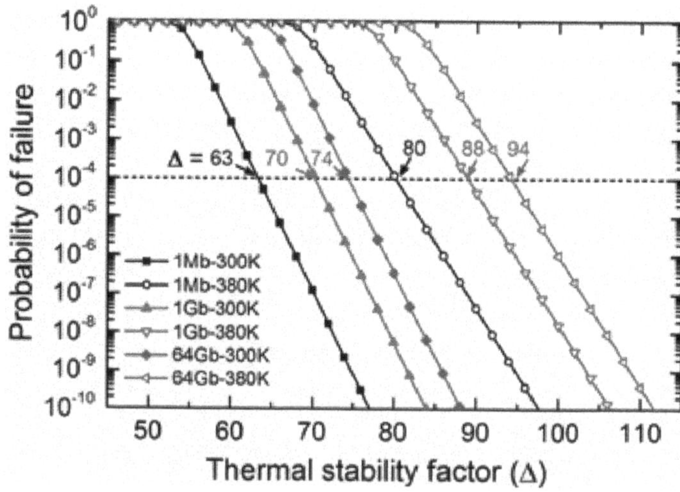

FIGURE 14.9
Bit failure rate at room temperature and 80°C during 10 years as a function of thermal stability factor at room temperature for different capacities of memory array.

for Heusler alloys to get these properties at room temperature. However, until now no successful results were obtained combining high TMR, perpendicular anisotropy, and low damping at room temperature. At present, FeCoB alloys are the standard material for storage layer, which can provide high TMR and also reasonably low damping constant [26–28]. Another approach for reducing the I_{co} is to increase the STT efficiency by configuring the stack in double MTJs where the storage layer is sandwiched between two reference layers [29,30]. In this configuration, spin-transfer torque comes from both interfaces of the storage layer with MgO. A factor of 10 reduction of switching current of DMTJ compared to single MTJ has been predicted by analytical calculation [31]. Another strategy is to reduce the anisotropy by thermal effect [32] or by electric field effect while writing. The second effects also known as voltage-controlled anisotropy (VCMA) can potentially reduce the anisotropy to write the cell with very low current [33,34].

The reliable and faster reading of a memory cell depends on the TMR values. A large TMR value provides a larger signal to separate out the high and low resistance states. In order to discriminate between the two resistances states, the two distributions must be separated by at least 12σ as shown in Figure 14.10, where σ is the standard deviation of the resistance distributions. Moreover, for reliable detection between the bits of low and high states with higher resistance and lower resistance of the distributions, there must be an even larger margin of about 20σ. This margin increases with smaller diameter, as the distribution might be broader due to relatively larger edge damage during etching. Therefore, high TMR is necessary for fast and reliable reading. Magnetic tunnel junction with MgO as a barrier can provide a large TMR ratio. A maximum TMR ratio of 604% was already obtained using in-plane MTJ [36]. The TMR ratio for pMTJs is limited by the thickness of the FeCoB layer as the PMA of thicker FeCoB diminishes. However, 350% TMR was already demonstrated in pMTJs, which ensures faster reading of pSTT-MRAM devices [26]. In addition, for practical industrial MRAM chips, uniformity of the barrier at the wafer level is very important since a small drift in the thickness yields exponential variation of MTJ's resistance.

The endurance of STT-MRAM is defined by the number of read-write cycles that the cell can withstand before dielectric breakdown. The writing voltage of a memory cell should be largely separated from breakdown voltage to ensure an endurance of $>10^{15}$ cycle. Therefore, endurance can be improved by reducing the writing voltage. In practical STT-MRAM devices, the write voltage is typically set in the range of 0.4–0.5V while the

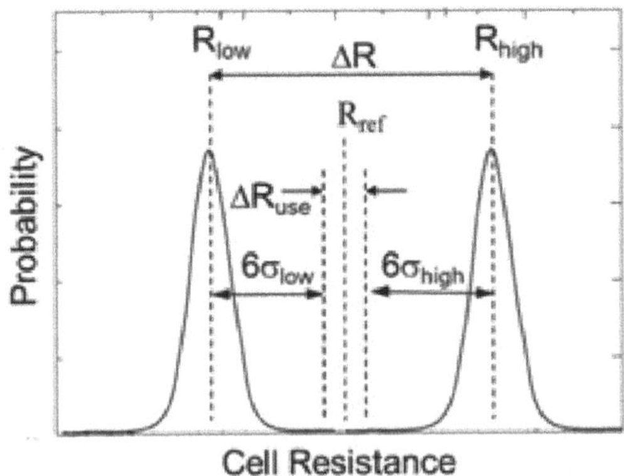

FIGURE 14.10
The distribution of high resistance and low resistance states of a memory array. (From Apalkov, D. et al., *Proc. IEEE*, 104, 685–697, 2016.)

dielectric breakdown voltage is in the range 1.2–1.5V. Improving the quality of MgO barrier by reducing the number of defects (such as trapped water molecules, vacancies, dislocation, etc.) increases the breakdown voltage. Similarly, the integration process should be optimized to reduce the edge damages while etching the MTJ stack. In the STT-MRAM industry, mastering the barrier quality is well controlled. However, the nano-patterning process for ultra-small technology nodes (sub-20 nm) and small pitch is still a remaining challenge. Recently, Jimmy J. Kan et al. reported endurance of more than 10^{15} cycle of 50 ns pulse for 1 Mb perpendicular STT-MRAM [8].

The last criterion, though not the least, is the compatibility of the stack with the STT-MRAM fabrication process. The memory is integrated with CMOS in the back-end-of-line (BEOL) fabrication. Therefore, the stack should be able to endure the high thermal budget, about 400°C, without compromising the magneto-electric properties [37,38]. The stack should be designed and deposited in such a way that the interfacial roughness at the interfaces between the MgO tunnel barrier and ferromagnetic electrodes remains well below the thickness of each individual layers. Otherwise, pin holes and defects in the MgO layer will significantly reduce TMR, PMA, and breakdown voltage of memory cells. The pMTJ stacks for STT-MRAM typically consist of multilayers comprising a few tens of layers of thickness ranging between 0.1 nm and a few nm. The stack should be as thin as possible to reduce the nonvolatile etch product during etching, which will improve the yield and decrease the dot-to-dot variability [35].

14.1.5 Spin-Orbit Torque-Induced Switching

Recently, a novel spin torque was discovered in ferromagnetic materials adjacent to a heavy metal (e.g., Pt, Ta, W) originating due to strong spin-orbit interaction in heavy metals. In the case of STT switching, the conduction electrons are polarized within a ferromagnetic material whereas in SOT switching, the angular momentum is transferred from crystal lattice to the electrons due to spin-orbit coupling. An in-plane current flowing through a heavy metal can manipulate the magnetization of a ferromagnet adjacent to it via spin-orbit torque effect. The basic requirement for the observation of the spin-orbit torque phenomenon is the structural inversion asymmetry in which the ferromagnetic materials should be sandwiched between two different heavy metals or the same heavy metal with different thicknesses. The structural inversion asymmetry helps to induce a net spin-orbit field on the ferromagnet from both the interfaces, which would otherwise be negated in the case of the same heavy metals with the same thickness. One can also induce spin-orbit field by exchanging one of the heavy metals with an oxide layer. The spin-orbit torque can be originated from one of these or in a combination of two physical phenomenon called Rashba effect [10,39,40] and Spin Hall effect [11,41,42], which will be elaborated on in the subsequent sections. To our knowledge, no method has been demonstrated so far to separate the individual contribution from each of these two phenomena in spin-orbit torque-induced magnetization dynamics.

14.1.5.1 Rashba Effect

In the systems with structural inversion asymmetry (e.g., Pt/Co/AlOx), the spin-orbit coupling in heavy metal induces a strong crystal electric field at the interface between the heavy metal and the ferromagnetic layer. The role of the oxide layer is to break the structural inversion symmetry to induce an effective spin-orbit field. When the current is applied, conduction electrons flowing at the interface in the ferromagnetic layer

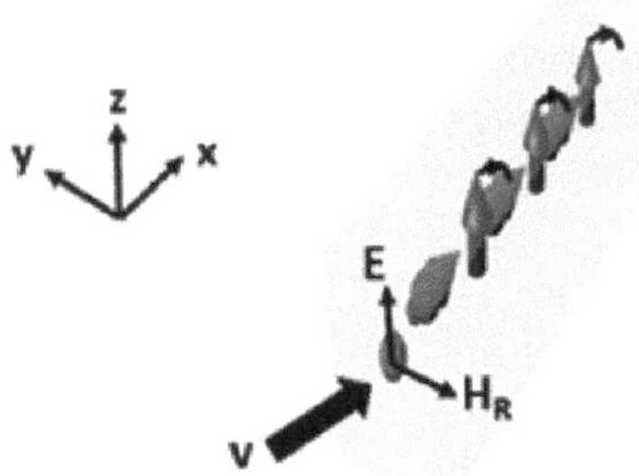

FIGURE 14.11
Schematic illustration of Rashba effect. The electrons with velocity v experience the electric field due to crystal potential, which acts as a Rashba magnetic field (H_R) in their rest frame. This tilts the moments perpendicular (along-X) to both the electric and magnetic fields.

experience the electric field as an effective magnetic field in their rest frame of reference as shown in Figure 14.11. This effective magnetic field polarizes the conduction electrons in the direction orthogonal to both the electric field and their velocity. Thus, there will be an effective spin accumulation due to Rashba field at the interface defined by the Equation 14.26. The exchange interaction between these accumulated spins at the interface and the local moments in the ferromganetic layer induces magnetization switching. The Rashba spin-orbit torque that is responsible for the switching of the magnetization is due to an effective Rashba field (H_R) induced at the interface, which is expressed as [43]

$$H_R = -\frac{2\alpha_R m_e}{\hbar|e|M_s} \mathrm{P}|J_e|(\hat{z}\times\hat{J}_e) \tag{14.26}$$

where α_R is a material parameter that quantifies the strength of spin-orbit coupling, m_e is the rest mass of the electron, and J_e is the current density. The Rashba field is along the transverse direction to the strip orthogonal to the direction of electron flow and the effective electric field (along z-direction) due to spin-orbit coupling. I. M. Miron et al. [39] reported the first observation of asymmetry in domain wall nucleation in ferromagnetic nanowires due to Rashba effect. In this experiment, 20 parallel wires were patterned from the Pt/Co/AlOx, and domain wall nucleation was observed by Kerr microscopy when the current was applied. For a given current direction, the domain wall nucleation disappeared for one magnetic field direction and the nucleation was enhanced for the other direction of the magnetic field. By reversing the current direction, similar behavior is witnessed but for the opposite sign of magnetic field. This observation confirms the presence of effective magnetic field due to the current flow in the heavy metal, thus torque exerted by Rashba effect is often called a "field-like torque." The Rashba effect due to current alone cannot switch out of plane magnetization of a nanodot as it can only induce a precession due to Rashba field. Therefore, an in-plane magnetic field collinear to the current direction is necessary for the magnetization switching [10].

The field-like torque acting on the magnetization due to Rashba field H_R can be expressed as

$$\tau_{R,\mathrm{FL}} = -|\gamma|\, m \times H_R \tag{14.27}$$

14.1.5.2 Spin Hall Effect

An alternative mechanism was proposed by Liu et al. [11,41] for the current-induced magnetization switching in ferromagnetic materials sandwiched between heavy metals. Due to spin-dependent scattering of the electrons in heavy metal, electrons with opposite spins move in opposite directions generating a spin current orthogonal to the charge current, as shown in Figure 14.12. This phenomenon is called spin Hall effect. The long spin diffusion lengths (~nm) in heavy metals allow the spins to be accumulated at the interface with the ferromagnetic layer. These accumulated spins exert a torque on the local magnetization of the ferromagnet to initiate the switching. The discovery of spin Hall effect suggests the novel concept of generation of spin currents in nonmagnetic materials and injection into a ferromagnet to manipulate the magnetization. The torque induced by spin Hall effect is a damping-like torque and is effectively expressed from the spin Hall angle Θ_H, which is a measure of spin current induced from the charge current.

The efficiency of the switching due to the Spin Hall effect is quantified by a Spin Hall angle (Θ_H), which is a parameter that shows conversion yield of the spin current (J_s) from a charge current (J_e).

$$\Theta_H = \frac{J_s}{J_e} \tag{14.28}$$

The sign and magnitude of the spin Hall angle are the material properties of heavy metal in which spin current is generated. Pt [44] possesses a positive spin Hall angle where as Ta and W posses negative spin Hall angles.

The spin current injected into the ferromagnetic layer from the heavy metal exerts Spin Hall torque, which has symmetry of damping-like torque. The torque due to Spin Hall effect can be expressed as [45]

$$\tau_{SH} = -|\gamma| m \times (m \times (\hat{z} \times \hat{J}_e) \tag{14.29}$$

$H_{SH} = (m \times (\hat{z} \times \hat{J}_e)$ is an effective field due to Spin Hall effect, which is parameterized by Spin Hall angle (Θ_H)

$$H_{SH} = \frac{\mathrm{h}\Theta_H |J_e|}{2|e| M_s t_F} \tag{14.30}$$

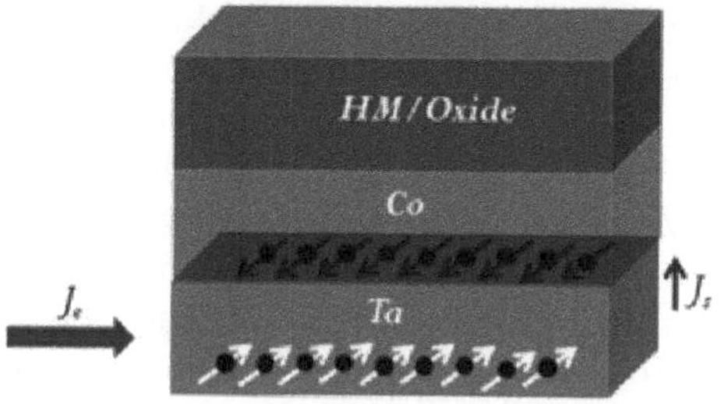

FIGURE 14.12
Schematic to show the spin Hall effect in heavy metal in which electrons of opposite spins scatter in opposite directions generating a spin current J_s orthogonal to charge current J_e.

where J_e is the charge current density, M_s, and t_F are the saturation magnetization and the thickness of the ferromagnetic layer into which the spin current is injected from the heavy metal.

14.1.5.3 Spin-Orbit Torque MRAM

The spin-orbit torque effect has opened a new avenue for the manipulation of magnetization by in-plane current, inspiring the proposals for novel and efficient spintronic devices. The proposal of spin-orbit torque MRAM has come with a lot of advantages to eliminate the intrinsic issues that are identified in STT-MRAM for ultrafast applications. In STT-MRAM, the current is flown through the tunnel barrier for both reading and writing cycles. Though the current required for reading is very small, there is a probability that an accidental switching can lead to the correction of the data. Moreover, for fast writing, one needs to apply a short pulse with large current density, which can cause junction breakdown leading to device failure. In spin-orbit torque MRAM (SOT-MRAM), the read and write paths are decoupled as shown in Figure 14.13. For reading, a small bias current is applied across the junction whereas for the writing, an in-plane current is applied through a heavy metal adjacent to the free layer. Hence, the problem of dielectric breakdown of the tunnel barrier does not exist anymore. This lifts the limit on the switching speed and also eliminates the chance of accidental writing as mentioned earlier. However, SOT-MRAM is a three-terminal device and therefore the footprint is relatively larger than two-terminal STT-MRAM. Thus, SOT-MRAM will be useful in nonvolatile memories that require faster data rates but no constraints on density therefore promising for SRAM applications.

The first proof of concept of a three-terminal SOT-MRAM device was demonstrated in 2014. M. Cubukcu et al. [46] have fabricated a perpendicular FeCoB/MgO/FeCoB MTJ on top of a Ta track, as shown in Figure 14.14, to demonstrate the reading and writing operations by TMR effect and SOT effect, respectively. They have observed switching with a current pulse of 50 ns with a current density of 5×10^{11} A/m^2. Currently, a lot of effort is devoted to improving the SOT-induced switching speed for ultrafast recording applications (e.g., SRAM). K. Garello et al. [47] have shown an ultrafast switching of a magnetic dot by an in-plane current of pulse width of 180 ps. However, in SOT-MRAM, the threshold current density may increase with a reduction in pulse width for faster switching.

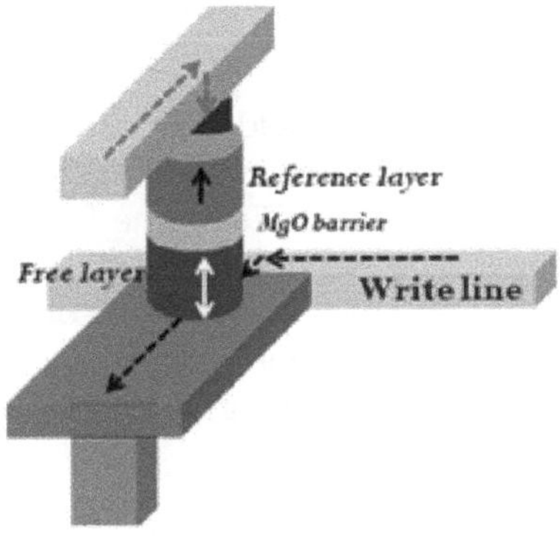

FIGURE 14.13
Schematic to show the three-terminal SOT-MRAM device with decoupled read and write paths.

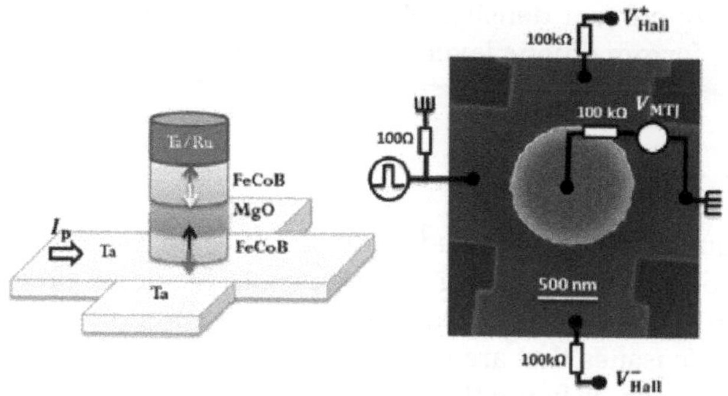

FIGURE 14.14
Schematic showing an MTJ cell placed on a Ta track to demonstrate the proof of concept of SOT-MRAM. (Adapted from Cubukcu, M. et al., *Appl. Phys. Lett.*, 104, 2014.)

14.1.6 Domain Wall Dynamics in Ferromagnetic Nanostructures

In general, in ferromagnetic nanostructures with relatively large dimensions (few tens of nm), magnetization reversal is mediated via domain wall (DW) nucleation and motion. The underlying physics behind the DW motion in the presence of a magnetic field or current is different from the coherent precession that occurs in magnetic nanostructures with small dimensions (typically less than 30 nm). Therefore, it is imperative to understand the DW dynamics under different driving mechanisms not only for fundamental understanding but also to realize DW-based memory and logic devices. Thus, this part of the chapter describes the configuration of DW types that are stable in ferromagnets as well as their dynamics under magnetic field and current along with their possible applications in memory devices. In a ferromagnetic material, exchange interaction aligns neighboring spins in the same direction. However, it breaks into several domains with different magnetization directions to reduce the magnetostatic energy. These domains are separated by DWs in which spins gradually rotate from one domain magnetization to another to minimize the exchange energy. Two kinds of DWs, Bloch and Néel type, are stable in ferromagnetic nanostructures depending on the dimensions. In Néel walls, magnetization rotates in plane whereas in Bloch walls the magnetization rotates out of plane.

14.1.6.1 Field-Induced Domain Wall Motion

The DW motion induced by the magnetic field is well known in magnetism as it acts as a medium for the magnetization reversal process in magnetic films. When the magnetic field is applied to a magnetic material with DWs, the field expands the domains of one orientation at the cost of other domains of opposite direction to minimize the Zeeman energy. Effectively, it means the boundaries between the domains are in motion due to the effective fields (internal and external magnetic field).

The motion of a DW in the presence of a magnetic field can be described by using the Landau-Lifshitz-Gilbert (LLG) equation.

$$\frac{d\vec{M}}{dt} = -\gamma\left(\vec{M}\times\vec{H}_{\text{eff}}\right)+\frac{\alpha}{M_s}\left(\vec{M}\times\frac{d\vec{M}}{dt}\right) \tag{14.31}$$

This is the basic equation and same as the one described earlier in 14.20 but without the STT term. The first term is the precession term that lifts the magnetic moments in the DW out of plane, which generates a demagnetizing field, H_d, as shown in Figure 14.15. The demagnetizing field exerts a torque to drive the DW forward. For device applications, the key parameter during the DW motion is the DW mobility, which is the rate of change of velocity as a function of the applied magnetic field. The mobility of the DW defines the limiting operating velocity of a DW-based device. The DW mobility has two regimes as a function of magnetic field. When the external magnetic field is low, the canting of the magnetic moments in the DW out of plane is small enough that demagnetizing field H_d drives the DW linearly, depicted as linear regime of domain wall velocity versus field plot, shown in Figure 14.16. The velocity of the DW in linear regime is $v = \mu H$, where μ is the mobility and H is the external magnetic field. The mobility is defined as $\mu = |\gamma|\delta/\alpha$, where γ is the gyromagnetic ratio, δ is the DW width, and α is the damping parameter.

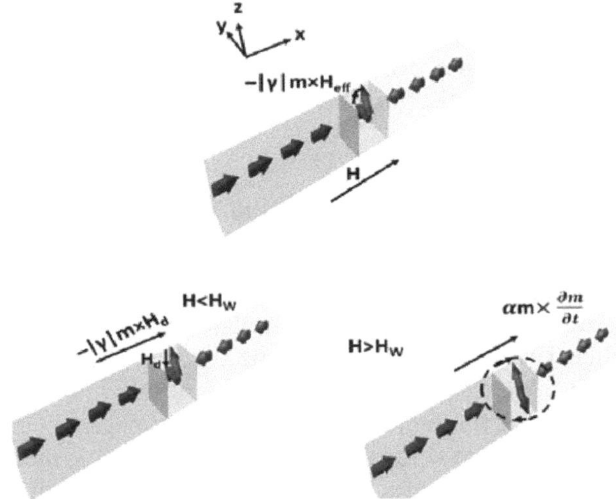

FIGURE 14.15
Schematics to describe the DW motion in the presence of a magnetic field. The top shows the precession of DW out of plane due to an external magnetic field. The bottom shows the resultant motion when the magnetic field is below and above Walker breakdown.

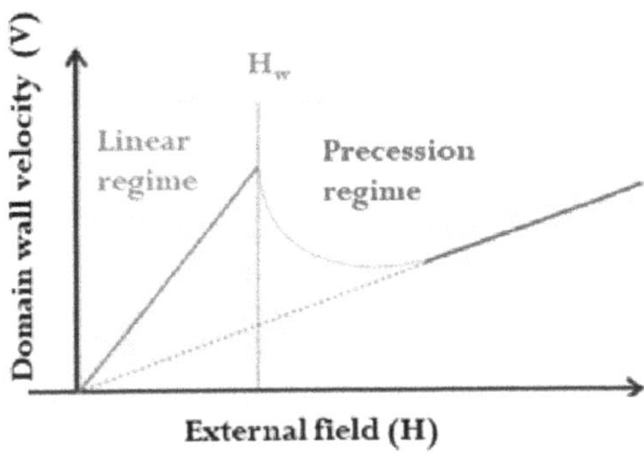

FIGURE 14.16
The domain wall velocity as a function of magnetic field showing linear and precession regimes separated by Walker field H_w.

The DW speed scales linearly with the external magnetic field until a threshold field called Walker field (H_w). When the magnetic field is larger than this threshold ($H > H_w$), the DW continuously precesses around the nanowire axis and now damping term drives the DW. This regime is called precession regime. The DW velocity in precession regime is defined as $v = \alpha^2 \mu H$, where α is the damping parameter, which is very small (<<1). The velocity of the DW drops significantly after Walker field, as a portion of the DW energy is used for the precession of DW rather than to move it. The continuous precession of the DW causes periodic transformation of the DW between the Bloch and Néel configurations. This phenomenon is called Walker breakdown [48], which limits the DW mobility.

Though field-induced DW motion is very interesting to understand the fundamental properties of DW motion, it is not very promising for application point of view due to the following reasons. The generation of magnetic field for on-chip application is not energy efficient and poses scalability issues. Moreover, adjacent DWs in the ferromagnetic layer always move in opposite directions causing either shrinking or expansion of domains. For memory applications to store the data as domains in a track, the DW motion needs to be unidirectional. Individual control of DWs by generating localized magnetic fields is a cumbersome process and power inefficient. Therefore, one needs an alternative method for reliable control of the DWs in magnetic nanowires to employ them in device applications.

14.1.6.2 Spin Transfer Torque-Induced Domain Wall Motion

Current-induced DW motion due to STT is promising because of the unidirectional DW motion irrespective of the type of the DW. When current is applied to a ferromagnetic material, the conduction electrons become spin-polarized due to spin-dependent scattering as discussed earlier. These conduction electrons are then aligned according to the local magnetization direction because of the angular momentum transfer from local moments. To conserve the total angular momentum, the conduction electrons exert a torque on the local moments; this phenomenon is called spin-transfer torque, which is explained earlier in Section 14.3. This transfer is very efficient when the spins and moments are noncollinear as in the case of a DW in ferromagnets. The moments in the DW are rotated toward the spin by the torque induced by the spin angular momentum, and the incoming spins align along the direction of moment of the DW. Thus, this leads to DW motion along the direction of electron flow, as illustrated in Figure 14.17.

STT-induced DW motion has attracted lots of interest to describe the equation of motion of the DW in the presence of current [49,50]. The most general agreement among all the theoretical modeling shows that the interaction of electric currents and the local magnetizations can be explained by considering two different spin-transfer torque terms, which are termed as adiabatic spin-transfer torque and non-adiabatic spin-transfer torque.

The modified LLG equation with the addition of these two terms can be written as [49]

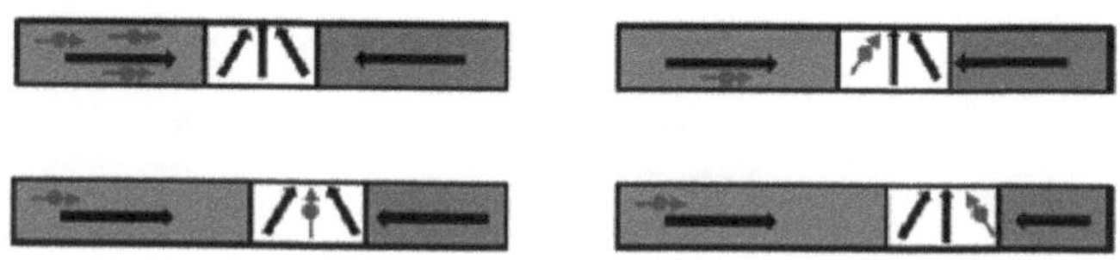

FIGURE 14.17
Schematic to depict the DW motion under current-induced spin-transfer torque effect.

$$\frac{d\vec{M}}{dt} = -\gamma\left(\vec{M}\times\vec{H}_{\text{eff}}\right) + \frac{\alpha}{M_s}\left(\vec{M}\times\frac{d\vec{M}}{dt}\right) + \text{STT}_1 + \text{STT}_2 \quad (14.32)$$

where M is the magnetization, M_s is the saturation magnetization of material, u is the electron drift velocity, γ_0 is the gyromagnetic ratio, and α is the Gilbert damping constant. The first two terms account for the torque by the effective field H_{eff} and the Gilbert damping torque, parameterized by a dimensionless α as described previously. As proposed by Slonczewski, STT_1 describes the angular momentum transfer between the conduction electrons and the local spins of the material. In this case, it is assumed that when the electrons move across the DW, they adjust their spin orientations adiabatically to the direction of the local moments while transferring angular momentum to the DW.

The adiabatic STT_1 term can be written as

$$\text{STT}_1 = (\boldsymbol{u}.\nabla)\vec{M} \quad (14.33)$$

where u is the conduction electron velocity quantified by magnitude of the spin-transfer torque given by

$$u = \frac{g\mu_B p}{2eM_s} \quad (14.34)$$

for which g, μ_B, and e are the Landè factor, Bohr magnetron, and electron charge, respectively; M_s is the saturation magnetization; and p is the efficiency of spin polarization of conduction electrons, which depends on the ferromagnetic material. As explained earlier for the case of field-induced DW dynamics, the DW only precesses when the torque due to the demagnetizing field and the current-induced adiabatic torque are compensated. In this case, there will be no effective displacement of the DW. This is called intrinsic pinning, and the current applied should overcome certain threshold current density for the adiabatic torque to drive the DW. Theoretical predictions [51] have shown that the threshold current density required to drive the DWs by adiabatic STT are close to 10^{13} A/m^2. However, initial experiments report a critical current density of the order of 10^{12} A/m^2 for DW motion [52–54]. This has led to proposal of an additional term called non-adiabatic torque term STT_2 to the LLG equation.

The STT_2 term can be defined as

$$\text{STT}_2 = \frac{\beta}{M_s}\vec{M}\times\left[(\boldsymbol{u}.\nabla)\vec{M}\right] \quad (14.35)$$

β is called a non-adiabatic constant. The non-adiabatic spin torque is always orthogonal to the adiabatic torque. It acts as an effective field on the system often referred as field-like torque. The speed of the DW depends on the relative ratio between β and α. When $\beta < \alpha$, the adiabatic torque drives the DW with a speed increasing with current density. When $\beta > \alpha$, non-adiabatic field-like torque induces precessional motion of the DW leading to Walker breakdown where the DW speeds significantly drops at relatively high current densities. It is generally believed that the non-adiabatic term originates from the spatial mistracking of the conduction electron spins and the local moments, where the spin-polarized conduction electrons are scattered (reflected), when it cannot follow the local moments. Zhang and Li [50] proposed a model for the case of slight mistracking between the electron spin and the local magnetization direction during the spin transfer phenomenon. This means that the spin of the electron is not exactly aligned in the direction of the local magnetic moment when it moves along the ferromagnetic material. This mistracking generates nonequilibrium spin accumulation across the magnetic DW. The accumulated spins are then relaxed through a spin-flip scattering process toward the local magnetization direction. The Zhang–Li model eventually then leads to both adiabatic and nonadiabatic spin-transfer torque terms.

Initially, the DW motion induced by spin-transfer torque was widely studied in nanowires of in-plane soft magnetic material such as $Ni_{81}Fe_{19}$ (permalloy). Yamaguchi et al. [54] first showed the real-space observation of the back-and-forth motion of vortex DW by microsecond current pulses with current density of the order of 10^{12} A/m^2. Recently, with the developments in lithography techniques, relatively high DW velocity of 150 m/s [55] was demonstrated by IBM, which makes the implementation of DW based memory and logic applications to be very promising.

14.1.6.3 Domain Wall Racetrack Memory

The DW-based racetrack memory was proposed by S.S.P. Parkin et al. [24] in 2008, which has attracted a lot of interest in the field of DW dynamics driven by the spin-transfer torque phenomenon. In racetrack memory, the binary bits are stored in the domains with opposite magnetizations as shown in Figure 14.18 (red and blue regions). These data bits are separated by DWs and can be moved along with the motion of DWs in the racetrack. The data bits can be written by a localized magnetic field, created by flowing current through a strip line (vertical green nanowire) connected with one end of the racetrack (horizontal nanowire). The direction of the current defines the orientation of the magnetization above the strip line thus writing “1” and “0” with currents of opposite directions. Magnetic tunnel junctions are patterned having an interface with the racetrack to read the binary bits saved in the domains. Here, the magnetic domains of nanowire act as a storage layer. The magnetic tunnel junction reads the magnetization direction of the free layer relative to the reference layer by low (parallel) and high

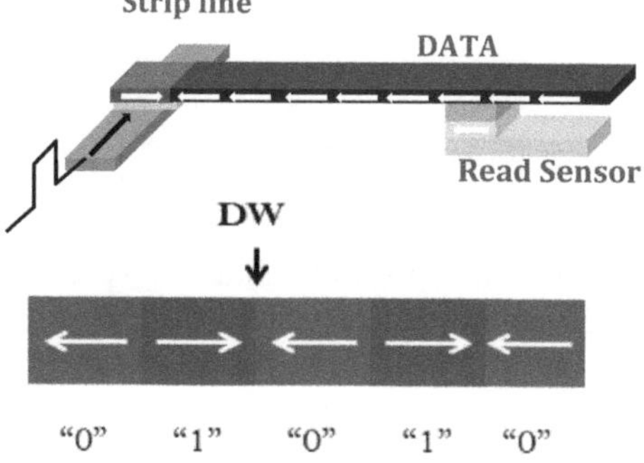

FIGURE 14.18
Schematic of the DW racetrack memory with read and write heads proposed by IBM for storing the data as direction of magnetization of domains. (From Parkin, S. S. P. et al., *Science*, 320, 190–195, 2008.)

magnetoresistance (anti-parallel) states like the way described in other MRAM devices. For reading, the bits need to be pushed back and forth by applying current pulses through the racetrack. One can use artificial pinning sites [56–58] such as notches to confine the data bits by pinning the DWs. Initially, IBM demonstrated the racetrack memory on 2D patterned nanowires. However, to increase the density by many folds, one can construct it as 3D standing nanowires by growing the cylindrical nanowires using template-assisted growth [59] or by creating trenches on top of Si substrate by using FIB or chemical methods. However, the fabrication and integration of these 3D nanowires are extremely challenging to realize. Thus, most of the studies are only focused on 2D patterned nanowires. A robust DW shift register operation with three bits and two DWs in a permalloy nanowire, in which by applying current pulse with opposite signs, back-and-forth motion of DWs with speeds up to 150 m/s was demonstrated by Hayashi et al. [55]. Much of the initial developments on 2D racetrack memory are focused primarily on permalloy nanowires. However, permalloy faces several obstacles to be employed in applications. First, the critical current density to drive the DWs with a speed of ~100 m/s requires relatively large current densities of ~10^{12} A/m^2 [60]. These high currents can lead to joule heating, which is a drawback for power efficiency and the retention of the stored data. Moreover, the widths of the domain walls in permalloy nanowire are typically of an order of the nanowire width (~100 nm). As DWs are occupying a large space in a racetrack, it can be detrimental for the storage density. Similar to the field-induced dynamics, the DWs in the racetrack can undergo the Walker breakdown process, in which a periodic transformation between transverse and vortex (anti-vortex) configurations limits the speed of the device. Therefore, the dynamics in permalloy are not compatible for reliable device applications. To replace the permalloy, the researchers have been looking for a different class of materials with perpendicular magnetic anisotropy (PMA). In multilayer systems, such as Co/Ni [61] and Co/Pt [62], interfacial anisotropy overpowers the shape anisotropy to bring the magnetization out of plane. The widths of the DWs found in PMA nanowires are much smaller, ~10 nm. The threshold current density to drive these DWs is substantially smaller in PMA materials ($J_c = 2 \times 10^{11}$ A/m^2) [63]. The DWs in PMA nanostrips also undergo the periodic transformation between Bloch and Néel configurations, but the internal structure of the DW remains intact. For the aforementioned reasons, the DWs in PMA materials are viable for reliable device applications.

Although the DW-based racetrack memory is a strong candidate for universal memory due to its high read/write speeds and scalability, it suffers from some practical limitations. For instance, high current density is required to de-pin and move the DWs. There is a possibility of mutual annihilation of multiple DWs during motion leading to a reduction in packing density. The dipolar interaction between successive DWs also limits the packing density by affecting neighboring DWs. Solutions have been proposed to overcome these shortcomings, which include patterning notches to control the data flow, pulsed current to move the DW, and using synthetic antiferromagnetic coupling to reduce the stray field [64]. More recently, heavy metal/ferromagnet/oxide stacks have been employed that include Dzyaloshinskii-Moriya interaction (DMI) field and SOT for efficient DW motion. This has reduced the current density for propagation [65–67]. In SOT, the anti-damping torque is always perpendicular to the magnetization, hence the incubation delay of the switching process is less, which has led to high-speed device performance [47]. Another advantage of SOT-based driving is that the nanowire dimensions can be made longer to increase device density without suffering from high resistivity problem, since current flows through the heavy metal conductor. Recently, Lorentz transmission electron microscopy (L-TEM) imaging showed that the existence of DMI makes the DWs topologically protected, preventing the walls from mutual annihilation [68]. This property would permit more DWs to be closely packed thereby increasing the device density.

14.1.7 Review of Spintronic Logic Devices

Charge-based logic schemes utilize the presence or absence of charge to distinguish between logic states 0 and 1. The switching device—e.g., CMOS—can be considered as two potential wells separating the charge using an adjustable barrier. Theoretical studies have found the minimum switching energy between the two states as k_BTln2 (23 meV) [69]. However, the projected gate switching energy for 10 nm gate length is 15 eV [70]. The information stored in the electron spin orientation is devoid of such switching energies. The non-volatility also eliminates the delay and the energy required to obtain information from the microprocessor while it is in the stand-by state, thus saving power needed to periodically turn the processor on or off. Numerous spintronic-based logic devices have been proposed; however, practical limitations such as direct cascading and small signal-to-noise ratio prevent successful commercialization of the technology [71]. In this section, we review some of the proposed schemes of the spintronic-based logic devices.

14.1.7.1 Nano-Magnetic Logic

Nanomagnetic logic (NML) is an alternative logic scheme to the existing CMOS-based technologies with promising features such as nonvolatile computing states and low power consumption. Magnetic quantum cellular automata (MQCA) or nanomagnetic logic (NML) are built from identical magnetic dots that are elongated along one direction to form a group of elliptical nanomagnets. These nanomagnets interact through electromagnetic interactions, which are either exchange or primarily dipole interactions. Room temperature demonstration of MQCA-based logic was performed by Cowburn et al. [72]. Majority gate logic application of MQCA was demonstrated by Imre et al. [73]. The neighboring magnets would adopt anti-parallel magnetization due to dipolar coupling. The singular magnet placed horizontally on the right corner would act as the control bit, which can control the magnetization direction of the other magnets. By different arrangement of the nanomagnets, logic gate operations can be obtained. However, all the logic operations require the need of an external magnetic field for clocking, which is an issue for on-chip integration and scalability.

Recently, Bhowmik et al. [74] proposed the utilization of Spin Hall effect to control the logic bits of NML with perpendicular anisotropy. The researchers showed that if the current is made to flow through a thick Ta layer underneath the nanomagnets, it could induce the switching through spin-orbit torque phenomenon. The current flow direction needs to be designed such that it does not exert torque on the input nanomagnet. The limitation of this approach is the excess joule heating caused by the current flow through the highly resistive Ta layer. Kiermaier et al. [75] proposed a current-controlled Oersted field generation in perpendicular anisotropy Co/Pt multilayers to influence the magnetic states of the nano-magnets. The researchers modeled the switching behavior of nanomagnets by employing Sharrock's formalism [76]. They also took into account the joule heating in the current driven wire. They estimated a current pulse requirement of about 7.8×10^{11} A/m^2 influencing the nanomagnets separated from the current carrying wire by 150 nm. The power dissipation in NML was computed through micromagnetic simulations by Csaba et al. [77]. They estimated the power consumption to be around a few 10 milli-electronvolts, which is close to the theoretical limit of any computation [78]. Although NML technology offers fast switching speeds, the dipole interaction between the magnets is sensitive to shape. The single domain states also become unstable due to thermal effects.

14.1.7.2 STT-MTJ-Based Logic

Many variations of STT-MTJ-based logic schemes have been proposed. An all-metallic logic scheme, referred to as mLogic, was proposed by Morris et al. [79]. It consists of two GMR or MTJ stacks for reading the output state, which is coupled to a free region carrying the input data. The input is applied by passing the current across the free layer through two pinned layers. DW exists at one end, which carries the information. When a current flows from the positive input to the negative input, the DW propagates toward the left and switches the free region of the spin valve to an up-magnetized state. Since the pinned region is also up-magnetized, the resistance is low in this state and corresponds to logic bit "0." On the other hand, when the current flows in the opposite direction, the resistance remains in the high state and corresponds to logic bit "1." Different combinations can be realized by changing the initial spin orientation of the pinned layer. The low-resistance path in the all-metallic path allows operation at low voltages, but it also provides a leakage path that could worsen the energy efficiency as compared to CMOS logic.

STT-based logic schemes would have practical issues such as high power consumption due to the requirement of large critical current to switch the magnetization. The reliability is also a concern due to breakdown of tunnel barrier on repeated switching cycles. To overcome these issues, Spin Hall effect (SHE)-based MTJ logic devices have been proposed [46,80]. In a three-terminal SHE-MTJ device, the free layer is in direct contact with the heavy metal layer. This overcomes the high write current and tunnel barrier degradation since the read and write paths are separate. However, it is difficult to cascade this three-terminal device for pipeline computing. This issue can be solved by considering a four-terminal device, proposed by Kang et al. [81]. Compared to a three-terminal device, the four- terminal device has an additional read terminal. The fully decoupled read and write paths eliminate the feedback to the input, hence enabling direct cascading. The issue with SHE-MTJ logic schemes is low density due to limitation in scaling down the device geometry and additional overheads.

14.1.7.3 All Spin Logic

The logic devices that combine charge-based and spin-based mechanisms, such as semiconductor spintronic logic devices, suffer from dynamic power dissipation due to frequent conversion from electrical to magnetic state. All-spin logic (ASL) overcomes the aforementioned challenges since it computes and stores the information in the magnetic state [82]. ASL employs a nonlocal spin signal to switch a nanomagnet, constituting input to the next stage. The input and output nanomagnets are free layers and can switch left or right representing the binary data bits. The magnetization of the input magnet generates a spin current that propagates in the channel toward the output nanomagnet. The output nanomagnet can in turn be switched by this spin current due to the application of spin torque. Thus, by applying a negative voltage to the input, the majority spins can be injected, and the information is "copied" to the output. By reversing the polarity of the input voltage to positive, the majority spins are attracted, and their absence aligns the output nanomagnet antiparallel to the input nanomagnet, thus providing a "NOT" gate operation. The conducting channel could be a metal or semiconductor. Although the semiconductors exhibit longer spin coherence lengths [83–86], the injection efficiency is low. Another important aspect is that the switching process should be nonreciprocal to prevent the output state from affecting the input state. This can be accomplished to have an input side with a tunneling interface, which would enhance the injection efficiency in a metal or a semiconductor [60,87–92], and an output side with a low resistance interface to suppress the back injection. In NML, explained previously, the nonreciprocity has been achieved using Bennett clocking [93]; however, it requires additional clocking circuitry as compared to CMOS-based logic. A full adder circuit has also been proposed recently based on the ASL technique [94].

An advantage of ASL over CMOS logic is that the logic gates can be implemented more compactly. There are some practical limitations of the ASL device. For instance, the limitation of spin-flip length in the metallic channel requires the use of repeaters such as nanomagnets to transmit information. These can consume significant power in the actual design. Scaling down ASL can increase the efficiency of spin-torque on the output due to less spin-flip scattering; however, as the input and output become closer, the dipole interaction between the magnets would start to create interference. There is a leakage current in the channel while cascading, which would increase on scaling down the device further. The spins could also flow into the ground due to reduction of the shunt path.

14.1.7.4 Domain Wall Logic

The first demonstration of a domain wall (DW)-based logic device [95] was shown by Allwood et al. The shape anisotropy of the nanostructure constrains the magnetization to follow the path of the conduit. Time-varying rotating magnetic fields were applied along the vertical and horizontal directions, and the MOKE signal was traced at various junctions. The bifurcation of DWs at the fan-out and their combination gives rise to logic operations—for instance, the MOKE signal is high at IV only when it is high at both II and III. Thus, output at IV is logical AND of inputs at II and III.

Recently, Murapaka et al. [96] demonstrated a programmable logic device based on controlled motion of DW through a U-shaped branch structure. The structure exploited geometrical asymmetry to constrain the DW motion along a particular branch as proposed by Sethi et al. [97]. The structure adopted a "pull-up" configuration in which the "u-branch" was displaced along the vertical direction (+ y-direction). This constrained the DWs to propagate along the lower branch by virtue of lower pinning strength and Zeeman energy, irrespective of their chiralities. To propel DW motion along the upper branch, a current carrying stripline was patterned in a second lithography step to partially overlap with the horizontal nanowire. The overlap was kept only 25% obtained using COMSOL simulations. This ensured that only specific DW chiralities were influenced by the gate magnetic field. The magnetization directions of the horizontal and the vertical nanowires served as the two inputs to the logic gates and the magnetization direction of the "u-branches" served as the output of the logic gates. The propagation of DW along a particular branch changed its magnetization direction and hence the output. Since the magnetization directions of the two branches are always opposite, it is possible to realize simultaneously the output and its complement—e.g., if the output of the upper branch corresponded to

NAND gate output, the lower branch would correspond to the AND gate output. Through magnetic force microscopy imaging and anisotropic magnetoresistance measurements, the researchers were able to demonstrate all basic logic operations. The magnetic field required can be generated using an inductor in the form of a meander making a possible on-chip integration. The DW motion can also be realized using spin-transfer torque (STT) with a magnetic tunnel junction read sensor integrated at the output making the device compatible with CMOS technology.

14.1.7.5 Spin-Wave Logic

Spin waves are collective excitations of spin lattice and can be used for dynamic computation or wave-based computing. Spin-wave-based logic schemes can find applications in non-Boolean devices. Since they operate in GHz range, their wavelength can be scaled down to nanometer range, hence they have excellent scaling potential. The versatility in their dispersion relation makes them useful for a variety of device applications. Kostylev et al. [98] utilized the principle of Mach-Zehnder-type spin-wave interferometers [99] to propose and experimentally demonstrate spin-wave logic device. The initial prototype was used to obtain XOR and NOT gate, which was later modified to obtain other logic functionalities [100]. A current controlled interferometer was used to construct the logic functionalities. A microwave is propagated through a power splitter and propagates through the respective branches of the interferometer. The phase shift is introduced using a controllable phase shifter (CPS) attached to the two branches, and the signals are then recombined using a mixer. In analogy to this, the researchers proposed the spin-wave logical gate where the components can be realized using a patterned ferromagnetic film of yttrium iron garnet (YIG) acting as the waveguide, and the CPS is controlled via a current carrying stripline. The carrier wave number of the spin waves would depend on the constant magnetic field applied to the ferromagnetic film, which was controlled by the current-carrying stripline attached to the ferromagnetic film. The magnetostatic carrier waves are excited in the stripe by using a transducer, and a second transducer placed a certain distance away is used for detection of the spin waves. The current direction in the stripline would dictate whether the total field is either locally enhanced or reduced resulting in the upward shift or the downward shift of the carrier wave number, respectively. When the phase shift between the CPS is π, an XOR gate is realized. The disadvantage of this method is the large current required for achieving the functionality and the large dimensions of the interferometer device. The dimensions can be reduced by using patterned permalloy (NiFe) instead of YIG.

Recently, an experimental prototype of spin-wave majority gate was proposed by Fischer et al. [101] utilizing the interference of spin waves in YIG waveguide structures. The prototype consists of three inputs that combine to give one output. The logic “0” corresponds to a certain phase $\phi(0)$, and the logic “1” corresponds to phase $\phi(1) = \phi(0) + \pi$. The RF currents passing through the copper striplines were used to excite the spin waves. The waves traveling in different waveguides are mixed in a single waveguide where they superpose to generate the output. The spin waves in the output waveguide generate an electrical signal in the copper striplines detected using a rectifying diode. The device is operated in the backward volume spin waves [102] based on the values of FMR linewidth obtained, and the magnetic field is applied parallel to the [103] waveguide. Due to the difference in the output voltages for different input voltage combinations, the spin-wave-based majority gates cannot be cascaded. To address this issue, parametric amplification or nonlinear magnon phenomena [104] may be used.

TABLE 14.1

Comparison of Device Parameters for CMOS and Spintronic-Based Logic Schemes

Device Name	Energy (fJ)	Delay (ps)	Active Power (W)	Stand-By Power (W)	Power Density (W/cm²)	Throughput (/s-cm²)
CMOS high performance	10^2	10^3	10^{-4}	10^{-5}–10^{-6}	10	10^2
CMOS low voltage	10	10^4	10^{-6}	10^{-8}–10^{-7}	1	10–10^2
All-spin logic	10^4	10^5–10^6	10^{-5}	10^{-8}	10	1
STT-MTJ/domain-wall	10^2	10^5–10^6	10^{-5}–10^{-6}	10^{-8}	1–10	1–10
Nanomagnetic logic	10^2	10^5–10^6	10^{-7}	10^{-10}	1–10^{-1}	10

Source: Nikonov, D. E., and Young, I. A., *IEEE J. Explor. Solid-State Comput. Devices Circuits*, 1, 3–11, 2015.

Table 14.1, shows a comparison of device parameters for CMOS and spintronic-based logic schemes. The benchmarking was carried out by Nikonov et al. [105] by adopting an electrical model for spin-based designs in 32-bit arithmetic logic unit (ALU) and adder circuits. The numerical values in the table indicate the computed order of magnitude of the device parameters for respective technologies. The data reveals that the spintronic devices in general switch slower and consume more energy per switching than the CMOS-based devices. The advantages are low dynamic and static power consumption, which compensates for the low throughput. Thus, although the spintronic logic has potential, more research is required to make its performance competitive with that of CMOS-based schemes.

References

1. M. N. Baibich et al., "Giant magnetoresistance of (001)Fe/(001)Cr magnetic superlattices," *Phys. Rev. Lett.*, vol. 61, no. 21, pp. 2472–2475, 1988.
2. G. Binasch, P. Grünberg, F. Saurenbach, and W. Zinn, "Enhanced magnetoresistance in layered magnetic structures with antiferromagnetic interlayer exchange," *Phys. Rev. B*, vol. 39, no. 7, pp. 4828–4830, 1989.
3. M. Jullière, "Tunneling between ferromagnetic films," *Phys. Lett. A*, vol. 54, no. 3, pp. 225–226, 1975.
4. J. S. Moodera, L. R. Kinder, T. M. Wong, and R. Meservey, "Large magnetoresistance at room temperature in ferromagnetic thin film tunnel junctions," *Phys. Rev. Lett.*, vol. 74, no. 16, pp. 3273–3276, 1995.
5. W. Butler, X.-G. Zhang, T. Schulthess, and J. MacLaren, "Spin-dependent tunneling conductance of Fe|MgO|Fe sandwiches," *Phys. Rev. B*, vol. 63, no. 5, pp. 1–12, 2001.
6. S. Yuasa, T. Nagahama, A. Fukushima, Y. Suzuki, and K. Ando, "Giant room-temperature magnetoresistance in single-crystal Fe/MgO/Fe magnetic tunnel junctions," *Nat. Mater.*, vol. 3, no. 12, pp. 868–871, 2004.
7. S. S. P. Parkin et al., "Giant tunnelling magnetoresistance at room temperature with MgO (100) tunnel barriers," *Nat. Mater.*, vol. 3, no. 12, pp. 862–867, 2004.
8. J. J. Kan et al., "A study on practically unlimited endurance of STT-MRAM," *IEEE Trans. Electron Devices*, vol. 64, no. 9, pp. 3639–3646, 2017.
9. A. Manchon and S. Zhang, "Theory of nonequilibrium intrinsic spin torque in a single nanomagnet," *Phys. Rev. B - Condens. Matter Mater. Phys.*, vol. 78, no. 21, pp. 1–4, 2008.
10. I. M. Miron et al., "Perpendicular switching of a single ferromagnetic layer induced by in-plane current injection," *Nature*, vol. 476, no. 7359, pp. 189–193, 2011.

11. L. Liu, C. F. Pai, Y. Li, H. W. Tseng, D. C. Ralph, and R. A. Buhrman, "Spin-torque switching with the giant spin hall effect of tantalum," *Science (80-.).*, vol. 336, no. 6081, pp. 555–558, 2012.
12. S. S. P. Parkin, M. Hayashi, and L. Thomas, "Magnetic domain-wall racetrack memory," *Science (80-.).*, vol. 320, pp. 190–195, 2008.
13. N. F. Mott, "The resistance and thermoelectric properties of the transition metals," *Proc. R. Soc. A Math. Phys. Eng. Sci.*, vol. 156, no. 888, pp. 368–382, 1936.
14. J. C. Slonczewski, "Conductance and exchange coupling of two ferromagnets separated by a tunneling barrier," *Phys. Rev. B*, vol. 39, no. 10, pp. 6995–7002, 1989.
15. J. Mathon and A. Umerski, "Theory of tunneling magnetoresistance of an epitaxial Fe/MgO/Fe(001) junction," *Phys. Rev. B*, vol. 63, no. 22, p. 220403, 2001.
16. J. C. Slonczewski, "Current-driven excitation of magnetic multilayers," *J. Magn. Magn. Mater.*, vol. 159, no. 1–2, pp. L1–L7, 1996.
17. L. Berger, "Emission of spin waves by a magnetic multilayer traversed by a current L.," *Phys. Rev. B*, vol. 54, no. 13, pp. 9353–9358, 1996.
18. M. Tsoi et al., "Excitation of a magnetic multilayer by an electric current," *Phys. Rev. Lett.*, vol. 80, no. 19. pp. 4281–4284, 1998.
19. J. Katine, F. Albert, R. Buhrman, E. Myers, and D. Ralph, "Current-driven magnetization reversal and spin-wave excitations in Co/Cu/Co pillars," *Phys. Rev. Lett.*, vol. 84, no. 14, pp. 3149–52, 2000.
20. S. Fujita, H. Noguchi, K. Ikegami, S. Takeda, K. Nomura, and K. Abe, "Novel memory hierarchy with e-STT-MRAM for near-future applications," *2017 International Symposium on VLSI Design, Automation and Test, VLSI-DAT 2017*, pp. 3–4, 2017.
21. J. M. Slaughter et al., "Technology for reliable spin-torque MRAM products," *Technical Digest - International Electron Devices Meeting IEDM*, pp. 21.5.1–21.5.4, 2017.
22. S. W. Chung et al., "4Gbit density STT-MRAM using perpendicular MTJ realized with compact cell structure," *Technical Digest - International Electron Devices Meeting IEDM*, pp. 27.1.1–27.1.4, 2017.
23. B. N. Engel et al., "A 4-Mb toggle MRAM based on a novel bit and switching method," *IEEE Trans. Magn.*, vol. 41, no. 1, pp. 132–136, 2005.
24. S. S. P. Parkin, M. Hayashi, and L. Thomas, "Magnetic domain-wall racetrack memory," *Science (80-.).*, vol. 320, no. 5873, pp. 190–194, 2008.
25. H. Sato et al., "Properties of magnetic tunnel junctions with a MgO/CoFeB/Ta/CoFeB/MgO recording structure down to junction diameter of 11nm," *Appl. Phys. Lett.*, vol. 105, no. 6, pp. 1–5, 2014.
26. D. E. M. Krounbi, V. Nikitin, D. Apalkov, J. Lee, X. Tang, R. Beach, "Status and challenges in spin-transfer torque MRAM technology," *ECS Trans.*, vol. 69, no. 3, pp. 119–126, 2015.
27. D. B. Gopman et al., "Enhanced ferromagnetic resonance linewidth of the free layer in perpendicular magnetic tunnel junctions Enhanced ferromagnetic resonance linewidth of the free layer in perpendicular magnetic tunnel junctions," *AIP Adv.*, vol. 7, no. 5, p. 055932, 2017.
28. M. A. W. Schoen et al., "Ultra-low magnetic damping of a metallic ferromagnet," *Nat. Phys.*, vol. 12, no. 9, pp. 839–842, 2016.
29. Z. Diao et al., "Spin transfer switching in dual MgO magnetic tunnel junctions," *Appl. Phys. Lett.*, vol. 90, no. 13, 2007.
30. G. Hu et al., "STT-MRAM with double magnetic tunnel junctions," *2015 IEEE International Electron Devices Meeting (IEDM)*, Washington, DC, 2015.
31. D. C. Worledge, "Theory of spin torque switching current for the double magnetic tunnel junction," *IEEE Magn. Lett.*, vol. 8, pp. 1–6, 2017.
32. S. Bandiera et al., "Spin transfer torque switching assisted by thermally induced anisotropy reorientation in perpendicular magnetic tunnel junctions," *Appl. Phys. Lett.*, vol. 99, no. 20, pp. 2011–2014, 2011.
33. T. Maruyama et al., "Large voltage-induced magnetic anisotropy change in a few atomic layers of iron," *Nat. Nanotechnol.*, vol. 4, no. 3, pp. 158–161, 2009.
34. W. Kang, L. Chang, Y. Zhang, and W. Zhao, "Voltage-controlled MRAM for working memory: Perspectives and challenges," *Proceedings of the 2017 Design, Automation and Test in Europe DATE 2017*, pp. 542–547, 2017.

35. D. Apalkov, B. Dieny, and J. M. Slaughter, "Magnetoresistive random access memory," *Proc. IEEE*, vol. 104, no. 10, pp. 685–697, 2016.
36. S. Ikeda et al., "Tunnel magnetoresistance of 604% at 300K by suppression of Ta diffusion in CoFeB/MgO/CoFeB pseudo-spin-valves annealed at high temperature," *Appl. Phys. Lett.*, vol. 93, no. 8, p. 082508, 2008.
37. S. H. Kang and K. Lee, "Emerging materials and devices in spintronic integrated circuits for energy-smart mobile computing and connectivity," *Acta Mater.*, vol. 61, no. 3, pp. 952–973, 2013.
38. J. Chatterjee, R. C. Sousa, N. Perrissin, S. Auffret, C. Ducruet, and B. Dieny, "Enhanced annealing stability and perpendicular magnetic anisotropy in perpendicular magnetic tunnel junctions using W layer," *Appl. Phys. Lett.*, vol. 110, no. 20, p. 202401, 2017.
39. I. M. Miron et al., "Current-driven spin torque induced by the Rashba effect in a ferromagnetic metal layer," *Nat. Mater.*, vol. 9, no. 3, pp. 230–234, 2010.
40. U. H. Pi et al., "Tilting of the spin orientation induced by Rashba effect in ferromagnetic metal layer," *Appl. Phys. Lett.*, vol. 97, no. 16, 2010.
41. L. Liu, O. J. Lee, T. J. Gudmundsen, D. C. Ralph, and R. A. Buhrman, "Current-induced switching of perpendicularly magnetized magnetic layers using spin torque from the spin hall effect," *Phys. Rev. Lett.*, vol. 109, no. 9, pp. 1–5, 2012.
42. C. Pai et al., "Spin transfer torque devices utilizing the giant spin Hall effect of tungsten," *Appl. Phys. Lett.*, vol. 122404, pp. 1–5, 2012.
43. A. Manchon et al., "Analysis of oxygen induced anisotropy crossover in Pt/Co/MOx trilayers," *J. Appl. Phys.*, vol. 104, no. 4, 2008.
44. S. Woo, M. Mann, A. J. Tan, L. Caretta, and G. S. D. Beach, "Enhanced spin-orbit torques in Pt/Co/Ta heterostructures," *Appl. Phys. Lett.*, vol. 105, no. 21, 2014.
45. A. V. Khvalkovskiy et al., "Matching domain-wall configuration and spin-orbit torques for efficient domain-wall motion," *Phys. Rev. B - Condens. Matter Mater. Phys.*, vol. 87, no. 2, pp. 2–6, 2013.
46. M. Cubukcu et al., "Spin-orbit torque magnetization switching of a three-terminal perpendicular magnetic tunnel junction," *Appl. Phys. Lett.*, vol. 104, no. 4, 2014.
47. K. Garello et al., "Ultrafast magnetization switching by spin-orbit torques," *Appl. Phys. Lett.*, vol. 105, no. 21, 2014.
48. G. S. D. Beach, C. Nistor, C. Knutson, M. Tsoi, and J. L. Erskine, "Dynamics of field-driven domain-wall propagation in ferromagnetic nanowires," *Nat. Mater.*, vol. 4, no. 10, pp. 741–744, 2005.
49. A. Thiaville, Y. Nakatani, J. Miltat, and Y. Suzuki, "Micromagnetic understanding of current-driven domain wall motion in patterned nanowires," *Europhys. Lett.*, vol. 69, no. 6, pp. 990–996, 2005.
50. S. Zhang and Z. Li, "Roles of nonequilibrium conduction electrons on the magnetization dynamics of ferromagnets," *Phys. Rev. Lett.*, vol. 93, no. 12, pp. 1–4, 2004.
51. G. Tatara and H. Kohno, "Theory of current-driven domain wall motion: Spin transfer versus momentum transfer," *Phys. Rev. Lett.*, vol. 92, no. 8, 2004.
52. J. Grollier et al., "Switching a spin valve back and forth by current-induced domain wall motion," *Appl. Phys. Lett.*, vol. 83, no. 3, pp. 509–511, 2003.
53. M. Tsoi, R. E. Fontana, and S. S. P. Parkin, "Magnetic domain wall motion triggered by an electric current," *Appl. Phys. Lett.*, vol. 83, no. 13, pp. 2617–2619, 2003.
54. A. Yamaguchi, T. Ono, S. Nasu, K. Miyake, K. Mibu, and T. Shinjo, "Real-space observation of current-driven domain wall motion in submicron magnetic wires," *Phys. Rev. Lett.*, vol. 92, no. 7, 2004.
55. M. Hayashi, L. Thomas, R. Moriya, C. Rettner, and S. S. P. Parkin, "Current-Controlled Magnetic Domain-Wall Nanowire Shift Register," *Science (80-.).*, vol. 320, pp. 209–211, 2008.
56. M. Chandra Sekhar, S. Goolaup, I. Purnama, and W. S. Lew, "Depinning assisted by domain wall deformation in cylindrical NiFe nanowires," *J. Appl. Phys.*, vol. 115, no. 8, pp. 1–7, 2014.
57. M. C. Sekhar, S. Goolaup, I. Purnama, and W. S. Lew, "Crossover in domain wall potential polarity as a function of anti-notch geometry," *J. Phys. D. Appl. Phys.*, vol. 44, no. 23, 2011.

58. D. W. Wong, I. Purnama, G. J. Lim, W. L. Gan, C. Murapaka, and W. S. Lew, "Current-induced three-dimensional domain wall propagation in cylindrical NiFe nanowires," *J. Appl. Phys.*, vol. 119, no. 15, 2016.
59. M. Chandra Sekhar, H. F. Liew, I. Purnama, W. S. Lew, M. Tran, and G. C. Han, "Helical domain walls in constricted cylindrical NiFe nanowires," *Appl. Phys. Lett.*, vol. 101, no. 15, pp. 1–6, 2012.
60. C. Guite, I. S. Kerk, M. C. Sekhar, M. Ramu, S. Goolaup, and W. S. Lew, "All-electrical deterministic single domain wall generation for on-chip applications," *Sci. Rep.*, vol. 4, pp. 1–5, 2014.
61. P. Sethi, C. Murapaka, G. J. Lim, and W. S. Lew, "In-plane current induced domain wall nucleation and its stochasticity in perpendicular magnetic anisotropy Hall cross structures," *Appl. Phys. Lett.*, vol. 107, no. 19, 2015.
62. J. H. Franken, H. J. M. Swagten, and B. Koopmans, "Shift registers based on magnetic domain wall ratchets with perpendicular anisotropy," *Nat. Nanotechnol.*, vol. 7, no. 8, pp. 499–503, 2012.
63. T. Koyama et al., "Observation of the intrinsic pinning of a magnetic domain wall in a ferromagnetic nanowire," *Nat. Mater.*, vol. 10, no. 3, pp. 194–197, 2011.
64. S. H. Yang, K. S. Ryu, and S. Parkin, "Domain-wall velocities of up to 750 m s-1driven by exchange-coupling torque in synthetic antiferromagnets," *Nat. Nanotechnol.*, vol. 10, no. 3, pp. 221–226, 2015.
65. S. Emori, U. Bauer, S. M. Ahn, E. Martinez, and G. S. D. Beach, "Current-driven dynamics of chiral ferromagnetic domain walls," *Nat. Mater.*, vol. 12, no. 7, pp. 611–616, 2013.
66. K.-S. Ryu, L. Thomas, S.-H. Yang, and S. Parkin, "Chiral spin torque at magnetic domain walls," *Nat. Nanotechnol.*, vol. 8, no. 7, pp. 527–533, 2013.
67. P. P. J. Haazen, E. Murè, J. H. Franken, R. Lavrijsen, H. J. M. Swagten, and B. Koopmans, "Domain wall depinning governed by the spin Hall effect," *Nat. Mater.*, vol. 12, no. 4, pp. 299–303, 2013.
68. M. J. Benitez et al., "Magnetic microscopy and topological stability of homochiral Néel domain walls in a Pt/Co/AlO x trilayer," *Nat. Commun.*, vol. 6, pp. 1–7, 2015.
69. R. Landauer, "Irreversibility and heat generation in the computational process," *IBM J. Res. Dev.*, vol. 5, pp. 183–191, 1961.
70. ITRS, "International Technology Roadmap for Semiconductors 2.0: Executive report," International Technology Roadmap for Semiconductors, p. 79, 2015.
71. J. S. Friedman and A. V. Sahakian, "Complementary magnetic tunnel junction logic," *IEEE Trans. Electron Devices*, vol. 61, no. 4, pp. 1207–1210, 2014.
72. R. P. Cowburn, "Room temperature magnetic quantum cellular automata," *Science (80-.).*, vol. 287, no. 5457, pp. 1466–1468, 2000.
73. A. Imre, G. Csaba, L. Ji, A. Orlov, H. Brenstein, and W. Porod, "Majority logic gate for magnetic quantum-dot cellular automata," *Science (80-.).*, vol. 311, pp. 206–208, 2006.
74. D. Bhowmik, L. You, and S. Salahuddin, "Spin hall effect clocking of nanomagnetic logic without a magnetic field," *Nat. Nanotechnol.*, vol. 9, no. 1, pp. 59–63, 2014.
75. J. Kiermaier, S. Breitkreutz, G. Csaba, D. Schmitt-Landsiedel, and M. Becherer, "Electrical input structures for nanomagnetic logic devices," *J. Appl. Phys.*, vol. 111, no. 7, pp. 100–103, 2012.
76. M. P. Sharrock, "Time dependence of switching fields in magnetic recording media (invited)," *J. Appl. Phys.*, vol. 76, no. 10, pp. 6413–6418, 1994.
77. G. Csaba, P. Lugli, and W. Porod, "Power dissipation in nanomagnetic logic devices," *4th IEEE Conference on Nanotechnology*, no. 3, pp. 346–348, 2004.
78. W. Porod, R. O. Grondin, D. K. Ferry, and G. Porod, "Dissipation in Computation," *Phys. Rev. Lett.*, vol. 52, no. 3, pp. 232–235, 1984.
79. D. Morris, D. Bromberg, J. Zhu, and L. Pileggi, "mLogic: Ultra-low voltage non-volatile logic circuits using STT-MTJ devices," *Design Automation Conference* (DAC), *2012 49th ACM/EDAC/IEEE*, pp. 486–491, 2012.
80. M. Yamanouchi et al., "Three terminal magnetic tunnel junction utilizing the spin Hall effect of iridium-doped copper," *Appl. Phys. Lett.*, vol. 102, no. 21, 2013.

81. W. Kang, Z. Wang, Y. Zhang, J.-O. Klein, W. Lv, and W. Zhao, "Spintronic logic design methodology based on spin Hall effect–driven magnetic tunnel junctions," *J. Phys. D. Appl. Phys.*, vol. 49, no. 6, p. 065008, 2016.
82. B. Behin-Aein, D. Datta, S. Salahuddin, and S. Datta, "Proposal for an all-spin logic device with built-in memory," *Nat. Nanotechnol.*, vol. 5, no. 4, pp. 266–270, 2010.
83. B. Huang, D. J. Monsma, and I. Appelbaum, "Coherent spin transport through a 350 micron thick silicon wafer," *Phys. Rev. Lett.*, vol. 99, no. 17, pp. 1–4, 2007.
84. B. Huang, H. J. Jang, and I. Appelbaum, "Geometric dephasing-limited Hanle effect in long-distance lateral silicon spin transport devices," *Appl. Phys. Lett.*, vol. 93, no. 16, pp. 2013–2016, 2008.
85. B. Huang and I. Appelbaum, "Spin dephasing in drift-dominated semiconductor spintronics devices," *Phys. Rev. B - Condens. Matter Mater. Phys.*, vol. 77, no. 16, pp. 1–6, 2008.
86. H. J. Jang, J. Xu, J. Li, B. Huang, and I. Appelbaum, "Non-ohmic spin transport in n-type doped silicon," *Phys. Rev. B - Condens. Matter Mater. Phys.*, vol. 78, no. 16, pp. 1–6, 2008.
87. E. I. Rashba, "Theory of electrical spin injection: Tunnel contacts as a solution of the conductivity mismatch problem," *Phys. Rev. B - Condens. Matter Mater. Phys.*, vol. 62, no. 24, pp. 267–270, 2000.
88. G. Schmidt, L. W. Molenkamp, A. T. Filip, and B. J. van Wees, "Basic obstacle for electrical spin-injection from a ferromagnetic metal into a diffusive semiconductor," *Phys. Rev.*, vol. 62, no. 8, pp. 4790–4793, 1999.
89. I. Appelbaum, B. Huang, and D. J. Monsma, "Electronic measurement and control of spin transport in silicon," *Nature*, vol. 447, no. 7142, pp. 295–298, 2007.
90. B. T. Jonker, G. Kioseoglou, A. T. Hanbicki, C. H. Li, and P. E. Thompson, "Electrical spin-injection into silicon from a ferromagnetic metal/tunnel barrier contact," *Nat. Phys.*, vol. 3, no. 8, pp. 542–546, 2007.
91. X. Lou et al., "Electrical detection of spin transport in lateral ferromagnet-semiconductor devices," *Nat. Phys.*, vol. 3, no. 3, pp. 197–202, 2007.
92. S. P. Dash, S. Sharma, R. S. Patel, M. P. De Jong, and R. Jansen, "Electrical creation of spin polarization in silicon at room temperature," *Nature*, vol. 462, no. 7272, pp. 491–494, 2009.
93. J. Atulasimha and S. Bandyopadhyay, "Bennett clocking of nanomagnetic logic using multiferroic single-domain nanomagnets," *Appl. Phys. Lett.*, vol. 97, no. 17, pp. 1–4, 2010.
94. Q. An, L. Su, J. O. Klein, S. Le Beux, I. O'Connor, and W. Zhao, "Full-adder circuit design based on all-spin logic device," *Proceedings of the 2015 IEEE/ACM International Symposium on Nanoscale Architectures NANOARCH 2015*, pp. 163–168, 2015.
95. D. A. Allwood, G. Xiong, C. C. Faulkner, D. Atkinson, D. Petit, and R. P. Cowburn, "Magnetic domain-wall logic," *Science (80-.).*, vol. 309, no. 5741, pp. 1688–1692, 2005.
96. C. Murapaka, P. Sethi, S. Goolaup, and W. S. Lew, "Reconfigurable logic via gate controlled domain wall trajectory in magnetic network structure," *Sci. Rep.*, vol. 6, pp. 1–11, 2016.
97. P. Sethi, C. Murapaka, S. Goolaup, Y. J. Chen, S. H. Leong, and W. S. Lew, "Direct observation of deterministic domain wall trajectory in magnetic network structures," *Sci. Rep.*, vol. 6, pp. 1–8, 2016.
98. M. P. Kostylev, A. A. Serga, T. Schneider, B. Leven, and B. Hillebrands, "Spin-wave logical gates," *Appl. Phys. Lett.*, vol. 87, no. 15, pp. 1–3, 2005.
99. Y. K. Fetisov and C. E. Patton, "Microwave bistability in a magnetostatic wave interferometer with external feedback," *Magn. IEEE Trans.*, vol. 35, no. 2, pp. 1024–1036, 1999.
100. T. Schneider, A. A. Serga, B. Leven, B. Hillebrands, R. L. Stamps, and M. P. Kostylev, "Realization of spin-wave logic gates," *Appl. Phys. Lett.*, vol. 92, no. 2, 2008.
101. T. Fischer et al., "Experimental prototype of a spin-wave majority gate," *Appl. Phys. Lett.*, vol. 110, no. 15, 2017.
102. A. A. Serga, A. V. Chumak, and B. Hillebrands, "YIG magnonics," *J. Phys. D. Appl. Phys.*, vol. 43, no. 26, 2010.

103. T. Brächer, P. Pirro, and B. Hillebrands, "Parallel pumping for magnon spintronics: Amplification and manipulation of magnon spin currents on the micron-scale," *Phys. Rep.*, vol. 699, pp. 1–34, 2017.
104. A. V. Chumak, A. A. Serga, and B. Hillebrands, "Magnon transistor for all-magnon data processing," *Nat. Commun.*, vol. 5, pp. 1–8, 2014.
105. D. E. Nikonov and I. A. Young, "Benchmarking of beyond-CMOS exploratory devices for logic integrated circuits," *IEEE J. Explor. Solid-State Comput. Devices Circuits*, vol. 1, pp. 3–11, 2015.

15

Fundamentals, Modeling, and Application of Magnetic Tunnel Junctions

Ramtin Zand, Arman Roohi, and Ronald F. DeMara

CONTENTS

Aggressive Metal Oxide Semiconductor (MOS) technology scaling in digital circuits has resulted in important challenges including a significant increase in leakage currents, short-channel effects, and drain saturation growth while reducing the power supply voltage for digital applications. Furthermore, by extensions to sub-10-nm regimes, error resiliency has become a major challenge for the microelectronics industry, particularly mission-critical systems, e.g., space and terrestrial applications. Therefore, emerging devices and technologies have attracted considerable attention in recent years as an alternative to CMOS-based technologies such as spintronic [1–6], resistive random access memory (RRAM) [7–10], phase-change memory (PCM) [11,12], and quantum cellular automata (QCA) [13–18]. Among promising devices, the 2014 Magnetism Roadmap [19] identifies nanomagnetic devices as capable post-CMOS candidates, of which magnetic tunnel junctions (MTJs) are considered one of the most promising technologies spanning both logic [20–22] and memory functionalities [23–25]. MTJs are characterized by non-volatility, near-zero standby power, high integration density, and radiation hardness as a technology progression from CMOS. Moreover, MTJs can be readily integrated at the back-end process of the CMOS fabrication due to their vertical structure [26,27].

15.1 Fundamentals and Modeling of Magnetic Tunnel Junctions

Figure 15.1 depicts the vertical structure of an MTJ [28,29], which consists of two ferromagnetic (FM) layers: (1) *Fixed Layer*, that is magnetically-pinned and utilized as a reference layer, and (2) *Free Layer*, that is, its magnetic orientation can be switched.

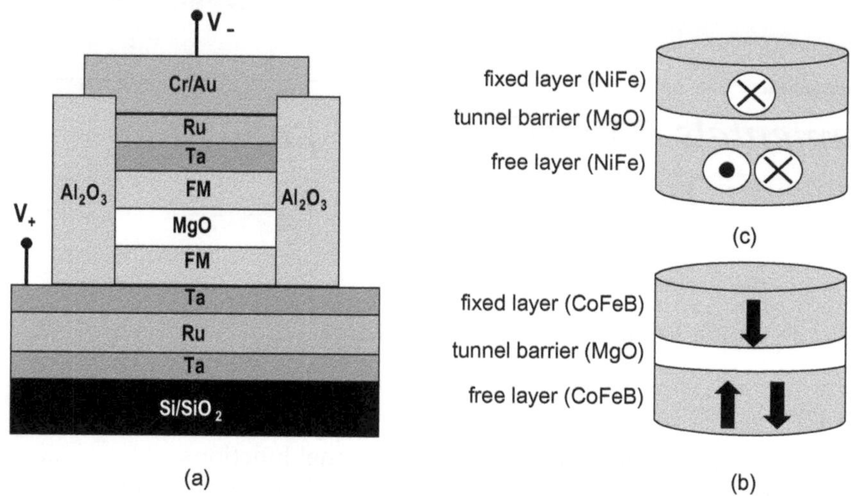

FIGURE 15.1
(a) MTJ vertical structure, (b) perpendicular MTJ (PMTJ), and (c) in-plane MTJ (IMTJ).

These two FM layers are separated by a thin oxide barrier, e.g., MgO [30]. The FM layers can have two different magnetization configurations called *parallel* (P) and *antiparallel* (AP), according to which MTJ resistance changes between R_P and R_{AP}, respectively. MTJ resistance is determined by the angle (θ) between the magnetization orientations of the fixed layer and free layer due to the tunnel magnetoresistance (TMR) effect. The MTJ resistance in P (θ = 0°) and AP (θ = 180°) states is expressed by the following equations [31–33]:

$$R(\theta) = 2R_{\text{MTJ}} \times \frac{1+\text{TMR}}{2+\text{TMR}+\text{TMR}.\cos\theta} = \begin{cases} R_p = R_{\text{MTJ}} & , \quad \theta = 0^\circ \\ R_{\text{ap}} = R_{\text{MTJ}}(1+\text{TMR}), & \theta = 180^\circ \end{cases} \tag{15.1}$$

$$R_{\text{MTJ}} = \frac{t_{\text{ox}}}{\text{Factor} \times \text{Area}.\sqrt{\varphi}} \exp\left(1.025 \times t_{\text{ox}}.\sqrt{\varphi}\right) \tag{15.2}$$

$$\text{TMR} = \text{TMR}(0)/1+\left(\frac{V_b}{V_h}\right)^2 \tag{15.3}$$

where V_b is the bias voltage, $V_h = 0.5\text{V}$ is the bias voltage when TMR is half of the TMR_0, t_{ox} is the oxide thickness of MTJ, *Factor* is obtained from the resistance-area product value of the MTJ that relies on the material composition of its layers, *Area* is the surface of MTJ, and φ is the oxide layer energy barrier height. The energy barrier between P and AP configurations of MTJ is in a range such that it can switch between configurations while also retaining thermal stability. The magnetic direction of MTJ layers can be in the film plane or out of the film plane, referred to as in-plane MTJ (IMA) and perpendicular MTJ (PMA) structure, respectively.

Two of the conventional switching methods used for changing the magnetization orientation of free layers are field-induced magnetic switching (FIMS) [34] and thermally assisted switching (TAS) [35]. In the aforementioned approaches, a current source with an amplitude in range of milliampere (mA) was required to generate the magnetic field, which should be applied to switch the MTJ state. Thus, these approaches are not appropriate for low-power integrated circuits due to the significantly high-switching energy consumption. In 1996, Slonczewski [36] proposed a spin-transfer torque (STT) switching method, which is known as a promising alternative for changing the MTJ states.

Based on the STT approach, the magnetic orientation of the free layer can be switched using a bidirectional spin-polarized current (I_{MTJ}) passing through the MTJ device. The MTJ states can be switched when the I_{MTJ} becomes higher than a critical current (I_C), as shown in Figure 15.2. First, the spin of the electrons is polarized by the fixed layer, then the spin of the electrons that flow through the MTJ free layer will be aligned with the magnetization direction of the nanomagnet by undergoing an exchange field. This phenomenon is called *spin-filtering effect*. The conservation of the angular momentum results in the exertion of an opposite sign torque with equal magnitude on the free layer, which eventually changes its magnetization direction. The direction of the current that flows through MTJ defines its *P* or AP configuration. The required bidirectional current could be readily produced by simple MOS-based circuits.

STT switching behavior can be categorized into two main regions based on the relation between I_{MTJ} and I_C: (1) *precessional region* ($I_{MTJ} > I_C$) described by the Sun model [37], where MTJ experiences a rapid precessional switching, and (2) *thermal activation region* ($I_{MTJ} < I_C$) defined by the Neel-Brown model [38], in which the switching can occur with a long input current pulse due to the thermal activation. The switching duration in the precessional and thermal activation regions is described by Equations (15.4) and (15.5), respectively [27],

$$\frac{1}{\langle \tau_{STT} \rangle} = \left[\frac{2}{C + \ln(\pi^2 \Delta)} \right] \frac{\mu_B P}{em(1+P^2)} (I_{MTJ} - I_C) \tag{15.4}$$

$$\frac{1}{\langle \tau_{STT} \rangle} = \tau_0 \exp\left(\Delta \times \left(1 - \frac{I_{MTJ}}{I_C} \right) \right) \tag{15.5}$$

where τ_{STT} is the mean switching duration, $C = 0.577$ is the Euler's constant, $\Delta = E/4k_BT$ is the thermal stability factor, m is the free layer magnetic moment, and τ_0 is the attempt period. In practice, MTJ is normally designed to work in the precessional region with an input current amplitude larger than the critical current to achieve high switching speed.

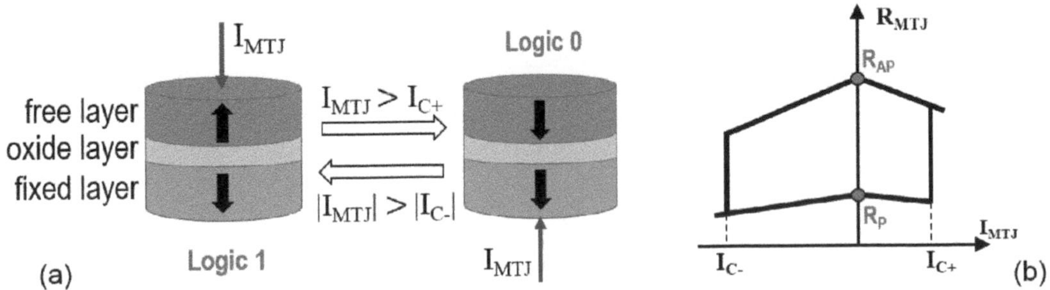

FIGURE 15.2
(a) MTJ state changes from *P* to AP due to the positive current $I_{MTJ} > I_{c+}$ condition, rather than a negative current $I_{MTJ} > |I_{c-}|$ condition, where $|I_{c-}| > I_{c+}$ and (b) MTJ resistance hysteresis curve.

Equations (15.6) and (15.7) express the switching critical current for IMA ($I_{c\text{-IMA}}$) [39] and PMA ($I_{c\text{-PMA}}$) [28] devices, respectively,

$$I_{c\text{-IMTJ}} = 2\alpha e M_s V\left(H_C + \frac{H_{\text{eff}}}{2}\right) / g(\theta)P\bar{h} \tag{15.6}$$

$$I_{c\text{-PMTJ}} = \alpha\gamma e M_s H_k V / \mu_B g(\theta) \tag{15.7}$$

where α is the magnetic damping constant, μ_B is the Bohr magneton, γ is the gyromagnetic ratio, e is the electric charge, V is the volume of the free layer, M_s is the saturation magnetization, $\bar{h}$ is the reduced Planck's constant, H_C is the in-plane coercive field, H_{eff} is the effective out-of-plane demagnetization field, and H_k is the anisotropy field. The effective demagnetization field in IMA is approximately equal to the saturation magnetization, which is normally larger than the anisotropy field in PMA. Thus, the switching current for PMA is smaller than that of the IMA devices according to Equations (15.6) and (15.7). Moreover, the spin polarization efficiency factor, $g(\theta)$, is a function of the angle between the free layer and fixed layer magnetization directions (θ) and is obtained by Equations (15.8) and (15.9) for IMA [40] and PMA [33] devices, respectively.

$$g_{\text{IMTJ}} = \left[-4 + \left(P^{-1/2} + P^{1/2}\right) \times \left(3 + \cos\theta\right)/4\right]^{-1} \tag{15.8}$$

$$g_{\text{PMTJ}} = g_{\text{SV}} \pm g_{\text{tunnel}} \tag{15.9}$$

$$g_{\text{SV}} = \left[-4 + \left(P^{-1/2} + P^{1/2}\right)^3 \times \left(3 + \cos\theta\right)/4\right]^{-1}$$

$$g_{\text{tunnel}} = \left(P/2\right) / \left(1 + P^2\cos\theta\right)$$

where P is the spin polarization percentage of the tunnel current, g_{SV} is the spin polarization efficiency in a spin valve, and g_{tunnel} is the spin polarization efficiency in tunnel junction nanopillars.

The dynamics of the magnetic moment of free layer in an STT-MTJ device can be defined by the Landau–Lifshitz–Gilbert (LLG) equation [41,42]

$$\frac{\partial\vec{m}}{\partial t} = -\gamma\mu_0\left(\vec{m} \times \vec{H}_{\text{eff}}\right) + \alpha\left(\vec{m} \times \frac{\partial\vec{m}}{\partial t}\right) - c_{\text{STT}}\left(\vec{m} \times \vec{m} \times \vec{M}\right) \tag{15.10}$$

where $\vec{m}$ and $\vec{M}$ are the unit vectors of the free layer and fixed layer magnetizations, respectively. The STT coefficient is $c_{\text{STT}} = \eta^{\mu_B I_{\text{MTJ}}}/_{eV}$, where η is spin-torque efficiency depending on the current direction. The Equation (15.10) can be expressed by a nonlinear system of two differential equations as shown in (15.11), where θ and φ are polar and azimuthal angles, respectively, and K is the anisotropic constant [43]:

$$\begin{cases} \dfrac{d\theta}{dt} = \alpha\dfrac{d\varphi}{dt} + \dfrac{c_{\text{STT}}}{M_s} \\ \dfrac{d\varphi}{dt} = -\gamma\mu_0 H_{\text{eff}}\sin\theta - \dfrac{2\gamma K}{M_s}\sin\theta\cos\theta \end{cases} \tag{15.11}$$

While the STT approach offers significant advantages in terms of read energy and speed, a significant incubation delay due to the pre-switching oscillation [44,45] incurs high switching energy. Consequently, Spin Hall Effect (SHE) and Rashba effect have been investigated recently to achieve an alternative low power-switching approach [46–48]. Recently, SHE-MTJ was introduced as an alternative for 2-terminal MTJs, which provides separate paths for read and write operations, while expending significantly less switching energy [49–51].

It is shown in [47] that a spin-polarized current can be generated in nanomagnetic devices through the SHE, which can be utilized to produce the required torque for switching the magnetization directions of the free layer in MTJs. In [50], Manipatruni et al. have provided the physical equations of the 3-terminal SHE-MTJ device behavior. Figure 15.3 shows the structure of the SHE-MTJ device, in which the magnetic orientation of the free layer changes by passing a charge current through a heavy metal (HM). The MTJ free layer is directly connected to the HM, which is normally made of β-tantalum (β-Ta) [47], β-tungsten (β-W) [48], or Pt [52]. The MTJ logic state is determined by the direction of the applied charge current. The spin-orbit coupling in the HM deflects the electrons with different spins in opposite directions, which results in a spin injection current transverse to the applied charge current. The injected current produces the required spin torque for aligning the magnetic direction of the free layer. The ratio of the injected spin current to the applied charge current, called Spin Hall injection efficiency (SHIE), is defined by Equation (15.12):

$$\text{SHIE} = \frac{I_{sz}}{I_{cx}} = \frac{(\pi/4).w_{\text{MTJ}}.l_{\text{MTJ}}}{t_{\text{HM}}.w_{\text{HM}}}\theta_{\text{SHE}}\left[1-\text{sech}\left(\frac{t_{\text{HM}}}{\lambda_{sf}}\right)\right] \quad (15.12)$$

where w_{MTJ} is the width of the MTJ, l_{MTJ} is the length of the MTJ, t_{HM} is the thickness of the HM, w_{HM} is the width of the HM, λ_{sf} is the spin-flip length in the HM, and $\theta_{\text{SHE}} = J_s/J_c$ is the Spin Hall angle, where J_c is the applied charge current density and J_s is the spin current density generated by SHE [50]. An interesting feature of the SHE phenomenon that can be observed through this equation is how it can lead to spin currents that are larger than the inducing charge currents. For instance, the SHIE value of 1.73 is achieved using the experimental parameters mentioned in Table 15.1. Thus, the generated spin current is larger than the applied charge current. In contrast, the spin injection efficiency of a 2-terminal MTJ is normally less than one, resulting in a favorable write switching energy

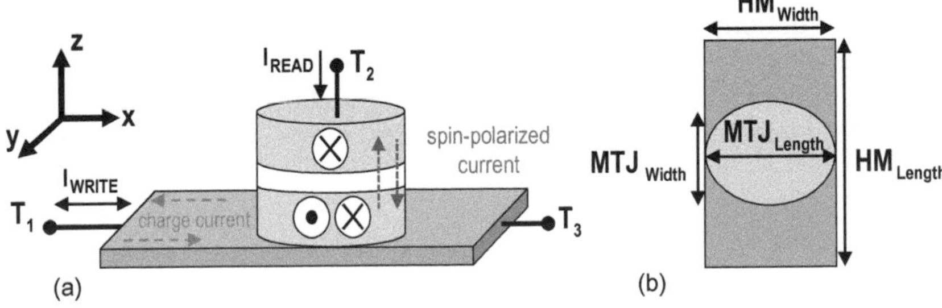

FIGURE 15.3
(a) SHE-MTJ vertical structure. Positive current along $+x$ induces a spin injection current $+z$ direction. The injected spin current produces the required spin torque for aligning the magnetic direction of the free layer in $+y$ directions and vice versa. (b) SHE-MTJ Top view.

TABLE 15.1

Experimental Parameters of STT/SHE-MTJ Devices

Parameter	Description	Value
HM Volume	$l_{HM} \times w_{HM} \times t_{HM}$	100 nm × 60 nm × 3 nm
MTJ_{Area}	$MTJ_{Length} \times MTJ_{Width} \times \pi/4$	60 nm × 30 nm × π/4
MTJ_{Area}	Reference MTJ surface	50 nm × 25 nm × π/4
t_{ox}	Thickness of oxide barrier	0.85 nm
α	Gilbert damping factor	0.007
μ_B	Bohr magneton	9.27e-24 $J{\cdot}T^{-1}$
M_s	Saturation magnetization	7.8e5 $A{\cdot}m^{-1}$
P	Spin polarization	0.52
γ	Gyromagnetic ratio	1.76e7 $(Oe.s)^{-1}$
R_{AP}, R_P	MTJ resistances	2.8 KΩ, 5.6 KΩ
R_P	Reference MTJ resistance	4.12 KΩ
TMR_0	TMR ratio	100%
H_k	Anisotropy field	80 Oe
μ_0	Permeability of free space	1.25663e-6 T.m/A
θ_{SHE}	Spin Hall angle	0.3
ρ_{HM}	HM resistivity	200 μΩ.cm
φ	Potential barrier height	0.4 V
Λsf	Spin-flip length	1.5 nm
E	Electric charge	1.602e-19 C
$\hbar$	Reduced Planck's constant	6.626e-34/2π J.s
$I_{C\text{-}SHE}$	SHE-MTJ critical current	108 μA
$I_{P\text{-}AP}$	STT-MTJ critical current For P to AP switching	37 μA
$I_{AP\text{-}P}$	STT-MTJ critical current For AP to P switching	18 μA

Source: Zand, R. et al., *IEEE Trans. Nanotechnol.*, 16, 32–43, 2017.

for SHE-MTJs. The critical spin current required for switching the free layer magnetization orientation is expressed by Equation (15.13) [53]:

$$I_{S,\text{critical}} = 2q\alpha M_S V_{\text{MTJ}} \left(H_k + 2\pi M_S\right)/\hbar \tag{15.13}$$

where V_{MTJ} is the MTJ free layer volume. Thus, the SHE-MTJ critical charge current can be calculated using Equations (15.12) and (15.13).

The relation between the switching time (τ_{SHE}) and the applied charge current (I_{SHE}) is shown in (15.14), in which v_c is the critical switching voltage, τ_0 is the characteristic time, and R_{HM} is the HM resistance, which are given by Equations (15.15) through (15.17), respectively [50],

$$\tau_{\text{SHE}} = [\tau_0.\ln\left(\pi/2\theta_0\right)]/[\left(\frac{R_{\text{HM}}.I_{\text{SHE}}}{v_c}\right) - 1] \tag{15.14}$$

$$v_c = 8\rho.I_c \left\{\theta_{\text{SHE}}.\left[1 - \text{sech}\left(\frac{t_{\text{HM}}}{\lambda_{\text{sf}}}\right)\right].\pi.l_{\text{HM}}\right\}^{-1} \tag{15.15}$$

$$\tau_0 = M_s \,.\, V_{\mathrm{HM}} \,.\, q/I_c \,.\, P \,.\, \mu_B \tag{15.16}$$

$$R_{\mathrm{HM}} = (\rho_{\mathrm{HM}} \,.\, l_{\mathrm{HM}})/(w_{\mathrm{HM}} \,.\, t_{\mathrm{HM}}) \tag{15.17}$$

where θ_0 is the effect of stochastic variation, l_{HM} is the length of the HM, and I_C is the critical charge current for spin-torque induced switching.

The dynamic behavior of a free layer in SHE-MTJ devices can be described using a modified Landau–Lifshitz–Gilbert (LLG) equation, as expressed in Equations (15.18) through (15.21) [49,54]

$$\frac{\partial \vec{m}}{\partial t} = -\gamma\mu_0\left(\vec{m}\times\vec{H}_{\mathrm{eff}}\right) + \alpha\left(\vec{m}\times\frac{\partial \vec{m}}{\partial t}\right) - c_{\mathrm{SHE}}\left(\vec{m}\times\vec{\sigma}_{\mathrm{S\hat{H}E}}\times\vec{m}\right) \tag{15.18}$$

$$\vec{m} = \sin\theta\cos\varphi\vec{x} + \sin\theta\sin\varphi\vec{y} + \cos\theta\vec{z} \tag{15.19}$$

$$\vec{\sigma}_{\mathrm{SHE}} = -\vec{x} \tag{15.20}$$

$$\vec{H}_{\mathrm{eff}} = H_k\sin\theta\cos\varphi\vec{x} - M_s\cos\theta\,\vec{z} \tag{15.21}$$

where $\vec{m}$ is the unit vector of free layer magnetization, γ is electron gyromagnetic ratio, μ_0 is vacuum permeability, and H_{eff} is the effective magnetic field. The Spin Hall torque coefficient is $c_{\mathrm{SHE}} = J_s\theta_{\mathrm{SHE}}\bar{h}\gamma/(2qd)$, where $\bar{h}$ is the reduced Planck constant and d is the free layer thickness. The Equation (15.14) can be solved assuming that $\vec{x}$, $\vec{y}$, and $\vec{z}$ are the unit vecors along X, Y, and Z axes of the Cartesian coordinate system, respectively. Moreover, θ and φ are polar and azimuthal angles, respectively. Substituting Equations (15.19) through (15.21) into Equation (15.18) results in a nonlinear system of two differential equations expressed in Equation (15.22) [49,54]:

$$\begin{cases} \dfrac{d\theta}{dt} = \dfrac{\begin{array}{c}\gamma\mu_0 H_k\sin\theta\cos\varphi(-\sin\varphi + \alpha\cos\theta\cos\varphi) \\ +\alpha\gamma\mu_0 M_s\cos\theta\sin\theta - c_{\mathrm{SHE}}\left(\cos\varphi\cos\theta + \alpha\sin\varphi\right)\end{array}}{\left(1+\alpha^2\right)} \\[2ex] \dfrac{d\varphi}{dt} = \dfrac{\gamma\mu_0\begin{pmatrix}-H_k\sin\theta\cos\varphi\cos\theta\cos\varphi - \alpha H_k\sin\theta\cos\varphi\sin\varphi \\ -M_s\cos\theta\sin\theta\end{pmatrix} + c_{\mathrm{SHE}}\left(\sin\varphi - \alpha\cos\varphi\cos\theta\right)}{\left(1+\alpha^2\right)\sin\theta} \end{cases} \tag{15.22}$$

The free layer magnetization dynamics can be described using the θ and φ coordinates, which can be obtained via solving Equation (15.22), in which the initial θ induced by the thermal distribution at a finite temperature can be calculated as $\theta_0 = \sqrt{k_B/2E_b}$, where $E_b = (1/2)\mu_0 M_s H_k V_{\mathrm{MTJ}}$ is the thermal barrier of the magnet [55].

Figure 15.4 shows the block diagram of the approach proposed in [21,56] to model the behavior of STT-MTJ and SHE-MTJ devices, in which a Verilog-AMS model is developed using the aforementioned equations. Then, the model is leveraged in a SPICE circuit simulator to design hybrid CMOS/spin-based circuits and validate their functionality using experimental parameters such as the ones listed in Table 15.1.

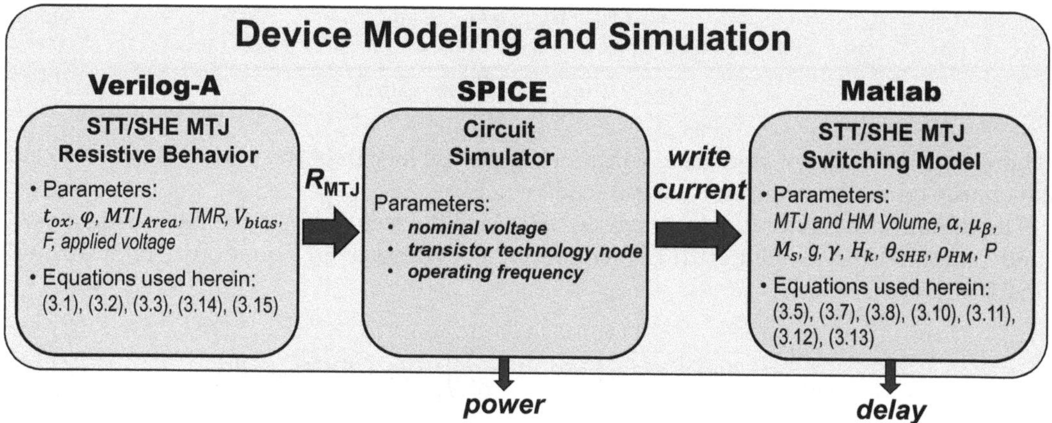

FIGURE 15.4
Modeling and simulation process of STT/SHE MTJ devices. (From Zand, R. et al., *IEEE Trans. Nanotechnol.*, 16, 32–43, 2017.)

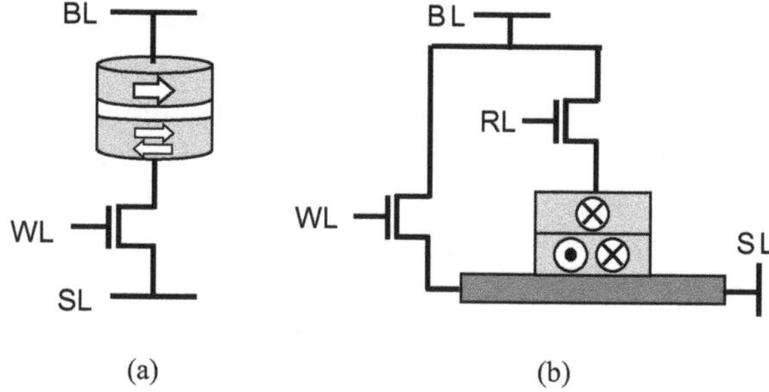

FIGURE 15.5
(a) 2-terminal MTJ bitcell and (b) SHE-MTJ bitcell. (From Zand, R. et al., *IEEE Trans. Nanotechnol.*, 16, 32–43, 2017.)

Figure 15.5a and b show the CMOS-based bitcell of the 2-terminal STT-MTJ and SHE-MTJ, respectively. In SHE-MTJ, the spin current can be significantly larger than the applied charge current. Therefore, the transistor utilized in the bitcell of the 2-terminal MTJ should be larger than that of the SHE-MTJ to be able to provide equal switching delay. Thus, although a SHE-MTJ bitcell requires two MOS transistors, its integration density is comparable to the 2-terminal MTJs. Increasing the transistor size in 2-terminal MTJs may also impact the reliability of the tunneling oxide barrier, which is improved in SHE-MTJ, since the current does not flow through it during the write operation.

15.2 Application of Magnetic Tunnel Junctions in Field Programmable Reconfigurable Fabrics

The main motivations for embracing reconfigurable fabrics, namely field-programmable gate arrays (FPGAs), could be divided into two categories: (1) *flexibility and accessibility*, enabling realization of logic elements at medium and fine granularities while incurring low non-recurring engineering and rapid deployment to market, and (2) *resiliency and runtime adaptability:* reconfigurable fabrics have been demonstrated to provide a viable solution for process-voltage-temperature variation-induced problems and could be utilized effectively for fault recovery [57–75] and aging mitigation [76–80]. However these advantages are achieved at a cost of increased fabric area and power consumption, as well as a decrease in performance compared to the application-specific integrated circuits (ASICs). Innovations using emerging devices within reconfigurable fabrics have been sought to bridge the gaps needed to provide these benefits.

Currently, static random access memory (SRAM) cells are the basis for most of the commercial FPGAs and can be found in the well-known Xilinx and Altera products. In FPGAs, SRAM cells are employed within programmable switching blocks to control the interconnection between logic building blocks. Moreover, they are utilized in lookup tables (LUTs) to store the logic function configuration data, which constitute the primary components in reconfigurable fabrics. In particular, a LUT is a memory with 2^m cells in which the truth table of an m-input Boolean function is stored [81]. The reprogrammability of the SRAM cells, and the fact that they can be implemented by highly scaled CMOS technology, have made the SRAM-based FPGAs the most popular reconfigurable fabric on the market. However, SRAM cells also have some limiting attributes that have caused FPGAs to have a niche market share of ASICs.

In [82], Kuon and Rose have provided a comprehensive comparison between SRAM-based FPGAs and ASICs in terms of area, performance, and power consumption. They have reported that in order to achieve the same functionality and performance in an FPGA as an ASIC, FPGA requires a significantly larger area while consuming approximately 14 times more power. This is mainly due to the crucial drawbacks of the SRAM cells such as: (1) *high static power*—This is due to the existence of intrinsic leakage current, which is significantly increasing by technology scaling; (2) *volatility*—SRAM is volatile and, all functions must be reprogrammed upon each power-up; consequently, an external nonvolatile memory is required to be integrated into the chip either in the same package or on the printed circuit board level; and (3) *low logic density*—SRAM consists of six transistors, which limits the logic density.

As depicted in Figure 15.6, FPGA fabrics continue their transition toward embracing the benefits of increased heterogeneity along several cooperating dimensions [83]. Since the inception of the first field-programmable devices, various granularities of general-purpose reconfigurable logic blocks and dedicated function-specific computational units have been added to configurable logic block (CLB) structures. These have resulted in increased computational functionality compared to homogeneous CLBs [84–86]. Over the last 10 years, reprogrammable fabrics have further embraced highly dedicated special-purpose co-processing units to handle complex floating-point computations [81,87]. Some of the standard co-processing units that appear within many contemporary FPGAs are digital signal processing (DSP) blocks, multiplier accumulators (MACs), and multi-bit block RAMs, as well as processor hard cores, which are commonly embedded within the fabric of many leading commercially available reconfigurable devices.

The upper right-hand corner of Figure 15.6 depicts that emerging devices could advance new transformative opportunities for exploiting technology-specific advantages, which we refer to as *technology heterogeneity.* Technology heterogeneity recognizes the cooperating advantages of CMOS devices for their rapid switching capabilities, while simultaneously embracing emerging devices for their non-volatility, near-zero standby power, high integration density, and radiation hardness [88,89]. Realization of technology heterogeneity in a field-programmable fabric enables synthesis time codesign and dynamic runtime adaptability among device technologies. Thus, we propose exciting feasible research toward utilizing the proven spin-based devices to complement CMOS devices.

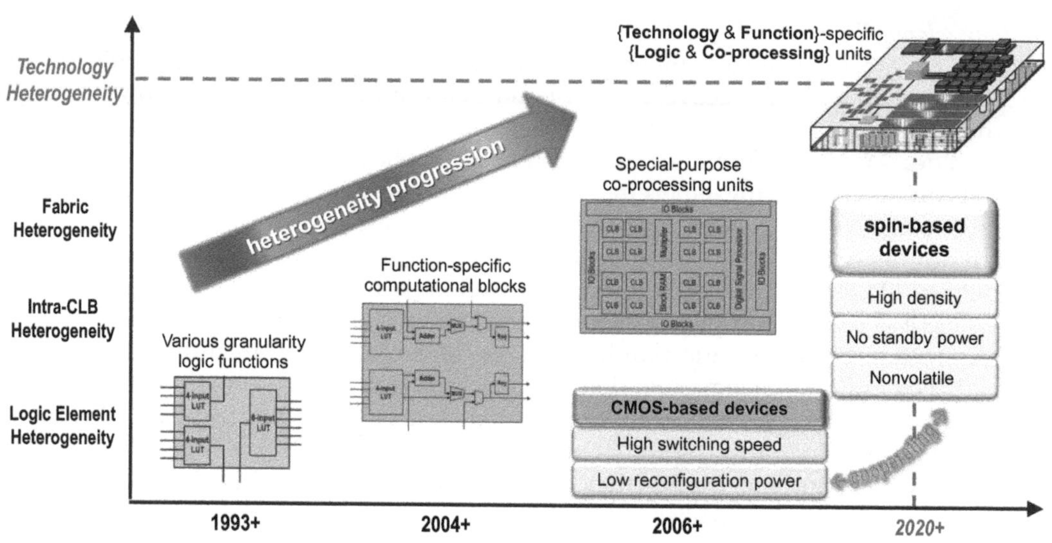

FIGURE 15.6
Escalation of heterogeneity within cronological and structural contexts. (From DeMara, R. F. et al., Heterogeneous technology configurable fabrics for field programmable co-design of CMOS and spin-based devices, in *IEEE International Conference on Rebooting Computing* (*ICRC-2017*), Washington, DC, 2017.)

LUTs are the building blocks of reconfigurable computing fabrics, providing a $2^m \times 1$ bit memory capable of realizing an m-input Boolean logic function. Currently, SRAM-based LUTs are a primary constituent for logic realization in most reconfigurable fabrics. However, SRAM's drawbacks such as high static power consumption, volatility, and restricted logic density [90,91] have motivated exploration of alternative LUT designs. One of the alternatives is based on nonvolatile flash-based LUTs; however, it targets a niche market due to its low reconfiguration endurance [92]. Higher endurance nonvolatile LUTs can be enabled by emerging resistive technologies, such as spintronic storage elements [1–6], resistive random access memory (RRAM) [7–10], and phase change memory (PCM) [11,12]. Although PCM can offer non-volatility, its considerable reconfiguration power and high write latency can significantly exceed that of a SRAM LUT. Spintronic devices offer non-volatility, near-zero static power, and high integration density [88,89]. Two of the spin-based devices, which have been previously proposed for use in reconfigurable fabrics, are magnetic tunnel junctions (MTJs) [1–4,56,93,94] and domain wall (DW)-based racetrack memory (RM) [5,6]. RM is effective for non-volatility and area density, although previous designs can incur significant delay and energy cost due to excessive shift activities to configure the implemented logic function [5]. Hence, MTJ-based LUTs are proposed to be placed at critical points of a large-scale digital circuit to implement various logic functions as a runtime adaptable fabric under middleware control. Radiation immunity of MTJ devices decrease the susceptibility of the design to radiation-induced errors [95]. Moreover, the magnetic LUTs provide the fabric with sufficient reconfigurability features to mitigate process variations. The fabric will be leveraged for fault detection and recovery using the adaptive self-healing approaches. MTJs comprising the storage elements in the adaptable LUTs are vertically integrated as a back-end process of typical CMOS fabrication, which significantly reduces the area cost of the redundancy.

Three types of energy consumption profiles can be identified in FPGA LUTs. First, there is an initial write energy consumption incurred at LUT configuration time. Second, the

LUTs comprising active logic paths will consume read energy, which may constitute only certain sub areas within the high gate equivalent capacity of contemporary FPGA chips. Third, the standby energy consumed by the remaining significant quantity of the LUTs comprising the fabric may be inactive. It is not possible to power-gate LUT islands, as they must retain the stored configuration. It has been estimated in [96] that if the combined effect of these three modes can be mitigated with a suitable SRAM alternative, then typical power consumption can be reduced up to 81% under representative applications based on measurements of fabricated devices. In [96], Suzuki et al. have fabricated a nonvolatile FPGA with 3,000 6-input STT-MTJ-based LUTs under 90 nm CMOS and 75 nm perpendicular MTJ technologies. They have utilized the LUT designs introduced in [2,3]. In addition to the aforementioned energy savings, they also achieved 56% area reduction.

In [93], Zand et al. have introduced a new STT-MTJ-based LUT design (STT-LUT). They provided a comparison between their proposed LUT and the design introduced by Suzuki et al. in [3], which was the basis of the aforementioned fabricated MTJ-based FPGA. The comparison results show an additional 42% reduction in active power consumption for read operations. Therefore, even more power reduction is expected in comparison to what is achieved in [96] by fabricating a nonvolatile FPGA based on the LUT design introduced in [93]. It is worth noting that the aforementioned total power reductions are achieved using MTJs that utilize the STT switching approach for reconfiguration operation, which consumes approximately 10 times greater power than that of the conventional SRAM cells [97]. Thus, extensions to accommodate the increased write energy consumed by MTJ-based LUTs are sought to be addressed. Herein, we will study two of the MTJ-based LUT designs previously developed by the author that are suitable candidates to be leveraged within the structure of the field programmable reconfigurable fabrics.

15.2.1 Design of Spin-Transfer Torque (STT)-Magnetic Tunnel Junction (MTJ)-Based LUT Circuits

A 4-input STT-LUT [93] is shown in Figure 15.7, which consists of read and write circuits. The *write circuit* includes two transmission gates (TGs) that provide the desired charge current for STT switching [98], while the *read circuit* is composed of a pre-charge sense amplifier (SA) [99–101], a TG-based multiplexer (MUX), and a reference tree. Each MTJ cell in the LUT circuit could be accessed according to the input signals—*A*, *B*, *C*, and *D*—through the MUX, which employs TGs instead of pass transistors (PTs). TGs are characterized by near optimal full-swing switching behavior, resulting in less delay, as well as more resiliency to process variation compared to PT-based designs [102].

The reference tree in the read circuit includes four TGs in a series configuration to compensate for the select tree active resistance. Reference MTJ resistance is designed in a manner such that its value in parallel configuration is between low resistance, R_P, and high resistance, R_{AP}, of the LUT MTJ cells as shown in the following equation:

$$R_{P\text{-reference MTJ}} \cong \frac{1}{2}\left(R_{\text{AP-LUT MTJ}} + R_{P\text{-LUT MTJ}}\right) \tag{15.23}$$

According to Equation 15.2, resistance of MTJ can be altered by changing the oxide barrier thickness, t_{ox}, or MTJ area. Oxide thickness could only be changed between 0.7 nm and 2.5 nm to keep the low resistance value and also show the TMR effect. Additionally, as established in [97], fabricating MTJs with various oxide thicknesses requires a different

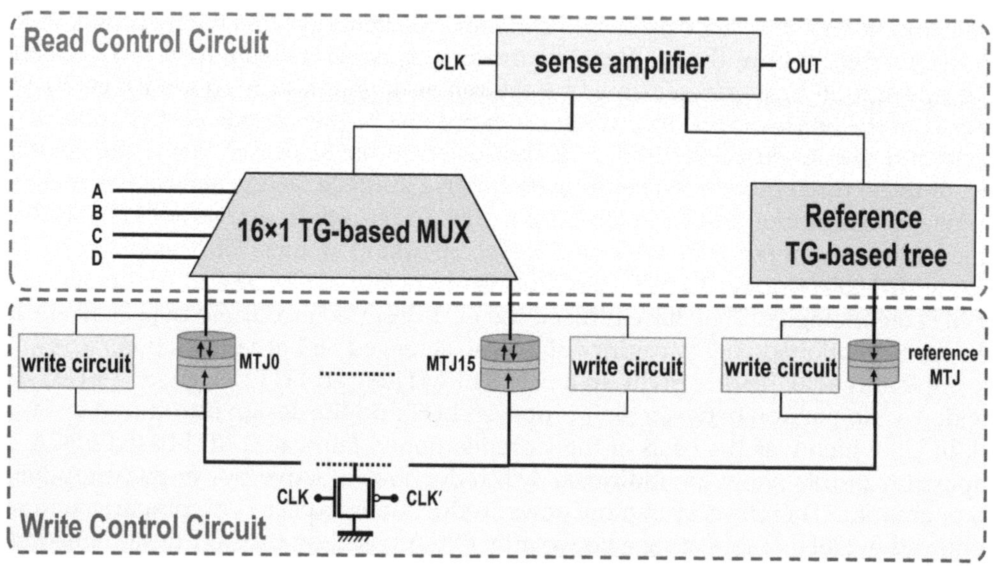

FIGURE 15.7
4-input STT-LUT functional diagram. (From Zand, R. et al., *IEEE Trans. Circuits Syst. II: Exp. Briefs*, 63, 678–682, 2016.)

magnetic process, which leads to a significant increase in fabrication cost. Thus, *Area* can be changed to determine the desired value of reference MTJ resistance.

In [93], an STT-LUT circuit is implemented utilizing both PTs and TGs. The performances of our STT-LUT implementations are compared with SRAM-LUT [103] and two aforementioned MTJ-based LUTs. The STT-LUT provides high speed and ultra-low power circuits with improved power-delay product (PDP) values. Furthermore, TG-based STT-LUT exhibits least PDP value while it leverages a larger number of MOS transistors compared to PT-based STT-LUT, which is the optimum choice from the area efficiency point of view. In [93], it is exhibited that PDP values and the number of LUT inputs are linearly proportional with a low slope, which validates the STT-LUT scalability. This capability led to the proposition of a 4-input adaptive LUT (A-LUT), as shown in Figure 15.8.

The proposed 4-input A-LUT could be configured to operate as different LUTs in seven independent modes: four 2-input STT-LUTs, two 3-input STT-LUTs, and one 4-input STT-LUT. Output of each configuration is individually connected to the SA through a *mode selector*, which includes PTs to choose between different operational modes, described in Table 15.2. For example, bitstream = `10'h104` configures A-LUT to operate as a 2-input STT-LUT based on the logic function stored in MTJ4 to MTJ7.

Figure 15.9 shows the layout of the A-LUT that occupies a cell area of 13.5 μm × 15.75 μm in a 90 nm process. A five-metal layer design is depicted. The MTJ cell has a vertical structure that could be readily integrated at the back-end process of CMOS fabrication. Generally, an n-input A-LUT PDP is smaller than the n-input STT-LUT PDP when performing 2-input to (n-1)-input Boolean functions.

Despite the aforementioned advantages of conventional MTJ devices, their main challenge is relatively high delay and power consumption. Moreover, they are 2-terminal devices that can experience occasional read/write disturbances due to sharing a common path for read and write operations. Recently, a 3-terminal Spin Hall Effect (SHE)-based MTJ has been introduced as an alternative for conventional 2-terminal MTJs. SHE-MTJ provides separate paths for read and write operations, while expending significantly

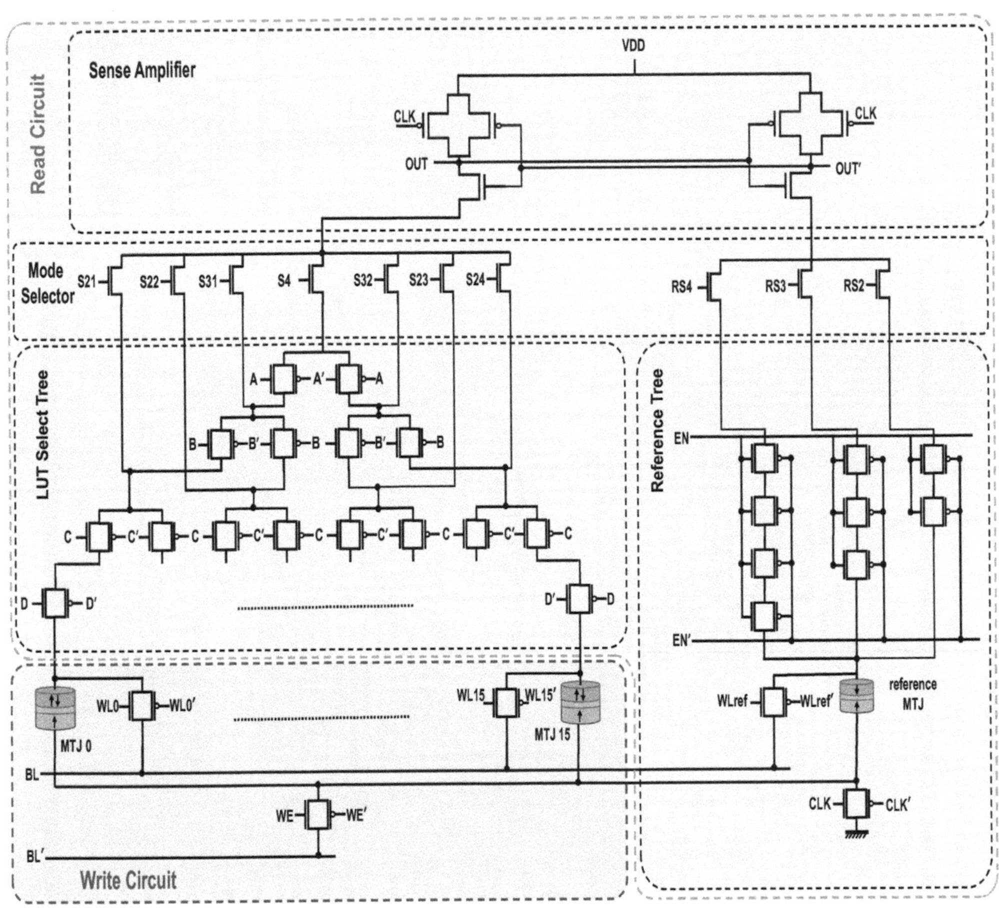

FIGURE 15.8
Circuit view of A-LUT schematic. (From Zand, R. et al., *IEEE Trans. Circuits Syst. II: Exp. Briefs*, 63, 678–682, 2016.)

TABLE 15.2
Configuration Specifications and MTJ Usage for 2-Input through 4-Input LUT Organization

	S21	S22	S23	S24	S31	S32	S4	RS2	RS3	RS4	bitstream	MTJ Usage	Description
mode 0	1	0	0	0	0	0	0	1	0	0	**10'h204**	0–3	2-input STT-LUT
mode 1	0	1	0	0	0	0	0	1	0	0	**10'h104**	4–7	2-input STT-LUT
mode 2	0	0	1	0	0	0	0	1	0	0	**10'h84**	8–11	2-input STT-LUT
mode 3	0	0	0	1	0	0	0	1	0	0	**10'h44**	12–15	2-input STT-LUT
mode 4	0	0	0	0	1	0	0	0	1	0	**10'h22**	0–7	3-input STT-LUT
mode 5	0	0	0	0	0	1	0	0	1	0	**10'h12**	8–15	3-input STT-LUT
mode 6	0	0	0	0	0	0	1	0	0	1	**10'h9**	0–15	4-input STT-LUT

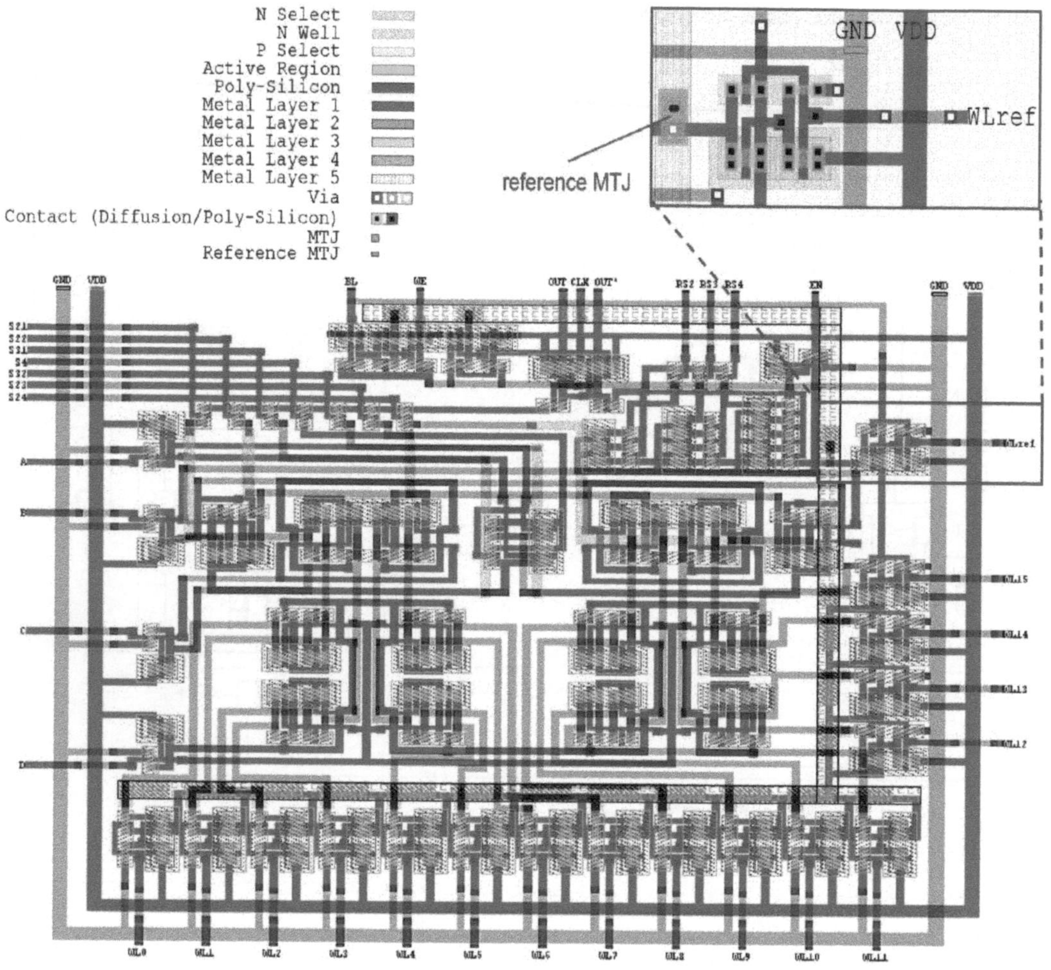

FIGURE 15.9
13.5 μm × 15.75 μm 4-input A-LUT layout. (From Zand, R. et al., *IEEE Trans. Circuits Syst. II: Exp. Briefs*, 63, 678–682, 2016.)

less switching energy [46–48]. A detailed comparison between the SHE-MTJ-based LUT (SHE-LUT) and 2-terminal MTJ-based LUTs is provided in [56].

15.2.2 Design of Spin Hall Effect (SHE)-Magnetic Tunnel Junction (MTJ)-Based LUT Circuits

The structure of a SHE-LUT circuit is shown in Figure 15.10, in which SHE-MTJ is utilized as a storage element. In general, data is stored in resistive memory cells in the form of different resistance levels, e.g., high resistance state stands for logic "1" and vice versa. Therefore, the SA is required to distinguish the resistive state of the memory cell.

Figure 15.10 shows a pre-charge sense amplifier (PCSA) circuit that includes four PMOS transistors connected to the VDD, two NMOS transistors that connect the PMOS transistors to the select trees and data storage cells, and one NMOS transistor that connects the circuit to ground (GND). Moreover, a *TG-based reference tree* including four TGs in series configuration is utilized in our designs to compensate for the select tree resistance.

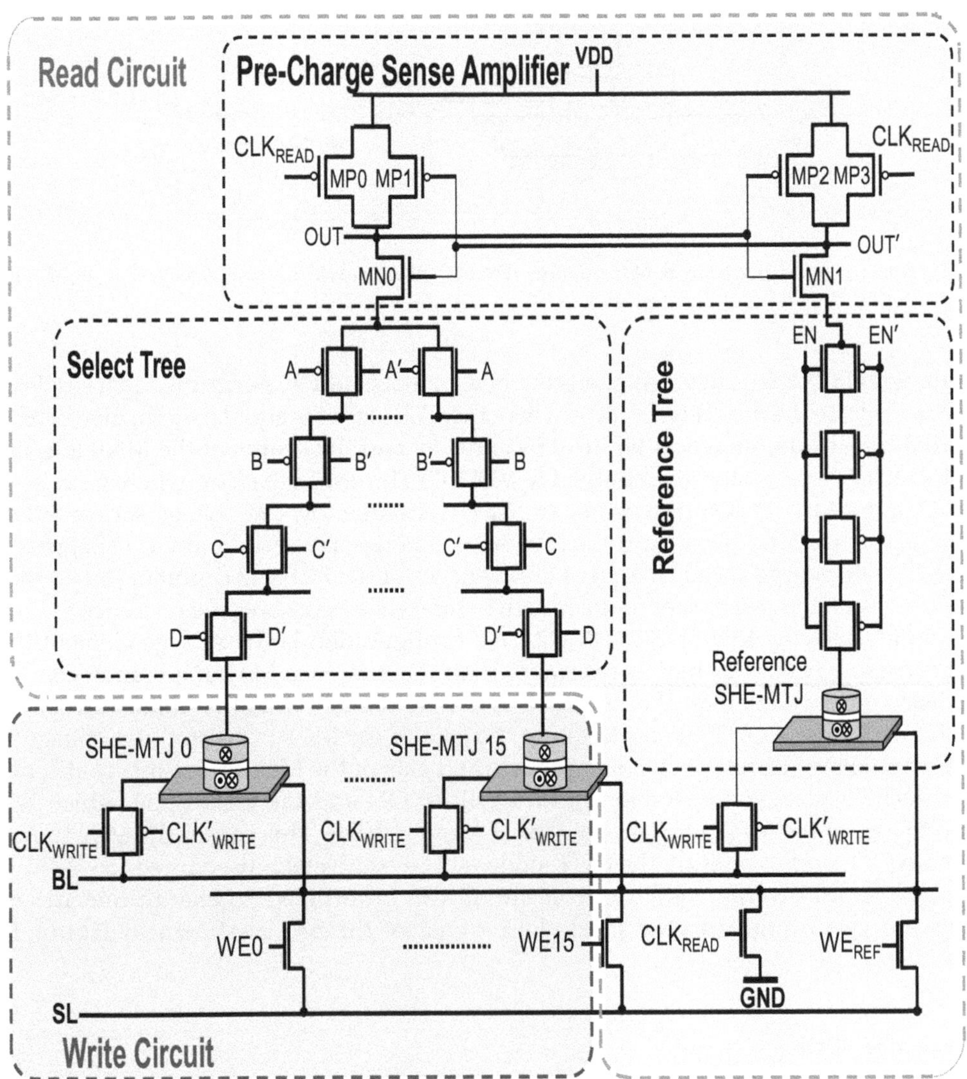

FIGURE 15.10
Circuit-level design of proposed SHE-LUT. (From Zand, R. et al., *IEEE Trans. Nanotechnol.*, 16, 32–43, 2017.)

Reference MTJ dimensions are designed in a manner such that their resistance value in parallel configuration is between low resistance, R_{Low}, and high resistance, R_{High}, of the SHE-based MTJ cells as shown in Figure 15.11 and elaborated by the following equations:

$$R_{P\text{-reference MTJ}} \cong \frac{1}{2}\left(R_{\mathrm{Low}} + R_{\mathrm{High}}\right) = \left(R_{\mathrm{AP\text{-}LUT\ MTJ}} + R_{P\text{-LUT MTJ}}\right)/2 + R_{\mathrm{HM}}/2 \qquad (15.24)$$

where,

$$R_{\mathrm{Low}} = \left(R_{P\text{-LUT MTJ}} + R_{\mathrm{HM}}/2\right)$$

$$R_{\mathrm{High}} = \left(R_{\mathrm{AP\text{-}LUT\ MTJ}} + R_{\mathrm{HM}}/2\right)$$

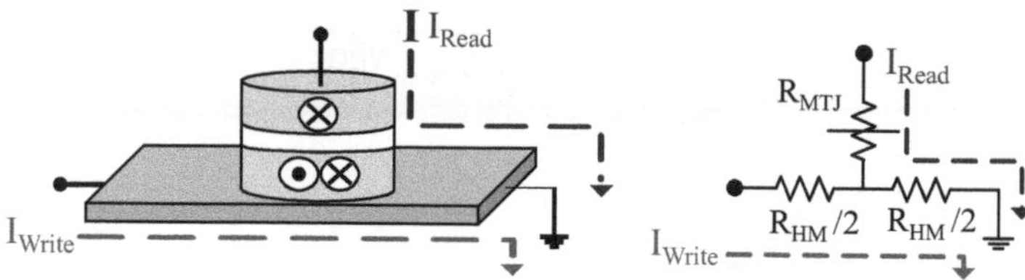

FIGURE 15.11
SHE-MTJ read and write path equivalent resistances. (From Zand, R. et al., *IEEE Trans. Nanotechnol.*, 16, 32–43, 2017.)

Sensing with PCSA requires two operating phases that could be performed in a single clock (CLK) period: (1) The *pre-charge phase* is where the CLK signal is equal to zero; therefore, MP0 and MP3 transistors, shown in Figure 15.10, are ON and the drains of the MN0 (OUT) and the MN1 (OUT′) transistors are charged to VDD. (2) The *discharge phase*, where CLK is equal to VDD and all the PMOS transistors are OFF. Consequently, the voltage source (VDD) is disconnected from the circuit and the pre-charged nodes, i.e., OUT and OUT′, begin to discharge. The discharge speed in each of the branches of the PCSA is different due to the difference between the resistances of the resistive storage elements and the reference SHE-MTJ. For example, assume that SHE-MTJ0 with AP configuration is the storage element that is being sensed. Since it has higher resistance than the reference MTJ, the branch connected to it discharges slower than the reference SHE-MTJ branch, thus the voltage drops faster in OUT′ node. Since OUT′ is connected to the gate of the MP1 transistor, the voltage drop results in a voltage difference between source and gate of the MP1 transistors that is higher than threshold voltage. Consequently, MP1 will be ON and the OUT node, which is connected to gate of the MP2 transistor, will be charged to VDD. This causes the MP2 transistor to remain OFF, and as a result, the OUT′ node will be completely discharged to GND.

In practice, an external synchronizer circuit can be utilized to ensure that the input signals are synchronous to the local clock signal of the SA, as shown in Figure 15.12.

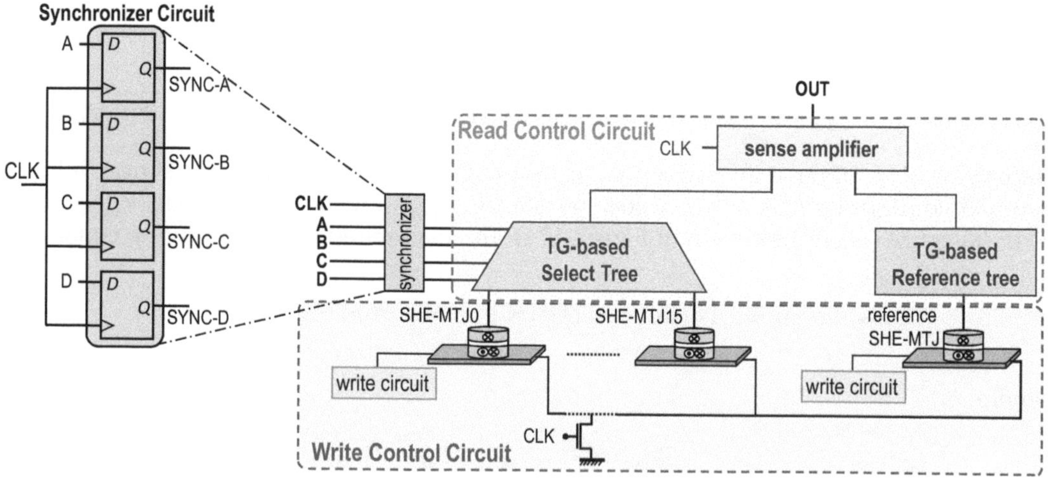

FIGURE 15.12
Schematic of 4-input SHE-based MTJ-LUT along with an external synchronizer circuit.

The synchronizer circuit for an n-input LUT includes *n* flip-flops that sample the inputs at each clock cycle. The pre-charge state of the SA should be sufficiently long to meet the required setup and hold times of the flip-flops to avoid metastability. The probability of synchronization failure caused by a flip-flop staying in the metastable state exponentially decreases with time [104].

In [56], a TG is utilized in the SHE-MTJ write circuit, as shown in Figure 15.13a. TGs are composed of one NMOS and one PMOS transistor, and they are characterized by their near optimal full-swing switching behavior. A TG-based write circuit provides a symmetric switching behavior, i.e., the generated write current amplitude for P to AP switching equals the current amplitude produced for switching from AP to P state. Moreover, TGs are capable of producing a current amplitude that is sufficiently large to ensure the complete switching of the MTJ devices.

Figure 15.13b shows the TG-based write circuit layout view. To address the feasibility of integrating SHE-MTJ with TGs, the three-dimensional (3D) cross-sectional view is provided in Figure 15.13c, which depicts the SHE-MTJ integration at the back-end process of CMOS fabrication. The required current for switching the SHE-MTJ passes through the heavy metal structure, which is built in the second metal layer. The MTJ stack is integrated between the second and forth metal layers, and it occupies the space for the third via and metal layer as well as the fourth via. Although, TG-based designs necessitate the availability of both CLK and inverse CLK signals, it is common and reasonable to assume access to both signal conditions within typical integrated circuits.

Most of the modern FPGAs, especially XILINX products such as Virtex-5 [105] and Spartan-6 [106], utilize fracturable 6-input LUTs in their design thus enabling six independent inputs and two separated outputs. The fracturable 6-input LUT can implement any six-input Boolean functions, as well as two five-input Boolean functions with common inputs [107]. In [56], a fracturable 6-input LUT is developed using SHE-MTJ devices, in which two PCSAs and two reference trees are utilized to ensure the independency of the outputs. Five NMOS transistors and two select signals, i.e., S5 and S6, are added to the LUT circuit to control the 5-input and 6-input operation modes of the SHE-based fracturable LUT. The structure of the proposed 6-input SHE-based structure LUT is shown in Figure 15.14. It provides significantly higher functional flexibility at the expense of slightly more area and power consumption.

Finally, Table 15.3 provides a qualitative comparison between aforementioned spin-based LUTs emphasizing their strengths and limitations relative to SRAM-based LUTs.

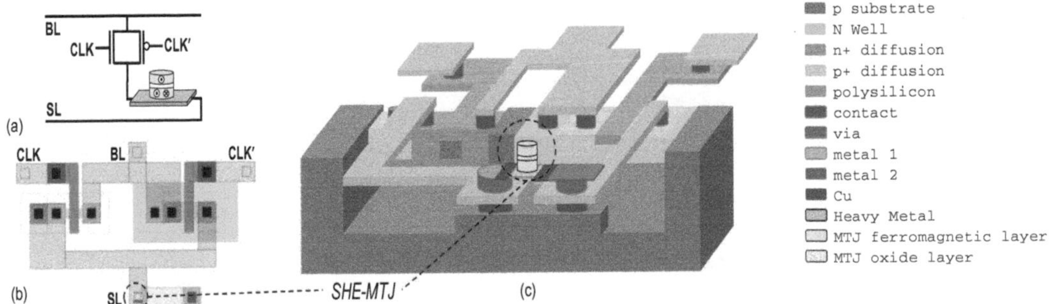

FIGURE 15.13
(a) Proposed transmission gate-based write circuits. (b) TG-based write circuit layout view. (c) Three-dimensional (3D) cross-sectional schematic of SHE-MTJ integration at the back-end process of TG fabrication. (From Zand, R. et al., *IEEE Trans. Nanotechnol.*, 16, 32–43, 2017.)

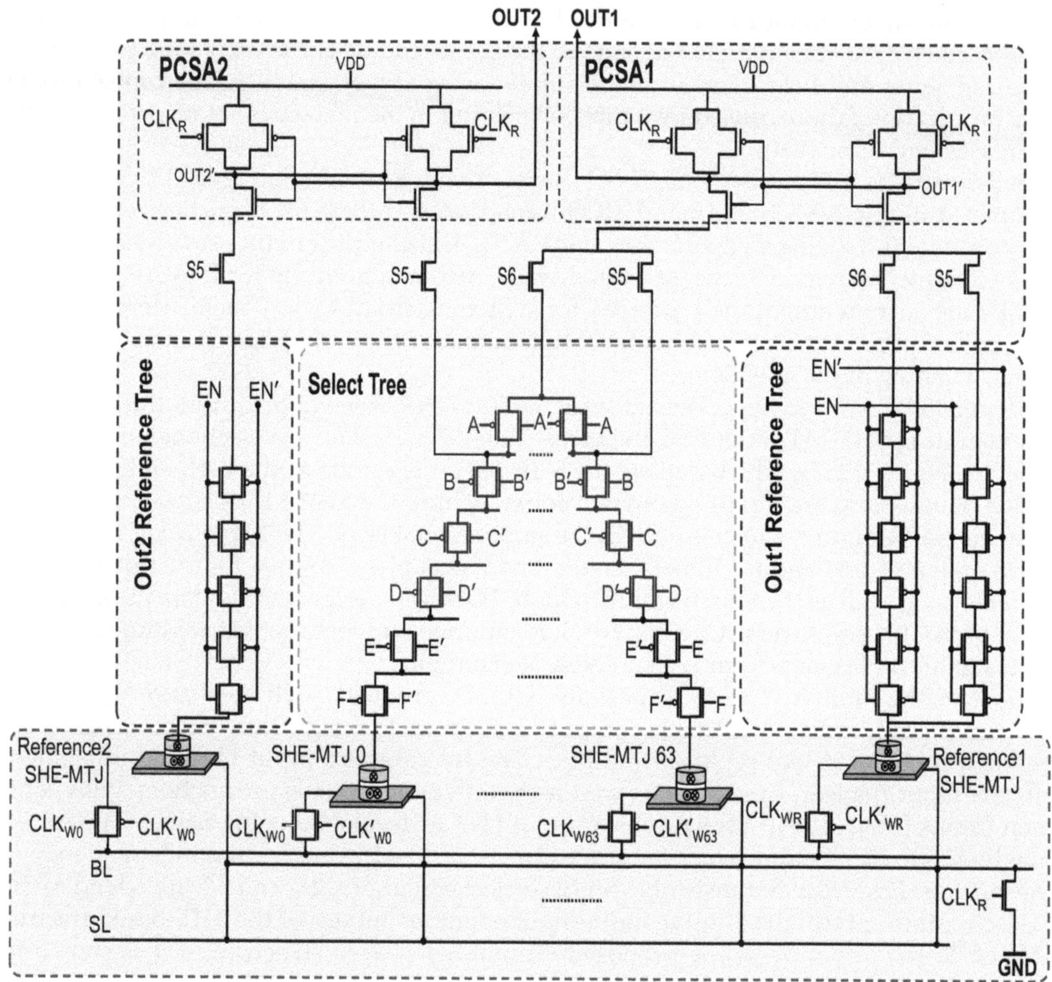

FIGURE 15.14
The structure of the 6-input SHE-based fracturable LUT. (From Zand, R. et al., *IEEE Trans. Nanotechnol.*, 16, 32–43, 2017.)

TABLE 15.3
Characteristics of Spin-Based LUT Circuits

Parameter		STT-LUT	A-LUT	SHE-LUT	Fracturable SHE-LUT
Power consumption	Static	✓✓	✓✓	✓✓	✓✓
	Read	—	-	—	-
	Reconfiguration	--	--	-	-
Speed	Read	—	-	—	-
	Reconfiguration	--	--	-	-
Functional flexibility		—	—	✓	—
Complexity		—	—	-	—
Scalability		✓	✓	✓	✓

Each "✓" or "-" indicates strength/limitation relative to SRAM-based LUTs.

15.3 Application of Magnetic Tunnel Junctions in Logic-In-Memory Circuits

The unifying computational mechanism underlying all of the MTJ devices is accumulation-mode operation, which enables realization of majority logic functions as basic computational building blocks [108,109]. Therefore, this section focuses on using MTJ-based majority gates (MGs) to implement Boolean logic circuits. Figure 15.15a shows the schematic of 3-input MG that is designed using SHE-MTJ devices, in which PCSA is utilized to sense its state. The minimum current required for switching the state of the s-MTJ devices is called the critical current (I_C), which is relative to the dimensions of the device. In an n-input SHE-MTJ-based MG, the device is designed such that at least $(n - 1)/2$ of the input transistors should be ON to produce a switching current amplitude greater than the critical current. MGs can be cascaded to realize conjunctive/disjunctive Boolean gate realizations. For instance, by affixing one (or two) of the three (or five) input transistors in ON or OFF states upon demand during the circuit operation, then a 2(or 3)-input OR gate or a 2(or 3)-input AND gate can be realized, respectively. For instance, the functionalities of 2-input OR and 3-input AND gates implemented by SHE-MTJ- based MGs are verified by circuit simulation using the SPICE-based model of a SHE-MTJ device developed in [110], and the parameters listed in Table 15.4, as shown in Figure 15.15b.

Binary addition is the most fundamental mathematical operation. It is worth mentioning that other operations in computer arithmetic such as subtraction and multiplication are usually implemented by adders, and this importance has motivated alternative designs for full adder (FA) structures. The logic functions of the FA can be expressed as follows:

$$\begin{aligned} \text{SUM} &= A \oplus B \oplus C \\ C_{\text{OUT}} &= AB + AC + BC \end{aligned} \tag{15.25}$$

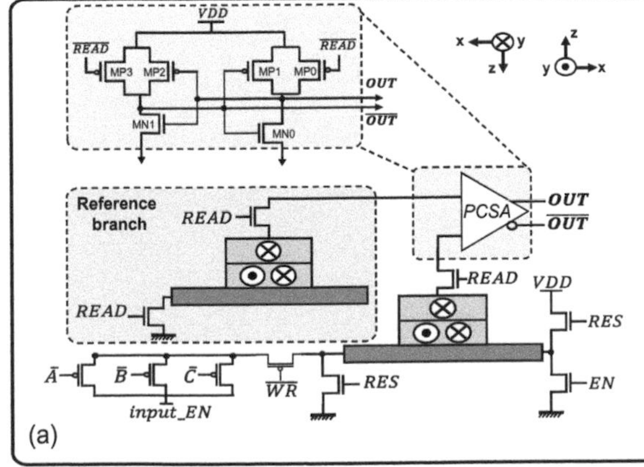

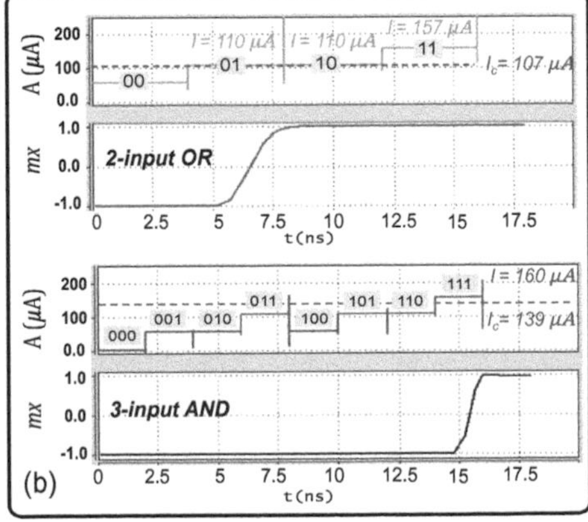

FIGURE 15.15
(a) s-MTJ based 3-input PG, (b) 2-input OR logic, and 3-input AND logic using PG3 and PG5, respectively.

TABLE 15.4
Proof-of-Concept Parameters

Parameter	Value
Nominal voltage	1.2 V
MOS technology node	45 nm
3-input MG	
HM volume	$100 \times 60 \times 3$ nm³
MTJ area	$60 \times 30 \times \pi/4$ nm³
Critical current	107 μA
5-input MG	
HM volume	$150 \times 80 \times 3$ nm³
MTJ area	$80 \times 40 \times \pi/4$ nm³
Critical current	139 μA
Spin Hall angle	0.3
HM resistivity	200 μΩ.cm

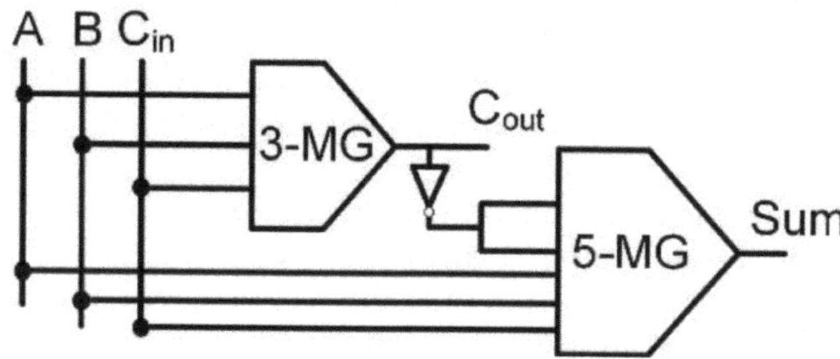

FIGURE 15.16
Schematic of a 1-bit full adder using 3- and 5-input MGs.

SUM and C_{OUT} logic functions can be implemented using MGs as shown in Figure 15.16 and expressed by Equations 15.26 and 15.27, respectively:

$$\begin{aligned} \text{SUM} &= \boldsymbol{M5}\left(A,B,C,\overline{M3},\overline{M3}\right) = \boldsymbol{M5}\left(A,B,C,\overline{C_{OUT}},\overline{C_{OUT}}\right) \\ &= ABC + AB\overline{C_{OUT}} + AC\overline{C_{OUT}} + BC\overline{C_{OUT}} + A\overline{C_{OUT}} + B\overline{C_{OUT}} \\ &\quad + C\overline{C_{OUT}} = ABC + \overline{C_{OUT}}\left(A+B+C\right) \\ &= ABC + \left(\overline{AB+AC+BC}\right)\left(A+B+C\right) \\ &= ABC + \left(\overline{AB}.\overline{AC}.\overline{BC}\right)\left(A+B+C\right) \\ &= ABC + \left(\overline{A}.\overline{B} + \overline{A}.\overline{C} + \overline{B}.\overline{C}\right)\left(A+B+C\right) \\ &= ABC + \overline{A}\overline{B}C + \overline{A}B\overline{C} + A\overline{B}\overline{C} = \boldsymbol{A+B+C} \end{aligned} \tag{15.26}$$

$$C_{OUT} = A.B + A.C + B.C = \text{3-input MG}\left(A,B,C\right) = \boldsymbol{M3}\left(A,B,C\right) \tag{15.27}$$

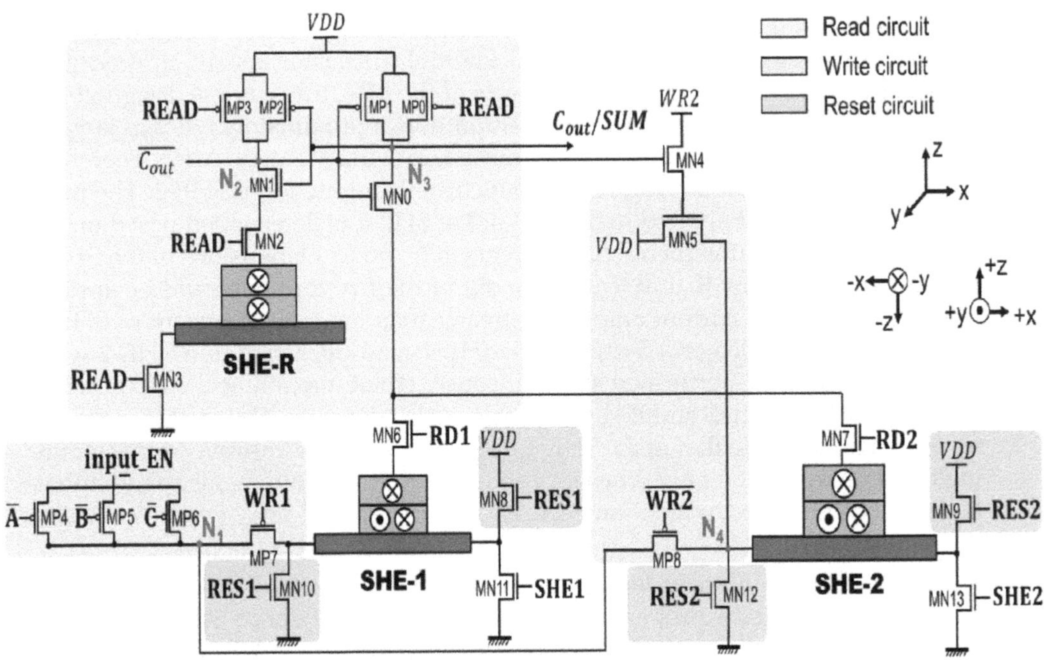

FIGURE 15.17
Circuit-view of SHE-based FA design. SHE-1 functions as a 3-input MG, while SHE-2 performs 5-input MG function. (From Roohi, A. et al., *IEEE Trans. Comput.-Aided Des. Integr. Circuits Syst.*, 36, 2134–2138, 2017.)

In [21], a nonvolatile (NV) FA is developed using SHE-MTJ devices, which is composed of 23 MOS transistors and 3 SHE-MTJs. Two of the SHE-MTJs function as MGs, and the other one is utilized as a reference element to sense the output of the FA. The switching behavior and functionality of the proposed circuit is verified using a SPICE circuit simulator. Figure 15.17 depicts the schematic of our proposed FA, which consists of two main parts as described below.

Write/Reset Circuit: For SHE-MTJ write operation, a charge current should be applied to the HM to produce a spin current greater than the critical switching spin current of the MTJ. In our SHE-based FA design, three PMOS transistors are leveraged to produce the input charge current according to the three inputs of the circuit—A, B, and C_{in}. The magnitude of the driven current for SHE-1 is determined based on the conservation of current on the N_1 node shown in Figure 15.17. The dimensions of the SHE-1 are designed in a manner such that the switching critical current is higher than a charge current produced by one of the input PMOS transistors (MP_4, MP_5, and MP_6). In order for the C_{out} to become "1," the SHE-1 state should change to anti-parallel. Hence, at least two of the three input transistors are required to be ON to switch the SHE-1 state. Therefore, the three PMOS transistors and SHE-1 device together function as a 3-input MG. To perform the SHE-1 write operation, RES1, WR1, and SHE1 signals should be "0," "1," and "1," respectively. For reset operation, two NMOS transistors (MN_8 and MN_{10}) are assigned to reset the SHE-1 state and prepare it for the next operation. Herein, reset operation means writing on SHE-MTJs in the -x direction to change their configuration to P state.

To implement the 5-input MG required for producing the SUM output, $\overline{C_{out}}$ is obtained through a sense amplifier, and the MN_5 transistor is used to produce a current based on the obtained $\overline{C_{out}}$. The size of the MN_5 transistor is designed in a manner such that it generates a current amplitude approximately twice as large as the currents produced by input PMOS transistors (MP_4, MP_5, and MP_6). Therefore, it can be assumed that there are five input currents injected into SHE-2. The magnitude of the current applied to the HM of the SHE-2 is determined based on the conservation of the aforementioned currents in N_1 node. Dimensions of the SHE-2 are designed in a way that at least three out of the five inputs should be applied to HM to produce a current amplitude greater than the critical current of SHE-2. Thus, SHE-2 functions as a 5-input MG. SHE-1 read operation and SHE-2 write operation should be performed simultaneously. Therefore, all of the RD1, SHE1, WR2, and SHE-2 signals should be "1" during this operation. The reset mechanism for SHE-2 is similar to that of the SHE-1. However, it can be improved by performing the reset operation only when C_{out} equals "1." Thus, unnecessary reset operations will be removed, which can decrease the energy and delay overhead caused by the reset scheme.

Read Circuit: The main components of the reading scheme are pre-charge sense amplifiers. To perform proper sensing operation, the reference SHE-MTJ device (SHER) is designed in a way that its resistance value in parallel configuration is between low resistances (R_Ps) and high resistances (R_{AP}s) of the SHE-1 and SHE-2 cells. Table 15.5 elaborates the required signaling for performing the write, read, and reset operations.

Figure 15.18 shows the functionality of our proposed SHE-based FA, in which the applied inputs are ABC = "010." There are three phases for one complete operation cycle of SHE-FA. In phase I, shown in Figure 15.18a, write and reset transistors for SHE-1 and SHE-2 are enabled, respectively. The produced input charge current according to ABC = "010" is smaller than SHE-1 critical current; therefore, the FL magnetization direction of SHE-1 remains in P state. Simultaneously, the SHE-2 reset transistors generate a current amplitude in $-x$ direction, which can reset SHE-2 to P state.

Figure 15.18b depicts the second phase of the SHE-FA operation including reading SHE-1 and writing SHE-2 devices at the same time. Since the SHE-1 input current for ABC = "010" is not sufficient to switch its states, C_{out} and $\overline{C_{out}}$ equal "0" and "1," respectively. Therefore,

TABLE 15.5

Required Signaling for 1-Bit SHE-Based FA [20]

Operation	Device	Signaling
WRITE	SHE-1	READ = "0", WR1 = SHE1 = "1"
	SHE-2	WR2 = SHE2 = SHE1 = RD1 = READ = "1"
READ[a]	SHE-1	RD1 = SHE1 = READ = "1"
	SHE-2	RD2 = SHE1 = READ = "1"
	SHE-3	READ = "1"
RESET	SHE-1	RES1 = "1"
	SHE-2	RES2 = "1"

[a] When READ is set ("1"), node N_2 in Figure 15.17 is connected to the ground via SHE-R, which is in parallel configuration.

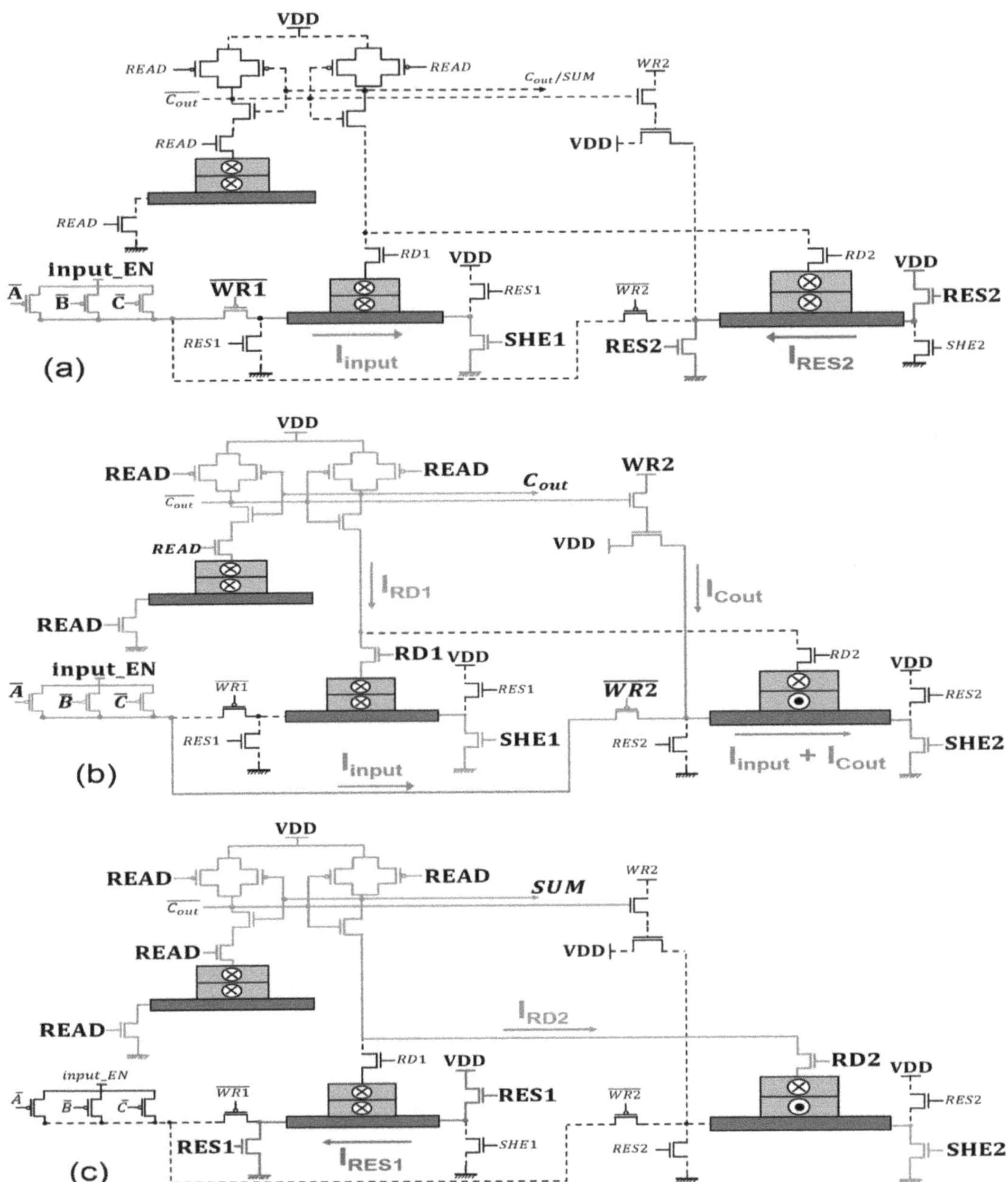

FIGURE 15.18
SHE-based functionality for input ABC = "010," (a) write and reset operations for SHE-1 and SHE-2 occurred, respectively; hence, FL of SHE-1 remains in *P* state, then (b) read and write operation for SHE-1 and SHE-2 perform simultaneously; therefore, FL of SHE-2 changes to AP state; and finally, (c) SHE-1 is reset along with reading SHE-2 state. (From Roohi, A. et al., *IEEE Trans. Comput.-Aided Des. Integr. Circuits Syst.*, 36, 2134–2138, 2017.)

MN_4 and MN_5 transistors are ON and $I_{\overline{Cout}}$ will be generated. Input currents and $I_{\overline{Cout}}$ are accumulated in N_4 node and produce the SHE-2 write current. In this example, the magnitude of the injected current is greater than the SHE-2 critical current; therefore, the state of the SHE-2 device changes to AP configuration.

In the third phase, the reset and read operations are performed for SHE-1 and SHE-2, respectively. Due to the difference between the resistances of the SHE-1 and SHE-2 HMs, the produced reset current for SHE-1 is different from that of the SHE-2. In this phase, SHE1 and SHE2 signals are equal to "0" and "1," respectively. Thus, during the read operation, PCSA senses the state of the SHE-2 device that is the SUM output. As previously mentioned, SHE-2 was configured to AP state in the second phase; therefore, the output of the PCSA equals "1." Simulation results including timing diagram and SHE-MTJs' magnetization directions are depicted in Figure 15.19, which validates the functionality of our proposed FA for two sets of inputs, ABC = "001" and ABC = "111."

In [21], the authors have examined the functionality of an n-bit SHE-FA to verify the concatenatability of our SHE-based FA. Figure 15.20 shows the schematic and timing diagram for a 4-bit SHE-based FA. To obtain the SUM output for each adder block, C_{out} of their previous block is required to be applied as one of the three input signals. Hence, each C_{out} bit in level n is utilized to obtain (1) SUM output in level n and (2) C_{out} bit in level $n+1$. Therefore, the C_{out} in each level should remain unchanged for a sufficient duration to ensure the correct operation of an n-bit SHE-FA. This can be achieved by SHE-MTJ devices without any additional energy consumption, due to their non-volatility feature. The timing limitations are considered in the timing diagram shown in Figure 15.20. To decrease the propagation delay of an n-bit SHE-FA, the independent operations are designed to be performed simultaneously. Namely, C_{out} write operation for the second adder block is independent of the SUM write operation of the first block, thus both are operated in the second time step.

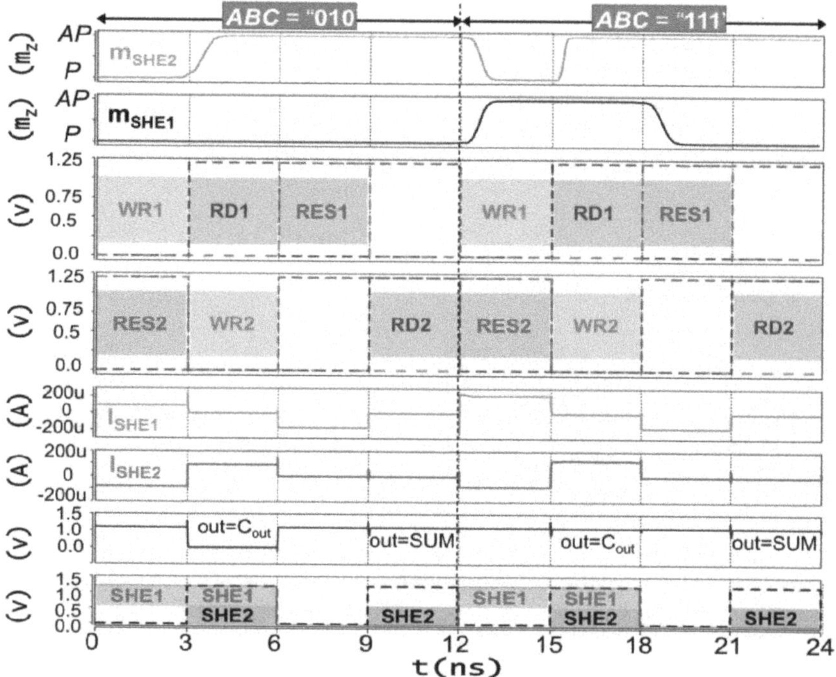

FIGURE 15.19
Simulation results of 1-bit SHE-based FA for two input sequences, "010" and "111." (From Roohi, A. et al., *IEEE Trans. Comput.-Aided Des. Integr. Circuits Syst.*, 36, 2134–2138, 2017.)

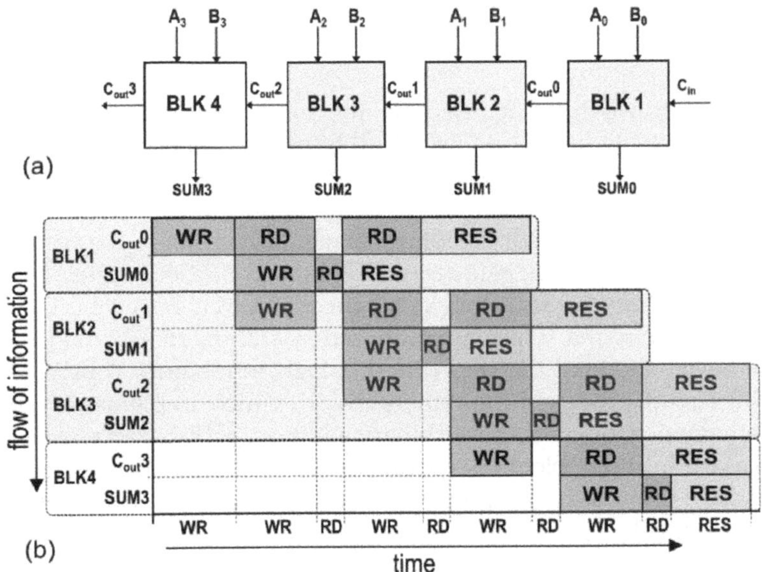

FIGURE 15.20
(a) Schematic of 4-bit SHE-based FA. (b) 4-bit SHE-based FA timing diagram. (From Roohi, A. et al., *IEEE Trans. Comput.-Aided Des. Integr. Circuits Syst.*, 36, 2134–2138, 2017.)

Power consumption of an n-bit SHE-FA relies on the number of write, read, and reset operations that are required to be executed for a complete addition cycle. For instance, there are eight SHE-MTJs in a 4-bit SHE-FA; therefore, eight write and reset operations should be performed in a complete addition operation. Moreover, as shown in Figure 15.20, 11 read operations are performed in a complete cycle, 8 operations to output the SUM and C_{out} values, and 3 operations for switching SHE-2 states. In general, the total propagation delay and power consumption of our proposed n-bit SHE-based FA can be calculated using the following equations

$$D_{n\text{-FA}} = \text{DWR}_{\text{SHE-1}} + N \times (\text{DWR}_{\text{SHE-2}}) + N \times (D_{\text{SA}}) + \text{DRES}_{\text{SHE-2}} \tag{15.28}$$

$$P_{n\text{-FA}} = 2N \times [\max(\text{PWR}_{\text{SHE-2}})] + 2N \times (\text{PRES}_{\text{SHE-1}}) + (2(N-1) + (N+1)) \times (\text{PRD}_{\text{SHE-1}}) \tag{15.29}$$

where N is number of bits, and $\text{DWR}_{\text{SHE-1}}$ and $\text{DWR}_{\text{SHE-2}}$ are write operation delays of SHE-1 and SHE-2, respectively. DSA is PCSA delay, and $\text{DRES}_{\text{SHE-2}}$ is reset operation delay of SHE-2. $\text{PWR}_{\text{SHE-2}}$, $\text{PRES}_{\text{SHE-1}}$, and $\text{PRD}_{\text{SHE-1}}$ are the power consumption of SHE-2 write, SHE-1 reset, and SHE-1 read operations, which are the worst-case values for each quantity.

The introduced SHE-based FA was examined using a SPICE circuit simulator, which indicated roughly 75% improvement in terms of power consumption compared to the most energy-efficient logic-in-memory designs using conventional MTJ devices [111–113]. Due to the scalability and voltage-based operation of our proposed 1-bit SHE-FA, it can be readily concatenated to constitute an n-bit SHE-based adder or an n-bit SHE-based ALU with significantly low area and energy consumption as depicted by the pipeline analysis.

15.4 Summary

In this chapter, first we have focused on the fundamentals and modeling of both spin-transfer torque (STT)-based and Spin Hall Effect (SHE)-based magnetic tunnel junctions (MTJs), which have recently attracted significant attentions as promising alternatives for CMOS devices in both memory and logic applications. Next, MTJ-based lookup table (LUT) circuits were studied as the primary building blocks for reconfigurable fabrics, namely field-programmable gate arrays (FPGAs). In particular, detailed descriptions were provided regarding the structure of adaptive STT-MTJ-based LUT and fracturable SHE-MTJ-based LUT circuits, which are proposed by authors in [93] and [56], respectively. Moreover, we have explained the structure of a logic-in-memory (LIM) SHE-MTJ-based full adder (FA) circuit proposed by authors in [21], which implements one of the most important mathematical operations, i.e., binary addition. SHE-based majority gates (MGs) were utilized as the building blocks of the proposed SHE-FA. SHE-MTJ- based MGs can be cascaded to realize conjunctive/disjunctive Boolean gate realizations enabling the implementation of larger scale logic designs. Herein, we have described the application of MTJs in reconfigurable fabrics and LIM architectures, and we have provided some examples of the corresponding MTJ-based circuits. However, the application of MTJ devices are not limited to these examples, and there are a wide range of applications that have been studied in recent years, including neuromorphic computing [114–118], associative computing [51,119,120], and secure and intermittent computing for energy-harvesting-powered Internet-of-Things (IoT) devices [121–123]. Readers are referred to the following references for additional information regarding the recent application of MTJ devices.

References

1. W. Zhao, E. Belhaire, C. Chappert, F. Jacquet, and P. Mazoyer, "New non-volatile logic based on spin-MTJ," *Physica Status Solidi (a)*, vol. 205, pp. 1373–1377, 2008.
2. D. Suzuki, M. Natsui, T. Endoh, H. Ohno, and T. Hanyu, "Six-input lookup table circuit with 62% fewer transistors using nonvolatile logic-in-memory architecture with series/parallel-connected magnetic tunnel junctions," *Journal of Applied Physics*, vol. 111, p. 07E318, 2012.
3. D. Suzuki, M. Natsui, and T. Hanyu, "Area-efficient LUT circuit design based on asymmetry of MTJ's current switching for a nonvolatile FPGA," in *Circuits and Systems (MWSCAS), 2012 IEEE 55th International Midwest Symposium on*, 2012, pp. 334–337.
4. X. Yao, J. Harms, A. Lyle, F. Ebrahimi, Y. Zhang, and J. P. Wang, "Magnetic tunnel junction-based spintronic logic units operated by spin transfer torque," *IEEE Transactions on Nanotechnology*, vol. 11, pp. 120–126, 2012.
5. W. Zhao, D. Ravelosona, J. Klein, and C. Chappert, "Domain wall shift register-based reconfigurable logic," *IEEE Transactions on Magnetics*, vol. 47, pp. 2966–2969, 2011.
6. W. Zhao, N. Ben Romdhane, Y. Zhang, J.-O. Klein, and D. Ravelosona, "Racetrack memory based reconfigurable computing," in *Faible Tension Faible Consommation (FTFC), 2013 IEEE*, 2013, pp. 1–4.
7. X. Tang, G. Kim, P. E. Gaillardon, and G. D. Micheli, "A study on the programming structures for RRAM-based FPGA architectures," *IEEE Transactions on Circuits and Systems I: Regular Papers*, vol. 63, pp. 503–516, 2016.

8. P. E. Gaillardon, X. Tang, J. Sandrini, M. Thammasack, S. R. Omam, D. Sacchetto, Y. Leblebici, and G. D. Micheli, "A ultra-low-power FPGA based on monolithically integrated RRAMs," in *2015 Design, Automation & Test in Europe Conference & Exhibition (DATE)*, 2015, pp. 1203–1208.
9. Y. C. Chen, H. Li, and W. Zhang, "A novel peripheral circuit for RRAM-based LUT," in *2012 IEEE International Symposium on Circuits and Systems*, 2012, pp. 1811–1814.
10. X. Xue, J. Yang, Y. Lin, R. Huang, Q. Zou, and J. Wu, "Low-power variation-tolerant nonvolatile lookup table design," *IEEE Transactions on Very Large Scale Integration (VLSI) Systems*, vol. 24, pp. 1174–1178, 2016.
11. K. Huang, Y. Ha, R. Zhao, A. Kumar, and Y. Lian, "A low active leakage and high reliability phase change memory (PCM) based non-volatile FPGA storage element," *IEEE Transactions on Circuits and Systems I: Regular Papers*, vol. 61, pp. 2605–2613, 2014.
12. C. Y. Wen, J. Li, S. Kim, M. Breitwisch, C. Lam, J. Paramesh, and L. T. Pileggi, "A non-volatile look-up table design using PCM (phase-change memory) cells," in *VLSI Circuits (VLSIC), 2011 Symposium on*, 2011, pp. 302–303.
13. A. Roohi, R. Zand, S. Angizi, and R. F. Demara, "A parity-preserving reversible QCA gate with self-checking cascadable resiliency," *IEEE Transactions on Emerging Topics in Computing*, pp. 1–1, 2016.
14. A. M. Chabi, A. Roohi, H. Khademolhosseini, S. Sheikhfaal, S. Angizi, K. Navi, and R. F. DeMara, "Towards ultra-efficient QCA reversible circuits," *Microprocessors and Microsystems*, vol. 49, pp. 127–138, 2017.
15. A. M. Chabi, A. Roohi, R. F. DeMara, S. Angizi, K. Navi, and H. Khademolhosseini, "Cost-efficient QCA reversible combinational circuits based on a new reversible gate," in *2015 18th CSI International Symposium on Computer Architecture and Digital Systems (CADS)*, 2015, pp. 1–6.
16. A. Roohi, H. Thapliyal, and R. DeMara, "Wire crossing constrained QCA circuit design using bilayer logic decomposition," *Electronics Letters*, vol. 51, pp. 1677–1679, 2015.
17. A. Roohi, R. F. DeMara, and N. Khoshavi, "Design and evaluation of an ultra-area-efficient fault-tolerant QCA full adder," *Microelectronics Journal*, vol. 46, pp. 531–542, 2015.
18. A. Roohi, S. Sayedsalehi, H. Khademolhosseini, and K. Navi, "Design and evaluation of a reconfigurable fault tolerant quantum-dot cellular automata gate," *Journal of Computational and Theoretical Nanoscience*, vol. 10, pp. 380–388, 2013.
19. R. L. Stamps, S. Breitkreutz, J. Åkerman, A. V. Chumak, Y. Otani, G. E. Bauer, J.-U. Thiele, M. Bowen, S. A. Majetich, and M. Kläui, "The 2014 magnetism roadmap," *Journal of Physics D: Applied Physics*, vol. 47, p. 333001, 2014.
20. A. Roohi, R. Zand, and R. F. DeMara, "A tunable majority gate-based full adder using current-induced domain wall nanomagnets," *IEEE Transactions on Magnetics*, vol. 52, pp. 1–7, 2016.
21. A. Roohi, R. Zand, D. Fan, and R. F. DeMara, "Voltage-based concatenatable full adder using spin Hall effect switching," *IEEE Transactions on Computer-Aided Design of Integrated Circuits and Systems*, vol. 36, pp. 2134–2138, 2017.
22. S. D. Pyle, H. Li, and R. F. DeMara, "Compact low-power instant store and restore D flip-flop using a self-complementing spintronic device," *Electronics Letters*, vol. 52, pp. 1238–1240, 2016.
23. N. Khoshavi, S. Salehi, and R. F. DeMara, "Variation-immune resistive non-volatile memory using self-organized sub-bank circuit designs," in *2017 18th International Symposium on Quality Electronic Design (ISQED)*, 2017, pp. 52–57.
24. X. Chen, N. Khoshavi, R. F. DeMara, J. Wang, D. Huang, W. Wen, and Y. Chen, "Energy-aware adaptive restore schemes for MLC STT-RAM cache," *IEEE Transactions on Computers*, vol. 66, pp. 786–798, 2017.
25. X. Chen, N. Khoshavi, J. Zhou, D. Huang, R. F. DeMara, J. Wang, W. Wen, and Y. Chen, "AOS: Adaptive overwrite scheme for energy-efficient MLC STT-RAM cache," *Presented at the Proceedings of the 53rd Annual Design Automation Conference*, Austin, TX, 2016.

26. X. Fong, Y. Kim, K. Yogendra, D. Fan, A. Sengupta, A. Raghunathan, and K. Roy, "Spin-transfer torque devices for logic and memory: Prospects and perspectives," *Computer-Aided Design of Integrated Circuits and Systems, IEEE Transactions on*, vol. 35, pp. 1–22, 2016.
27. W. Kang, Y. Zhang, Z. Wang, J.-O. Klein, C. Chappert, D. Ravelosona, G. Wang, Y. Zhang, and W. Zhao, "Spintronics: Emerging ultra-low-power circuits and systems beyond MOS technology," *ACM Journal on Emerging Technologies in Computing Systems (JETC)*, vol. 12, p. 16, 2015.
28. S. Ikeda, K. Miura, H. Yamamoto, K. Mizunuma, H. Gan, M. Endo, S. Kanai, J. Hayakawa, F. Matsukura, and H. Ohno, "A perpendicular-anisotropy CoFeB–MgO magnetic tunnel junction," *Nature Materials*, vol. 9, pp. 721–724, 2010.
29. E. Deng, W. Kang, Y. Zhang, J.-O. Klein, C. Chappert, and W. Zhao, "Design optimization and analysis of multicontext STT-MTJ/CMOS logic circuits," *IEEE Transactions on Nanotechnology*, vol. 14, pp. 169–177, 2015.
30. B. Behin-Aein, J.-P. Wang, and R. Wiesendanger, "Computing with spins and magnets," *MRS Bulletin*, vol. 39, pp. 696–702, 2014.
31. W. Zhao, E. Belhaire, Q. Mistral, C. Chappert, V. Javerliac, B. Dieny, and E. Nicolle, "Macro-model of spin-transfer torque based magnetic tunnel junction device for hybrid magnetic-CMOS design," in *Behavioral Modeling and Simulation Workshop, Proceedings of the 2006 IEEE International*, 2006, pp. 40–43.
32. Z. Xu, C. Yang, M. Mao, K. B. Sutaria, C. Chakrabarti, and Y. Cao, "Compact modeling of STT-MTJ devices," *Solid-State Electronics*, vol. 102, pp. 76–81, 2014.
33. Y. Zhang, W. Zhao, Y. Lakys, J.-O. Klein, J.-V. Kim, D. Ravelosona, and C. Chappert, "Compact modeling of perpendicular-anisotropy CoFeB/MgO magnetic tunnel junctions," *IEEE Transactions on Electron Devices*, vol. 59, pp. 819–826, 2012.
34. S. Wolf, D. Awschalom, R. Buhrman, J. Daughton, S. Von Molnar, M. Roukes, A. Y. Chtchelkanova, and D. Treger, "Spintronics: A spin-based electronics vision for the future," *Science*, vol. 294, pp. 1488–1495, 2001.
35. I. Prejbeanu, M. Kerekes, R. Sousa, H. Sibuet, O. Redon, B. Dieny, and J. Nozieres, "Thermally assisted MRAM," *Journal of Physics: Condensed Matter*, vol. 19, p. 165218, 2007.
36. J. C. Slonczewski, "Current-driven excitation of magnetic multilayers," *Journal of Magnetism and Magnetic Materials*, vol. 159, pp. L1–L7, 1996.
37. R. Koch, J. Katine, and J. Sun, "Time-resolved reversal of spin-transfer switching in a nanomagnet," *Physical Review Letters*, vol. 92, p. 088302, 2004.
38. W. F. Brown Jr, "Thermal fluctuations of a single-domain particle," *Journal of Applied Physics*, vol. 34, pp. 1319–1320, 1963.
39. L. Liu, T. Moriyama, D. Ralph, and R. Buhrman, "Reduction of the spin-torque critical current by partially canceling the free layer demagnetization field," *Applied Physics Letters*, vol. 94, p. 122508, 2009.
40. L.-B. Faber, W. Zhao, J.-O. Klein, T. Devolder, and C. Chappert, "Dynamic compact model of spin-transfer torque based magnetic tunnel junction (MTJ)," in *Design & Technology of Integrated Systems in Nanoscal Era, 2009. DTIS'09. 4th International Conference on*, 2009, pp. 130–135.
41. J. Z. Sun, "Spin-current interaction with a monodomain magnetic body: A model study," *Physical Review B*, vol. 62, p. 570, 2000.
42. J.-B. Kammerer, M. Madec, and L. Hébrard, "Compact modeling of a magnetic tunnel junction—Part I: Dynamic magnetization model," *IEEE Transactions on Electron Devices*, vol. 57, pp. 1408–1415, 2010.
43. Z. Xu, K. B. Sutaria, C. Yang, C. Chakrabarti, and Y. Cao, "Compact modeling of STT-MTJ for SPICE simulation," in *Solid-State Device Research Conference (ESSDERC), 2013 Proceedings of the European*, 2013, pp. 338–341.
44. T. Devolder, C. Chappert, J. Katine, M. Carey, and K. Ito, "Distribution of the magnetization reversal duration in subnanosecond spin-transfer switching," *Physical Review B*, vol. 75, p. 064402, 2007.

45. H. Zhao, B. Glass, P. K. Amiri, A. Lyle, Y. Zhang, Y.-J. Chen, G. Rowlands, P. Upadhyaya, Z. Zeng, and J. Katine, "Sub-200 ps spin transfer torque switching in in-plane magnetic tunnel junctions with interface perpendicular anisotropy," *Journal of Physics D: Applied Physics*, vol. 45, p. 025001, 2011.
46. L. Liu, C.-F. Pai, D. Ralph, and R. Buhrman, "Magnetic oscillations driven by the spin Hall effect in 3-terminal magnetic tunnel junction devices," *Physical Review Letters*, vol. 109, p. 186602, 2012.
47. L. Liu, C.-F. Pai, Y. Li, H. Tseng, D. Ralph, and R. Buhrman, "Spin-torque switching with the giant spin Hall effect of tantalum," *Science*, vol. 336, pp. 555–558, 2012.
48. C.-F. Pai, L. Liu, Y. Li, H. Tseng, D. Ralph, and R. Buhrman, "Spin transfer torque devices utilizing the giant spin Hall effect of tungsten," *Applied Physics Letters*, vol. 101, p. 122404, 2012.
49. W. Kang, Z. Wang, Y. Zhang, J.-O. Klein, W. Lv, and W. Zhao, "Spintronic logic design methodology based on spin Hall effect–driven magnetic tunnel junctions," *Journal of Physics D: Applied Physics*, vol. 49, p. 065008, 2016.
50. S. Manipatruni, D. E. Nikonov, and I. A. Young, "Energy-delay performance of giant spin Hall effect switching for dense magnetic memory," *Applied Physics Express*, vol. 7, p. 103001, 2014.
51. D. Fan, S. Maji, K. Yogendra, M. Sharad, and K. Roy, "Injection-locked spin Hall-induced coupled-oscillators for energy efficient associative computing," *IEEE Transactions on Nanotechnology*, vol. 14, pp. 1083–1093, 2015.
52. L. Liu, T. Moriyama, D. Ralph, and R. Buhrman, "Spin-torque ferromagnetic resonance induced by the spin Hall effect," *Physical Review Letters*, vol. 106, p. 036601, 2011.
53. S. Rakheja and A. Naeemi, "Graphene nanoribbon spin interconnects for nonlocal spin-torque circuits: Comparison of performance and energy per bit with CMOS interconnects," *IEEE Transactions on Electron Devices*, vol. 59, pp. 51–59, 2012.
54. Z. Wang, W. Zhao, E. Deng, J.-O. Klein, and C. Chappert, "Perpendicular-anisotropy magnetic tunnel junction switched by spin-Hall-assisted spin-transfer torque," *Journal of Physics D: Applied Physics*, vol. 48, p. 065001, 2015.
55. T. Aoki, Y. Ando, D. Watanabe, M. Oogane, and T. Miyazaki, "Spin transfer switching in the nanosecond regime for CoFeB/MgO/CoFeB ferromagnetic tunnel junctions," *Journal of Applied Physics*, vol. 103, p. 103911, 2008.
56. R. Zand, A. Roohi, D. Fan, and R. F. DeMara, "Energy-efficient nonvolatile reconfigurable logic using spin Hall effect-based lookup tables," *IEEE Transactions on Nanotechnology*, vol. 16, pp. 32–43, 2017.
57. A. Alzahrani and R. F. DeMara, "Fast online diagnosis and recovery of reconfigurable logic fabrics using design disjunction," *IEEE Transactions on Computers*, vol. 65, p. 1, 2016.
58. R. A. Ashraf and R. F. DeMara, "Scalable FPGA refurbishment using netlist-driven evolutionary algorithms," *IEEE Transactions on Computers*, vol. 62, pp. 1526–1541, 2013.
59. R. Al-Haddad, R. S. Oreifej, R. Zand, A. Ejnioui, and R. F. DeMara, "Adaptive mitigation of radiation-induced errors and TDDB in reconfigurable logic fabrics," in *Test Workshop (NATW), 2015 IEEE 24th North Atlantic*, 2015, pp. 23–32.
60. R. S. Oreifej, R. Al-Haddad, R. Zand, R. A. Ashraf, and R. F. DeMara, "Survivability modeling and resource planning for self-repairing reconfigurable device fabrics," *IEEE Transactions on Cybernetics*, vol. 48, pp. 1–13, 2017.
61. R. A. Ashraf and R. F. DeMara, "Scalable FPGA refurbishment using netlist-driven evolutionary algorithms," *IEEE Transactions on Computers*, vol. 62, pp. 1526–1541, 2013.
62. N. Imran, J. Lee, and R. F. DeMara, "Fault demotion using reconfigurable slack (FaDReS)," *IEEE Transactions on Very Large Scale Integration (VLSI) Systems*, vol. 21, pp. 1364–1368, 2013.
63. N. Imran, J. Lee, Y. Kim, M. Lin, and R. F. DeMara, "Fault-mitigation by adaptive dynamic reconfiguration for survivable signal-processing architectures," *International Journalof Control and Automation*, vol. 6, 2013.
64. J. Huang, M. Parris, J. Lee, and R. F. Demara, "Scalable FPGA-based architecture for DCT computation using dynamic partial reconfiguration," *ACM Transactions on Embedded Computing Systems*, vol. 9, pp. 1–18, 2009.

65. K. Zhang, N. Khoshavi, J. M. Alghazo, and R. F. De Mara, "Organic embedded architecture for sustainable FPGA soft-core processors," in *Reliability and Maintainability Symposium (RAMS), 2015 Annual*, 2015, pp. 1–6.
66. R. S. Oreifej, R. Al-Haddad, R. A. Ashraf, and R. F. DeMara, "Sustainability assurance modeling for SRAM-based FPGA evolutionary self-repair," in *2014 IEEE International Conference on Evolvable Systems*, 2014, pp. 17–22.
67. R. F. DeMara, J. Lee, R. N. Al-Haddad, R. S. Oreifej, R. Ashraf, B. Stensrud, and M. Quist, "Dynamic partial reconfiguration approach to the design of sustainable edge detectors," in *ERSA*, 2010, pp. 49–58.
68. N. Imran and R. F. DeMara, "Distance-ranked fault identification of reconfigurable hardware bitstreams via functional input," *International Journal of Reconfigurable Computing*, vol. 2014, 2014.
69. N. Imran, L. Jooheung, K. Youngju, L. Mingjie, and F. Ronald, "Amorphous slack methodology for autonomous fault-handling in reconfigurable devices," *International Journal of Multimedia & Ubiquitous Engineering*, vol. 7, 2012.
70. M. G. Parris, C. A. Sharma, and R. F. Demara, "Progress in autonomous fault recovery of field programmable gate arrays," *ACM Computing Surveys (CSUR)*, vol. 43, p. 31, 2011.
71. R. F. DeMara, K. Zhang, and C. A. Sharma, "Autonomic fault-handling and refurbishment using throughput-driven assessment," *Applied Soft Computing*, vol. 11, pp. 1588–1599, 2011.
72. R. F. DeMara and K. Zhang, "Autonomous FPGA fault handling through competitive runtime reconfiguration," in *Evolvable Hardware, 2005. Proceedings. 2005 NASA/DoD Conference on*, 2005, pp. 109–116.
73. J. Lohn, G. Larchev, and R. DeMara, "Evolutionary fault recovery in a Virtex FPGA using a representation that incorporates routing," in *Parallel and Distributed Processing Symposium, 2003. Proceedings. International*, 2003, p. 8.
74. J. Lohn, G. Larchev, and R. DeMara, "A genetic representation for evolutionary fault recovery in Virtex FPGAs," in *ICES*, 2003, pp. 47–56.
75. S. D. Pyle, V. Thangavel, S. M. Williams, and R. F. DeMara, "Self-scaling evolution of analog computation circuits with digital accuracy refinement," in *Adaptive Hardware and Systems (AHS), 2015 NASA/ESA Conference on*, 2015, pp. 1–8.
76. R. A. Ashraf, A. Al-Zahrani, N. Khoshavi, R. Zand, S. Salehi, A. Roohi, M. Lin, and R. F. DeMara, "Reactive rejuvenation of CMOS logic paths using self-activating voltage domains," in *2015 IEEE International Symposium on Circuits and Systems (ISCAS)*, 2015, pp. 2944–2947.
77. N. Khoshavi, R. A. Ashraf, and R. F. DeMara, "Applicability of power-gating strategies for aging mitigation of CMOS logic paths," in *2014 IEEE 57th International Midwest Symposium on Circuits and Systems (MWSCAS)*, 2014, pp. 929–932.
78. N. Khoshavi, R. A. Ashraf, R. F. DeMara, S. Kiamehr, F. Oboril, and M. B. Tahoori, "Contemporary CMOS aging mitigation techniques: Survey, taxonomy, and methods," *Integration, the VLSI Journal*, vol. 59, pp. 10–22, 2017.
79. R. A. Ashraf, N. Khoshavi, A. Alzahrani, R. F. DeMara, S. Kiamehr, and M. B. Tahoori, "Area-energy tradeoffs of logic wear-leveling for BTI-induced aging," in *Proceedings of the ACM International Conference on Computing Frontiers*, 2016, pp. 37–44.
80. R. A. Ashraf, R. Oreifej, and R. F. DeMara, "Scalability of sustainable self-repair to mitigate aging induced degradation in SRAM-based FPGA devices," in *Presentations at the ReSpace/MAPLD 2011 Conference*, 2011.
81. I. Kuon, R. Tessier, and J. Rose, "FPGA architecture: Survey and challenges," *Foundations and Trends in Electronic Design Automation*, vol. 2, pp. 135–253, 2008.
82. I. Kuon and J. Rose, "Measuring the gap between FPGAs and ASICs," *Computer-Aided Design of Integrated Circuits and Systems, IEEE Transactions on*, vol. 26, pp. 203–215, 2007.
83. R. F. DeMara, A. Roohi, R. Zand, and S. D. Pyle, "Heterogeneous technology configurable fabrics for field programmable co-design of CMOS and spin-based devices," in *IEEE International Conference on Rebooting Computing (ICRC-2017)*, Washington, DC, 2017.
84. J. He and J. Rose, "Advantages of heterogeneous logic block architecture for FPGAs," in *Custom Integrated Circuits Conference, 1993. Proceedings of the IEEE 1993*, 1993, pp. 7.4.1–7.4.5.

85. J. Cong and S. Xu, "Delay-optimal technology mapping for FPGAs with heterogeneous LUTs," in *Design Automation Conference, 1998. Proceedings*, 1998, pp. 704–707.
86. A. Koorapaty, V. Chandra, K. Tong, C. Patel, L. Pileggi, and H. Schmit, "Heterogeneous programmable logic block architectures," in *Proceedings of the Conference on Design, Automation and Test in Europe-Volume 1*, 2003, p. 11118.
87. M. P.-E. Gaillardon, "Reconfigurable logic architectures based on disruptive technologies," Citeseer, 2012.
88. J. Kim, A. Paul, P. A. Crowell, S. J. Koester, S. S. Sapatnekar, J.-P. Wang, and C. H. Kim, "Spin-based computing: Device concepts, current status, and a case study on a high-performance microprocessor," *Proceedings of the IEEE*, vol. 103, pp. 106–130, 2015.
89. D. E. Nikonov and I. A. Young, "Overview of beyond-CMOS devices and a uniform methodology for their benchmarking," *Proceedings of the IEEE*, vol. 101, pp. 2498–2533, 2013.
90. J. H. Anderson and Q. Wang, "Area-efficient FPGA logic elements: Architecture and synthesis," in *Proceedings of the 16th Asia and South Pacific Design Automation Conference*, 2011, pp. 369–375.
91. N. S. Kim, T. Austin, D. Baauw, T. Mudge, K. Flautner, J. S. Hu, M. J. Irwin, M. Kandemir, and V. Narayanan, "Leakage current: Moore's law meets static power," *Computer*, vol. 36, pp. 68–75, 2003.
92. P. Pavan, R. Bez, P. Olivo, and E. Zanoni, "Flash memory cells-an overview," *Proceedings of the IEEE*, vol. 85, pp. 1248–1271, 1997.
93. R. Zand, A. Roohi, S. Salehi, and R. F. DeMara, "Scalable adaptive spintronic reconfigurable logic using area-matched MTJ design," *IEEE Transactions on Circuits and Systems II: Express Briefs*, vol. 63, pp. 678–682, 2016.
94. M. K. G. Krishna, A. Roohi, R. Zand, and R. F. DeMara, "Heterogeneous energy-sparing reconfigurable logic: Spin-based storage and CNFET-based multiplexing," *IET Circuits, Devices & Systems*, vol. 11, pp. 274–279, 2017.
95. R. Zand and R. Demara, "Radiation-hardened MRAM-based LUT for non-volatile FPGA soft error mitigation with multi-node upset tolerance," *Journal of Physics D: Applied Physics*, vol. 50, p. 505002, 2017.
96. D. Suzuki, M. Natsui, A. Mochizuki, S. Miura, H. Honjo, H. Sato, S. Fukami, S. Ikeda, T. Endoh, and H. Ohno, "Fabrication of a 3000-6-input-LUTs embedded and block-level power-gated nonvolatile FPGA chip using p-MTJ-based logic-in-memory structure," in *VLSI Circuits (VLSI Circuits), 2015 Symposium on*, 2015, pp. C172–C173.
97. W. Zhao, E. Belhaire, C. Chappert, and P. Mazoyer, "Spin transfer torque (STT)-MRAM–based runtime reconfiguration FPGA circuit," *ACM Transactions on Embedded Computing Systems (TECS)*, vol. 9, p. 14, 2009.
98. R. Zand, A. Roohi, and R. F. DeMara, "Energy-efficient and process-variation-resilient write circuit schemes for spin Hall effect MRAM device," *IEEE Transactions on Very Large Scale Integration (VLSI) Systems*, vol. 25, pp. 2394–2401, 2017.
99. W. Zhao, C. Chappert, V. Javerliac, and J.-P. Nozière, "High speed, high stability and low power sensing amplifier for MTJ/CMOS hybrid logic circuits," *IEEE Transactions on Magnetics*, vol. 45, pp. 3784–3787, 2009.
100. S. Salehi, D. Fan, and R. F. Demara, "Survey of STT-MRAM cell design strategies: Taxonomy and sense amplifier tradeoffs for resiliency," *ACM Journal on Emerging Technologies in Computing Systems (JETC)*, vol. 13, p. 48, 2017.
101. S. Salehi and R. F. DeMara, "Process variation immune and energy aware sense amplifiers for resistive non-volatile memories," in *2017 IEEE International Symposium on Circuits and Systems (ISCAS)*, 2017, pp. 1–4.
102. A. Alzahrani and R. F. DeMara, "Process variation immunity of alternative 16nm HK/MG-based FPGA logic blocks," in *Circuits and Systems (MWSCAS), 2015 IEEE 58th International Midwest Symposium on*, 2015, pp. 1–4.
103. Y. Zhou, S. Thekkel, and S. Bhunia, "Low power FPGA design using hybrid CMOS-NEMS approach," in *Proceedings of the 2007 International Symposium on Low Power Electronics and Design*, 2007, pp. 14–19.

104. J. U. Horstmann, H. W. Eichel, and R. L. Coates, "Metastability behavior of CMOS ASIC flip-flops in theory and test," *IEEE Journal of Solid-State Circuits*, vol. 24, pp. 146–157, 1989.
105. Xilinx, "Virtex-5 FPGA user guide," http://www.xilinx.com/support/documentation/user_guides/ug190.pdf, March 2012.
106. Xilinx, "Spartan-6 FPGA configurable logic block user guide," http://www.xilinx.com/support/documentation/user_guides/ug384.pdf, February 2010.
107. A. Percey, "Advantages of the Virtex-5 FPGA 6-Input LUT architecture," *White Paper: Virtex-5 FPGAs, Xilinx WP284 (v1.0)*, 2007.
108. B. Behin-Aein, D. Datta, S. Salahuddin, and S. Datta, "Proposal for an all-spin logic device with built-in memory," *Nature Nanotechnology*, vol. 5, pp. 266–270, 2010.
109. D. E. Nikonov, G. I. Bourianoff, and T. Ghani, "Proposal of a spin torque majority gate logic," *IEEE Electron Device Letters*, vol. 32, pp. 1128–1130, 2011.
110. K. Y. Camsari, S. Ganguly, and S. Datta, "Modular approach to spintronics," *Scientific Reports*, vol. 5, 2015.
111. S. Matsunaga, J. Hayakawa, S. Ikeda, K. Miura, H. Hasegawa, T. Endoh, H. Ohno, and T. Hanyu, "Fabrication of a nonvolatile full adder based on logic-in-memory architecture using magnetic tunnel junctions," *Applied Physics Express*, vol. 1, p. 091301, 2008.
112. S. Matsunaga, J. Hayakawa, S. Ikeda, K. Miura, T. Endoh, H. Ohno, and T. Hanyu, "MTJ-based nonvolatile logic-in-memory circuit, future prospects and issues," in *Proceedings of the Conference on Design, Automation and Test in Europe*, 2009, pp. 433–435.
113. E. Deng, Y. Zhang, J.-O. Klein, D. Ravelsona, C. Chappert, and W. Zhao, "Low power magnetic full-adder based on spin transfer torque MRAM," *IEEE Transactions on Magnetics*, vol. 49, pp. 4982–4987, 2013.
114. A. Sengupta and K. Roy, "Spin-transfer torque magnetic neuron for low power neuromorphic computing," in *2015 International Joint Conference on Neural Networks* (*IJCNN*), 2015, pp. 1–7.
115. G. Srinivasan, A. Sengupta, and K. Roy, "Magnetic tunnel junction based long-term short-term stochastic synapse for a spiking neural network with on-chip STDP learning," *Scientific Reports*, vol. 6, p. 29545, 2016.
116. A. F. Vincent, J. Larroque, W. S. Zhao, N. B. Romdhane, O. Bichler, C. Gamrat, J. O. Klein, S. Galdin-Retailleau, and D. Querlioz, "Spin-transfer torque magnetic memory as a stochastic memristive synapse," in *2014 IEEE International Symposium on Circuits and Systems* (*ISCAS*), 2014, pp. 1074–1077.
117. W. Zhao, D. Querlioz, J. O. Klein, D. Chabi, and C. Chappert, "Nanodevice-based novel computing paradigms and the neuromorphic approach," in *2012 IEEE International Symposium on Circuits and Systems*, 2012, pp. 2509–2512.
118. R. Zand, K. Y. Camsari, I. Ahmed, S. D. Pyle, C. H. Kim, S. Datta, and R. F. DeMara, "R-DBN: A resistive deep belief network architecture leveraging the intrinsic behavior of probabilistic devices," *arXiv preprint arXiv:1710.00249*, 2017.
119. Q. Guo, X. Guo, R. Patel, E. Ipek, and E. G. Friedman, "AC-DIMM: Associative computing with STT-MRAM," *SIGARCH Computer Architecture News*, vol. 41, pp. 189–200, 2013.
120. H. Jarollahi, N. Onizawa, V. Gripon, N. Sakimura, T. Sugibayashi, T. Endoh, H. Ohno, T. Hanyu, and W. J. Gross, "A nonvolatile associative memory-based context-driven search engine using 90 nm CMOS/MTJ-hybrid logic-in-memory architecture," *IEEE Journal on Emerging and Selected Topics in Circuits and Systems*, vol. 4, pp. 460–474, 2014.
121. A. Roohi, L. Wang, S. Kose, and R. F. DeMara, "Secure intermittent-robust computation for energy harvesting device security and outage resilience," in *14th IEEE International Conference on Advanced and Trusted Computing* (*ATC 2017*), San Francisco, CA, 2017.
122. K. Ma, Y. Zheng, S. Li, K. Swaminathan, X. Li, Y. Liu, J. Sampson, Y. Xie, and V. Narayanan, "Architecture exploration for ambient energy harvesting nonvolatile processors," in *2015 IEEE 21st International Symposium on High Performance Computer Architecture* (*HPCA*), 2015, pp. 526–537.
123. R. Aitken, V. Chandra, J. Myers, B. Sandhu, L. Shifren, and G. Yeric, "Device and technology implications of the Internet of Things," in *2014 Symposium on VLSI Technology* (*VLSI-Technology*): *Digest of Technical Papers*, 2014, pp. 1–4.

16

*RRAM Devices: Underlying Physics, SPICE Modeling, and Circuit Applications**

Firas Odai Hatem, T. Nandha Kumar, and Haider A. F. Almurib

CONTENTS

16.1 Introduction to the Resistive Nonvolatile Memories

Over the last decade, several fast and scalable NVM technologies are being explored and researched for data storage applications, neuromorphic networks, and other digital logic and circuits applications [4–6]. These emerging memories have been attracting an increasing research interest as a potential technology for replacing the conventional semiconductor memories [7,8] such as the existing nonvolatile (NV) flash memory, static random access memory (SRAM), and dynamic random access memory (DRAM) [9]. Examples of

* *This chapter is a reproduction of the mathematical and SPICE RRAM models published in [1] and [2], respectively. All figures, equations, analysis, etc., in this chapter are extracted from [1], [2], and the PhD thesis [3]. Note that [1], [2], and [3] are authored by the same authors of this chapter.*

such NVMs are RRAM [10,11], spin-transfer torque memories (STTRAMs) [12], and phase-change random access memory (PRAM) [13,14]. Each of these memories presents advantages in terms of scaling and speed [15].

A common feature of all these potential NVM technologies is that they are resistive memories where applying electrical pulses on these devices can alter different physical processes, which in turn changes the device resistance [15]. For example, applying electrical pulses on RRAM devices alters the state of the conductive filament (CF), which changes the resistance of the device. On the other hand, resistance change in the PCM device depends on the phase of the active material while in STTRAM; the resistance relies on the magnetic polarization of the ferromagnetic layer in the magnetic tunnel junction (MTJ) [15].

Currently, the leading companies and research institutes in the field are still carrying out more explorations and extensive experiments on most of these promising NVM candidates. However, the metal-oxide bipolar RRAM that employs a resistive switching (RS) mechanism has emerged to be a possible candidate for the next-generation NVMs [7,16,17] and their various applications [18,19]. One of the main advantages of the RRAM technology against PCM and SSTRAM is the simple device structure (an insulator layer sandwiched between two or more metallic layers [15,20]). RRAM can be considered as a specific type of the memristor, which can describe the bipolar resistive switching (BRS) behavior [17,21]. RRAM technology demonstrates various characteristics and attributes that make it a useful element for logic circuit design, neuromorphic systems, and 3D memory applications. For example, RRAM provides NV characteristics, fast write and read time, and it has a non-destructive reading mechanism where the stored data is ideally not altered (or slightly altered) during the reading mechanism [22]. RRAM exhibits the non-destructive behavior because its state variable changes slightly in the case of low reading current while the change due to high current is significantly larger. Thus, RRAM state variable has a nonlinear dependence on the charge [22]. In addition, RRAM provides a high R_{OFF}/R_{ON} ratio to store distinct Boolean data where R_{OFF} and R_{ON} are the RRAM resistances at high-resistance state (HRS) and low-resistance state (LRS), respectively. In general, when storing a digital state into a certain memory device, it is important for the stored data to be distinct. This means that the differences between the stored data are considerably large. By doing this, we guarantee that the digital stored state is not sensitive to the changes in both the operating conditions and the parameters [22]. There are also other applications developed using RRAM technology such as NV lookup tables (LUTs) for field-programmable gate array (FPGA) [5,6,23–26].

However, despite that RRAM exhibits all these useful characteristics, which makes it an important device for the potential circuits applications, the final suitable practical applications for the RRAM memory technology are not yet confirmed. There are still technology limits (e.g., leakage current in crossbar structure), reliability issues (e.g., variability), and understanding limitations (e.g., the underlying physics), which must be overcome before there can be a move toward the industry implementation of the RRAM technology [15].

This chapter focuses on highlighting the underlying physics behind the RRAM operating mechanism by taking one type of RRAM device (bi-layered RRAM) and its available analytical and circuit models as a case study. The physics-based equations and the simulation results obtained from these models are reproduced and combined in this chapter to get better insight into the current conduction (static behavior) and the RS (dynamic behavior) mechanisms of the RRAM devices.

16.2 The First Fabricated Memristor: HP Labs Memristor

In 1971, Leon Chua argued that the memristor is the fourth circuit element that exists besides the other three fundamental circuit elements (resistor, inductor, and capacitor) [27]. Until 2008, Chua's idea about the memristor had not been investigated, and no physical memristor model had been fabricated to demonstrate the actual operation of the device. However, in 2008, a research group from HP Labs revealed for the first time that the memristor element exists naturally in the nanometer scale electronic systems where both the ionic conduction and the solid-state electronics are used to describe the memristor equations [17].

The first memristor cell structure announced by HP Labs consisted of two layers of TiO_2 metal-oxide sandwiched between two metal electrodes (platinum electrodes). A schematic diagram of the HP memristor is shown in Figure 16.1. One of the oxide layers is highly doped with oxygen vacancies, which behave as a semiconductor while the second layer is undoped and has higher resistivity (behaves as an insulator layer). In Figure 16.1, D represents the constant total length of the oxide layer, and w is the variable length of the doped region. Also, w is known as the state variable of the memristor.

Depending on the amount of the electric charge flowing in the device (obtained by applying external stimulus voltage V), the boundary between the two oxide regions shown in Figure 16.1 is moving in the same direction of the passing current I [17]. This movement is caused by the oxygen ions/vacancies movement. The flow of mobile dopants increases and decreases the state variable width w of the doped region. Therefore, the total resistance result is equal to the sum of the series resistances of the doped and undoped regions of the memristor and is given by [17].

$$V = \left(R_{ON} \frac{w}{D} + R_{OFF} \left(1 - \frac{w}{D} \right) \right) \cdot I \tag{16.1}$$

where R_{ON} and R_{OFF} are the resistances of the semiconductor layer for the memristor when $w = 1$ (ON state) and 0 (OFF state), respectively.

If the simple case of ionic electronic conduction and linear ionic drift in a uniform field with average ions mobility of μ_V is assumed as in [17], then

$$\frac{dw}{dt} = \mu_V \frac{R_{ON}}{D} \cdot I \tag{16.2}$$

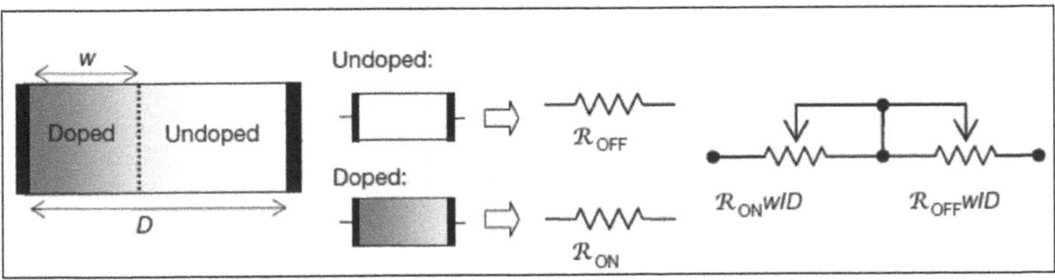

FIGURE 16.1
A simple equivalent circuit of HP Labs' memristor announced in 2008. (From Strukov, D.B. et al., *Nature*, 453, 80–83, 2008.)

where μ_V is the average carriers (oxygen ions/vacancies) mobility. Equations (16.1) and (16.2) reflect the nonlinear relationship between the integrals of the voltage and current. As mentioned previously, RRAM can be considered a specific type of the memristor that can describe the BRS behavior.

16.3 Electroforming Process and the RS Mechanism of RRAM Devices

RRAM devices require an irreversible one-step process before the electrical switching mechanism. This step is referred to as electroforming [28]. The ideal irreversible electroforming process and the reversible RS behavior of the RRAM devices are shown in Figure 16.2 [28]. The inset of Figure 16.2 shows the typical metal-insulator-metal (MIM) RRAM device structure of $Si/SiO_x/Ti$ 5 nm/Pt 15 nm/TiO_2 25–50 nm/Pt 30 nm.

Before the ON/OFF switching states, the device is in its virgin (almost insulator) state. The repeatable ON/OFF switching states shown in Figure 16.2 can be reached after the electroforming process is applied; this is done by applying a high negative voltage (shown in green color in Figure 16.2) or high positive voltage (shown in red color in Figure 16.2). At the electroforming voltage, the device characteristics will change abruptly to a Higher/Lower current and Lower/Higher voltage for negative and positive forming voltages, respectively. The abrupt change in the current indicates the occurrence of the forming process and a decrease in the resistance of the device (several orders in magnitude). After the electroforming process, the initial switching state depends on the polarity of the voltage of the electroforming state. The device can then be switched ON and OFF by applying negative and positive voltages, respectively, on the top electrode (TE). These switching

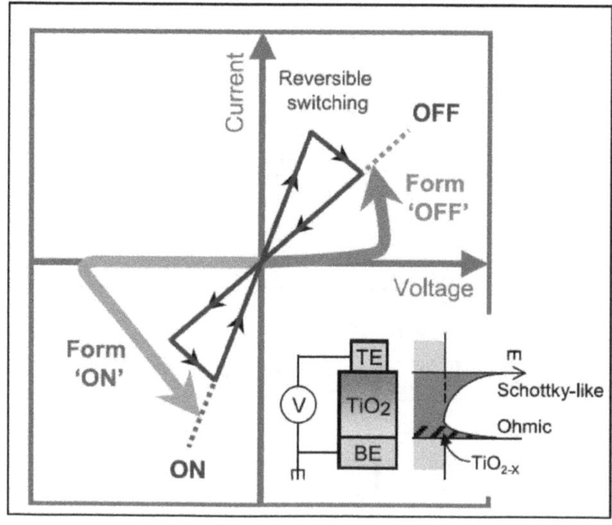

FIGURE 16.2
The ideal reversible BRS process and the required forming voltage polarity for the MIM RRAM. (From Yang, J.J. et al., *Nanotechnology*, 20, 215201, 2009.)

polarities depend on the asymmetry of the interfaces after the device fabrication. In the device shown in Figure 16.2, the top interface is a Schottky barrier-like interface while the bottom interface is ohmic [28]. The physics behind the electroforming process are readily available in many previous articles [28,29].

16.4 Bi-Layered RRAM: Device Structure, Operating Mechanism, and Device Physics

Based on the device structure, RRAM can be classified into MIM [20] and metal-insulator-semiconductor-metal (MISM) RRAM devices [7,16] (hereafter will be referred to as bi-layered RRAM). A typical tantalum oxide-based MISM and MIM RRAM physical devices are shown in Figure 16.3a and b, respectively [30].

Bi-layered RRAM has been widely studied as one of the potential candidates for the NVM [1,7,16,20,30–36], and several RS oxide materials have been explored for the bi-layered RRAM devices such as: CuO/ZnO [34], AlO_x/WO_x [33], HfO_x/AlO_x [35], TiO_x/HfO_x [36], and Ta_2O_5/TaO_x [1,7,16,30–32]. Among these materials is the Ta_2O_5 metal oxide, which showed surprising performance. It exhibited an extreme cycling endurance of over 10^{12}; a retention time of more than 10 years; an abrupt switching time (< 1 ns); multilevel states; two stable phases of TaO_x and Ta_2O_5 [29,37]; and good scalability [7,38–40].

Due to these promising RS characteristics of the Ta_2O_5 oxide material, the researchers have adopted the $Pt/Ta_2O_5/TaO_x/Pt$ as one of the potential bi-layered RRAM devices where TaO_x acts as a bulk material and Ta_2O_5 as an insulator layer, forming a MISM structure [7,30,38] (see Figure 16.3a). Therefore, this bi-layered RRAM will be used in this chapter as a case study to further explore the RRAM devices in terms of the underlying physics behind the device operating mechanism, device modeling, and integrating RRAM devices in circuit applications.

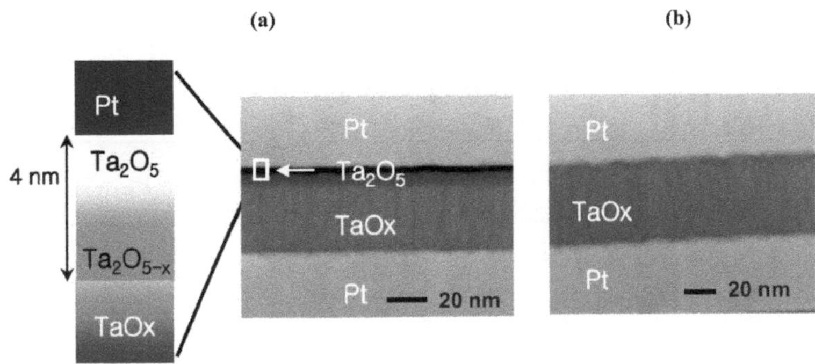

FIGURE 16.3
Transmission electron microscopy (TEM) image of (a) $Pt/Ta_2O_5/TaO_x/Pt$ bi-layered RRAM film and (b) $Pt/TaO_x/Pt$ MIM RRAM film. (From Hur, J.H. et al., *Phys. Rev. B—Condens. Matter Mater. Phys.*, 82, 1–5, 2010.)

16.4.1 The Detailed Bi-Layered RRAM Structure and the RS Mechanism

The detailed structure of the oxide-based Pt/Ta_2O_5/TaO_x/Pt RRAM cell is shown in Figure 16.4 (extracted from [1]), which consists of a high-resistivity oxide layer (top layer) fixed on a TaO_x metal-rich less resistive layer (bulk layer) sandwiched between two pt electrodes. The bulk layer is highly doped with oxygen vacancies (V_O) and its resistance R_B is assumed to be constant during the switching cycles. D is the thickness of the Ta_2O_5 layer, and R_I is the resistance of this layer. The resistance R_I is variable due to the drifting of oxygen vacancies/ions in and out of the layer during the RS mechanism. The thickness of these layers varies from 2 to 5 nm and 10 to 50 nm for the top and base layers, respectively [31]. In this chapter, the thickness of the insulating layer Ta_2O_5 is 3–4 nm [1]. The term w is used to refer to the length of the undoped region in the CF. The doped region is the CF region with high concentration of donor-like oxygen vacancies (the low-resistance region of the CF). The BE is grounded, and the applied voltages and simulation measurements will take place on the TE.

Low-Resistance State: The device reaches its LRS when a negative voltage is applied as in Figure 16.4a. During switching into LRS, oxygen ions in the Ta_2O_5 layer and in the area around the interface region will hop toward the TaO_x layer. The hopped ions will leave behind donor-like oxygen vacancies, which in turn dope the high resistive Ta_2O_5 layer [7], creating a metal-rich doped region and reducing the barrier height $\varnothing_b$ simultaneously. The term $\varnothing_b$ refers to the potential barrier height for electrons moving from TE→TaO_x (Schottky-like barrier). When the total length of the Ta_2O_5 layer is doped by oxygen vacancies ($w = 0$), the device reaches its LRS, and the Schottky barrier-type interface is

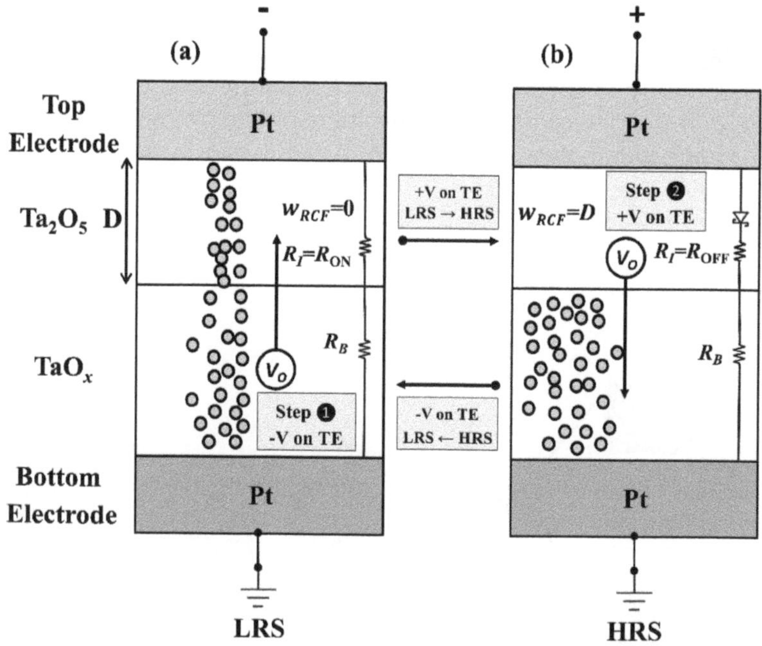

FIGURE 16.4
A schematic representation of the bi-layered RRAM model and its BRS mechanism. (a) The LRS and (b) the HRS. (From Hatem, F.O. et al., A SPICE model of the Ta2O5TaOx Bi-Layered RRAM, *IEEE Transactions on Circuits and System I: Regular Paper*, 2016.)

changed into an ohmic interface. This is shown in Figure 16.4a by step 1 where the arrow inside the cell signifies V_o movement direction.

High-Resistance State: The device reaches HRS by applying a positive voltage as shown in Figure 16.4b. In this case, oxygen ions hop from the TaO_x layer toward the Ta_2O_5 layer and the interface region. Thus, the Ta_2O_5 layer is undoped, and the barrier height is increased (the ohmic interface is changed to Schottky barrier-type interface). As a result, the RRAM state is changed into HRS when all the length of the CF layer is undoped ($w = 1$). This state change is indicated in Figure 16.4b by step 2.

16.4.2 Resistive Switching: Underlying Physics and Modeling

For all values of the applied voltage V, the growth rate dw/dt of the undoped is given by the average ionic drift velocity of oxygen ions $v(t)$ as follows [1,2]:

$$v(t) \approx \frac{dw}{dt} \approx \begin{cases} v_{1,2} \cdot a \cdot c \cdot \exp^{\left(-\frac{U}{KT}\right)} \cdot \sinh \frac{x_{1,2}E}{E_0}, & 0 < w < D \\ 0, & 0 \geq w \geq D \end{cases} \tag{16.3}$$

where v_1 and v_2 are fitting parameters account for the vertical growth/dissolution velocity; a is the effective hopping distance; c is the attempt-to-escape frequency; U is the intrinsic barrier for ion hopping; KT is the thermal energy; x_1 and x_2 are the enhancement fitting factors of the electric field E dependence; and $E_0 = 2KT/qa$, where q is the electron charge. The full derivation of (16.3) is available in [1]. Equation (16.3) reflects the oxygen ions hopping mechanism, which is derived using the Mott and Gurney rigid point ion model [41,42]. The term E represents the electric field in the undoped region, which is the main factor that specifies the oxygen ions' migration rate [32,43]. E can be calculated by dividing the voltage drop on the undoped region of the CF by w, which can be reduced to [1]

$$E = \frac{V_D}{w + R_r(D - w)} \tag{16.4}$$

where $R_r = R_{ON}/R_{OFF}$.

The term V_D represents the voltage across the Ta_2O_5 switching layer, and it can be calculated after considering all the voltage drops across the RRAM cell. These voltages are TE/Ta_2O_5 contact voltage (V_S) which acts like a variable Schottky barrier (represented by Schottky diode), the bulk layer voltage (IR_B), and the insulator layer voltage V_D which is given by [1]:

$$V_D = V - IR_B - V_S. \tag{16.5}$$

$$= \left(R_{ON}\left(1 - \frac{w}{D}\right) + R_{OFF} \cdot \frac{w}{D}\right) \cdot I = R_I \cdot I. \tag{16.6}$$

where R_{ON} and R_{OFF} are the resistance of Ta_2O_5 layer for the LRS and HRS, respectively (see Figure 16.4).

16.4.3 Current Conduction: Underlying Physics and Modeling

Schottky barrier modulation and tunneling mechanism are assumed to be the dominant conduction mechanism of the bi-layered Ta_2O_5/TaO_x RRAM device. The *I–V* equation based on this mechanism is given by [1,44,45]

$$I = AA^{*}T^{2}\exp^{-(n_{1,2}\varnothing_b)/V_T}\left(\exp^{\frac{(V-IR_{\text{series}})}{\eta V_T}}-1\right)\left(\exp^{-\sqrt{\xi}w\times 10^{10}}\right). \tag{16.7}$$

where A is the electrode area, A^{*} is Richardson constant, T is the temperature, n_1 and n_2 are fitting parameters that determine the influence of $\varnothing_b$ on I, V_T is the thermal voltage that is equal to ($kT/q \approx 26\,\text{mV}$) where k is the Boltzmann constant, $R_{\text{series}} = R_I + R_B$, and η is the ideality factor that determines the deviation of the conduction from the pure thermionic emission mechanism of the Schottky barrier. The term ξ is the effective barrier height that is used as a fitting parameter, and $w \times 10^{10}$ is the thickness of the insulator volume in Å. For simplification, T_0 is assumed to be the device temperature when it is not in the switching stage; hence, T_0 is used in (16.7).

The term $\left(\exp^{-\sqrt{\xi}w\times 10^{10}}\right)$ in (16.7) is the tunneling probability factor (TPF) that reflects the effect of tunneling current on the RRAM device [1,2,44,45]. This tunneling depends on D. Tunneling is possible when D is thicker than 1 nm. However, it has been theorized in [44] that the tunneling mechanism occurs in an insulator layer of a few nanometers thickness only. According to [31], the typical thickness of the insulator layer in the Ta_2O_5/TaO_x bi-layered RRAM is 2–5 nm, which is in the reasonable range for the tunneling to occur [44].

The ideality factor η is given by [1,2]

$$\eta = m\left[(\eta_{\text{LRS}})\left(1-\frac{w}{D}\right)+(\eta_{\text{HRS}})\frac{w}{D}\right] \tag{16.8}$$

where m is a fitting parameter, and η_{LRS} and η_{HRS} are the values of η at LRS and HRS, respectively. These values can be calculated by fitting (16.8) into (16.7) to get the best *I–V* fitting. Equation (16.8) reflects the effect of the continuous charging and discharging of the interface traps densities reported in [1] and [2]. When a bias is applied for RS, the Fermi level moves continuously up or down with respect to the interface traps levels where the degree of bending is proportional to the applied bias [44]. This movement results in a continuous change of the charge of the interface traps [46], which in turn affects the magnitude of the interface traps density continuously.

The barrier height $\varnothing_b$ in (16.7) is given by [1]

$$\varnothing_b = \varnothing_{\text{Bn0}}\left(\frac{w}{D}\right)-\left[\frac{q^3 N\varphi_s}{8\pi^2{\varepsilon_s}^3}\right]^{1/4}. \tag{16.9}$$

where $\varnothing_{\text{Bn0}}$ is the ideal Schottky barrier height when an n-type semiconductor is used, $\Delta\varnothing$ is the image-force lowering factor, N is the bulk layer charge density, and ε_s is the permittivity of the bulk layer. The value of φ_s is given by $\varphi_s \approx \varnothing_{\text{Bn}}$ [13]. Equation (16.9) reflects the effect of the image-force lowering factor on the barrier height $\varnothing_b$ reported in [1].

16.4.4 Temperature Modeling

During the RS process, the formation and rupture of the CF are sensitive to the temperature. Thus, the temperature effect is important during the RS behavior and should be considered in the CF growth rate equation. The temperature in the simulation is the local temperature around the doped region, which can be raised during the RS due to the Joule heating effect [47]. This enhances the ions migration process and thus it should be used in the growth rate equation. For the simplification of the model, the approximate value of the temperature used in this model is based on the simple analysis in [43] and [47]. The doped region temperature is given by [47]

$$T = T_0 + I^2 \cdot R \cdot R_{\text{th}} \tag{16.10}$$

where T_0 is the ambient temperature, R_{th} is the effective thermal resistance, R is the doped region electrical resistance, and I is the current through that region.

16.5 Bi-Layered RRAM SPICE Modeling

The equations representing w evolution and the current conduction are summarized in Table 16.1. These equations are implemented as a RRAM circuit model in LTSPICE. Table 16.2 contains the parameters used in LTSPICE modeling with their values. The structure of the SPICE model is shown in Figure 16.5, which determines the time-dependent current flow at LRS, HRS, and the transitional state (switching stage).

TABLE 16.1

Summary of the Bi-layered RRAM Model Equations

Bi-Layered RRAM	
$\frac{dw}{dt} = v(t) \approx \begin{cases} v_{1,2} \cdot a \cdot c \cdot \exp^{\left(-\frac{U}{KT}\right)} \cdot \sinh \frac{x_{1,2}E}{E_0}, & 0 < w < D, \\ 0, & 0 \geq w \geq D \end{cases}$	(16.3)
$E = \frac{V_D}{w + R_r(D - w)}$	(16.4)
$V_D = V - IR_B - V_{S.}$	(16.5)
$V_D = \left(R_{\text{ON}}\left(1 - \frac{w}{D}\right) + R_{\text{OFF}} \cdot \frac{w}{D}\right) \cdot I = R_I \cdot I$	(16.6)
Mathematical Modeling of the I–V Behavior and the Temperature	
$I = AA^* T_0^2 \exp^{-(m_{1,2}\varnothing_b)/V_T} \left(\exp^{\frac{(V_S)}{\eta V_T}} - 1\right)\left(\exp^{-\sqrt{\xi}w \times 10^{10}}\right)$	(16.7)
$\eta = \text{m}\left[(\eta_{\text{LRS}})\left(1 - \frac{w}{D}\right) + (\eta_{\text{HRS}})\frac{w}{D}\right]$	(16.8)
$\varnothing_b = \varnothing_{\text{Bn}} - \Delta\varnothing = \varnothing_{\text{Bn0}}\left(\frac{w}{D}\right) - \left[\frac{q^3 N \varphi_s}{8\pi^2 \varepsilon_s{}^3}\right]^{1/4}$	(16.9)
$T = T_0 + I^2 \cdot R_I \cdot R_{th}$	(16.10)

Source: Hatem, F.O. et al., A SPICE model of the Ta_2O_5TaOx Bi-Layered RRAM, *IEEE Transactions on Circuits and System I: Regular Paper*, 2016.

TABLE 16.2

Parameters Used in the SPICE Bi-layered RRAM Model Simulation for $D = 4$ and 3 nm

Parameter	Value When (D = 4/3 nm)
T_0	300K
D	4 nm [30,1]/3 nm [31,1]
V_T	26 mV
A	9×10^{-6} cm^2 [30,1]
ε_S	$27\times\varepsilon_0$
ε_0	8.85×10^{-12} F·m^{-1}
U	1eV [32]
a	1nm [32,48]
k	1.38×10^{-23} J·K^{-1}
q	1.6×10^{-19} C
A^*	120×10^4 A·m^{-2}·K^{-2} [44]
N	1×10^{23} C·m^{-3}
ξ	0.00175/0.0031
R_{OFF}	40kΩ/20kΩ
R_{ON}	1.7kΩ
R_B	12kΩ
c	1×10^{13} Hz [48]
$\varnothing_{Bn0}$	0.6eV [30,1]/0.45 eV
x_1	215/95
x_2	0.4/0.03
v_1	1
v_2	0.6e−6
n_1	0.1852/0.1905
n_2	0.0117
m	6e6

Source: Hatem, F.O., *Semicond. Sci. Technol.*, 30, 115009, 2015; Hatem, F.O. et al., A SPICE model of the Ta_2O_5TaOx Bi-Layered RRAM, *IEEE Transactions on Circuits and System I: Regular Paper*, 2016.

16.5.1 Current Path SPICE Modeling

It can be seen in Figure 16.5a that the single Ta_2O_5/TaO_x RRAM device is represented by the two-terminal SPICE subcircuit. The current port equation of the subcircuit is composed of two elements connected in series as shown in the top part of Figure 16.5a. The Schottky barrier tunneling element (the upper-left corner of Figure 16.5a) contains a parallel connection of two voltage-controlled current sources (VCCS) GItunON and GItunOFF (G-type source in LTSPICE), which model the tunneling current when $V < 0$ and $V > 0$, respectively. The correct VCCS is selected by using the applied bias V as a parameter in the STP function in LTSPICE. In this case, GItunON will be delivering

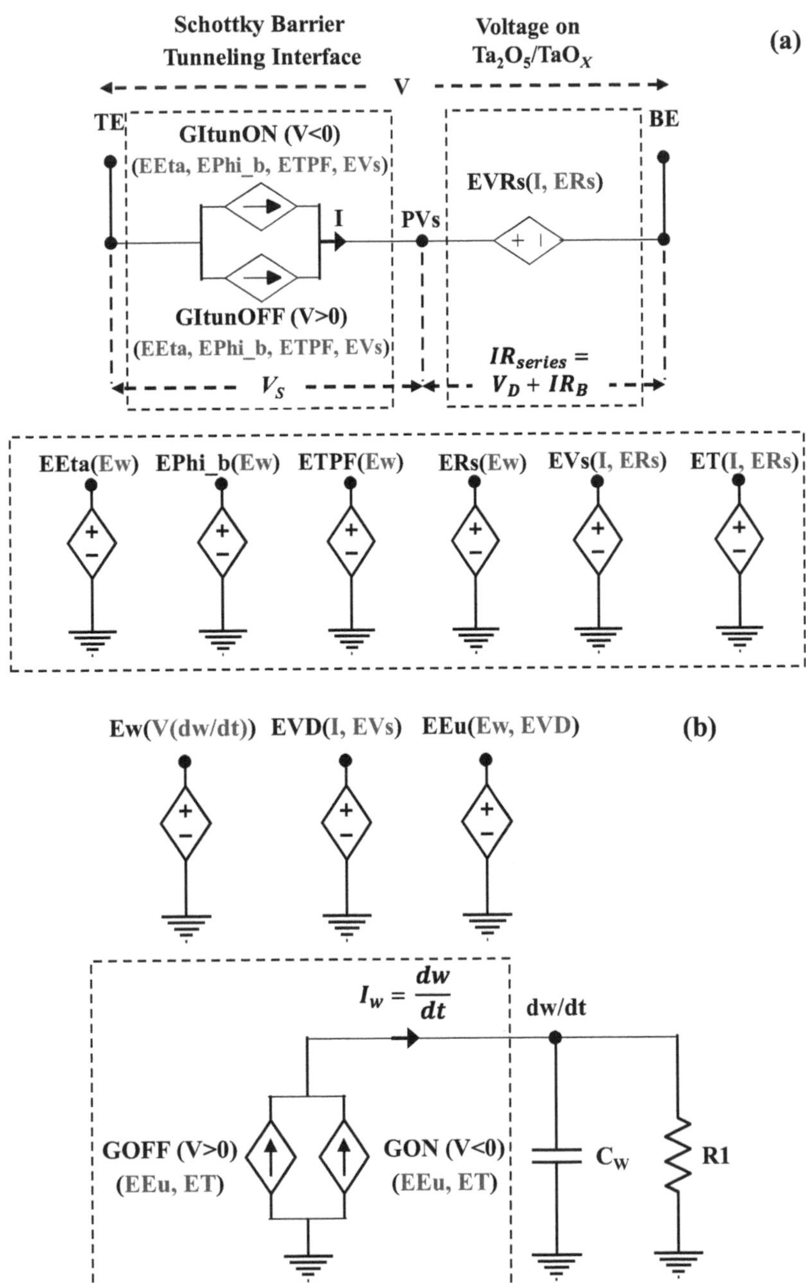

FIGURE 16.5
LTSPICE implementation of the SPICE model. (a) The two-terminal (current path) SPICE implementation of the single Pt/Ta_2O_5/TaO_x/Pt device and the implementation of the related parameters $\varnothing_b$, η, TPF, and V_S. (b) The SPICE subcircuit implementation of w evolution, E, and the self-limiting effect. (From Hatem, F.O. et al., A SPICE model of the Ta_2O_5TaOxBi-Layered RRAM, *IEEE Transaction Circuits and System: I Regular Paper*, 2016.)

a charge into the point PVs when $V < 0$ whereas GItunOFF is activated when $V > 0$. The function STP(x) is a unit step that returns 1 and 0 for $x \geq 0$ and $x < 0$, respectively. The current I passing through the point PVs is the total current through the RRAM device as given by (16.7). The expression for the reverse bias condition GItunOFF is similar to GItunON except that the polarity of Schottky barrier potential and the current direction are reversed. The voltage drop V_S on the tunneling element represents Schottky barrier voltage as shown in Figure 16.5a. The second element in the port equation (the upper-right corner of Figure 16.5a) is a voltage-controlled voltage source (VCVS) EVRs (E-type source in LTSPICE), which represents the total voltage drop on the device (IR_{series} or $V_D + IR_B$) except that of V_S. However, in order to produce the total output current I from GItunON and GItunOFF, four parameters—$\varnothing_b$, η, TPF, and V_S—are modeled using VCVSs: EPhi_b, EEta, ETPF, and EVs, respectively. The terminal voltage of the first three voltage sources is controlled by the instantaneous value of w. The value of w is produced using the VCVS Ew (Figure 16.5b), which is discussed in the next section. The fourth VCVS source EVs is implemented based on the device current and the value of R_{series}. The resistance R_{series} is implemented as ERs as shown in Figure 16.5a.

16.5.2 *w* and *E* SPICE Modeling

Figure 16.5b shows the SPICE modeling of w and E evolution dynamic. Two VCVS, EEu and EVD, are used to calculate E based on V_D. EEu is used to implement two VCCSs, GON and GOFF, which are used to produce the current $I_w = dw/dt$ using (16.7) (see Figure 16.5b). Based on the applied bias polarity, the function STP (see the subcircuit listing in [2]) is used to switch on GON (I_w = GON) when $V < 0$ (switching into LRS) while GOFF is used (I_w = GOFF) when $V > 0$ (switching into HRS). VCVS Ew is used to calculate the instantaneous value of w. This is achieved by integrating I_w inside a nested ternary function. The integration is performed by taking the voltage V(dw/dt) across the capacitor C_w (see Figure 16.5b). The capacitor C_w has a value of 1 F to keep the units of w unchanged (in nanometer). The nested ternary function is also used to model the self-limiting effect of the device as in (see the subcircuit listing in [2]). The initial state of the device is assumed to be in the ON states. This is implemented by setting the initial voltage on C_w to 0; hence, initial w value is also 0 nm. Furthermore, for the simulation to determine the initial DC operating point, a dc path from point dw/dt to the ground is provided using a resistor R1 with a very large value.

16.6 The Agreement with the *I–V* Characteristics for Different Values of *D*

16.6.1 Nonlinear Ionic Drift Mechanism

The validity of the SPICE model is first verified by comparing the obtained simulation results in LTSPICE with the experimental data from $Pt/Ta_2O_5/TaO_x/Pt$ RRAM devices with $D = 4$ and 3 nm.

Using the LTSPICE subcircuit provided in [2] and the associated parameters adjusted in Table 16.2, Figure 16.6 shows the semi-log scale simulated *I*–*V* characteristics for the SPICE model (solid line) and the experimental data from [30] for $D = 4$ nm. The simulated and the experimental data in Figure 16.6 are obtained by using the same bias protocol of a 100 Hz sine wave, SET voltage of −2 and 3 V RESET voltage. The model simulation results match with the measured data and show excellent agreements, both qualitatively and quantitatively except for a small discrepancy in the simulated current magnitude while switching into HRS. It is observed that the LRS switching voltage (LSV) = −1.24 V, I_{LSV} = 0.25 μA, HRS switching voltage (HSV) = 1.9 V, and I_{HSV} = 70 μA. I_{LSV} and I_{HSV} are the device current at LSV and HSV, respectively.

The SPICE model is used to reproduce the *I*–*V* characteristics from the experimental data with smaller Ta_2O_5 layer thickness ($D = 3$ nm) [31]. However, due to the change in the Ta_2O_5 layer thickness and its resistance at HRS, some of the used parameters must be tuned to reflect this change as follows. It is assumed that $\varnothing_{Bn0}$ is reduced compared to its initial height when $D = 4$ nm, following the reduction in the insulating volume where the smaller insulating volume results in smaller bandgap [31,49]. Thus, $\varnothing_{Bn0}$ is approximated to be reduced by around 25% to $\varnothing_{Bn0} \approx 0.45$ eV, following (16.9), which is similar to the value used in [31] ($\varnothing_{Bn0} = 0.4$ eV for $D = 3$ nm). Besides the change in $\varnothing_{Bn0}$, changes will occur to R_{OFF}, X_1, X_2, ξ, and n_1 in which $R_{OFF} = 20\,\text{k}\Omega\text{s}$, $X_1 = 95$, $X_2 = 0.03$, $\xi = 0.0031$, and $n_1 = 0.1905$. The simulation results of applying the proposed model into

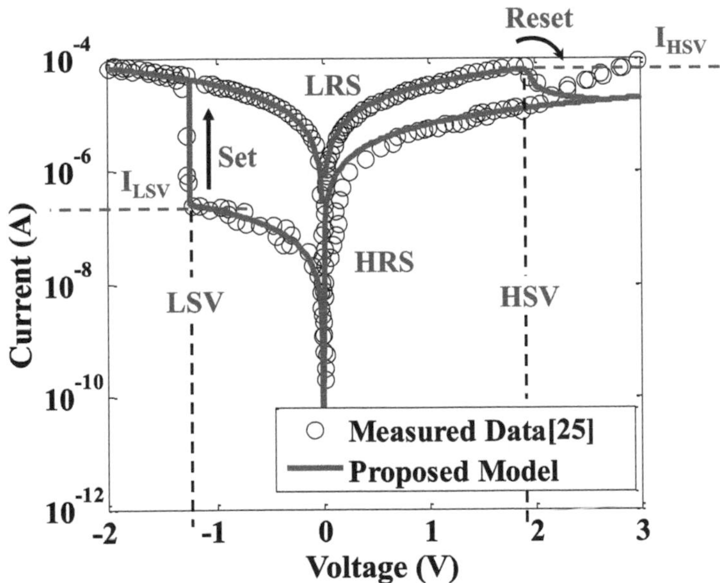

FIGURE 16.6
Experimental measurements [30] and the semi-log scale plot of the SPICE model simulation I–V characteristics with D = 4 nm, $\varnothing_{Bn0} = 0.6$ eV, $\xi = 0.00175$ eV, $R_{OFF} = 40$ kΩ, $x_1 = 215$, $x_2 = 0.4$, $n_1 = 0.1852$, and a −2/3 V 100 Hz sine wave bias signal. (From Hatem, F.O. et al., A SPICE model of the Ta_2O_5TaOx Bi-Layered RRAM, *IEEE Transactions on Circuits and System I: Regular Paper*, 2016.)

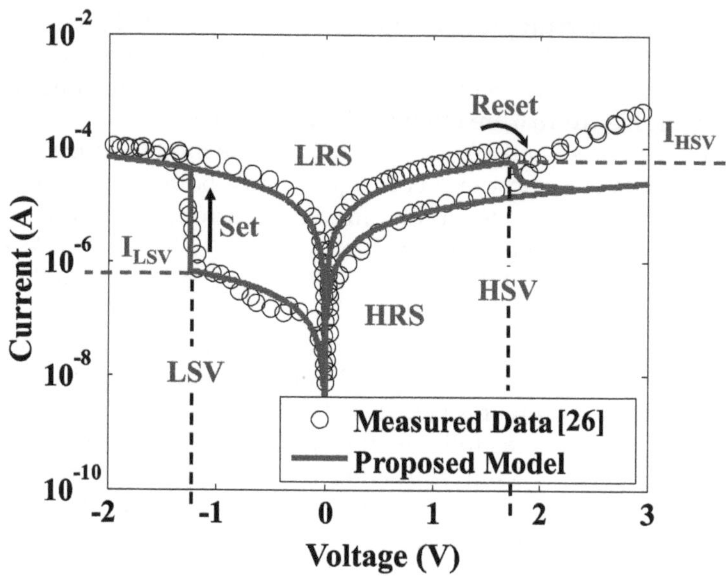

FIGURE 16.7
Experimental measurements [31] and the semi-log scale plot of the SPICE model simulation I–V characteristics with D = 3 nm, $\varnothing_{Bn0}$ = 0.45 eV, ξ=0.0031eV, R_{OFF} = 20 kΩ, x_1 = 95, x_2 = 0.03, n_1 =0.1905, and a −2/3 V 0.2 Hz sine wave bias signal. (From Hatem, F.O. et al., A SPICE model of the Ta_2O_5TaOx Bi-Layered RRAM, *IEEE Transactions on Circuits and System I: Regular Paper*, 2016.)

Ta_2O_5/TaO_x RRAM with $D = 3$ nm are shown in Figure 16.7. The measured and calculated *I–V* curves in Figure 16.7 are obtained by applying a sine wave voltage of 5 s period with voltages of −2 and 3 V for the SET and RESET, respectively. The simulation results in Figure 16.7 are consistent with the experimental data, which shows that the model emphasizes the dependency of the device behavior on the change of *D*. This dependency is achieved by integrating the physics involved when *D* is changed (i.e., integrating *D* and/or *w* into *E*, $\varnothing_b$, TPF, and η), providing that *D* is still within the allowable range for tunneling. It can be seen that LSV = −1.24 V, I_{LSV} = 0.67 μA, HSV = 1.7 V, and I_{HSV} = 70 μA.

16.6.2 Ideal State—Linear Dopant Drift

Figure 16.8 shows the simulated semi-log scale plot of the *I–V* characteristic using the simple ideal ionic drift equation (16.2) given in [17]. Similar to the case of the nonlinear ionic drift mechanism, it can be seen that the device is switching ON and OFF periodically by a 100 Hz sine wave $V = 2\sin(\omega_0 t)$ but does not exhibit a good agreement with the results of the experimental measurement for the same RRAM cell structure fabricated in previous publications [30], especially at the switching regions. However, the device shows the correct LRS and HRS current levels. These results show the importance of using the correct physics involved during the RS process of the bi-layered RRAM (nonlinear ions hopping mechanism).

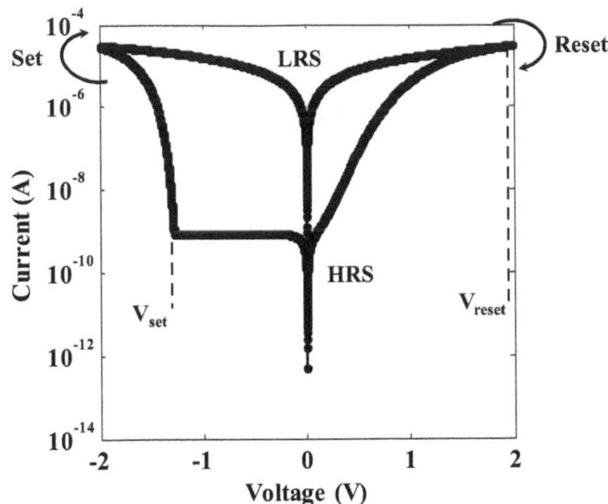

FIGURE 16.8
Semi-log scale plot of the *I–V* characteristics of the proposed RRAM mathematical model with the addition of TPF and the continuous variation of the interface traps densities but with use of a linear dopant drift model.

16.7 The Dependency of HSV and LSV on *D*

Figure 16.9 shows the SPICE model simulation results for $D = 3$ and 4 nm plotted together. These results have already been verified with the experimental data in Figures 16.6 and 16.7. By comparing the two devices' characteristics, it can be seen that the SPICE model can

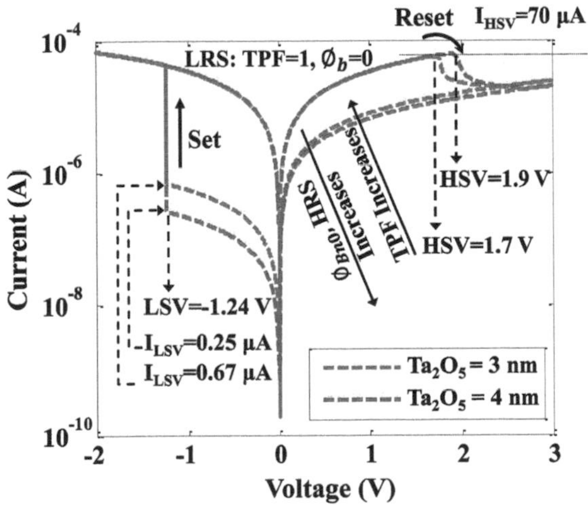

FIGURE 16.9
The semi-log scale plot of the SPICE model simulation *I–V* characteristics with $D = 3$ nm, $\varnothing_{Bn0} = 0.45$ eV (blue line), and $D = 4$ nm, $\varnothing_{Bn0} = 0.6$ eV (red line). Both curves are obtained using −2/3 V Hz sine wave bias signal with 0.2 Hz (for $D = 3$) and 100 Hz (for $D = 4$). (From Hatem, F.O. et al., A SPICE model of the Ta_2O_5TaOx Bi-Layered RRAM, *IEEE Transactions on Circuits and System I: Regular Paper*, 2016.)

capture the change in HSV, which decreases for the 3 nm device compared with that of 4 nm from around 1.9–1.7 V. With HSV as the required voltage for E to reach $E_{LRS\to HRS}$, the SPICE model can capture the change in HSV as follows. When a positive RESET bias is applied and before HSV is reached, E follows (16.11) and hence, two factors are contributing to the value of E at this stage, D and V_D. Due to the nanoscale nature of the term D in (16.11), a reduction of D by 1 nm will have a great influence on E_{LRS}. Hence, a smaller voltage V is required to reach $E_{LRS\to HRS}$ when D is reduced from 4 to 3 nm, which explains the smaller HSV obtained in Figure 16.9. It can be seen in Figure 16.9 that during LRS, the parameters TPF and $\varnothing_b$ maintain their maximum and minimum values (1 and around 0, respectively), irrespective of the value of D. Thus, HSV is not affected by these two parameters when D is changed.

$$E_{\mathrm{LRS}} = \frac{V_D}{R_r D} = \frac{V - IR_B - V_S}{R_r D} \tag{16.11}$$

Besides HSV, the SPICE model can also demonstrate the correct LSV when D is changed. According to the experiments in [16], [30], and [31], the Ta_2O_5/TaO_x RRAM maintains its LSV despite the change in D. Figure 16.9 shows that the SPICE model simulation captures this intrinsic feature and shows the same LSV when D is changed from 3 to 4 nm. These results verify that due to the correct modelling of E, the SPICE model can successfully demonstrate the dependency of HSV and LSV on D.

16.8 The Effects of Changing D on the Values of LRS and HRS

The SPICE model can successfully demonstrate the change in LRS and HRS values when D and the value of R_{OFF} are changed. According to the experiments in [16], [31], and [30], HRS for Ta_2O_5/TaO_x RRAM decreases by decreasing D while the LRS is insensitive to the change in D. Figure 16.9 shows that the SPICE model can successfully capture this feature where HRS is lowered when D is decreased from 4 to 3 nm whereas LRS remains unchanged. The SPICE model can predict this intrinsic phenomenon as follows.

HRS: While at HRS, the device resistance is determined by $R_B + R_{\mathrm{OFF}} + V_S / I$ [1]. The interface resistance V_S/I at HRS is influenced by TPF and $\varnothing_b$. At HRS, TPF and $\varnothing_b$ are proportional to $w = D$ where decreasing D results in larger and smaller values of TPF and $\varnothing_b$ ($\varnothing_{\mathrm{Bn0}}$ is smaller), respectively. As a result, I is increased according to (16.7) and the interface voltage V_S is decreased. Consequently, the resistance V_S/I is reduced for smaller D. Besides V_S/I, R_{OFF} is also reduced for smaller D (due to the smaller insulating volume). It is found that $R_{\mathrm{OFF}} = 20$ kΩs provides the best fitting to the experimental results when $D = 3$ nm. Therefore, the reduction in V_S/I and R_{OFF} when D is reduced leads to smaller HRS. However, the resistance R_B is insensitive to the change in D.

LRS: Following the same explanation for HRS, the device resistance at LRS is determined by: $R_B + R_{\mathrm{ON}} + V_S/I$. TPF and $\varnothing_b$ at LRS have the values of 1 and around 0, respectively, regardless of the value of D (see Figure 16.9) which makes the resistance V_S/I at LRS insensitive to these two parameters (V_S/I reaches its minimum value at LRS). Also, R_B and R_{ON} are selected to satisfy the condition $R_B \gg R_{\mathrm{ON}}$ as in [7]. Hence, the small change in R_{ON} due to the change of D is negligible (R_{ON} is assumed to be fixed in the SPICE model), and the total current while at LRS is limited by the relatively large R_B. This explains why the change in D has no effect on the LRS. These results show that by integrating TPF and $\varnothing_b$, the proposed SPICE model can manifest the dependency of LRS and HRS on D.

16.9 The Intrinsic Schottky Barrier and Its Effect on HRS during SET Switching

Using the proposed SPICE model, a comprehensive analysis and simulation of a Schottky barrier-like tunneling interface (Pt/Ta_2O_5) reveal that the device resistance during HRS is not fixed but depends on the bias polarity.

Figure 16.10a shows the simulated device resistance $R_{series} + V_S/I$ and the current I as a function of time for $D = 4$nm for a complete RS cycle while Figure 16.10b compares the corresponding variation in $\varnothing_b$, TPF, V_S, and V. Assuming LRS as the initial state (point A), it can be seen that the device first starts switching into HRS when HSV = 1.9 V is reached (point B). The device reaches its HRS at the end of the switching period (end of BC). However, the imperative remark that can be observed is that once a negative SET bias is applied (point D), the resistance switches from a few hundreds of kilo ohms (HRS value) to a few mega ohms, and the current tends to align with 0 A. The device maintains this very high resistance until LSV = –1.24 V is reached (point E) where the resistance drops significantly and an abrupt increase in the current is observed, indicating LRS. This behavior is attributed to the coexistence of the reversed-biased Schottky barrier associated with TPF as explained next. The device current is determined by the multiplication of three terms—T1, T2, and TPF—as follows:

$$I = \underbrace{AA^{*}T_0^2 \exp^{-(\varnothing_b)/V_T}}_{T1} \underbrace{\left(\exp^{\frac{V_S}{\eta V_T}} - 1 \right)}_{T2} \underbrace{\left(\exp^{-10^{10} \times \sqrt{\xi w}} \right)}_{TPF} \tag{16.12}$$

(1) While at LRS (AB), I is mainly influenced by T2 and varies according to V. TPF equals 1 and has no effect on suppressing I at LRS (no tunneling current). Similarly, $\varnothing_b \approx 0$ at LRS (ohmic contact) and T1 is fixed at its maximum value and does not reduce I. **(2)** Over time, HSV is reached (point B), RESET starts, and all the three terms in (16.12) will have a great influence on I during the switching. TPF starts to decrease, reducing I (indicates tunneling). Eventually, TPF switches to its minimum value at $w = D$ (point C). Simultaneously, $\varnothing_b$ is ramped to its maximum value at $w = D$ (point C) and tunneling Schottky barrier is formed. **(3)** While the device is at HRS, I at the positive bias region (CD) is suppressed by means of TPF and a forward biased tunneling Schottky barrier (positive V_S) composed of T1×T2 where T2 ≥ 0. In contrast, I at the negative bias region (DE) is suppressed by the same value of TPF and T1 but with Schottky barrier in T2 being reversed biased (negative V_S) where T2 ≤ 0. As a result, I in segment DE is getting closer to zero compared to that in the segment CD because $|T2|$ is smaller for negative V_S compared to positive V_S (($\exp^y - 1$) is smaller

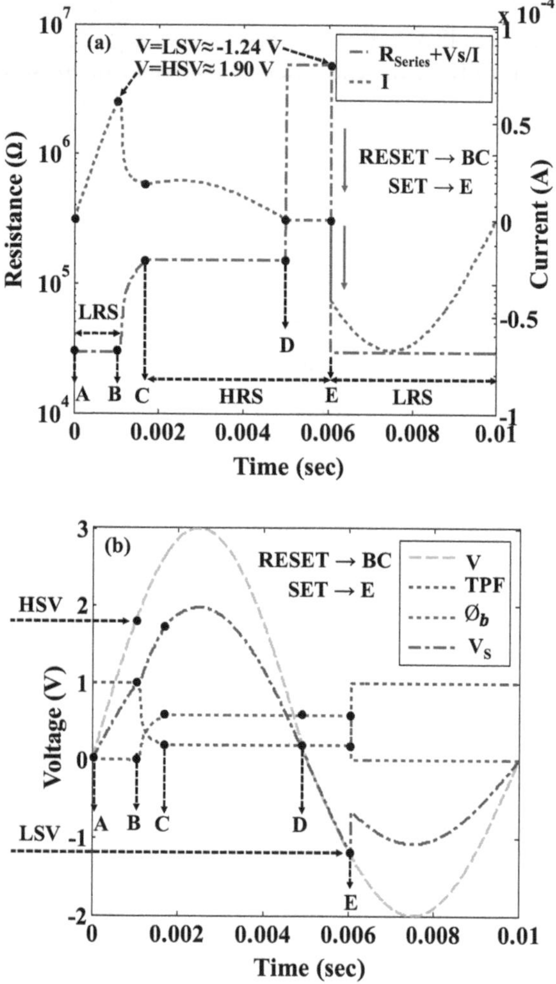

FIGURE 16.10
(a) The SPICE model simulation of the device total resistance $R_{series} + V_S/I$ (blue line) and the total current I (purple line) as a function of time for the 4 nm Ta_2O_5 layer thickness and $\varnothing_{Bn0} = 0.6$ eV. The curves plotted are obtained for a complete RS cycle using 2/3 V 100 Hz sine wave bias signal: (b) The corresponding variation in $\varnothing_b$, TPF, V_S, and V. (From Hatem, F.O. et al., A SPICE model of the Ta_2O_5TaOx Bi-Layered RRAM, *IEEE Transactions on Circuits and System I: Regular Paper*, 2016.)

for negative y). Consequently, the very small I forces V_S/I and the total resistance ($R = R_{series} + V_S/I$) to switch to the temporal high value observed in Figure 16.10a. **(4)** Once LSV is reached (point E), $\varnothing_b$ drops to 0 eV (the contact is changed into ohmic), forcing T1 to increase considerably. This is associated with TPF increases again to 1. The increase in T1 and TPF eliminates the effect of reverse biased Schottky barrier in T2; therefore, I increases again, which in turn forces V_S/I to drop to its minimum value, and LRS is reached. These results show that the proposed SPICE model can successfully capture the asymmetric current profile at HRS and highlight the dependency of the HRS on the bias polarity, making the model more reliable and predictive for potential circuit application.

16.10 RRAM-Based Nonvolatile D-Latch

As an application of the proposed Ta_2O_5/TaO_x RRAM SPICE model, a RRAM-based NV D-latch circuit is designed. Figure 16.11 shows the NV D-latch circuit design where the proposed RRAM model acts as an NV element that retains the latched data in the event of the power interruption. The SPICE simulation results of the NV behavior of the D-latch is illustrated in Figure 16.12a.

It can be seen in Figure 16.12a that between t = 0 and 0.1 ms, G is at low ($\bar{G}$ is high) and the input data D_{in} is kept at high. Therefore, the transmission gate TG1 is turned-off and TG2 is turned-on. The data in the latch during this period is invalid or depends on the previous state of the latch. During t = 0.1–0.3 ms, G is high and that turns-on TG1 and turns-off TG2. Now, the data D is written into the back-to-back connected inverter. Hence, D is high and $\bar{D}$ is low.

At the same duration, the data is also written into the RRAM. It can be observed from the simulation results in Figure 16.12a that the current *I* flowing through the RRAM decreases from 80 μA (at 0.1 ms) to 18 μA (at 0.3 ms); this is due to the resistance of the RRAM, which is initially at LRS and then changes to HRS. At t = 0.3 ms, G goes to low, hence, TG2 is turned-on and TG1 is turned-off. This retains the data in the latch and the HRS in the RRAM. When a power interruption occurs at t = 0.4 ms (V_{DD} = 0 V), it can be observed that D, $\bar{D}$, and the current through the RRAM are low. Then at t = 0.42 ms, when the power resumes (V_{DD} = 3 V), the current through the high-resistance RRAM (18 μA) retains the original value of D (to high) and $\bar{D}$ (to low). Thus, the data is successfully retained in the latch.

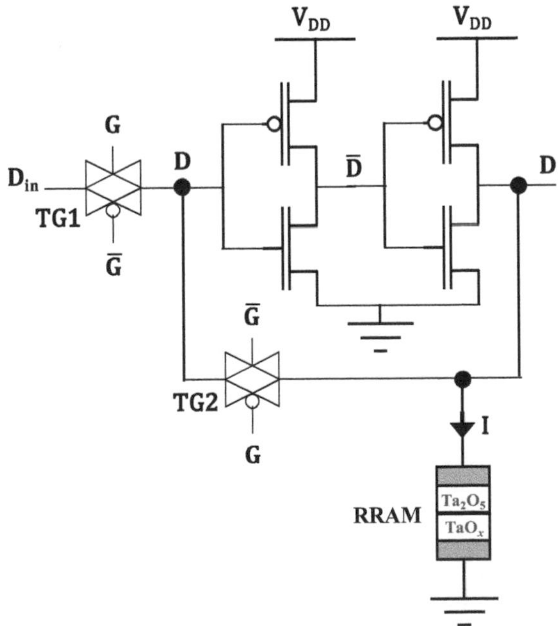

FIGURE 16.11
Schematic of the RRAM-based NV D-latch with the Ta_2O_5/TaO_x RRAM SPICE model integrated. (From Hatem, F.O. et al., A SPICE model of the Ta_2O_5TaOx Bi-Layered RRAM, *IEEE Transactions on Circuits and System I: Regular Paper*, 2016.)

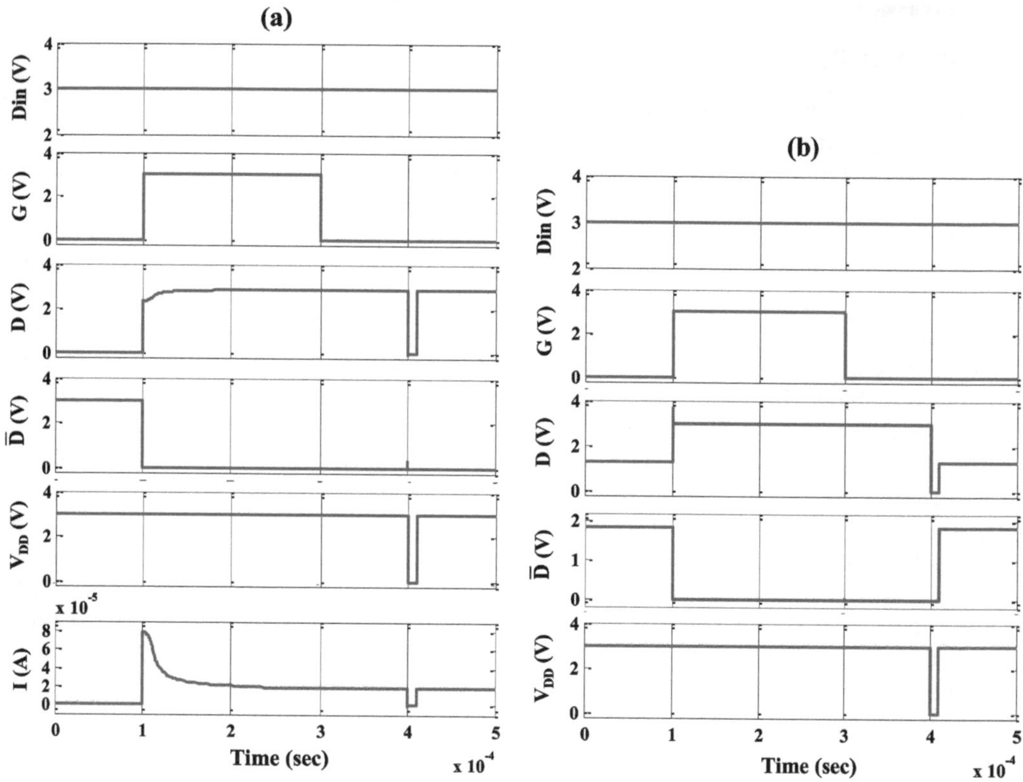

FIGURE 16.12
The SPICE simulation of the RRAM-based NV D-latch (a) with the Ta_2O_5/TaO_x RRAM SPICE model integrated and (b) without the RRAM connected. (From Hatem, F.O. et al., A SPICE model of the Ta_2O_5TaOx Bi-Layered RRAM, *IEEE Transactions on Circuits and System I: Regular Paper*, 2016.)

Next, RRAM is removed from the D-latch circuit, and the simulation is performed on the circuit under the similar simulation environment setup of the NV D-latch. The simulation results without the RRAM connected are shown in Figure 16.12b. It can be observed from the results that at t = 0.42 ms, when the power resumes, D and $\bar{D}$ do not resume to their original states after the power interruption and hence, the latched data is corrupted. Thus, an application of the proposed Ta_2O_5/TaO_x RRAM SPICE model as a nonvolatile element in the RRAM-based NV D-latch is demonstrated successfully.

References

1. F. O. Hatem, P. W. C. Ho, T. N. Kumar, and H. A. F. Almurib, "Modeling of bipolar resistive switching of a nonlinear MISM memristor," *Semicond. Sci. Technol.*, vol. 30, no. 11, 115009, 2015.
2. F. O. Hatem, T. N. Kumar, and H. A. F. Almurib, "A SPICE Model of the Ta_2O_5TaOxBi-Layered RRAM," *IEEE Transactions Circuits System: I Regular Paper*, 2016.
3. F. O. Hatem, *"Bipolar Resistive Switching of Bi-Layered Pt/Ta_2O_5/TaOx/Pt RRAM—Physics-Based Modelling, Circuit Design and Testing,"* Semenyih, Malaysia: The University of Nottingham Malaysia Campus, 2017.

4. S. H. Jo, T. Chang, I. Ebong, B. B. Bhadviya, P. Mazumder, and W. Lu, "Nanoscale memristor device as synapse in neuromorphic systems," *Nano Lett.*, vol. 10, no. 4, pp. 1297–1301, 2010.
5. T. N. Kumar, H. A. F. Almurib, and F. Lombardi, "Design of a memristor-based look-up table (LUT) for low-energy operation of FPGAs," *Integr. VLSI J.*, vol. 55, pp. 1–11, 2016.
6. H. A. F. Almurib, F. Lombardi, and T. N. Kumar, "Design and evaluation of a memristor-based look-up table for non-volatile field programmable gate arrays," *IET Circuits, Devices Syst.*, vol. 10, no. 4, pp. 292–300, 2016.
7. M.-J. Lee, C. B. Lee, D. Lee, S. R. Lee, M. Chang, J. H. Hur, Y.-B. Kim et al., "A fast, high-endurance and scalable non-volatile memory device made from asymmetric Ta_2O_{5-x}/TaO_{2-x} bilayer structures," *Nat. Mater.*, vol. 10, no. 8, pp. 625–630, 2011.
8. K. Eshraghian, O. Kavehei, K. R. Cho, J. M. Chappell, A. Iqbal, S. F. Al-Sarawi, and D. Abbott, "Memristive device fundamentals and modeling: Applications to circuits and systems simulation," in *Proceedings of the IEEE*, 2012, vol. 100, no. 6, pp. 1991–2007.
9. K. Kim and G. Koh, "The prospect on semiconductor memory in nano era," *Proc. 7th Int. Conf. Solid-State Integr. Circuits Technol. 2004*, vol. 1, pp. 662–667, 2004.
10. P. W. C. Ho, F. O. Hatem, H. A. F. Almurib, and T. N. Kumar, "Enhanced SPICE memristor model with dynamic ground," in *Proceeding—2015 IEEE International Circuits and Systems Symposium, ICSyS 2015*, 2016.
11. P. W. C. Ho, F. O. Hatem, H. A. F. Almurib, and T. N. Kumar, "Comparison between $Pt/TiO_2/Pt$ and $Pt/TaO_X/TaO_Y/Pt$ based bipolar resistive switching devices," *J. Semicond.*, vol. 37, no. 6, p. 64001, 2016.
12. Y. Chong Jian, F. O. Hatem, T. N. Kumar, and H. A. F. Almurib, "Compact SPICE modeling of STT-MTJ device," in *2015 IEEE Student Conference on Research and Development, SCOReD 2015*, 2016.
13. N. H. El-Hassan, T. N. Kumar, and H. A. F. Almurib, "Implementation of time-aware sensing technique for multilevel phase change memory cell," *Microelectronics J.*, vol. 56, pp. 74–80, 2016.
14. N. H. El-Hassan, T. N. Kumar, and H. A. F. Almurib, "Phase change memory cell emulator circuit design," *Microelectronics J.*, vol. 62, pp. 65–71, 2017.
15. D. Ielmini, "Resistive switching memories based on metal oxides: Mechanisms, reliability and scaling," *Semicond. Sci. Technol.*, vol. 31, no. 6, p. 63002, 2016.
16. K. M. Kim, S. R. Lee, S. Kim, M. Chang, and C. S. Hwang, "Self-Limited switching in Ta_2O_5/TaOx memristors exhibiting uniform multilevel changes in resistance," *Adv. Funct. Mater.*, vol. 25, no. 10, pp. 1527–1534, 2015.
17. D. B. Strukov, G. S. Snider, D. R. Stewart, and R. S. Williams, "The missing memristor found.," *Nature*, vol. 453, no. 7191, pp. 80–3, 2008.
18. X. Tang, G. Kim, P.-E. Gaillardon, and G. De Micheli, "A study on the programming structures for RRAM-based FPGA architectures," *IEEE Trans. Circuits Syst. I Regul. Pap.*, vol. 63, no. 4, pp. 503–516, 2016.
19. Y. Zhang, Y. Shen, X. Wang, and L. Cao, "A novel design for memristor-based logic switch and crossbar circuits," *IEEE Trans. Circuits Syst. I Regul. Pap.*, vol. 62, no. 5, pp. 1402–1411, 2015.
20. P. W. C. Ho, F. O. Hatem, H. A. F. Almurib, and T. N. Kumar, "Comparison on TiO_2 and TaO_2 based bipolar resistive switching devices," in *2014 2nd International Conference on Electronic Design, ICED 2014*, Penang, Malaysia, 2011.
21. L. Chua, "Resistance switching memories are memristors," *Appl. Phys. A*, vol. 102, no. 4, pp. 765–783, Mar. 2011.
22. S. Kvatinsky, E. G. Friedman, A. Kolodny, C. Uri, U. C. Weiser, S. Member, and U. C. Weiser, "TEAM: ThrEshold adaptive memristor model," *Circuits Syst. I Regul. Pap. IEEE Trans.*, vol. 60, no. 1, pp. 211–221, 2013.
23. T. Nandha Kumar, H. A. F. Almurib, and F. Lombardi, "On the operational features and performance of a memristor-based cell for a LUT of an FPGA," in *2013 13th IEEE International Conference on Nanotechnology* (*IEEE-NANO 2013*), 2013, pp. 71–76.
24. H. A. F. Almurib, T. N. Kumar, and F. Lombardi, "A memristor-based LUT for FPGAs," in *The 9th IEEE International Conference on Nano/Micro Engineered and Molecular Systems* (NEMS), 2014, pp. 448–453.

25. T. N. Kumar, H. A. F. Almurib, and F. Lombardi, "A novel design of a memristor-based look-up table (LUT) for FPGA," in *2014 IEEE Asia Pacific Conference on Circuits and Systems (APCCAS)*, 2014, pp. 703–706.
26. P. W. C. Ho, H. A. F. Almurib, and T. N. Kumar, "Configurable memristive logic block for memristive-based FPGA architectures," *Integr. VLSI J.*, vol. 56, pp. 61–69, Jan. 2017.
27. L. O. Chua, "Memristor—The Missing Circuit Element," *IEEE Trans. Circuit Theory*, vol. 18, no. 5, pp. 507–519, 1971.
28. J. J. Yang, F. Miao, M. D. Pickett, D. A. A. Ohlberg, D. R. Stewart, C. N. Lau, and R. S. Williams, "The mechanism of electroforming of metal oxide memristive switches," *Nanotechnology*, vol. 20, no. 21, p. 215201, 2009.
29. T. H. Park, S. J. Song, H. J. Kim, S. G. Kim, S. Chung, B. Y. Kim, K. J. Lee, K. M. Kim, B. J. Choi, and C. S. Hwang, "Thickness-dependent electroforming behavior of ultra-thin Ta_2O_5 resistance switching layer," *Phys. Status Solidi—Rapid Res. Lett.*, vol. 9, no. 6, pp. 362–365, 2015.
30. J. H. Hur, M. J. Lee, C. B. Lee, Y. B. Kim, and C. J. Kim, "Modeling for bipolar resistive memory switching in transition-metal oxides," *Phys. Rev. B—Condens. Matter Mater. Phys.*, vol. 82, no. 15, pp. 1–5, 2010.
31. J.-H. Hur, K. M. Kim, M. Chang, S. R. Lee, D. Lee, C. B. Lee, M.-J. Lee, Y.-B. Kim, C.-J. Kim, and U.-I. Chung, "Modeling for multilevel switching in oxide-based bipolar resistive memory," *Nanotechnology*, vol. 23, no. 22, p. 225702, 2012.
32. S. Kim, S.-J. Kim, K. M. Kim, S. R. Lee, M. Chang, E. Cho, Y.-B. Kim, C. J. Kim, U.-I. Chung, and I.-K. Yoo, "Physical electro-thermal model of resistive switching in bi-layered resistance-change memory.," *Sci. Rep.*, vol. 3, p. 1680, 2013.
33. Y. Zhang, H. Wu, Y. Bai, A. Chen, Z. Yu, J. Zhang, and H. Qian, "Study of conduction and switching mechanisms in Al/AlOx/WO x/W resistive switching memory for multilevel applications," *Appl. Phys. Lett.*, vol. 102, no. 23, pp. 1–5, 2013.
34. F. Yang, M. Wei, and H. Deng, "Bipolar resistive switching characteristics in CuO/ZnO bilayer structure," *J. Appl. Phys.*, vol. 114, no. 13, 2013.
35. P. Huang, X. Y. Liu, B. Chen, H. T. Li, Y. J. Wang, Y. X. Deng, K. L. Wei et al., "A physics-based compact model of metal-oxide-based RRAM DC and AC operations, *IEEE Trans. Electron Devices*, vol. 60, no. 12, pp. 4090–4097, Dec. 2013.
36. M. H. Chiang, K. H. Hsu, W. W. Ding, and B. R. Yang, "A predictive compact model of bipolar RRAM cells for circuit simulations," *IEEE Trans. Electron Devices*, pp. 1–8, 2015.
37. T. H. Park, S. J. Song, H. J. Kim, S. G. Kim, S. Chung, B. Y. Kim, K. J. Lee et al., "Thickness effect of ultra-thin Ta_2O_5 resistance switching layer in 28 nm-diameter memory cell," *Sci. Rep.*, vol. 5, p. 15965, 2015.
38. S. B. Lee, H. K. Yoo, K. Kim, J. S. Lee, Y. S. Kim, S. Sinn, D. Lee, B. S. Kang, B. Kahng, and T. W. Noh, "Forming mechanism of the bipolar resistance switching in double-layer memristive nanodevices," *Nanotechnology*, vol. 23, no. 31, p. 315202, 2012.
39. J. J. Yang, M. X. Zhang, J. P. Strachan, F. Miao, M. D. Pickett, R. D. Kelley, G. Medeiros-Ribeiro, and R. S. Williams, "High switching endurance in TaOx memristive devices," *Appl. Phys. Lett.*, vol. 97, no. 23, 2010.
40. A. C. Torrezan, J. P. Strachan, G. Medeiros-Ribeiro, and R. S. Williams, "Sub-nanosecond switching of a tantalum oxide memristor," *Nanotechnology*, vol. 22, no. 48, p. 485203, 2011.
41. N. Cabrera and N. F. Mott, "Theory of the oxidation of metals," *Reports Prog. Phys.*, vol. 12, no. 1, pp. 163–184, 2002.
42. D. B. Strukov and R. S. Williams, "Exponential ionic drift: Fast switching and low volatility of thin-film memristors," *Appl. Phys. A Mater. Sci. Process.*, vol. 94, no. 3, pp. 515–519, 2009.
43. S. Yu and H. S. P. Wong, "Compact modeling of conducting-bridge random-access memory (CBRAM)," *IEEE Trans. Electron Devices*, vol. 58, no. 5, pp. 1352–1360, 2011.
44. S. M. Sze and K. K. Ng, *Physics of Semiconductor Devices*, 3rd ed., Hoboken, NJ: John Wiley & Sons, 2006.
45. S. M. Sze and M. K. Lee, *Semiconductor Devices, Physics and Technology*, New York: John Wiley & Sons, 2012.

46. H. C. Card and E. H. Rhoderick, "Studies of tunnel MOS diodes I. Interface effects in silicon Schottky diodes," *J. Phys. D. Appl. Phys.*, vol. 4, no. 10, pp. 1589–1601, 2002.
47. U. Russo, D. Ielmini, C. Cagli, and A. L. Lacaita, "Self-accelerated thermal dissolution model for reset programming in unipolar resistive-switching memory (RRAM) devices," *IEEE Trans. Electron Devices*, vol. 56, no. 2, pp. 193–200, 2009.
48. N. Mott, *Electronic Processes in Ionic Crystals*, New York: Dover, 1964.
49. Z. Wei, Y. Kanzawa, K. Arita, Y. Katoh, K. Kawai, S. Muraoka, S. Mitani et al., "Highly reliable TaO_x ReRAM and direct evidence of redox reaction mechanism," *2008 IEEE International Electron Devices Meetings*, pp. 1–4, 2008.

[illegible]

17

Evaluation of Nanoscale Memristor Device for Analog and Digital Application

Jeetendra Singh and Balwinder Raj

CONTENTS

17.1 Introduction

A memristor is basically a resistor accompanied by memory. It is a propitious nonlinear device that became the fourth fundamental electrical circuit component after resistor, capacitor, and inductor. Its nanoscale size and nonvolatile memorizing capability provide more potential to replace conventional data storage devices. In conventional memories such as dynamic random access memory (DRAM), static random access memory (SRAM), and flash memory, data are stored as charge on a capacitor that dissipates with time—and data is eventually lost. In memristor-based memory, the high resistance and low resistance states are stored unlike charge in conventional memory. Since it stores resistance value indefinitely, the memristor can be used as a nonvolatile memory. The brain can be developed with the help of implementing a memristor in an analog circuit since the memristor can also be used as synapses of neurons.

There are three well-known basic circuit elements, namely capacitor (C), resistor (R), and inductor (L). They are related by four basic electrical parameters, namely charge (Q), voltage (V), current (I), and magnetic flux (ϕ), and all of them are frequently used in circuit theory. The details of them are given in the following table.

S.N.	Circuit Element	Inventor	Invention Year	Relating Parameter	Relation	Symbol
1	Capacitor	Ewald Georg von Kleist	1745	Charge (q) and voltage (v)	$dq = Cdv$	⊣⊢
2	Resistor	George Simon Ohm	1827	Voltage (v) and current (i)	$dv = Rdi$	-WW-
3	Inductor	Michael Faraday	1831	Magnetic flux(ϕ) and current (i)	$d\varphi = Ldi$	-ꝏ-

Since there are four basic electrical parameters, and if these are combining in pairs, then $4C_2 = 6$ gives six possible relations of four parameters. And out of the six possible combinations, the well-known five relations are given in the following equations—but one relation is missing.

1. $q(t) = \int_{-\infty}^{t} i(\tau)d\tau$ or $i = \frac{dq}{dt}$ Basic law of electricity

2. $\varphi(t) = \int_{-\infty}^{t} v(\tau)d\tau$ or $v = \frac{d\varphi}{dt}$ Basic law of magnetism

   Also, other relations are defined by the three basic linear circuit components (axiomatic definition):

3. $R = \frac{v}{i}$ Resistance

4. $L = \frac{\varphi}{i}$ Inductance

5. $C = \frac{q}{v}$ Capacitance

The aforementioned relations can also be written on the basis of source control if the circuit elements are nonlinear:

Current-controlled nonlinear resistor $v = f_R(i)$ and
Voltage-controlled nonlinear resistor $i = g_R(v)$ or in combination

$$f(v_R, i_R) = 0 \tag{17.1}$$

Voltage-controlled nonlinear capacitor $q = f_C(v)$ and
Charge-controlled nonlinear capacitor $v = g_C(q)$ or

$$f(q_C, v_C) = 0 \tag{17.2}$$

Current-controlled nonlinear inductor $\varphi = f_L(i)$ and
Flux-controlled nonlinear inductor $i = g_L(\varphi)$ or

$$f(\varphi_L, i_L) = 0 \tag{17.3}$$

As mentioned previously, there should be one more relation between charge (q) and flux (φ); it is missing here [1] and shown in Figure 17.1. In 1971, L. Chua recognized the missing link and gave it the name memristor.

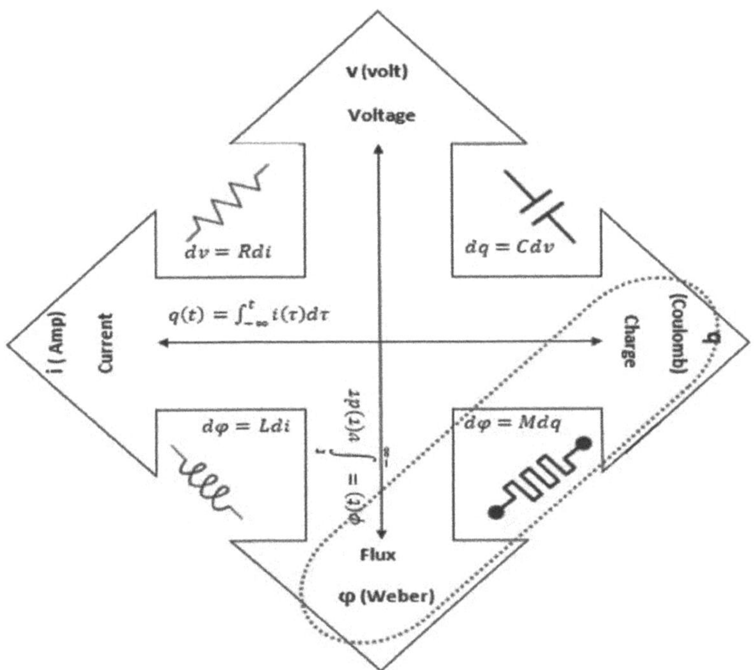

FIGURE 17.1
Explanation of four fundamental circuit elements with interlinked relations, memristor is encircled that relates flux and charge.

In this chapter, part 1 gave the introduction of the existing circuit element with a memristor, which explains the logic behind the existence of a memristor. Part 2 defines the memristor and basically explains what it is. Mathematically, a memristor is explained in part 3, and in part 4, the practical behavior of a memristor is given with a water pipe analogy. In part 5, the various types of the memristor are given, and the fabrication of all types is given in part 6. The explanation of the HP Labs memristor is explained in part 8, which details memristor structure, construction, and workings; its fabrication steps are also explained. A periodic table of 25 circuit family elements that was postulated by L. Chua is explained in part 9. Finally, the properties and advantages of memristor are explained in parts 10 and 11.

L. Chua invented the concept of the memristor in 1971. Before that, in 1960, Widrow unveiled a new device called a memister [2] (memory with register); it was made by electroplating a metal onto a resistive substrate. This device is a three-terminal device, the same as a transistor except the resistance between the two terminals is controlled by integral current instead of instantaneous current—i.e., controlled by the charge similar to a memristor.

17.2 Periodic Table of Circuit Elements

A theory of nonlinear circuit elements given by L. Chua has a family of circuit elements. He considered a circuit of two-terminal or one-port black box, shown in Figure 17.2, to predict this theory. This circuit is characterized by a constitutive relation in the $V^{(\alpha)}$ versus $i^{(\beta)}$ plane called an (α,β) element, where $V^{(\alpha)}$ and $i^{(\beta)}$ are complementary variables. These elements are derived from the voltage and current as in Equations (17.4) and (17.5), respectively [3]. A pair of variables $\left(V^{(\alpha)}i^{(\beta)}\right)$ are said to be complementary if $V^{(\alpha)}(t)$ is derived from a voltage signal $v(t)$, while $i^{(\beta)}(t)$ is derived from a current signal $i(t)$. Here, α and β are any integer (negative, positive, or zero) and need not be identical.

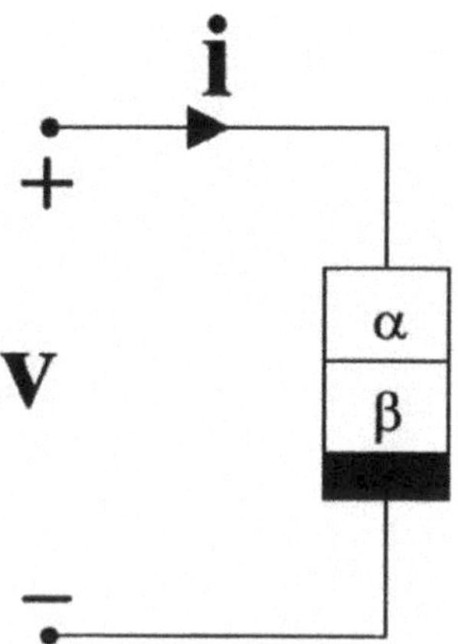

FIGURE 17.2
A Two terminal or one port black box (α, β) elements. (From Chua, L. O., *Proc. IEEE*, 91, 1830–1859, 2003.)

According to the Equations (17.4) and (17.5), for the various values of α and β, an infinite number of circuit elements are possible.

$$V^{(\alpha)}(t) \triangleq \begin{cases} \dfrac{d^{\alpha}v(t)}{dt^{\alpha}} & \text{if } \alpha = 1, 2, \ldots\ldots\ldots\infty \\ v(t) & \text{if } \alpha = 0 \\ \displaystyle\int_{-\infty}^{t} v(\tau)d\tau, & \text{if } \alpha = -1 \\ \displaystyle\int_{-\infty}^{t}\int_{-\infty}^{\tau|\alpha|} \cdots\cdots \\ \displaystyle\int_{-\infty}^{\tau_2} v(\tau)d\tau_1 d\tau_2 \ldots d\tau_{|\alpha|}, & \text{if } \alpha = -2, -3 \ldots \infty \end{cases} \tag{17.4}$$

$$i^{(\beta)}(t) \triangleq \begin{cases} \dfrac{d^{\beta}i(t)}{dt^{\beta}} & \text{if } \beta = 1, 2, \ldots\ldots\ldots\infty \\ i(t) & \text{if } \beta = 0 \\ \displaystyle\int_{-\infty}^{t} i(\tau)d\tau, & \text{if } \beta = -1 \\ \displaystyle\int_{-\infty}^{t}\int_{-\infty}^{\tau|\beta|} \cdots\cdots \\ \displaystyle\int_{-\infty}^{\tau_2} i(\tau)d\tau_1 d\tau_2 \ldots d\tau_{|\beta|}, & \text{if } \beta = -2, -3 \ldots \infty \end{cases} \tag{17.5}$$

If one constructs a hypothetical circuit using these circuit elements, then at a finite value of k such that $|\alpha| > k$ and $|\beta| > k$, the solution of the circuit does not exist due to the presence of the singularity [4]. Most of the real devices are modeled for $|\alpha| > 2$ and $|\beta| > 2$. Figure 17.3 shows the family of circuit elements for $|\alpha| = 2$ and $|\beta| = 2$. There are 25 circuit elements including 6 recognized circuit elements—i.e., resistor, capacitor, inductor, memristor, memcapacitor, and meminductor [5].

Circuit Elements		α and β Values
Resistor	→	$\alpha = 0, \beta = 0$
Capacitor	→	$\alpha = 0, \beta = -1$
Inductor	→	$\alpha = -1, \ \beta = 0$
Memristor	→	$\alpha = -1, \ \beta = -1$
Memcapacitor	→	$\alpha = -1, \ \beta = -2$
Memindutor	→	$\alpha = -2, \ \beta = -1$

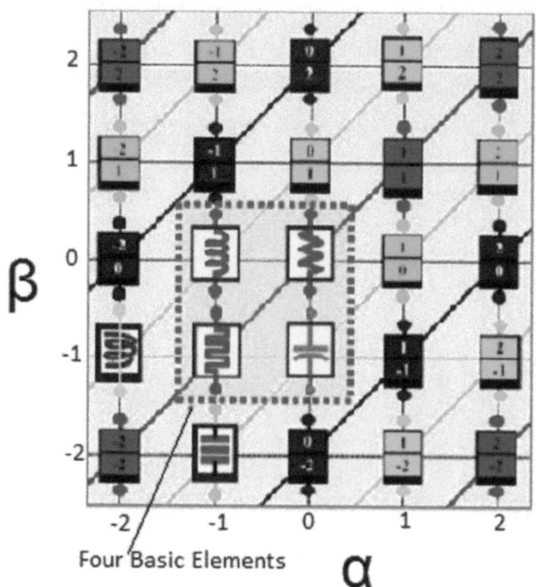

FIGURE 17.3
Family of 25 elements. (From Chua, L. O., *Proc. IEEE*, 91, 1830–1859, 2003.)

17.3 Definition of Memristor

In 1971, Leon Chua [1] had defined a relation and propounded a new nonlinear device called memristor in short (memory + resistor). It has two properties: one is dissipative resistance and the second is memory. It memorizes the last resistance or amount of charge flown through it, corresponding to the magnitude and direction of applied voltage across the device after removing applied voltage.

The flux and charge of the device can be related as

$$d\varphi = Mdq \tag{17.6}$$

There is no difference in memristance and resistance if the memristor is linear and M is constant. That is, the memristance is independent of a state variable charge, and memductance (inverse of memristance) is independent of state variable flux. If M is not constant—i.e., it is a dependent variable of charge q—then it will become nonlinear. A frequency-dependent (i, v) characteristic (hysteresis curve) will be obtained if the sinusoidal input is applied to the nonlinear memristor. Also, the relation between φ and q becomes nonlinear. Memristor is a fundamental element because the behavior of the memristor is unique and cannot be replicated by other fundamental elements; this was proved by Chua [1] and defined as:

Charge-controlled memristor

$$\varphi = f_M(q) \tag{17.7}$$

and Flux-controlled memristor

$$q = g_M(\varphi) \tag{17.8}$$

or in combination

$$f(\varphi_M, q_M) = 0 \tag{17.9}$$

Differentiating Equation (17.7) w.r.t time

$$\frac{d\varphi}{dt} = \frac{d}{dt} f_M(q)$$

$$\frac{d\varphi}{dt} = \frac{d}{dq} f_M(q).\frac{dq}{dt}$$

or

$$v(t) = \frac{df_M(q)}{dq}.i(t) \tag{17.10}$$

$$v(t) = M(q).i(t) \tag{17.11}$$

Where $M(q) = \frac{df_M(q)}{dq}$ incremental memristance unit Ohm (17.12)

Now differentiating Equation (17.8) w.r.t time

$$\frac{dq}{dt} = \frac{d}{dt} g_M(\varphi)$$

$$\frac{dq}{dt} = \frac{d}{d\varphi} g_M(\varphi).\frac{d\varphi}{dt}$$

or

$$i(t) = \frac{dg_M(\varphi)}{d\varphi}.v(t) \tag{17.13}$$

$$i(t) = W(\varphi).v(t) \tag{17.14}$$

Where $W(\varphi) = \frac{dg_M(\varphi)}{d\varphi}$ incremental memductance of unit siemens (17.15)

S.N.	Circuit Element	Inventor	Invention Year	Relating Parameter	Relation	Symbol
4.	Memristor	Leon Chua	1971	Magnetic flux (φ) and charge (q)	$d\varphi = Mdq$	

17.4 Mathematical Definition

Mathematically, we can realize the memristance of the memristor by allowing a transition metal oxide or semiconductor oxide thin film of thickness "D," which is sandwiched between two metal electrodes [6], as shown in Figure 17.4.

The thin semiconductor film consists of two regions: one with high or less in oxygen will act as a conductive or doped layer, and the other with equally proportionate oxygen with metal acts as an insulating layer. The doped region is considered R_{ON} since this region shows high conductivity because oxygen vacancies are created in this region by removing some oxygen from metal oxide. The undoped region will act as a high resistance region due to the presence of equal oxygen atoms in this region. This region is duplicated by R_{OFF}.

The total resistance of the device looked as the two variable resistance connected in series in Figure 17.4. The boundaries of the doped and undoped regions—i.e., the values of R_{ON} and R_{OFF}—will vary according to the magnitude and polarity of the applied voltage across the device and the time interval up to which voltage is applied. The applied voltage causes the oxygen vacancies to drift with dopant mobility μ_V.

17.4.1 Linear Drift Model

According to Figure 17.4 [6], total resistance of the device is written as

$$R_T = R_{ON}\frac{w(t)}{D} + R_{OFF}\left(1 - \frac{w(t)}{D}\right) \tag{17.16}$$

The voltage across the memristor is given by the following relation [6]:

$$v(t) = R_{ON}\frac{w(t)}{D} + R_{OFF}\left(1 - \frac{w(t)}{D}\right).i(t) \tag{17.17}$$

The linear relationship between drift-diffusion velocity and the net electric field is given by [7] Blanc and Staebler, 1971.

$$v = \mu_V E \tag{17.18}$$

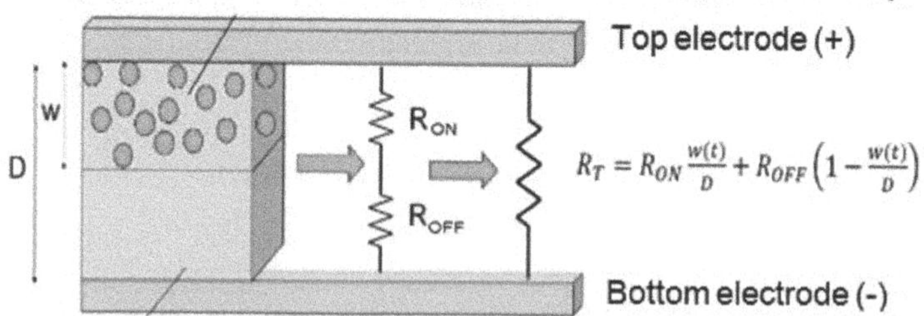

FIGURE 17.4
Structure of memristor with doped width W and undoped layer. (From Strukov, D. B. et al., *Nature*, 453, 80–83, 2008.)

or Equation (17.18) can be written as

$$\frac{dw(t)}{dt} = \mu_V . \frac{V}{D} \tag{17.19}$$

or

$$\frac{dw(t)}{dt} = \mu_V . \frac{R_{ON}}{D} i(t) \tag{17.20}$$

Integrating Equation (17.20)

$$\int dw(t) = \mu_V . \frac{R_{ON}}{D} \int i(t)dt \tag{17.21}$$

or

$$w(t) = \mu_V \frac{R_{ON}}{D} q(t) \tag{17.22}$$

Putting $w(t)$ from Equation (17.22) to Equation (17.17)

$$v(t) = \left[R_{ON}\mu_V \frac{R_{ON}}{D.D} q(t) + R_{OFF}\left(1 - \mu_V \frac{R_{ON}}{D.D} q(t)\right)\right] .i(t) \tag{17.23}$$

$$v(t) = R_{OFF}\left(1 - \mu_V \frac{R_{ON}}{D^2} q(t)\right)\left[1 + \frac{R_{ON}\mu_V \frac{R_{ON}}{D^2}}{R_{OFF}\left(1 - \mu_V \frac{R_{ON}}{D^2} q(t)\right)}\right] .i(t) \tag{17.24}$$

If $R_{OFF} \gg R_{ON}$

Then,

$$v(t) = R_{OFF}\left(1 - \mu_V \frac{R_{ON}}{D^2} q(t)\right) .i(t) \tag{17.25}$$

So,

$$M(q) = R_{OFF}\left(1 - \mu_V \frac{R_{ON}}{D^2} q(t)\right) \tag{17.26}$$

where:
μ_V is average dopants mobility in cm^2/volt-sec
$q(t)$ is the total charge passing through the device in time window t–t_o
R_{ON} and R_{OFF} are state resistances

Thus, the memristor is a function of charge q, mobility of dopant μ_V, and device thickness "D". Since memristance is inversely proportional to the device dimension, upshots of memristance are more as the dimension becomes as small as possible.

17.4.2 Nonlinear Ion Drift

If few volts are applied across a thin film (e.g., in nanometer), then an exponential electric field is developed inside the film [8]. This large electric field produces nonlinear ionic drifting across the thin film. The resistive switching, i.e., switching of doped and undoped interfaces at the boundaries of the memristor, is nonlinear because of the nonlinear drift-diffusion of ions. At boundaries, the drift velocity of the linear model does not become zero so the width "w" exceeds the limits of total thickness "D" of the device. Therefore, the linear model produces such boundaries problems that can be overcome by inserting the window function $f(w)$ at the right-hand side of Equation 17.20 [6] where w is a state variable, and further sticking of the boundaries is resolved by inserting the i of the memristor in the window function. The nonlinear ionic drift is considered, and the boundaries problem is resolved by including the window function.

$$\frac{dw(t)}{dt} = \mu_V . \frac{R_{ON}}{D} i(t) f(w,i) \tag{17.27}$$

The initial current, voltage, boundaries, and minimum and maximum "w" should be considered in simulation for the linear model; thus, large differences occur at output using the linear and nonlinear models.

17.4.2.1 Window Function

There are various window functions given in the literature to consider the nonlinear ionic drift and to resolve the boundaries problem of a memristor. The first window function is proposed, as given in Equation 17.28, which is a function of state variable w [6]

$$f(w) = \frac{w(1-w)}{D^2} \tag{17.28}$$

$$f(0) = 0 \tag{17.29}$$

$$f(D) = 0 \tag{17.30}$$

This function gave the zero values at the boundaries and fulfilled the boundary condition, but there are still problems with this function. First, there is a problem when the memristor is driven at boundaries $dw/dt \to 0$ because the memristor state can't be changed by the external field, and this is the basic problem of window function [9]. Second, a problem arises because nonlinearity is not fully considered since the window function abruptly goes zero at the boundaries.

The second window function is [10] alternative of the first window function, but it still has a sticking problem at the boundaries.

$$f(w) = x(1-x) \tag{17.31}$$

Here, $x = w/D$ instead of $x = w$ and also at the boundaries $f(w \to 0) = 0$ although this window function is considered the nonlinearity.

The linear behavior earns linear ionic drift in the range ($0 < w < D$), and nonlinear behavior nonlinear ionic drift at the boundaries is well approximated by another window function [11] expressed as

$$f(x) = 1 - (2x - 1)^{2p} \tag{17.32}$$

where $x = w/D$ and "p" is a positive integer that controls the linearity and nonlinearity of the window function and is known as the control parameter. This window provides 0 value at $x = 0$ and $x = 1$ and thus limits the boundaries between 0 and D. If $p \geq 4$, the state variable equation provides the linear drift for $f(o < x < 1) \approx 1$. This model also shows a problem at the terminal state; any external spur cannot change its state if either of the boundaries is reached [9].

A new window function including the memristor current "i" together with variable "x" and "p" is given as [9]

$$f(x) = 1 - (x - \text{sgn}(-i))^{2p} \tag{17.33}$$

$$\text{sgn}(i) = 1 \qquad \text{when } i \geq 0$$

$$\text{sgn}(i) = 0 \qquad \text{when } i \leq 0$$

Here, a positive current is correlated with increasing doped width, a negative current is correlated with decreasing doped width, and the ON and OFF states are brought out of the terminal state when the current is in the reverse direction. The problem with the window function is that it cannot provide the continuity at the boundaries—i.e., it has discontinuity problem at the boundaries. Also, it does not have a scaling factor, i.e., the maximum value can't be extended vertically.

Another window function is given by Prodromakis that removes the scaling problem [12]. The nonlinearity is considered by a quadratic equation $f(x) = ax^2 + bx + c = 0$, and the constants a, b, and c are easily calculated by applying boundary conditions such as $\left|df/dx\right|_{0.5} = 0$, $f(0.5) = 1$, and $f(0) = f(1) = 0$. The proposed window function is given in Equation 17.34 where j works as a vertical scalar and p as a lateral scalar, which determines the maximum value of $f(w)$, and p is a positive real number.

$$f(w) = j\left(1 - \left[\left(\frac{w}{D} - 0.5\right)^2 + 0.75\right]^p\right) \tag{17.34}$$

Although this elevates the boundary issues, the stuck problem of states is not resolved since this model ignores the nonlinear dependence of the state derivative on current.

To remove the ambiguity of the aforementioned window functions, a new window function is developed by mixing the window functions of Biolek and Prodromakis [13]. This window function considers the nonlinearity as well as the stuck state problem.

$$f(w) = j\left(1 - \left[0.25(x - \text{sgn}(-i))^2 + 0.75\right]^p\right)$$

$$\text{sgn}(i) = 1 \quad \text{when } i \geq 0$$

$$\text{sgn}(i) = 0 \quad \text{when } i \leq 0$$

All the window functions endure the same problem and show the dependency of the state variable; the memristor remembers all the charge passing through it.

17.4.3 Nonlinear Drift Model

In this model, the two metal/oxide interfaces of the device and the corresponding I-V graph are characterized by an ohmic interface (in case of heavy doping or $TiO_{2\text{-}X}$) and a rectifying interface (in case of low doping or TiO_2). A model that considers a highly nonlinear behavior is presented by the current-voltage relation given in Equation 17.35 [14]. First, the term of Equation 17.35 characterizes the ON state; second, the term is an estimation of the rectifier I-V expression and characterizes the OFF state of the memristor. Figure 17.5 shows the device structure consisting of $TiO_2/TiO_{2\text{-}X}$ layers between two platinum electrodes and corresponding equivalent circuit entailing a rectifier in parallel with the memristor [14].

$$i(t) = w(t)^n \beta \sinh\left(\alpha v(t)\right) + \chi\left[\exp\left(\gamma v(t)\right) - 1\right] \tag{17.35}$$

α, β, γ, and χ are fitting parameters, and n determines the influence of the state variable on current

$$\frac{dw}{dt} = a \cdot f(w) \cdot g(V) \tag{17.36}$$

where a is constant $f:[0,1] \rightarrow R$ a window function, and $g: R \rightarrow R$ is linear function.

17.4.4 Simmons's Tunnel Barrier Model

Simmons's tunnel barrier model consists of a thin TiO_2 layer in between the two platinum electrodes and uses an electroformation process; a $TiO_{2\text{-}X}$ layer is grown, as illustrated in Figure 17.6. This oxygen-deficient layer works as a high-conducting channel and is considered a series resistance R_s of value 215Ω. A small gap of width "w" remains of the TiO_2

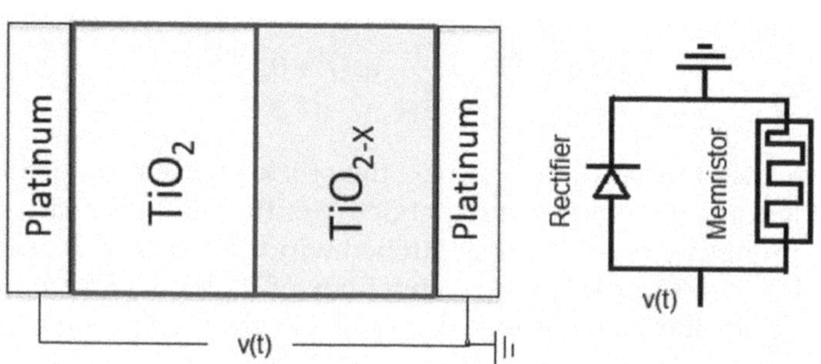

FIGURE 17.5
Device structure and equivalent circuit model. (From Yang, J. J. et al., *Nat. Nanotechnol.*, 3, 429–433, 2008.)

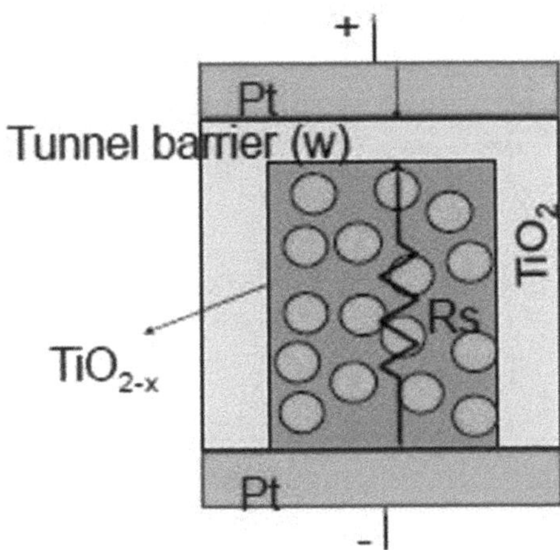

FIGURE 17.6
Simmons tunnel device structure. (From Pickett, M. D. et al., *J. Appl. Phys.*, 106, 1–6, 2009.)

during electroformation. Tunneling phenomena results in tunneling current "i" which works as voltage-controlled current source. The corresponding circuit model is shown in Figure 17.7 in which voltage across the resistance is assumed as V_R, and the current controlled voltage source is V_g [15]. In this model, the tunnel barrier of width "w" is considered with a series resistance instead of two series (R_{ON} and R_{OFF}) resistances as in the linear model. The tunnel width "w" increases on application of the positive bias since positively charged oxygen vacancies are repelled, and the width gap decreases on applying the negative bias. Therefore, the current is limited by this barrier gap width "w" and represented by Simmons's equations. The $\sinh(i/i_{\text{off}})\exp(w/w_c)$ term assumes nonlinear ionic drift and Joule heating of the interface, which gear up the drift of oxygen vacancies.

$$\frac{dw}{dt} = f_{\text{off}} \sinh\left(\frac{i}{i_{\text{off}}}\right) \exp\left[-\exp\left(\frac{w - a_{\text{off}}}{w_c} - \frac{|i|}{b}\right) - \frac{w}{w_c}\right] \quad i > 0 \tag{17.37}$$

$$\frac{dw}{dt} = f_{\text{on}} \sinh\left(\frac{i}{i_{\text{on}}}\right) \exp\left[-\exp\left(\frac{w - a_{\text{on}}}{w_c} - \frac{|i|}{b}\right) - \frac{w}{w_c}\right] \quad i < 0 \tag{17.38}$$

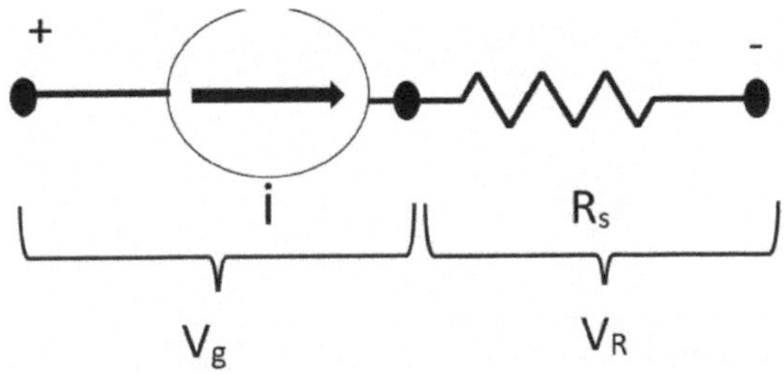

FIGURE 17.7
Memristor circuit model, current controlled voltage source is connected with series resistance. (From Pickett, M. D. et al., *J. Appl. Phys.*, 106, 1–6, 2009.)

The current is defined by Simmons's tunnel model [16]:

$$i = \frac{j_0 A}{\Delta w^2}\left\{\varphi_I e^{-B\sqrt{\varphi_I}} - \left(\varphi_I + e\left|v_g\right|\right)e^{-B\sqrt{\varphi_I + e\left|v_g\right|}}\right\} \tag{17.39}$$

where

$$j_0 = \frac{e}{2\pi h},\ w_1 = \frac{1.2\lambda w}{\varphi_0},\ \Delta w = w_2 - w_1$$

$$\varphi_I = \varphi_0 - e\left|v_g\right|\left(\frac{w_1 + w_2}{w}\right) - \left(\frac{1.15\lambda w}{\Delta w}\right)\ln\left(\frac{w_2(w - w_1)}{w_1(w - w_2)}\right)$$

$$B = \frac{4\pi\Delta w\sqrt{2m}}{h},\ \lambda = \frac{e^2\ln 2}{8\pi k\varepsilon_0 w}$$

$$w_2 = w_1 + w\left(1 - \frac{9.2\lambda}{(3\varphi_0 + 4\lambda - 2e\left|v_g\right|)}\right)$$

Voltage across the device:

$$v = v_g + i(t)R_s \tag{17.40}$$

17.4.5 Threshold Adaptive Memristor Model

The complications of the Simmons tunnel model were reduced by a simple model; the threshold adaptive memristor (TEAM) model [17] is introduced. State derivatives rely on polynomial rather than exponential, and the memristive device is modeled with the threshold current to consider the nonlinearity. Here, the state derivative equation is modeled by multiplying two polynomials—one is a function of the device current, and the other is a function of the state variable "w."

$$\frac{dw(t)}{dt} = \left\{k_{\text{off}}\left(\frac{i(t)}{i_{\text{off}}} - 1\right)^{\alpha_{\text{off}}} \cdot f_{\text{off}}(w); \qquad 0 < i_{\text{off}} < i \right. \tag{17.41}$$

$$\frac{dw(t)}{dt} = 0;\ i_{\text{off}} < i < i_{\text{on}} \tag{17.42}$$

$$\frac{dw(t)}{dt} = \left\{k_{\text{on}}\left(\frac{i(t)}{i_{\text{on}}} - 1\right)^{\alpha_{\text{on}}} \cdot f_{\text{on}}(w); \qquad i < i_{\text{on}} < 0 \right. \tag{17.43}$$

Here, i_{off} and i_{on} are threshold currents and k_{off}(+)ve, k_{on}(−)ve, $\alpha_{\text{off}}, \alpha_{\text{on}}$ are constants. $f_{\text{off}}(w)$ Also, $f_{\text{on}}(w)$ are working as a window function and are not necessarily the same.

The current and voltage relation is given as follows. If resistance is linear dependence of the state variable, then the current voltage relation is given in Equation (17.44). And if memristance exponentially depends on the state variable, then the current voltage relation is given by Equation (17.45).

$$v(t)=\left[R_{ON}+\frac{R_{OFF}-R_{ON}}{w_{off}-w_{on}}\left(w-w_{on}\right)\right]\cdot i(t) \tag{17.44}$$

$$v(t)=R_{ON}e^{\left(\frac{\lambda}{w_{off}-w_{on}}\right)(w-w_{on})}i(t) \tag{17.45}$$

Where $w\in\left[w_{on},w_{off}\right]$, λ is a fitting parameter and R_{ON} and R_{OFF} are bounded resistances and satisfy the following relation:

$$\frac{R_{OFF}}{R_{ON}}=e^{\lambda} \tag{17.46}$$

The window functions $f_{off}(w)$ and $f_{on}(w)$ can be expressed as [17]:

$$f_{off}(w)=\exp\left[-\exp\left(\frac{w-a_{off}}{w_c}\right)\right] \tag{17.47}$$

$$f_{on}(w)=\exp\left[-\exp\left(-\frac{w-a_{on}}{w_c}\right)\right] \tag{17.48}$$

17.4.6 Voltage Threshold Adaptive Memristor Model

The voltage threshold adaptive memristor (VTEAM) model is developed to fulfill the threshold voltage conditions required by many logic circuits and memory applications. It is an extension of the TEAM model, and the only difference is that it is voltage controlled whereas the TEAM model is current controlled. State variable derivatives for this model are given as [18]

$$\frac{dw(t)}{dt}=\left\{k_{off}\left(\frac{v(t)}{v_{off}}-1\right)^{\alpha_{off}}\cdot f_{off}(w);\qquad 0<v_{off}<v\right. \tag{17.49}$$

$$\frac{dw(t)}{dt}=0;\qquad v_{on}<v<v_{off} \tag{17.50}$$

$$\frac{dw(t)}{dt}=\left\{k_{on}\left(\frac{v(t)}{v_{on}}-1\right)^{\alpha_{on}}\cdot f_{on}(w);\qquad i<v_{on}<0\right. \tag{17.51}$$

$$i(t) = \left[R_{ON} + \frac{R_{OFF} - R_{ON}}{w_{off} - w_{on}} (w - w_{on}) \right]^{-1} \cdot v(t) \tag{17.52}$$

$$i(t) = \frac{e^{\left(\frac{\lambda}{w_{off} - w_{on}}\right)(w - w_{on})}}{R_{ON}} i(t) \tag{17.53}$$

The current voltage relation given in Equation (17.52) is valid when memristance linearly varies with state variable, and Equation (17.53) is valid for exponential variation of memristance with state variable. The window function for the VTEAM model is the same as that of the TEAM model given in Equations (17.47) and (17.48).

The drift velocity within the memristor is identified in two forms [8] and given as follows:

1. Linear

$$v = \mu E \qquad \text{where } E \ll E_o$$

2. Nonlinear

$$v = \mu E_o \exp^{\left(\frac{E}{E_o}\right)}$$

Where v = average drift velocity, E = applied electric field, μ = mobility, E_o = characteristic field for a particular mobile ion, typically $E_o = 1(\text{mV/cm})$.

The nonlinear properties of memristor are better described by another model that considers the memristive as well as capacitive effects. In this model, the memristive properties are realized by an infinite number of crystalline magnetic (Fe_3O_4) nanoparticles [19].

17.5 Experimental Definition of Memristor

We can explain the workings and behavior of the memristor by considering the resistor as a pipe of variable diameter and charge as the water that flows through the pipe [20]. The property of the pipe is that its diameter changes according to quantity and direction of water flow through it. In a particular direction of flow, the pipe's diameter prolongs, which means it allows more water to flow. In another direction, the diameter of the pipe expurgates, which means it restricts the flow of water. Similarly, the memristor in one direction permits charges to flow—i.e., shows low resistance—and in another direction it restrict charges to flow, i.e., shows high resistance. If the supply of the water is halted, then the diameter of the pipe is frost until the next supply is started.

The memristor remembers the last diameter of the pipe or resistance even if the water flow starts after many years. The previous explanation shows that the memristor is a device that has variable resistance with memory. This water-charge analogy, shown in Figure 17.8a, reveals a pipe of a certain diameter without any flow of water. In Figure 17.8b, water flows from left to right through the pipe, and the diameter of the pipe extends. In Figure 17.8c, when the flow of water is from right to left through the pipe, the diameter reduces.

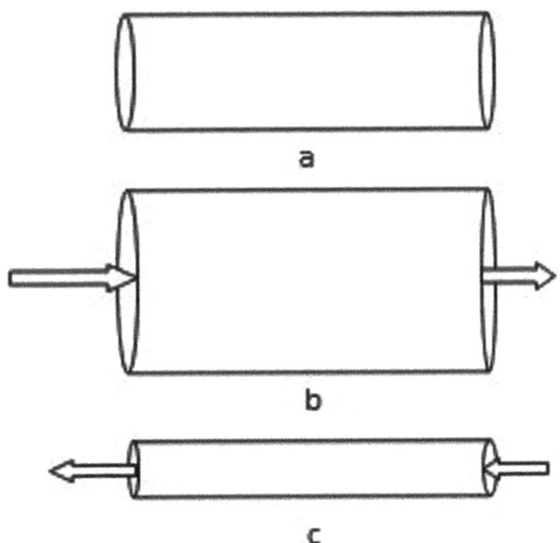

FIGURE 17.8
(a) A pipe of certain diameter (b) prolonged diameter of pipe when water flows left to right through pipe (c) expurgated diameter when water flows from right to left through pipe. (From Williams, S. R., *IEEE Spectr.*, 45, 28–35, 2008.)

17.6 Memristive Materials and Types

A memristor is classified in three parts as filament-type memristor, barrier-type memristor, and ferroelectric memristor. Further, the filament-type memristor is divided as valence change memristor (VCM) and electrochemical metallization memristor (ECM) [21], as shown in Figure 17.9. These classifications are based on the insulating materials and mechanism of conduction.

17.6.1 Filament Type Memristor

A filament-type memristor supports resistive switching in the insulating layer between the two electrodes. A conductive filament is formed in a particular direction of voltage supply, and in the reverse direction the conductive filament will rupture [21].

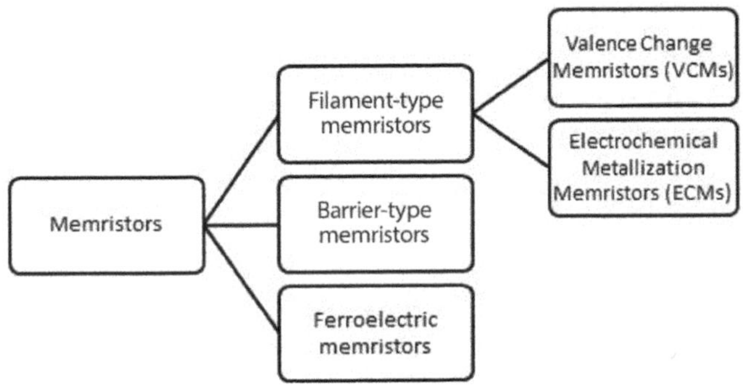

FIGURE 17.9
Different types of memristors.

17.6.1.1 Valence Change Memristor

The following oxides can be used as material for the insulating layer:

1. **Some binary metal oxides** such as AgO_x [22], MgO [23], TiO_x [24], FeO_x [25], ZnO_x [26], AlO_x [27], NiO_x [28], GeO_x [29], and SnO_x [30]
2. **Some complex metal oxides** such as LiXZn1-XO [31], $ZnFe_2O_4$ [32], FeZnO [33], and Zn_2SnO_4 [34]
3. **Some non-metal oxides** such as SiOx [35], and (Graphene Oxide) GO [36]
4. **Some metal nitride** such as ZrN [37]

Electrode materials: We can use the following relative inert material such as Au [38], Pt [38], Al [39], Ni [40], and Ge [41].

17.6.1.2 Electrochemical Metallization Memristor

This type of memristor is also known as atomic switches. The insulating material here is an electrolyte film with metallic cations as the mobile species, which are come from the electrode [21]. These mobile cations have high mobility. The insulating material here is made of the following materials:

1. **Some chalcogenides** such as Cu_2S [42], Ge_xS_x [43], and $Zn_xCd_{1-x}S$ [44]
2. **Some oxides** such as Al_2O_3 [45], SiO_2 [46], TiO_2 [47], CuO_x [48], and ZnO [49]
3. **Some halides** such as AlN [50] and SiC [51]
4. **Electrode materials** in this type of electrode that work as an anode are made of electrochemically active materials such as Cu [52], Ag [53], and Ni [40].

The other terminal, acting as a cathode, is made up of relatively active material such as Pt [38] and Ag [54].

17.6.2 Barrier-Type Memristor

In a barrier-type memristor, the following materials are used frequently:

1. **Some complex perovskite oxide** such as $BiFeO_3$ [54], $Bi_{0.9}Ca_{0.1}FeO_3$ [55], $Pr_{0.7}Ca_{0.3}MnO_3$ [56], and $CaMnO_3$ [56]
2. **Some binary oxides** such as GaO_x [57,58], ZnO [59], TiO_x [14,60], and CuO [61]
3. **Some sulfides and selenides** such as CdSe [62]

Here, electrode materials are metal, and alloy metals such as Pt, Au, Ti and semiconductors such as $SrTiO_3$ and $LaNo_3$ are generally used.

17.6.2.1 Schottky Barrier-Type Memristor

The barrier-type memristor also can be formed as a Schottky barrier-type memristor in which one interface is formed by joining the metal and oxide, also known as ohmic-metal junction; another interface is Schottky barrier, also known as Schottky-metal. The whole configuration is looked as Ohmic-metal/insulator/schottky-metal. At high resistive state (HRS), rectifying I-V behavior, and at low resistive state (LRS), ohmic conduction nature is shown by the Schottky barrier-type memristor [54,63]. Here, resistive switching behavior is affected by the dimension of the Schottky barrier—i.e., determined by height and width of the Schottky barrier. The barrier is altered by the redistribution of the oxygen vacancies [54,61,63].

17.6.3 Ferroelectric Memristor

In a ferroelectric memristor, an ultra-thin layer of ferroelectric material is used as an insulating layer that provides a ferroelectric tunnel junction [64,65], which is sandwiched between two electrodes. It is also a kind of memristor that shows resistive switching with a ferroelectric tunnel junction. Ferroelectric material such as $BiFeO_3$ [54], $BaTiO_3$ [66], and some organic ferroelectric materials [67] are generally used, and Pt and $La_{0.5}Sr_{0.5}CoO_3$ [68] are used as electrode materials. Tunneling resistive property of ferroelectric metal is the main characteristic of these materials. The disadvantage of ferroelectric memristors is low R_{ON}/R_{OFF} [69].

17.7 Resistive Switching Mechanism

The resistive switching may be unipolar if the polarity of the applied voltage is the same, and the magnitude is different or bipolar if the applied voltage is of the same and different polarity.

17.7.1 Working of VCM

Electric field and thermal effect occur on the application of the earmark voltage, and these effects cause movement of the anions. These anions will result in the formation and rupture of the conductive filament. The anions are in the form of the oxygen vacancy (positive charge depleted) that can be understood from the following equation:

$$O_o \rightarrow V_o^{2+} + O_i^{2-} \tag{17.54}$$

where:
- O_o is the oxygen atom on regular lattice
- V_o^{2+} is the oxygen vacancy on regular lattice
- O_i^{2-} is the oxygen atom not on regular lattice

These oxygen vacancies in pairs are created inside the insulating layer by the process of impact ionization [70]. Then, ions can be moved to the anode through the insulating layer by the applied electric field and Joule heating due to leakage current. The mobile anions get the ion diffusion path through defects in the structure as dislocation. The drift velocity of ions is exponentially related with the applied electric field as $v = \mu E \exp^{(E/E_o)}$, where E_o is a characteristic field of any moving atom. A high electric field will cause the high velocity; the Joule heating also increases the drift velocity of the mobile ions.

At the anode, the following oxidation reaction takes place:

$$O_i^{2-} \rightarrow 2e^- + \frac{1}{2}O_2 \tag{17.55}$$

Thus, oxygen bubbles are observed at the anode [24]. Oxygen vacancies are collected at the cathode as oxygen ions are moved to the anode; therefore, the oxygen-deficient region grows up and increases toward the anode.

When reverse bias is applied, the oxygen ions move back to the cathode and a reset process occurs or the conductive filament breaks; thus, the low resistance path will become the high resistive path. The equation of reset process is as follows:

$$O_i^{2-} + V_O^{2+} \rightarrow O_O^0 \tag{17.56}$$

This process is initiated at the middle of the conductive filament and then prolongs toward the anode but still remains at the cathode. If the insulating layer is made of p-type oxide, then the conductive filament will start to build up from the anode side. If the oxide is of n-type, the conductive filament starts from the cathode side [71].

17.7.2 Working of Electrochemical Metallization Memristor

In an electrochemical metallization memristor, the metallic cations are the mobile species and are created at the active anode as

$$M \rightarrow M^{Z+} + ze^- \tag{17.57}$$

These metallic cations move through the electrolyte toward the inert cathode and thus conductive filament is formed. The movement of the metallic cation is due to the electric field and concentration gradient. The formation of conductive filament consists of a metallic atom, and the device will turn from high resistance state to low resistance state on the application of opposite voltage at the inert electrode. Then, the conductive filament dissolves and the device turns from low-resistance to high resistance state. The reset process of the conductive filament is given by the following equation [21].

$$M^{Z+} + ze^- \rightarrow M \tag{17.58}$$

The resistive switching is caused by the electric field effect and thermal effect. The bipolar behavior is shown by the device if the electric field is dominated, and unipolar behavior is obtained when the thermal effect is dominated [72].

Conductance quantization effect is also observed in VCM and ECM. Slow voltage sweeping with appropriate pulse shows quantized conductance in the insulating layer [73,74]. If the conductive filament is very small, i.e., in nanometer scale, then the electron must flow via the quantized state. $G = nG_o$ where, n (integer [73,74] or half integer [75–77]) is the multiple quantized states and $G_o = 2(e^2/h)$ [78]. These multiple quantized states can be exploited as multiple bit storage. This means we can store many numbers between 0 and 1 unlike in recent memory devices. Therefore, this device shows great promise in the development of memory devices.

17.7.3 Working of Barrier-Type Memristor

In the barrier-type memristor, the interpolation of the barrier is done by redistribution of the oxygen vacancies [54,56,59–61]. Oxygen vacancies in a barrier-type memristor act as a donor when we apply negative voltage to the left electrode, as shown in Figure 17.10.

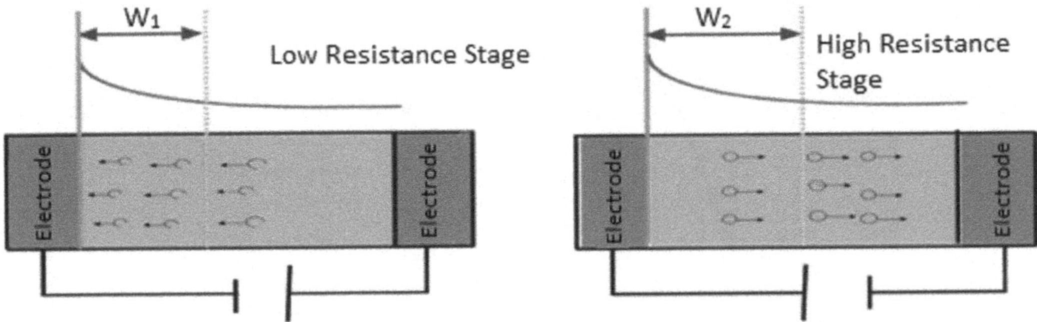

FIGURE 17.10
Re-distribution of the oxygen vacancies. (From Di Ventra, M. et al., *Proc. IEEE*, 97, 1717–1724, 2009.)

The donor oxygen vacancies are accumulated at the left electrode; thus, the donor density will increase. This means the depletion region will reduce and turn the memristor from high to low resistance states. When large positive voltage is applied to the left electrode, the oxygen vacancies will move away from this electrode; thus, the donor density decreases. This means the depletion region will widen and the transistor will turn from low resistance to high resistance state. Similarly, a mechanism is observed for *p*-type insulating material [59].

At the boundaries interface, the charge carriers trap and detrap, which causes alteration of the Schottky barrier. This also changes the height of the barrier and leads to alteration of resistance states from high to low and vice versa [79–81].

17.8 HP Memristor

In 1971, L. Chua gave the theory of the memristor. After a long delay, it was possible to fabricate this device in 2008. R. Stanley Williams and his colleagues at HP Labs succeeded in finally fabricating this device after a difficult endeavor.

17.8.1 Switch of Crossbar

The switch formed in the crossbar structure of the HP memristor consists of two layers of TiO_2 [20] including two top and bottom electrodes. One layer of Tio_2 has an exact 2:1 ratio of oxygen and titanium; thus, this layer behaves as a perfect insulator. However, the second layer has a deficit in oxygen, i.e., this layer has oxygen vacancies. These vacancies can conduct through this layer and make this layer conductive. The enlarged view of the switch is shown in Figure 17.11.

17.8.2 Crossbar Structure of Memristor

The crossbar structure is basically made by multiple horizontal and vertical wires crossing each other. If the crossbar structure is made of a memristor, then these wires are working as top and bottom electrodes, or these are platinum wires for the HP memristor [20] as shown in Figure 17.1 The upper and lower wires are separated by insulating material; thus,

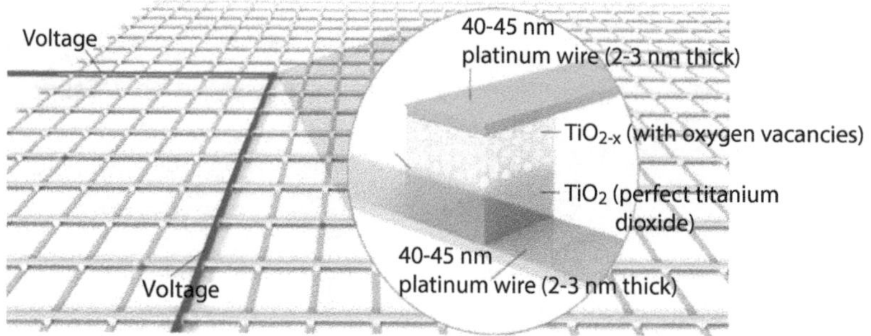

FIGURE 17.11
Crossbar array structure and switch configuration of HP memristor. (From Williams, S. R., *IEEE Spectr.*, 45, 28–35, 2008.)

the cross-section point of any two wires works as a switch, or this switch is configured as a memristor. In other words, the crossbar structure is made of a number of memristors. The switch or memristor turns ON if a positive bias is applied across the memristor, and the memristor turns off if reversed bias is applied.

17.8.3 Working

The switch operation is easily understood by Figure 17.12; the whole device is considered as two layers of TiO_2. The upper layer has oxygen deficiencies and acts as the conducting layer, whereas the bottom layer has an exact 2:1 oxygen-to-titanium ratio and acts as an insulating layer. If the positive voltage is applied to the upper layer, then these vacancies prolong toward the bottom insulating layer. Prolongation of the oxygen vacancies shows extension of the conductive width, and at a certain voltage the whole region becomes conductive. If the reverse voltage is applied at the upper layer, then the conducting layer shortens since oxygen vacancies move in a backward direction; thus, the whole region becomes highly resistive.

The most important property of this device is that the oxygen vacancies are frozen if the applied voltage is removed—either the positive or reverse voltage. Thus, the device

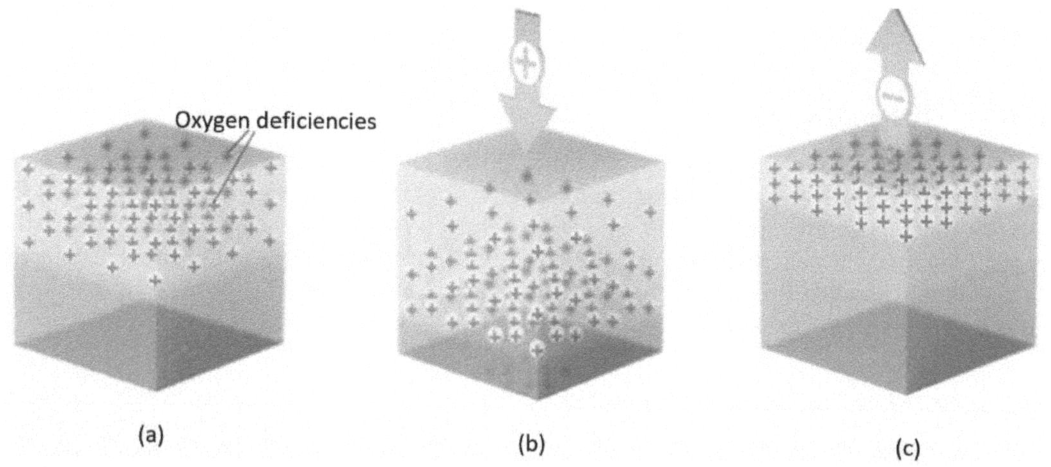

FIGURE 17.12
Variation of oxygen vacancies on application of biasing (a) no bias (b) positive biasing repels positively charged vacancies (c) negative biasing attracts positively charged oxygen vacancies.

remembers the last voltage applied, or the oxygen vacancies start from the last location when the next voltage was applied. It remembers the last applied voltage even if the next voltage is applied after several years.

17.8.4 Fabrication of Memristor

There are different layers in thin-film memristor as shown in Figure 17.13 [82]. The whole device is fabricated upon a silicon substrate. Two layers of TiO_2, one rich in oxygen and another with exact oxygen, are the two active layers and are fabricated one over another. Top and bottom electrodes made of Ti/Pt are fabricated above and below the two layers of the TiO_2 layers.

The fabrication of all the layers is done using standard photolithography processes. The electron gun evaporation technique is used to deposit the first layer of titanium and platinum (Ti/Pt) bi-layer, which works as a bottom electrode. RF magnetron sputtering is used to deposit the first layer of TiO_2, which works as the insulating layer. Then using the same process, the second layer of the TiO_2, which is rich in oxygen, is deposited and will work as the conducting layer. The excess oxygen atom can be added to the TiO_2 by flowing excess oxygen during the fabrication and making it non-stoichiometric. Then, the top layer of the electrode is deposited again by using the electron gun evaporation process; thus, the whole memristor is fabricated. The flow diagram of the fabrication process is shown in Figure 17.14 [83].

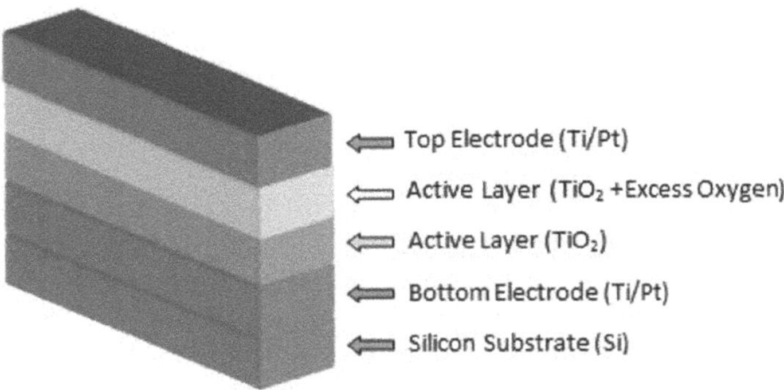

FIGURE 17.13
Different layers of thin film memristor. (From Mohanty, S. P., *IEEE Potentials*, 32, 34–39, 2013.)

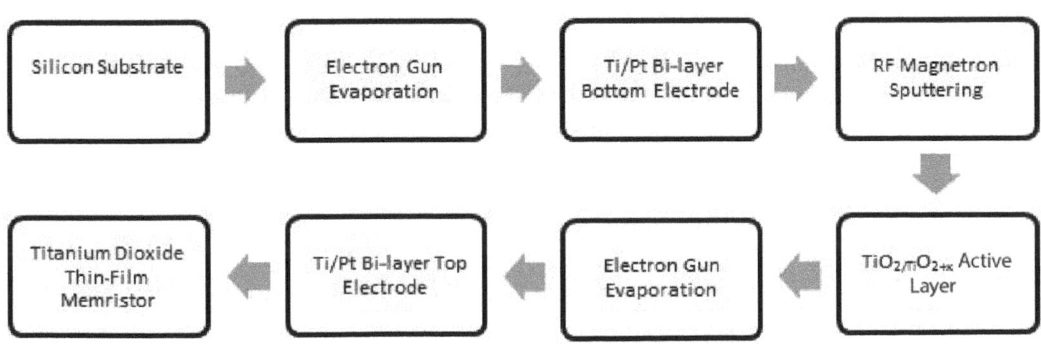

FIGURE 17.14
Fabrication steps of memristor. (From Mohanty, S. P., *IEEE Potentials*, 32, 34–39, 2013.)

17.9 Various Fabrication Technology of Insulating Layer

The memristor is realized as an insulating layer between two electrodes. The insulating layer can be fabricated using the following techniques [84]:

1. Atomic layer deposition (ALD)
2. ALD is the most anticipated technique for fabricating a memristor, and this technique is also helpful for using a memristor as a commercial device.
3. Sputter
4. Pulsed layer deposition (PLD)
5. Sol-gel process
6. Chemical solution deposition (CSD)
7. Chemical vapor deposition (CVD)
8. E-beam evaporation deposition
9. Molecular beam epitaxy (MBE)

There are various applications of a memristor according to the insulating layer property such as memory application prefers discrete resistance states, and neuromorphic application prefers continuous resistance states.

17.10 Properties of Memristor

The memorizing property of a memristor makes it capable to use in computer memory, and since it stores resistance value indefinitely, it can be used as nonvolatile memory. Memristors can be used as synapses of neurons, and the human brain can be developed with the help of a memristor implementing in an analog circuit.

Memristance is an inherent property of an electronic circuit. It is more significant at a small dimension. It is unobservable at millimeter scale, and the memristance is millions of times higher when we go from microscale to nanoscale because memristance is universally proportional to the square of dimension. As dimension becomes lesser, it is more noticeable.

When positive and negative voltage are applied, the memristor becomes conducting and nonconducting, respectively. When we turn off voltage, the device is frozen at that place; synapses work in the same manner as the memristor. The polarity and duration of the applied chemical electrical signal will determine the strength of the synaptic connection between neurons [20].

If there is asymmetry in applied bias, multiple continuous states will be obtained.

The equation $w(t) = \mu_v \frac{R_{ON}}{D} q(t)$ is valid when the value of w lies between $[0, D]$. The switching event should have a large amount of charge when the doped width "w" approaches the boundaries (o or D) at the threshold voltage to keep the ON and OFF states much longer; therefore, switching is essentially binary.

Since memristance dominates at nanoscale or more in less dimension, this device will provide high density. Another property of it is memorizing capabilities.

The nonvolatile nature of the memristor is comparable to flash. Memristors $\varphi - q$ curve shows two individual slopes that correspond to two values of $M(q)$ that fulfill the need of binary logics. This device is purely dissipative [85].

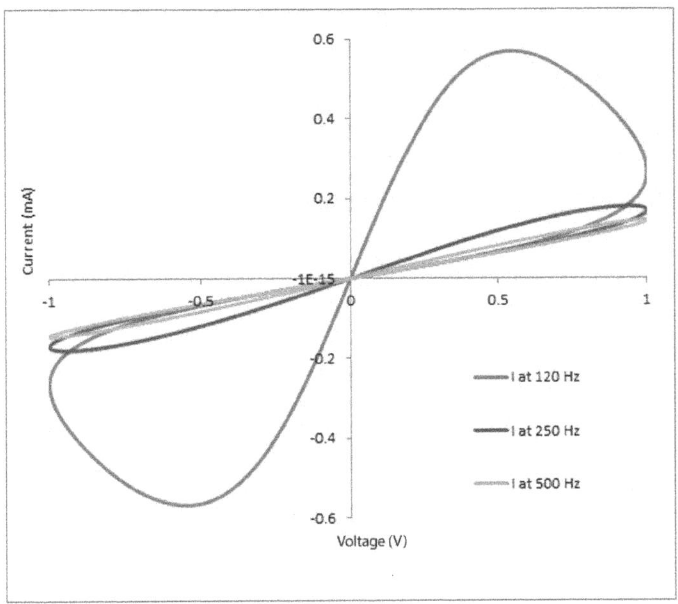

FIGURE 17.15
Pinched hysteresis loop of memristor. (From Chua, L. O., *IEEE Trans. Circuits Syst.*, 27, 1014–1044, 1980.)

Current-voltage characteristics of memristor will give pinched hysteresis loop and always pass through the origin for any bipolar periodic input voltage $v(t)$ [86].

Beyond a certain critical frequency, the area of pinched hysteresis lobe decreases monotonically as frequency of the input signal increases [86].

The shape of the pinched hysteresis loop varies with frequencies and shrinks to a single valued function through the origin as frequency tends to infinity [86].

$p(i) = M(q)(i(t))^2$ always give positive power for $M(q) \geq 0$, which implies a passive device.

Characteristic are shown in Figure 17.15.

The integral of current $q(t) = \int_{t_0}^{t} i(\tau)\,d\tau$ will work as a state variable and counts charge passing through the memristor instead of storage of charge unlike in the capacitor. The memristor is not a charge storage element. It is not merely a nonlinear resistor but is a nonlinear resistor with charge as a state variable [87].

The memristor will provide inverse relation between flux and frequency for a given periodic signal; that means it is merely a linear resistor at very high frequencies [85].

17.11 Advantages of Memristor

Since memristive behavior is dominating at a small size in the nanoscale range, its capability to memorize indefinably makes it popular in today's technology. Various advantages of memristor are explained as follows:

1. High scalability
2. Low power consumption

3. Excellent compatibility with CMOS, which is extremely useful for constructing modern integrating circuits including microprocessor, microcontroller, memories, logic circuits, and analog circuits such as sensors
4. The nanscale size of the switch in the crossbar array has a much higher density of switch than a comparable integrating circuit base on a transistor
5. The crossbar structure using the memristor will have the following advantages [88]:
    a. Inherent defect tolerance capability
    b. Simplicity
    c. Flexibility
    d. Maximum density

17.12 Conclusion

In this book chapter, the memristor is explored, starting with the missing link between charge and flux and then providing definition and theory as it is deployed mathematically and practically. The types of memristor—barrier, filament, and ferroelectric—are also explained. The workings and mechanisms of resistive switching are briefly discussed. The basic HP Labs model of the memristor, the device's structure, and its fabrication steps with techniques are also explained. A family of circuit elements postulated by L. Chua is provided. Then, the various advantages and important properties of the memristor are explained, which make it a popular and important circuit element.

References

1. Chua, L. O., "Memristor-the missing circuit element," *IEEE Transactions on Circuit Theory*, vol. 18, no. 5, pp. 507–519, 1971.
2. Widrow, B., *Adaptive "Adaline" Neuron Using Chemical "Memistors,"* Stanford Electronics Laboratories, Stanford, CA, 1960.
3. Chua, L. O., "Nonlinear circuit foundations for nano devices, Part I: The four-element torus," *Proceedings of the IEEE*, vol. 91, no. 11, pp. 1830–1859, 2003.
4. Chua, L. O., "Device modeling via nonlinear circuit elements," *IEEE Transactions on Circuits and Systems*, vol. 27, no. 11, pp. 1014–1044, 1980.
5. Di Ventra, M., Y. V. Pershin, and L. O. Chua, "Circuit elements with memory: Memristors, memcapacitors, and meminductors," *Proceedings of the IEEE*, vol. 97, no. 10, pp. 1717–1724, 2009.
6. Strukov, D. B., G. S. Snider, D. R. Stewart, and R. S. Williams, "The missing memristor found," *Nature*, vol. 453, no. 7191, pp. 80–83, 2008.
7. Blanc, J., and D. L. Staebler, "Electrocoloration in $SrTiO_3$: Vacancy drift and oxidation–reduction of transition metals," *Physical Review B*, vol. 4, no. 10, pp. 3548–3557, 1971.
8. Strukov, D. B., and R. S. Williams, "Exponential ionic drift: fast switching and low volatility of thin-film memristors," *Applied Physics A*, vol. 94, no. 3, pp. 515–519, 2009.

9. Biolek, D. B., and V. Biolkova, "Spice model of memristor with nonlinear dopant drift," *Radioengineering*, vol. 18, no. 2, pp. 210–214, 2009.
10. Benderli, S., and T. A. Wey, "On SPICE macromodelling of TiO_2 memristors," *Electronics Letters*, vol. 45, no. 7, pp. 377–379, 2009.
11. Joglekar, Y. N., and S. J. Wolf, "The elusive memristor: Properties of basic electrical circuits," *European Journal of Physics*, vol. 30, no. 4, p. 661, 2009.
12. Prodromakis, T., B. P. Peh, C. Papavassiliou, and C. Toumazou, "A versatile memristor model with nonlinear dopant kinetics," *IEEE Transactions on Electron Devices*, vol. 58, no. 9, pp. 3099–3105, 2011.
13. Zha, J., H. Huang, and Y. Liu, "A novel window function for memristor model with application in programming analog circuits," *IEEE Transactions on Circuits and Systems II: Express Briefs*, vol. 63, no. 5, pp. 423–427, 2016.
14. Yang, J. J., M. D. Pickett, X. Li, D. A. A. Ohlberg, D. R. Stewart, and R. S. Williams, "Memristive switching mechanism for metal/oxide/metal nanodevices," *Nature Nanotechnology*, vol. 3, no. 7, pp. 429–433, 2008.
15. Pickett, M. D., D. B. Strukov, J. L. Borghetti, J. J. Yang, G. S. Snider, D. R. Stewart, and R. S. Williams, "Switching dynamics in titanium dioxide memristive devices," *Journal of Applied Physics*, vol. 106, no. 7, pp. 1–6, 2009.
16. Simmons, J. G., "Generalized formula for the electric tunnel effect between similar electrodes separated by a thin insulating film," *Journal of Applied Physics*, vol. 34, no. 6, pp. 1793–1803, 1963.
17. Kvatinsky, S., E. G. Friedman, A. Kolodny, and U. C. Weiser, "TEAM: ThrEshold adaptive memristor model," *IEEE Transactions on Circuits and Systems I: Regular Papers*, vol. 60, no. 1, pp. 211–221, 2013.
18. Kvatinsky, S., M. Ramadan, E. G. Friedman, and A. Kolodny, "VTEAM: A general model for voltage-controlled memristors," *IEEE Transactions on Circuits and Systems II: Express Briefs*, vol. 62, no. 8, pp. 786–790, 2015.
19. Kumar, S., and B. Raj, "Estimation of stability and performance metric for inward access transistor based 6T SRAM cell design using n-type/p-type DMDG-GDOV TFET," *IEEE VLSI Circuits and Systems Letter*, vol. 3, no. 2, 2017.
20. Williams, S. R., "How we found the missing memristor," *IEEE Spectrum*, vol. 45, no. 12, pp. 28–35, 2008.
21. Waser, R., R. Dittimann, G. Staikov, and K. Szot, "Redox-based resistive switching memories-nono mechanism, prospects and challenges," *Advance Materials*, vol. 21, no. 25–26, pp. 2632–2663, 2009.
22. Wei, L. L., J. Wang, Y. S. Chen, D. S. Shang, Z. G. Sun, B. G. Shen, and J. R. Sun, "Pulse-induced alternation from bipolar resistive switching to unipolar resistive switching in the $Ag/AgO_x/Mg_{0.2}Zn_{0.8}O/Pt$ device," *Journal of Physics D: Applied Physics*, vol. 45, no. 42, p. 425303, 2012.
23. Raj, B., A. K. Saxena, and S. Dasgupta, "NanoscaleFinFET based SRAM cell design: Analysis of performance metric, process variation, UnderlappedFinFET and temperature effect," *IEEE Circuits and System Magazine*, vol. 11, no. 2, pp. 38–50, 2011.
24. Jain, N., and B. Raj, "Thermal stability analysis and performance exploration of asymmetrical dual-k underlap spacer (ADKUS) SOI FinFET for security and privacy applications," *Journal of Information Privacy and Security (JIPS)*, vol. 23, pp. 1890–1901, 2017.
25. Muraoka, S., K. Osano, Y. Kanzawa, S. Mitani, S. Fujii, K. Katayama, Y. Katoh et al., "Fast switching and long retention Fe-O ReRAM and its switching mechanism," In *2007 IEEE International Electron Devices Meeting*, 2007.
26. Sharma, V. K., M. Pattanaik, and B. Raj, "PVT variations aware low leakage INDEP approach for nanoscale CMOS circuits," *Microelectronics Reliability*, vol. 54, no. 1, pp. 90–99, 2014.
27. Lin, C.-Y., C.-Y. Wu, C.-Y. Wu, C. Hu, and T.-Y. Tseng, "Bistable resistive switching in Al_2O_3 memory thin films," *Journal of the Electrochemical Society*, vol. 154, no. 9, pp. G189–G192, 2007.
28. Hu, S. G., Y. Liu, T. P. Chen, Z. Liu, M. Yang, Q. Yu, and S. Fung, "Effect of heat diffusion during state transitions in resistive switching memory device based on nickel-rich nickel oxide film," *IEEE Transactions on Electron Devices*, vol. 59, no. 5, pp. 1558–1562, 2012.

29. Singh, A., M. Khosla, and B. Raj, "Design and analysis of electrostatic doped Schottky barrier CNTFET based low power SRAM," *International Journal of Electronics and Communications (AEÜ)*, vol. 80, pp. 67–72, 2017.
30. Almeida, S., B. Aguirre, N. Marquez, J. McClure, and D. Zubia, "Resistive switching of SnO_2 thin films on glass substrates," *Integrated Ferroelectrics*, vol. 126, no. 1, pp. 117–124, 2011.
31. Lin, C.-C., Z.-L. Tseng, K.-Y. Lo, C.-Y. Huang, C.-S. Hong, S.-Y. Chu, C.-C. Chang, and C.-J. Wu, "Unipolar resistive switching behavior of $Pt/Li_XZn_{1-X}O/Pt$ resistive random access memory devices controlled by various defect types," *Applied Physics Letters*, vol. 101, no. 20, p. 203501, 2012.
32. Hu, W., X. Chen, G. Wu, Y. Lin, N. Qin, and D. Bao, "Bipolar and tri-state unipolar resistive switching behaviors in $Ag/ZnFe_2O_4/Pt$ memory devices," *Applied Physics Letters*, vol. 101, no. 6, p. 063501, 2012.
33. Raj, B., A. K. Saxena, and S. Dasgupta, "Analytical modeling for the estimation of leakage current and subthreshold swing factor of nanoscale double gate FinFET device," *Microelectronics International, UK*, vol. 26, pp. 53–63, 2009.
34. Dong, H., X. Zhang, D. Zhao, Z. Niu, Q. Zeng, J. Li, L. Cai et al., "High performance bipolar resistive switching memory devices based on Zn_2SnO_4 nanowires," *Nanoscale*, vol. 4, no. 8, pp. 2571–2574, 2012.
35. Makihara, K., M. Ikeda, H. Murakami, S. Higashi, and S. Miyazaki, "Evaluation of chemical composition and bonding features of $Pt/SiO_x/Pt$ MIM diodes and its impact on resistance switching behavior," *IEICE Transactions on Electronics*, vol. 96, no. 5, pp. 702–707, 2013.
36. Jeong, H. Y., J. Y. Kim, J. W. Kim, J. O. Hwang, J. E. Kim, J. Y. Lee, T. H. Yoon et al., "Graphene oxide thin films for flexible nonvolatile memory applications," *Nano Letters*, vol. 10, no. 11, pp. 4381–4386, 2010.
37. Kim, H.-D., H.-M. An, Y. M. Sung, H. Im, and T. G. Kim, "Bipolar resistive-switching phenomena and resistive-switching mechanisms observed in zirconium nitride-based resistive-switching memory cells," *IEEE Transactions on Device and Materials Reliability*, vol. 13, no. 1, pp. 252–257, 2013.
38. Sun, X., G. Li, X. Zhang, L. Ding, and W. Zhang, "Coexistence of the bipolar and unipolar resistive switching behaviours in Au/SrTiO3/Pt cells," *Journal of Physics D: Applied Physics*, vol. 44, no. 12, p. 125404, 2011.
39. Singh, A., M. Khosla, and B. Raj, "Analysis of electrostatic doped schottky barrier carbon nanotube FET for low power applications," *Journal of Materials Science: Materials in Electronics*, vol. 28, pp. 1762–1768, 2017.
40. Sun, J., Q. Liu, H. Xie, X. Wu, F. Xu, T. Xu, S. Long, H. Lv, Y. Li, L. Sun, and M. Liu, "In situ observation of nickel as an oxidizable electrode material for the solid-electrolyte-based resistive random access memory," *Applied Physics Letters*, vol. 102, no. 5, p. 053502, 2013.
41. Prakash, A., S. Maikap, S. Z. Rahaman, S. Majumdar, S. Manna, and S. K. Ray, "Resistive switching memory characteristics of Ge/GeOx nanowires and evidence of oxygen ion migration," *Nanoscale Research Letters*, vol. 8, no. 1, pp. 1–10, 2013.
42. Nayak, A., T. Ohno, T. Tsuruoka, K. Terabe, T. Hasegawa, J. K. Gimzewski, and M. Aono, "Controlling the synaptic plasticity of a Cu_2S gap-type atomic switch," *Advanced Functional Materials*, vol. 22, no. 17, pp. 3606–3613, 2012.
43. van den Hurk, J., V. Havel, E. Linn, R. Waser, and I. Valov, "$Ag/GeS_x/Pt$-based complementary resistive switches for hybrid CMOS/Nanoelectronic logic and memory architectures," *Scientific Reports*, vol. 3, 2013.
44. Wang, Z., P. B. Griffin, J. McVittie, S. Wong, P. C. McIntyre, and Y. Nishi, "Resistive switching mechanism in $Zn_xCd_{1-x}S$ nonvolatile memory devices," *IEEE Electron Device Letters*, vol. 28, no. 1, pp. 14–16, 2007.
45. Sleiman, A., P. W. Sayers, and M. F. Mabrook, "Mechanism of resistive switching in $Cu/AlO_x/W$ nonvolatile memory structures," *Journal of Applied Physics*, vol. 113, no. 16, p. 164506, 2013.
46. Sharma, V. K., M. Pattanaik, and B. Raj, "INDEP approach for leakage reduction in nanoscale CMOS circuits," *International Journal of Electronics*, vol. 102, no. 2, pp. 200–215, 2015.

47. Raj, B., A. K. Saxena and S. Dasgupta, "Quantum mechanical analytical modeling of nanoscale DG FinFET: Evaluation of potential, threshold voltage and source/drain resistance," *Elsevier's Journal of Material Science in Semiconductor Processing*, Elsevier, vol. 16, no. 4, pp. 1131–1137, 2013.
48. Li, Y., G. Zhao, J. Su, E. Shen, and Y. Ren, "Top electrode effects on resistive switching behavior in CuO thin films," *Applied Physics A*, vol. 104, no. 4, pp. 1069–1073, 2011.
49. Zhuge, F., S. Peng, C. He, X. Zhu, X. Chen, Y. Liu, and R.-W. Li, "Improvement of resistive switching in Cu/ZnO/Pt sandwiches by weakening the randomicity of the formation/rupture of Cu filaments," *Nanotechnology*, vol. 22, no. 27, p. 275204, 2011.
50. Chen, C., S. Gao, G. Tang, C. Song, F. Zeng, and F. Pan, "Cu-embedded AlN-based nonpolar nonvolatile resistive switching memory," *IEEE Electron Device Letters*, vol. 33, no. 12, pp. 1711–1713, 2012.
51. Lee, W., J. Park, M. Son, J. Lee, S. Jung, S. Kim, S. Park, J. Shin, and H. Hwang, "Excellent state stability of Cu/SiC/Pt programmable metallization cells for nonvolatile memory applications," *IEEEElectron Device Letters*, vol. 32, no. 5, pp. 680–682, 2011.
52. Yang, X., S. Long, K. Zhang, X. Liu, G. Wang, X. Lian, Q. Liu et al., "Investigation on the RESET switching mechanism of bipolar Cu/HfO_2/Pt RRAM devices with a statistical methodology," *Journal of Physics D: Applied Physics*, vol. 46, no. 24, p. 245107, 2013.
53. Pattanaik, M., B. Raj, S. Sharma, and A. Kumar, "Diode based trimode multi-threshold CMOS technique for ground bounce noise reduction in static CMOS adders," *Advanced Materials Research*, vol. 548, pp. 885–889, 2012.
54. Tang, X., X. Zhu, J. Dai, J. Yang, L. Chen, and Y. Sun, "Evolution of the resistive switching in chemical solution deposited-derived $BiFeO_3$ thin films with dwell time and annealing temperature," *Journal of Applied Physics*, vol. 113, no. 4, p. 043706, 2013.
55. Rubi, D., F. Gomez-Marlasca, P. Bonville, D. Colson, and P. Levy, "Resistive switching in ceramic multiferroic $Bi_{0.9}Ca_{0.1}FeO_3$," *Physica B: Condensed Matter*, vol. 407, no. 16, pp. 3144–3146, 2012.
56. Lee, H.-S., S. G. Choi, H.-J. Choi, S.-W. Chung, and H.-H. Park, "A study of resistive switching property in $Pr_{0.7}Ca_{0.3}MnO_3$, $CaMnO_3$, and their bi-layer films," *Thin Solid Films*, vol. 529, pp. 347–351, 2013.
57. Gao, X., Y. Xia, J. Ji, H. Xu, Y. Su, H. Li, C. Yang, H. Guo, J. Yin, and Z. Liu, "Effect of top electrode materials on bipolar resistive switching behavior of gallium oxide films," *Applied Physics Letters*, vol. 97, no. 19, p. 3501, 2010.
58. Raj, B., "Quantum mechanical potential modeling of FinFET," in *Towards Quantum FinFET*, Springer, Switzerland, vol. 17, pp. 81–97, 2014.
59. Wang, W., R. Dong, X. Yan, B. Yang, and X. An, "Memristive behavior of ZnO/Au film investigated by a TiN CAFM tip and its model based on the experiments," *IEEE Transactions onNanotechnology*, vol. 11, no. 6, pp. 1135–1139, 2012.
60. Raj, B., A. K. Saxena, and S. Dasgupta, "A compact drain current and threshold voltage quantum mechanical analytical modeling for FinFETs," *Journal of Nanoelectronics and Optoelectronics (JNO), USA*, vol. 3, no. 2, pp. 163–170, 2008.
61. Choi, S. J., G. S. Park, K. H. Kim, W. Y. Yang, H. J. Bae, K. J. Lee, H. I. Lee et al., "In situ observation of vacancy dynamics during resistance changes of oxide devices," *Journal of Applied Physics*, vol. 110, no. 5, p. 056106, 2011.
62. Jain, N., and B. Raj, "Device and circuit co-design perspective comprehensive approach on FinFET technology - A review," *Journal of Electron Devices*, vol. 23, no. 1, pp. 1890–1901, 2016.
63. Vishvakarma, S. K., V. Agrawal, B. Raj, S. Dasgupta, and A. K. Saxena, "Two dimensional analytical potential modeling of nanoscale symmetric double gate (SDG) MOSFET with ultra thin body (UTB)," *Journal of Computational and Theoretical Nanoscience*, vol. 4, no. 6, pp. 1144–1148, 2007.
64. Contreras, J. R., H. Kohlstedt, U. Poppe, R. Waser, ChBuchal, and N. A. Pertsev, "Resistive switching in metal–ferroelectric–metal junctions," *Applied Physics Letters*, vol. 83, no. 22, pp. 4595–4597, 2003.

65. Raj, B., J. Mitra, D. K. Bihani, V. Rangharajan, A. K. Saxena, and S. Dasgupta, "Process variation tolerant FinFET based robust low power SRAM cell design at 32nm technology," *Journal of Low Power Electronics (JOLPE), Academy Publisher, FINLAND*, vol. 7, no. 2, pp. 163–171, 2011.
66. Singh, A., M. Khosla, and B. Raj, "Circuit compatible model for electrostatic doped Schottky barrier CNTFET," *Journal of Electronic Materials*, vol. 45, no. 12, pp. 4825–4835, 2016.
67. Asadi, K., D. M. De Leeuw, B. De Boer, and P. W. M. Blom, "Organic non-volatile memories from ferroelectric phase-separated blends," *Nature Materials*, vol. 7, no. 7, pp. 547–550, 2008.
68. Choi, J., J.-S. Kim, I. Hwang, S. Hong, I.-S. Byun, S.-W. Lee, S.-O. Kang, and B. H. Park, "Different nonvolatile memory effects in epitaxial Pt/$PbZr_{0.3}Ti_{0.7}O_3$/LSCO heterostructures," *Applied Physics Letters*, vol. 96, no. 26, p. 262113, 2010.
69. Pantel, D., H. Lu, S. Goetze, P. Werner, D. J. Kim, A. Gruverman, D. Hesse, and M. Alexe, "Tunnel electroresistance in junctions with ultrathin ferroelectric Pb $(Zr_{0.2}Ti_{0.8})$ O_3 barriers," *Applied Physics Letters*, vol. 100, no. 23, p. 232902, 2012.
70. Omura, Y., and Y. Kondo, "Impact-ionization-based resistive transition model for thin TiO_2 films," *Journal of Applied Physics*, vol. 114, no. 4, p. 043712, 2013.
71. Singh, A., M. Khosla, and B. Raj, "Compact model for ballistic single wall CNTFET under quantum capacitance limit," *Journal of Semiconductors (JoS)*, vol. 37, pp. 104001-8, 2016.
72. Yang, J. J., D. B. Strukov, and D. R. Stewart, "Memristive devices for computing," *Nature Nanotechnology*, vol. 8, no. 1, pp. 13–24, 2013.
73. Raj, B., and S. Vaidyanathan, Analysis of dynamic linear memristor device models, in S. Vaidyanathan, and C. Volos (Eds.), *Advances in Memristors, Memristive Devices and Systems*, Springer, AG, 2017.
74. Tsuruoka, T., T. Hasegawa, K. Terabe, and M. Aono, "Conductance quantization and synaptic behavior in a Ta_2O_5-based atomic switch," *Nanotechnology*, vol. 23, no. 43, p. 435705, 2012.
75. Singh, A., D. K. Saini, D. Agarwal, S. Aggarwal, M. Khosla, and B. Raj, "Modeling and simulation of carbon nanotube field effect transistor and its circuit application," *Journal of Semiconductors (JoS)*, vol. 37, pp. 074001-6, 2016.
76. Mehonic, A., A. Vrajitoarea, S. Cueff, S. Hudziak, H. Howe, Christophe Labbe, R. Rizk, M. Pepper, and A. J. Kenyon, "Quantum conductance in silicon oxide resistive memory devices," *Scientific Reports*, vol. 3, 2013.
77. Bansal, P., and B. Raj, "Memristor modeling and analysis for linear dopant drift kinetics," *Journal of Nanoengineering and Nanomanufacturing*, vol. 6, pp. 1–7, 2016.
78. Singh, A., M. Khosla, and B. Raj, "Comparative analysis of carbon nanotube field effect transistor and nanowire transistor for low power circuit design," *Journal of Nanoelectronics and Optoelectronics*, vol. 11, pp. 388–393, 2016.
79. Sun, J., C. H. Jia, G. Q. Li, and W. F. Zhang, "Control of normal and abnormal bipolar resistive switching by interface junction on in/Nb:$SrTiO_3$ interface," *Applied Physics Letters*, vol. 101, no. 13, p. 133506, 2012.
80. Bansal, P., and B. Raj, "Memristor: A versatile nonlinear model for dopant drift and boundary issues," *Quantum Matter*, vol. 5, pp. 1–5, 2016.
81. Huang, H.-H., W.-C. Shih, and C.-H. Lai, "Nonpolar resistive switching in the Pt/MgO/Pt nonvolatile memory device," *Applied Physics Letters*, vol. 96, no. 19, p. 193505, 2010.
82. Prodromakis, T., K. Michelakis, and C. Toumazou, "Fabrication and electrical characteristics of memristors with TiO_2/TiO_{2+x} active layers," in *Proceedings of the IEEE International Symposium Circuits and Systems*, pp. 1520–1522, 2010.
83. Mohanty, S. P., "Memristor: From basics to deployment," *IEEE Potentials*, vol. 32, no. 3, pp. 34–39, 2013.
84. Hu, S. G., S. U. Wu, W. W. Jia, Q. Yu, L. J. Deng, Y. Q. Fu, Y. Liu, and T. P. Chen, "Review of nanostructured resistive switching memristor and its applications," *Nanoscience and Nanotechnology Letters*, vol. 6, no. 9, 2014.
85. Kavehei, O., Y.-S. Kim, A. Iqbal, K. Eshraghian, S. F. Al-Sarawi, and D. Abbott, "The fourth element: Insights into the memristor," In *Communications, Circuits and Systems, 2009. ICCCAS 2009. International Conference on*, pp. 921–927, IEEE, 2009.

86. Adhikari, S. P., M. Pd. Sah, H. Kim, and L. O. Chua, "Three fingerprints of memristor," *IEEE Transactions Circuits and Systems I: Regular Papers*, vol. 60, no. 11, pp. 3008–3021, 2013.
87. Jain, A., S. Sharma, and B. Raj, "Design and analysis of high sensitivity photosensor using cylindrical surrounding gate MOSFET for low power sensor applications", *Engineering Science and Technology, an International Journal*, vol. 19, no. 4, pp. 1864–1870, 2016.
88. Sharma, S., B. Raj, and M. Khosla, "A Gaussian approach for analytical subthreshold current model of cylindrical nanowire FET with quantum mechanical effects," *Microelectronics Journal*, vol. 53, pp. 65–72, 2016.

Index

Note: Page numbers in italic and bold refer to figures and tables respectively.

Milton Keynes UK
Ingram Content Group UK Ltd.
UKHW050456071024
449327UK00015B/399